# INDUSTRIAL GUIDE TO CHEMICAL AND DRUG SAFETY

# INDUSTRIAL GUIDE TO CHEMICAL AND DRUG SAFETY

T.S.S. Dikshith

Prakash V. Diwan

A JOHN WILEY & SONS, INC., PUBLICATION

Published by John Wiley & Sons, Inc., Hoboken, New Jersey.
Published simultaneously in Canada.



***Library of Congress Cataloging-in-Publication Data*:**

Dikshith, T.S.S.
Industrial guide to chemical and drug safety / T.S.S. Dikshith, Prakash V. Diwan.
p. cm.
Includes bibliographical references and index.
ISBN 0-471-23698-5(cloth)
1. Industrial toxicology. 2. Drugs—Toxicology. 3. Chemicals—Safety measures. I. Diwan, Prakash V. II. Title.
RA1229 .D535 2002
615.9′02—dc21 2002027404

Printed in the United States of America

10 9 8 7 6 5 4 3 2 1

# CONTENTS

## FOREWORD

Chemicals and drugs have become essential for the improvement of human health, control of diseases, growth and development of industries, and for quality enrichment of life. Associated with this, synthesis and manufacturing of chemicals and drugs in varieties and abundance have taken a new turn. The large-scale use, storage, and disposal of complex chemicals and drugs have been increasing over the decades around the world. Also, men, women, and children are exposed in multiple ways to chemicals and drugs at home and in the workplace environment. Improper handling of chemicals has resulted in chemical disasters and health effects. Working safe with the chemicals and drugs is, therefore, an important educational subject for different groups of human society.

A much-awaited guidebook, *Industrial Guide to Chemical and Drug Safety*, is devoted to discussing the safe handling of drugs and chemicals by individuals associated with their manufacture in laboratories and industry. In its 20 chapters, the reader will discern for himself that Dr. Dikshith and Dr. Diwan have revealed certain jewels. Indeed, these could form the beginnings of more research and new knowledge, which I hope the readers will be challenged to unfold for themselves.

A.S. PAINTAL

# PREFACE

Exposure to a variety of chemicals and drugs has become very common in industrial, laboratory, and household environments. Human concern for safe workplace and safe substance is only natural. Chemical and drug toxicology has registered its importance in research and management around the world. In fact, toxicology is an important discipline playing an essential role in the evaluation of a variety of chemicals and drugs. Human relationships are becoming closer. Global understanding regarding the management of toxic and hazardous substances is reaching uniformity, for instance, in the methods of handling, use, transportation, storage, and disposal. Management of toxic and hazardous substances is in no way different in a developing country than in a developed country. Furthermore, the manufacture, use, application, and disposal pattern of host of chemicals and drugs have crossed national and international boundaries.

The methods and manner of evaluation of chemicals must, therefore, be based on uniform, universally accepted, principles of testing. Any data generated based on quality and international acceptance will go a long way for use and product sale in the world market. The regulatory function of governments and approval agencies is to determine and predict the safety of the substances, products, and drugs based on the evidence of preclinical toxicity tests. To achieve this goal and in the interest of the general public, an impartial assessment becomes necessary, and the regulatory bodies need accurate data of quality and integrity.

Over the years global regulatory agencies have been responsible for the generation of quality data on chemicals and drugs for purposes of establishing human safety and the protection of other species of biological systems. In recent years concern has increased around the world regarding widespread misuse and abuse of drugs and related products. This is more so because of the development of drug dependence among children and adults. Also, the danger of the development of dependence from the therapeutic use of drugs, often resulting from improper prescription or overprescription has attracted the attention of regulatory agencies and governments. The World Health Organization has recognized the importance of conducting proper evaluation of chemicals and drugs using standard methods of testing.

It is heartening to note that regulatory authorities in different countries are very anxious to bring in line certain work carried out in their respective countries with other countries of the world and to reach a state of harmonization. Several countries have introduced GLP accreditation to establish quality and integrity of data, which is a must for global competition. The book covers important chemicals and drugs vis-à-vis relative safety and potential efficacy to contain adverse health effects to humans and to help for a judicious management to reach a pragmatic decision. It is hoped that the book will offer valuable guidance to a host of occupational workers who regularly handle hazardous chemicals and drugs. Furthermore, students, trainees, skilled workers, managers, and personnel associated with regulatory agencies will be able to know the good and the harmful aspects of chemicals and drugs before use. The book also gives scope for proper training of the workers and beginners to avoid human disorders associated with the misuse of chemicals and drugs.

T.S.S. Dikshith

Prakash V. Diwan

*Hyderabad, India*

# ACKNOWLEDGMENTS

We record here our sincere thanks to all authors and publishers from whom we have inadvertently quoted and/or paraphrased certain matter. We could not obtain their formal permission due to our inability to trace their source. We thank them for their cooperation and support. The authors express their gratitude to Dr. J. Stober, Executive Secretary; Dr. Maged Younes, IFCS; Dr. Tim Meredith, Coordinator, International Program on Chemical Safety, Publication Division, World Health Organization, Geneva, Switzerland, for granting permission to use the published information in this book.

It is a great pleasure to express our sincere thanks to Dr. Dian Turnheim and Dr. Tiffany Larsen, Environmental Health and Safety Division, Organization of Economic Cooperation and Development (OECD), Paris, France, for granting permission to use their published literature from the OECD Principles of Good Laboratory Practice documents in this book. We also thank Dr. Marianne Yaun, Director, ATSDR Information Center, Washington, D.C., for allowing us to use the published literature of the federal government of the United States. It is a great pleasure to thank Prof. A. S. Paintal, FRS, for the encouraging foreword to this book; and Dr. K. V. Raghavan, Director, Indian Institute of Chemical Technology (Council of Scientific and Industrial Research), Hyderabad, India, for encouragement in the publication of the book.

The idea for this book took shape when the first author was in Ann Arbor, Michigan; Waukesha, Wisconsin; and Hamden, Connecticut, during 1999–2000. It was with the cooperation and support of his family members that he was able to complete the project. The author expresses his deep sense of appreciation to Mrs. Saroja Dikshith, Mr. Narasimha Kramadhati, Mrs. Pratibha Kramadhati, Dr. Deepak Murthy, and Mrs. Prerana Murthy for providing help and facilities needed for his work and for sharing their thoughts about the book.

The active cooperation of Mr. Abhay Singh Chauhan, Mr. Kishore Babu Chalasani, Mrs. S. Sridevi, and Dr. S. Ramakrishna at different stages of the manuscript preparation offered very valuable support to the authors, and our sincere thanks and appreciation to them all. Our appreciation to Mrs. Rekha Diwan for cooperation and encouragement in this work. Last, but not least, we express our sincere thanks to Mr. N. Ramakrishna, who did excellent secretarial work and typing to give final shape to the manuscript.

The authors express their sincere thanks to Janet D. Bailey, Executive Publisher; Dr. Bob Esposito, Executive Editor; and Ayana Meade, Editorial Assistant,

John Wiley & Sons, Inc., Hoboken, New Jersey; and Joan Wolk, Project Manager, Joan Wolk Editorial Services, New York, New York, for very valuable cooperation and encouragement in the publication of the book.

CHAPTER 1

# Introduction to Chemical and Drug Safety

## INTRODUCTION

Humankind has continually developed and progressed over the centuries, and nowhere is this more evident than in recent modern history. The past few decades have demonstrated man's dependency on chemical and pharmaceutical products to achieve both material benefit and health. Indeed, the assimilation of chemicals into our daily activities (e.g., cleaning, washing, gardening, preparing of food and beverages) and maintenance has become commonplace. One may even state that today we are a chemical society!

The variety, number, and volume of chemicals available today are quite staggering, as is the manner of their production, use, and disposal. In fact, many chemicals are discharged directly into water bodies and have reached surface and groundwater systems through use in homes and agriculture fields, or from industrial disasters. The population and ecosystems of the United States are subject to possible exposure to more than 75,000 synthetic chemicals, most of which are poorly tested or untested for potential health effects. Of this large list, more than 70,000 such substances are in commercial use, and only a small fraction are known to be tested and evaluated. Although there exists some scientific information about the adverse biological effects of some substances (e.g., solvents, metals, and pesticides), newly emerging evidence on the toxicity of others has been ignored. Soon after entering into soil, water, or atmosphere, almost all chemicals change composition, and this sets off a chain of actions with organisms in the environment. In fact, chemical disasters, resulting from contaminant seepage into soil, runoff into rivers and sources of drinking water, and chemical explosions, have become a global problem during the past few decades.[1,2] This process cannot continue. In the enthusiasm of new product development, one cannot ignore aspects of human health, environmental contamination, and other subsequent hazards. Disasters such as animal and plant life destruction, genetic deformities, human immune system, damage and tumor growth, and many other disturbances demand us to be pragmatic in the management of chemicals. Man has synthesized

*Industrial Guide to Chemical and Drug Safety*, By T.S.S. Dikshith and Prakash V. Diwan
ISBN 0-471-23698-5 

and mixed chemical formulations for use in agriculture and forestry (pesticides), homes (colors, cosmetics, detergents, drugs, paints, and soaps), industry (metals, solvents, plastics), and improved quality of life. It is ironic that the very properties that make these chemicals desirable for use (e.g., toxicity to kill crop pests, persistence to eliminate worms and vectors of diseases) have caused harm to human health and the environment. To what is this due? Investigations have indicated that we have become complacent in chemical management, more so with toxic chemicals. All chemicals should be handled with great care, as they deserve to be handled, since they are toxic. This requires a thorough knowledge of the proper handling, use, and disposal of chemicals, as well as qualified and experienced personnel, whether in a laboratory, chemical industry, pest control operation, or transportation and disposal of waste. Basic knowledge of chemical safety affords a worker the ability to protect his health and the environment. This, in turn, provides the benefits of chemicals rather than disasters.

## CHEMICALS AND THEIR IMPACT

A wide variety of chemicals, ranging from mild to extremely toxic, have been used to improve the quality of human life in areas such as health, agriculture, and daily regiments. Examples of these chemical products are drugs and pharmaceuticals, pesticides and fertilizers, soaps, detergents, cosmetics, solvents, packaging materials, dyes and pigments, and many more. Due to past uncontrolled use, misuse, or accidental leakage from manufacture sources, chemicals can be found widespread throughout water, sediments, and even tissues of aquatic organisms. Concentrations of toxic chemicals in lakes, ponds, and the atmosphere — although decreasing in many areas — are still over prescribed levels of safety.

A number of complex chemical compounds (e.g., pesticides, vapors and fumes, automobile exhausts, petroleum distillates, foundry fumes, heterocyclic amines, solid particles) have caused adverse effects to humans and environment. Some persistent contaminants have originated from industrial sources. For example, DDT, mirex, PCBs, dioxin, and others have been traced in the atmosphere as toxic depositions causing concern to human health.[3]

According to the Bay Area Air Quality Management District in San Francisco, more than 5.5 million pounds of toxic chlorinated substances are released in the area annually. This includes approximately 13,000 pounds of chloroform, 1.4 million pounds of freon, 2 million pounds of perchloromethylene, and trace amounts of dioxin. (Dioxin is one of the most toxic chemicals known.) A report released in September 1994 by the U.S. Environmental Protection Agency (EPA) clearly describes dioxin as a serious public health threat. The public health impact of dioxin may rival the impact that DDT had on public health in the 1960s. According to the EPA report, not only does there appear to be no "safe" level of exposure to dioxin, but levels of dioxin and similar chemicals have been found in the U.S. population that are "at or near levels associated with adverse health effects." The EPA report also confirmed that dioxin is a cancer hazard, exposure to dioxin also can cause severe reproductive and developmental problems (at levels 100 times lower than those associated with its cancer-causing effects), and dioxin can cause immune system damage and interfere with regulatory hormones.

The Niagara River has been a focus since 1987 as a major source of environmental contaminants. Both the U.S. and Canadian environmental agencies have made rigorous efforts to reduce pollution of toxic substances. Further, the EPA and New York State Department of Environmental Conservation identified 26 U.S. hazardous sites. The Niagara River Toxic Management Plan (NRTMP) has named several toxic chemicals, 10 of which are considered of major concern (Table 1-1).

In view of these developments, attempts have been made to understand the impact of different chemicals on human health. The United States and Europe currently are actively associated with major studies relating human health to toxic chemicals and compounds. An international group on humans has been formed to study the combination effects of chemicals and to more precisely understand the phenomenon of synergism and antagonism. Investigations in Europe include the development and application of statistically designed experiments combined with multivariate data analysis. Further, the researches also are modeling in *in vitro* and *in vivo* studies on a variety of chemicals, including petroleum hydrocarbons, aldehydes, food contaminants, industrial solvents, and mycotoxins. The

**TABLE 1-1 Chemicals of Priority Under the Niagara River Toxic Management Plan**

| |
|---|
| Arsenic |
| Benz(a)anthracene* |
| Benzo(a)pyrene* |
| Benzo(b)fluoranthene* |
| Benzo(k)fluoranthene* |
| Chlordane |
| Chrysene |
| DDT |
| Dieldrin |
| Dioxins* |
| Hexachlorobenzene* |
| Lead |
| Mercury* |
| Mirex* |
| Octachlorostyrene |
| PCBs* |
| Tetrachloroethylene* |
| Toxaphene. |

*Source*: NRTMP[4], 1988.
*Indicates chemical of concern.

significance of these studies is to develop safety evaluation strategies such as the use of toxic equivalence factors or alternatives. Some of these approaches include, the question-and-answer strategy, fractionation followed by recombination of compounds with mixture design, and quantitative structure–activity relationship analysis combined with lumping analysis and physiologically based pharmacokinetic and pharmacodynamic modeling. Recently, Feron and coworkers[3,4] reported the hazard identification and risk assessment of complex chemical compounds. They have suggested the possibility of a consistent method for generating total volatile organic compound values for indoor air.[3,4]

It is known that the health effects from environmental toxins may be a more serious problem in third world countries than in developed countries because of economic conditions. This is because several factors modulate a chemical's toxicity; this will be discussed in Chapter 2. However, it may be noted that failure to enforce appropriate regulations is a universal factor in negligent management practices for both developed and third world countries. Occupational exposure to toxic chemicals is higher in third world countries than in developed countries due to lack of stringent regulations, lack of knowledge of the risks involved, and worker negligence. General pollution is another important issue; developed countries have established strict pollution regulations, and risky industrial processes are being exported to third world countries along with banned substances and dangerous wastes. However, it should be emphasized that stringent regulations in developed countries will not prevent exposures in the long term, because toxic substances released into the environment will ultimately affect all future generations.[5]

Much importance is being given to understanding the impact of mixtures of different chemicals and the risks to human health. Reports have shown that the United States and Europe are actively associated with major studies on human health and related aspects about chemicals and their mixtures. In fact, an international study group on combination effects of chemicals has been formed to understand more precisely the phenomenon of synergism and antagonism. Investigations in Europe include the development and application of statistically designed experiments combined with multivariate data analysis. Furthermore, the researchers are also modeling in *in vitro* and *in vivo* studies on a wide variety of chemicals such as petroleum hydrocarbons, aldehydes, food contaminants, industrial solvents, and mycotoxins to understand the intricacies of the phenomenon. Many approaches are being developed, for example, safety evaluation strategies for mixtures of chemicals. The use of toxic equivalence factors or alternatives, the question-and-answer approach, fractionation followed by recombination of the mixture in combination with a mixture design, and quantitative structure–activity relationship analysis combined with lumping analysis and physiologically based pharmacokinetic and/or pharmacodynamic modeling are some of the leads.[3]

With the extensive use of pest control chemicals in agriculture and industry, residues of organochlorine pesticides and polychlorinated biphenyls (PCBs) have been discovered to elicit toxicological effects on aquatic organisms and wildlife. It is well known that these compounds are lipophilic (meaning attraction to fat molecules) and may become chemically bound to organisms. The toxic chemicals then accumulate in the body of organisms through the process of bioaccumulation. Once absorbed by species of plankton and other aquatic organisms, the organochlorine molecules work their way up levels of the food web, from small plankton to large plankton, to crustaceans and other invertebrates, to small fish, and finally to larger predatory fish and fish-eating species like osprey, gulls, terns, cormorants, or humans.

At each step in the food web, organochlorine contaminants undergo biomagnification and remain firmly bound to lipid molecules. The term *biomagnification* is self-explanatory, whereby the minuscule levels of toxic chemicals in the water become concentrated in wildlife bodies, often ending up at high concentrations in organisms at the top of the food web.[6–8]

## Pesticides

Pesticides have played an important role both as protectors of crops and food storage and as pollutants of the environment, based on the use and disposal of pesticides. In the early 1960s, the pesticide debate was largely confined to the industrialized nations. Today, however, pesticides are produced and used more globally. For instance, in 1960, India used pesticides for application on 6 million hectares per year; by 1988, this number increased to 80 million hectares. Brazil, India, and Mexico are now among the largest pesticide consumers in the world. (The leading pesticide producers are the United States, Western Europe, and Japan.) The third world's use of pesticides increased greatly during Green Revolution in the 1960s and beyond. In a 1996 study, it was reported[8a] that the United States exported approximately 25 million pounds of pesticides that are banned for use in the United States. More than 344 million pounds of hazardous pesticides have been exported worldwide. It also has been reported that chlordane, which has restricted use and is banned in the United States and 47 other countries, is still exported to countries in Asia, as well as Argentina and Venezuela.

Many types of pesticides are designed to kill bugs, fungus, and crop pests. Laboratory studies have found these chemicals to cause cancer, mutations, nervous system disorders, or hormonal disruptions; epidemiological data, though inadequate, have caused concern. In view of the high lipid solubility and slow rate of metabolism, the compounds accumulate in the tissues and organs of animals. Some of these compounds undergo oxidative biotransformation. These risks are increased with children and growing infants, who are far more sensitive than adults. A faster and different metabolism, as well as rapid growth and development, are the basic reasons for an infant's increased vulnerability to any toxic substance. Human exposure to pesticides occurs mostly from ingestion of contaminated foods, such as root crops, fish, or seafood. In fact, the buildup of aldrin and dieldrin in the human body occurs after years of exposure and can damage the nervous system. The toxicological significance of pesticide residues through food, fruits, and drinking water is not known.

Heptachlor epoxide and dieldrin have been identified in the milk of cows that were given aldrin and heptachlor in their feed. As an extrinsic factor, sunlight modulates the decomposition of aldrin and dieldrin into photodieldrin. Use of a large number of organophosphorus insecticides has caused adverse effects including neurologic disturbances and cancer in humans. This calls for proper management procedures by qualified and experienced personnel.[9–17]

Fetal mice were exposed *in utero* to low doses of DES, *o,p′*-DDT and methoxychlor and examined during adulthood for the rate of territorial marking in a novel territory. Each chemical had a strong, dose-dependent effect on this element of behavior, which increased above the controls and has become evident at all tested levels. According to vom Saal et al.,[12] during fetal life, hormones have marked effects on subsequent social behavior. Therefore, perturbation of systems that differentiate under endocrine control may result in the disruption not only of organ function, but also of an individual's social interactions, and these effects on social behaviors may be dramatic. If animals within a population show changes in social-sexual behaviors, a marked disturbance in social structure can occur. Thus, it has been observed that many pesticides induce birth defects in animal species. It is a matter of concern that the farming populations in Minnesota showed a higher rate of birth defects, particularly in children of farmers who were directly exposed to toxins at work or who live in and around agricultural regions. These areas were known to use heavy amounts of fungicides and chlorophenoxy herbicides.[18,19]

## Toxic Chemicals and Eggshell Thinning

In the past, dangerous levels of toxic chemicals were found in osprey eggs near the Great Lakes region. These chemicals directly affected the thickness of the eggs and thus the safety of the osprey population in this region. Although a range of persistent toxic chemicals are still traced in the eggs and chicks of osprey around parts of the Great Lakes, management of this contamination problem has begun to yield positive results. Most organochlorine pesticide levels in osprey eggs from the Great Lakes basin have declined over the past two decades. With increased and stringent regulatory management, it has been found that few eggs contained more than 4 ppm of the chemical DDE or had eggshells more than 10% thinner than those prior to the introduction of the chemical DDT. However, the information, in contrast to organochlorines, is still very meager. Residues of chemicals like PCBs in eggs have declined only slightly and others (e.g., polychlorinated dibenzo-p-dioxins [PCDDs] and polychlorinated dibenzofurans [PCDFs]) may have much greater toxicity even though they occur at much smaller concentrations.

Recent studies[19a,19b] indicate that most contaminants occur below suspected critical levels in osprey eggs, but eggs and chicks in the Great Lakes region are usually more contaminated than those further inland. For example, studies carried out in the 1990s showed that the concentrations of the most toxic PCDD (2,3,7,8-TCDD) and PCDF (2,3,7,8-TCDF) were generally higher in osprey eggs from

Lake Huron than from eggs further inland. Fish and fish-eating birds show similar patterns of geographic variation with these contaminants, suggesting that ospreys are accumulating toxic chemicals through the local food web. Much higher levels of these compounds have been found recently in osprey eggs from the vicinity of pulp mills in British Columbia, and in eggs of herring gulls and bald eagles on the Great Lakes. During the 1970s, thinning eggshells in birds of prey became a major concern and was traced to organochlorine contamination of the aquatic systems in the United States and other countries. Eggshell thinning was the first reproductive problem related to contaminants identified in fish-eating birds on the Great Lakes. Eggshells are made of calcium carbonate, which is synthesized from a chemical reaction of calcium and carbon dioxide in a bird's body. Thus chemical reaction is disturbed by DDE, and the eggshells cannot deposit the required amount of calcium carbonate, thus becoming thinner. The fragility of the eggs leads to breakage during incubation. Further, the exchange of gases through the pores in the eggshell is also affected by DDE contamination. Reports[19c,19d] have proven the thickness of osprey eggs was significantly associated with the level of DDE in the eggshell. For instance, 2 ppm (wet weight) of DDE caused approximately 10% eggshell thinning, 4 ppm caused 15%, and 9 ppm caused 20%. Eggs more than 15% thinner due to DDE levels often suffered breakage before hatching. A range of persistent toxic chemicals are still found in osprey eggs and chicks from all parts of the Great Lakes breeding area. However, since the early 1970s, declining DDE levels have been associated with increasing eggshell thickness. The problem of broken eggs is no longer regularly observed in Great Lakes osprey nests.

## Heavy Metals

A metal is regarded as toxic if it injures the growth or metabolism of cells when it is present above a given concentration. Almost all metals are toxic at high concentrations, and some are severe poisons even at very low concentrations. Copper, for example, is a micronutrient, a necessary constituent of all organisms, but if copper intake is increased above the proper level, it becomes highly toxic. Like copper, each metal has an optimum range of concentration, in excess of which the element is toxic. The metals that have been of concern are cadmium, mercury, tin, lead, vanadium, chromium, manganese, cobalt, and nickel. The toxicity of a metal depends on its route of administration and the chemical compound with which it is bound. Combining a metal with an organic compound may either increase or decrease its toxic effects on cells. On the other hand, the combination of a metal with sulphur (to form a sulphide) results in a less toxic compound than the corresponding hydroxide or oxide, because the sulphide is less soluble in body fluids than the oxide. Toxicity generally results when an excessive concentration is exposed to an organism over a prolonged period of time; when the metal is presented in an unusual biochemical form; or when the metal is presented to an organism by way of an unusual route of intake. Less well understood, but perhaps of equal significance, are the carcinogenic and teratogenic properties of some metals. The association of lead with human activities and the resulting adverse effects need no emphasis except that lead has severely disturbed vital functions in adults and children.[17,20]

Humans have been exposed more and more to metallic contaminants in the environment, mostly from the products of industry. There are three main sources of metals in the environment. The most obvious are the processes of extraction and purification: mining, smelting, and refining. Another is the release of metals from fossil fuels (e.g., coal, oil), when these are burned. Cadmium, lead, mercury, nickel, vanadium, chromium, and copper are all present in these fuels, and considerable amounts enter the air or are deposited in ash. The third and most diverse source is the production and use of industrial products containing metals, which is increasing as new applications are found. The modern chemical industry, for example, uses many metals or metal compounds as catalysts; metal compounds are used as stabilizers in the production of many plastics, and metals are added to lubricants, which then find their way into the environment.[21]

Mercury is an especially dangerous compound. Mercury levels have changed relatively little over the past two decades. Mercury is now used, in increasing quantities, in parts of the Amazon basin where prospectors pan for gold along small streams and tributaries. Atmospheric deposition is now a major source of mercury as well in the Great Lakes ecosystem.

Arsenic is a powerful poison; at high levels, it can cause death or illness. Exposure to higher than average levels of arsenic happens mostly in the workplace, near hazardous waste sites, or in areas with high natural levels. Arsenic is organically found in nature at low levels. The inorganic arsenic compound combines with carbon and hydrogen in plants and animals and becomes organic arsenic. Compared to the inorganic form, organic arsenic is usually less harmful. Inorganic arsenic compounds are used for wood preservation and the formulation of insecticide products.

Another dangerous compound is cadmium. Exposure to cadmium happens mostly in the workplace where cadmium products are produced. The general population is exposed to cadmium most commonly from breathing cigarette smoke or eating cadmium-contaminated foods. Cadmium damages the lungs, can cause kidney disease, and may irritate the digestive tract.

### Endocrine-Disrupting Chemicals

Endocrine-disrupting chemicals (EDCs) have been defined as exogenous substances that alter function(s) of the endocrine system and consequently cause adverse health effects in the organism and the progeny. There is a growing concern that many substances interfere with the normal functioning of the body governed by the endocrine system. These have the potential to cause adverse affects in both humans and wildlife. Laboratory studies using species of animals have demonstrated that certain synthetic chemicals affect the immune system. The list of chemicals that disrupt the endocrine system includes the aromatic hydrocarbons, carbamates and other pesticides, heavy metals, organohalogens, organophosphates, organotins, oxidant air pollutants, such as ozone and nitrogen dioxide, polycyclic aromatic hydrocarbons, synthetic hormones, monomers

**TABLE 1-2 Causative Agents in Disrupting Endocrine Functions in Animals and Humans**

| |
|---|
| Pesticides |
| Heavy metals (cadmium, lead, mercury) |
| Organochlorines |
| Dioxins |
| Polyvinyl chloride |
| Pentachlorophenol |
| Plasticizers and surfactants |
| Phthalates |
| Polycarbonates, styrenes |
| Ethoxylates |

and additives used in the plastics industry, and detergent components. Some of the potential environmental EDCs are ubiquitous and persistent, and easily cross national boundaries (Table 1-2).

Many chemicals present in the work and general environments have the potential to disturb the immune system of wildlife and humans. The consequences of such interference on the developing immune system are not well understood, and the literature is still inadequate. New evidence is coming to light, which demonstrates the combined impacts of complex mixtures that are highly unpredictable and sometimes synergistic. Moreover, one of the paradoxes of endocrine disruption is that in some cases, higher doses do not cause adverse effects, whereas very low doses can. Reports have shown that chemicals have caused changes such as thyroid dysfunction in birds and fish; decreased fertility in birds, fish, shellfish, and mammals; decreased hatching success in birds, fish and turtles; gross birth deformities in birds, fish, and turtles; metabolic abnormalities in birds, fish, and mammals; and behavioral defects in birds. These effects are manifested as alterations in the immune system that may lead to a decreased quality of life. These alterations include immune modulation expressed as an increase or decrease in measured immune parameters, hypersensitivity, and autoimmunity. More studies are required on the development of the immune system of diverse animal species and the factors that lead to maturation and senescence. Further study also is needed to understand the mechanistic role of synthetic chemicals in the alteration of these processes. With the ubiquitous nature of environmental chemicals to which wildlife and human populations have unknown exposure, it becomes difficult to identify suitable control populations (i.e., populations with no exposure levels), indicating that "true" control populations for epidemiological studies are lacking.

Given the complexity of endocrine systems, it is not surprising that the range of substances thought to cause endocrine disruption is wide, and includes both natural and manufactured (synthetic) chemicals. Industrial, agricultural, and municipal wastes can expose organisms in the environment to unusually high concentrations of natural substances, such as sex hormones or phytoestrogens. Manufactured chemicals (e.g., pesticides, dioxins, PCBs) may be released intentionally, as by-products of industrial processes and waste disposal, or as discharges from industrial or municipal treatment systems (alkylphenols). The variety of sources and substances presents an enormous challenge to environmental managers in industry and government.[22,23]

Arnold et al. found 150- to 1600-fold synergistic interactions among weakly estrogenic pesticides using a test system in which yeast cells were genetically engineered to contain functional human estrogen receptors. They then exposed the yeast to low levels of endosulfan, dieldrin, toxaphene, and chlordane. The chemicals administered individually provoked only partial estrogenic reactions. While in combination, the estrogenic activity was increasingly significant than any one chemical alone. Chlordane increased the impact of other chemicals when put into combination, but alone it caused none.[24] The striking thing about these findings with the same yeast screening method is not the existence of synergistic effects, but rather their reported magnitude. However, these results could not be confirmed by other researchers,[25] which indicates inadequacy of the data and the need for more studies.

Endocrine systems are complex mechanisms, coordinating and regulating internal communication among cells. Endocrine systems release hormones that act as chemical messengers. The messengers interact with cell receptors to trigger responses and prompt normal biological functions such as growth, embryonic development, and reproduction. Scientists know that endocrine systems can be adversely affected by a variety of substances. A large number of compounds bioactivate and interfere with the normal function of body systems.[26] These disturb the communication between the messenger and cell receptors, so that the chemical message is not interpreted properly. Even subtle effects on the endocrine system can result in changes in growth, development, reproduction, or behavior that can affect the organism itself or its next generation.[24] The specific mechanisms by which substances disrupt the endocrine systems are complex and not yet completely understood. Global concern, expressed by international and U.S. authorities, about the potential adverse effects to human and ecological systems and the specific impact on endocrine systems prompted the World Health Organization (WHO) (IPCS) to take measures to address these issues on a larger scale.

### Polychlorinated Biphenyls, Polycyclic Aromatic Hydrocarbons, Dioxins, Toxic Gases, Vapors, and Their Pollutants

Polychlorinated biphenyls (PCBs) were first synthesized in 1864, and commercial use has been active since 1929. Over 1 million tons of PCBs have been produced commercially with different trade names, such as Aroclor, Clophen, Fenchlor, and Kanechlor. There are 209 PCB isomers or types, which differ from each other in the number and relative position of the chlorine atoms on the biphenyl molecular frame. A small number of these isomers are particularly toxic and are believed to account for the bulk of PCB-induced toxicity in animals. PCBs can produce large amounts of furans when they are burned, and the chemicals that are often mixed with PCBs for use in electrical equipment can produce dioxins. Fires involving transformers and capacitors have contaminated buildings, power stations, locomotives, and other locations with dioxins and furans. Many products and wastes that are contaminated with dioxins and furans will produce even larger amounts when burned. For example, when treated wood is burned, the chlorophenols that burn with it could be a widespread source of dioxins and furans. Wood treatment facilities often collect waste pentachlorophenol in ponds, and in the past have set fire to the ponds to reduce their volume. This practice generated significant amounts of dioxins and furans. PCBs and polybrominated biphenyls (PBBs) have been widely used as dielectrics in transformers and large capacitors; in heat transfer and hydaulic systems; in lubricating and cutting oils; and as flame retardants in textiles, carpets, and plastics. These materials are resistant to igniting but will burn in a building fire or incinerator. When they burn, they can be a source of dioxins and furans. PCBs are also in use as ink solvent/carriers in carbonless copy paper, as adhesives, and as sealants. PCBs with extremely stable molecules are desirable for industrial uses, but they also persist for a long time once released into the environment. The low flammability of PCBs make them useful as lubricating oils and fire retardants in insulating and heat-exchanging fluids used in electrical transformers and capacitors.[27,28] PCBs also were used as plasticizers and waterproofing agents. Industrial manufacturers voluntarily cut back PCB production in 1971. In Canada, the use of PCBs was regulated in 1977 under the Environmental Contaminants Act. PCBs have not been manufactured in North America since 1978, and the importation of all electrical equipment containing PCBs was banned after 1980, restricting their use to existing equipment.

Exposure to polycyclic aromatic hydrocarbons (PAHs) usually occurs by breathing air contaminated by wild fires or coal tar, or by eating foods that have been grilled. PAHs are a group of over 100 different chemicals that are formed during the incomplete burning of coal, oil and gas, garbage, or other organic substances like tobacco or charbroiled meat. PAHs are usually found as a mixture containing two or more of these compounds, such as soot. *Dioxin* is a general term that describes a group of hundreds of chemicals that are highly persistent in the environment. The most toxic compound is 2,3,7,8-tetrachlorodibenzo-p-dioxin (TCDD), but animal species vary considerably in their sensitivity to this chemical. Elevated levels of 2,3,7,8-TCDD in the environment are linked closely to effluent from previous 2,4,5-trichlorophenol manufacturing (for wood preservatives) and to toxic waste disposal sites associated with this manufacturing. Atmospheric deposition of 2,3,7,8-TCDD, either bound to dry airborne particles or in rain or snow, is now a major source of this and other organochlorine compounds, particularly in the upper Great Lakes regions. The toxicity of other dioxins and chemicals, such as PCBs, that act similar to dioxin are measured in relation to TCDD.

Dioxin is formed as a by-product of many industrial processes involving chlorine such as waste incineration, chemical and pesticide manufacturing, and pulp and paper bleaching. Dioxin was the primary toxic component in Agent

Orange, found at Love Canal in Niagara Falls, New York, and was the basis for evacuations at Times Beach, Missouri, and Seveso, Italy. Dioxin is also the popular name for a class of chlorinated hydrocarbon compounds known as polychlorinated dibenzo-p-dioxins (PCDDs). PCDDs, along with polychlorinated dibenzofurans (PCDFs), are produced as by-products during chemical reactions involving high temperatures in the presence of chlorine. The original sources of atmospheric PCDDs and PCDFs today include municipal incinerators, which burn a wide range of chlorinated compounds put out with the trash, and exhaust from vehicles burning leaded gasoline or diesel.

PCDDs, PCDFs, and PCBs are chemically classified as halogenated aromatic hydrocarbons. The chlorinated and brominated dibenzodioxins and dibenzofurans are classified as tricyclic aromatic compounds with similar physical and chemical properties, and both classes are similar structurally. Certain PCBs (the so-called coplanar or mono-ortho coplanar congeners) are also structurally and conformationally similar. The most widely studied of these compounds is TCDD; this compound, often called simply dioxin, represents the reference compound for this class.

Polyvinyl chloride (PVC) is a common plastic that can produce dioxins and furans when burned. PVC is often present in municipal waste in large amounts, and is believed to contribute to the dioxins and furans from incinerators. Many sources of combustion produce dioxins and furans. Incinerators, both municipal and industrial, are significant sources; dioxins and furans have been found in incinerator ash and in gases and tiny particles escaping through smokestacks. Power plants, smelters, steel mills, oil and wood stoves, and furnaces all emit dioxins and furans.

The internal combustion engine of petroleum-powered motor vehicles discharges carbon monoxide, lead, nitrogen oxides, aldehydes, ethylene, and other aliphatic hydrocarbons into the atmosphere only a few feet from the breathing zone of the population. Local concentrations of these substances reach appreciable levels, which are highest in urban centers where traffic density is greatest, at major intersections, and in so-called "canyon streets." Under conditions of poor natural ventilation and strong sunlight, a complex series of reactions take place between the nitrogen oxide and hydrocarbons, leading to the formation of ozone peroxyacyl nitrates (PANs) and other substances (usually grouped together as photochemical oxidants). This more extensive type of motor vehicle pollution can affect the air of an entire community (e.g., Los Angeles smog). Emissions of aliphatic hydrocarbons other than ethylene are not considered important, except in the existence of photochemical pollution. This may be significant regarding damage to forests and crops.[29–31]

## Greenhouse Gases

Gases, such as water vapor, carbon dioxide, tropospheric ozone, nitrous oxide, methane, and chloroflurocarbons (CFCs), are largely transparent to solar radiation but opaque to outgoing long-wave radiation. Their action is similar to that of glass in a greenhouse. Some of the long-wave (infrared) radiation is absorbed and reemitted by the greenhouse gases. The effect of this is to warm the surface and the lower atmosphere of the Earth. The CFCs are a family of inert, nontoxic, and easily liquefied chemicals used in refrigeration, air conditioning, packaging, and insulation, or as solvents or aerosol propellants. Because they are not destroyed in the lower atmosphere, they drift into the upper atmosphere, where, given suitable conditions, their chlorine components break down the ozone layer. Since the beginning of the Industrial Revolution, atmospheric concentrations of carbon dioxide have increased nearly 30%, methane concentrations have more than doubled, and nitrous oxide concentrations have risen by nearly 15%. These increases have enhanced the heat-trapping capability of the Earth's atmosphere. Sulfate aerosols, a common air pollutant, cool the atmosphere by reflecting light back into space; however, sulfates are short-lived in the atmosphere and vary regionally. In addition, other gases such as hydrogen sulfide fumes, ozone, and associated air-polluting gases have caused a variety of adverse effects in the respiratory system of animals and humans.[32–37]

### Ozone Depletion

Ozone depletion is the result of a complex set of circumstances and chemistry. Since the appearance of an "ozone hole" over the Antarctic in the early 1980s, Americans have become aware of the health threats posed by ozone depletion, which decreases our atmosphere's natural protection from the sun's harmful ultraviolet (UV) rays. The Earth's climate is predicted to change because of the human activities altering the chemical composition of the ozone, through the buildup of greenhouse gases—primarily carbon dioxide, methane, and nitrous oxide. The heat-trapping property of these gases is undisputed. Although uncertainty exists about how the Earth's climate responds to these gases, it is certain that global temperatures are rising. Worldwide concern about possible climate change and acceleration of sea-level rise—resulting from increasing concentrations of greenhouse gases—has led governments to consider international action to address the issue. An international task force was developed to do just that: the United Nations Framework Convention on Climate Change (UNFCCC). The United States, in cooperation with over 140 countries, is phasing out the production of ozone-depleting substances in an effort to safeguard the ozone layer.

The ozone layer around the Earth has provided us all a natural ring of protection from harmful UV radiation. However, multiple anthropogenic activities have acted as barriers and often caused damage to this shield. Less protection from UV light will, over time, lead to increased health problems and crop damage. Major health problems linked to overexposure to UV radiation by the depletion of ozone include skin cancer (melanoma and nonmelanoma), premature aging of the skin and other skin problems, cataracts and other eye damage, and suppression of normal immune system function.[29]

## Motor Vehicle–Emitted Pollutants

There are four recognized pollution sources from the ordinary automobile, namely, the exhaust pipe, the crank case, the carburetor, and the fuel tank. Tire and road dust and asbestos particles from brake linings are not generally included in any discussion of the problem; some pollution from these sources will certainly continue, even if all other emissions can be eliminated. The distribution of pollutants within the vehicle are many. For instance, (1) evaporation losses, tank and carburetor (20% of the hydrocarbons); (2) crank case blow-by (25% of the hydrocarbons); and (3) exhaust (55% of hydrocarbons and almost all of the lead, carbon monoxide, and nitrogen oxides). The gases from diesel engine exhaust contain negligible amounts of carbon monoxide, no lead, and somewhat lower amounts of lighter hydrocarbons per unit of fuel consumed than gasoline-powered vehicles. They are, nevertheless, recipients of much public criticism because they are offensive, malodorous, and have a high content of particulate matter. Because diesel vehicles are small in number compared with gasoline-powered vehicles, their overall contribution to pollution levels is not great. Nevertheless, because they often discharge pollutants in close proximity to people, they must be regarded seriously; in cities, diesels make a significant contribution to soiling buildings and materials. In other countries, the number of motor vehicles is much smaller, but owing to their concentration in cities, it is possible that pollution comparable to levels in the United States may occur in the future. Local concentrations of the characteristic pollutants are inevitable in any city, and the likelihood of photochemical pollution in widely scattered parts of the world must be recognized.

Air pollution from motor vehicles in third world countries does not yet present as great a problem as in highly industrialized countries. The number of cars in use is relatively small, so the pollution caused by them is much less than that from industrial complexes. A typical example is found in India, where large installations, such as chemical or petrochemical complexes, fertilizer, and power plants surround or are scattered among most large cities. The discharges from these installations are so great that, proportionally, pollution from automobiles is insignificant, except in a few large cities.

However, the effects from automobile exhausts in these cities are similar to those in industrial cities of advanced countries, because many vehicles have a high weight-to-horsepower ratio and are often old and poorly maintained. The horsepower of 85% of the cars in India is between 10 and 14, and 60% of all vehicles are more than 10 years old. Vehicle maintenance is poor because spare parts are expensive or unavailable, and technical competence is low. Consequently, pollution is out of all proportion to the number of cars in circulation. Carbon monoxide peaks of 100 ppm have been recorded at street level at major intersections. As the number of vehicles continues to increase, it is expected that oxidant pollution may become a problem in other cities if control measures are not introduced.

## NEED OF THE TIME

Industrial development is the order of the present day. During the enlargement of global technology of industries, humans are using and disposing of a variety of chemicals, such as solvents, metals, polymers, food coloring/dyes, drug detergents, and a large number of agrochemicals. In fact, these chemicals have become a common part of the workplace and the home. Increasing technologies, then requiring increased use, of chemicals, metals, and solvents in one measure or other, improper use, handling, storage, and waste disposal have left a trail of adverse effects both to the worker and the living environment. This must be properly managed to avoid further unfortunate occurrences. Workers must be properly educated about the risks and hazards of the chemicals that are being used. The need of the time is to educate the bench worker, the management, and others about the chemical properties, uses, and dangers to achieve human safety and contain chemical toxicity.

## CONCLUSION

By acknowledging that cancer may not be the only long-term health threat we face, we can modify the way we define toxicity and determine public health risk. Finding answers to increasingly complicated and interrelated problems that have environmental, health, economic, and political consequences is no easy matter. To understand the long-term health effects in a meaningful way, more research is needed. Regulatory agencies, industries, and universities worldwide are devoting time, money, and research to learn how environmental estrogens may affect wildlife and human health. More studies would provide answers to a number of questions, for instance, (1) to discern problems in the correct perspective; (2) to identify chemicals and develop reliable and sensitive test systems and methods; (3) to identify related health effects; (4) to understand how individual factors such as species, age, dose, length of exposure, and genetics determine health effects; and (5) to determine how multiple synthetic chemicals react with each other in the body and the environment.

The manner of using toxic chemicals (e.g., use of pesticides nearly 50 years ago) with overenthusiasm made us throw the elementary caution of safety to the wind. We followed the same trend of management with scores of newer and exciting chemicals. The goal for us now in the twenty-first century is to move forward and address the problem with pragmatism and precautionary principles to achieve maximum safety and improve quality of life.

## REFERENCES

1. T. Schettler, G. Solomon, M. Valenti, and A. Huddle, *Generations at Risk, Reproductive Health and the Environment*. The MIT Press, Cambridge, MA, 1999.

2. L. Guillette, Endocrine-disrupting environmental contaminants and developmental abnormalities in embryos. *Hum. Ecol. Risk Assess.* **1**:25, 1995.
3. V.J. Feron, F.R. Cassee, and J.P. Groten, Toxicology of chemical mixtures: International perspective. *Environ. Health Perspect.* **106**(suppl. 6):1281, 1998.
4. Anonymous. *The Report, U.S. Environmental Protection Agency and the New York State Department of Environmental Conservation*, U.S. Environmental Protection Agency, Washington, DC, November, 1998.
5. P. Ostrosky-Wegman, and M.E. Gonsebatt, *Environ. Health Perspect.* **104**(suppl. 3): 599, 1996.
6. G.C. Butler, ed., *Principles of Ecotoxicology*, Scope 12, John Wiley & Sons, New York, 1978.
7. R. Truhaut, Ecotoxicology objectives, principles and perspectives. In Hunter, W.J., and Smeets, J.G.M., eds., *Evaluation of Toxicological Data for the Protection of Public Health.* Pergamon Press, Oxford, 1977, 339.
8. D.C. Malins, and G. Ostrander, *Aquatic Toxicology: Molecular, Biochemical, and Cellular Perspectives*, CRC Press, Boca Raton, FL, 1994.

8a. Pesticide Action Network, Pesticide production in the South Linking production and trade. *Pesticides News* No. **26**, 7–10, PAN, London, United Kingdom, December 1994.

9. Agency for Toxic Substances and Disease Registry. *Toxicological profile for alpha-, beta-, gamma- and delta-hexachlorocyclohexane (update).* U.S. Department of Health and Human Services, Public Health Service, Atlanta, GA, 1994, PB/99/166662.
10. B. Eskenazi, A. Bradman, and R. Castorina, Exposure of children to organophosphate pesticides and their potential adverse health effects. *Environ. Health Perspect.* **107**(suppl. 3):409, 1999.
11. V.F. Garry, D. Schreinemachers, M.E. Harkins, and J. Griffith, Pesticide appliers, biocides and birth defects in rural Minnesota. *Environ. Health Perspect.* **104**(4):394, 1996.
12. F. vom Saal, S. Nagel, P. Palanza, M. Boechler, S. Parmigiani, and W. Welshons, Oestrogenic pesticides: Binding relative to estradiol in MCF-7 cells and effects of exposure during fetal life on subsequent territorial behavior in male mice. *Toxicol. Lett.* **77**:343, 1995.
13. L. Hardel, and M. Ericksson, A case control study of non-Hodgkinson lymphoma and exposure to pesticides. *Cancer* **85**(6): 1999.
14. M.B. Abou-Donia, The cytoskeleton as a target for organophosphorus ester-induced delayed neurotoxicity (OPIDN). *Chem. Biol. Interact.* **87**(1–3):383, 1993.
15. M.C.R. Alavanja, G. Akland, D. Baird, et al., Cancer and noncancer risk to women in agriculture and pest control: The agricultural health study (update). *J. Occup. Med.* **36**(11):1247, 1994, PB/99/166704.
16. Agency for Toxic Substances and Diseases Registry. *Toxicology profile for lead.* U.S. Department of Health Services, Atlanta, GA, 1999.
17. R.J. Levine, R.M. Moore, G.D. McLaren, W.F. Barthel, and P.J. Landrigan, Occupational lead poisoning, animal deaths, and environmental contamination at a scrap smelter. *Am. J. Pub. Health* **66**:548, 1976.

18. Department of Environment. *Endocrine disrupting substances in the environment.* Environment Canada, Canada, 1999.
19. L. Birnbaum, Endocrine effects of prenatal exposures to PCBs, dioxins, and other xenobiotics: Implications for policy and future research. *Environ. Health Perspect.* **102**(8):676, 1994.
19a. K. Michael, Great Lakes water clear, but contamination high, Chicago Tribune, April 2, 2002.
19b. Government of Canada, *Toxic chemicals in the Great Lakes and associated effects.* Environment Canada, Dept. of Fisheries and Oceans, Health and Welfare Ottawa, Canada, 1991.
19c. Government of Canada, *The Fall and rise of osprey populations in the Great Lakes basin.* Great Lakes Factsheet, Ontario, Canada, 2000.
19d. Pesticide Action Network, Contamination of the Great Lakes. *Pesticides New* No. **21**, 20, PAN, London, United Kingdom, September 1993.
20. S.F. Arnold, D.M. Klotz, B.M. Collins, P.M. Vonier, L.J. Guillette Jr., and J.A. McLachlan, Synergistic activation of estrogen receptor with combinations of environmental chemicals. *Science* **272**:1489, 1996.
21. K. Ramamoorthy, F. Wang, I.C. Chen, et al., Potency of combined estrogenic pesticides. *Science* **275**:405, 1997.
22. M.W. Anders, ed., *Bioactivation of Foreign Compounds*, Academic Press, Inc., New York, 1985.
23. J.L. Jacobson, S.W. Jacobson, and H. Humphrey, Effects of exposure to PCBs and related compounds on growth and activity in children. *Neurotoxicol. Teratol.* **12**:319, 1990.
24. J.L. Jacobson, and S.W. Jacobson, Intellectual impairment in children exposed to polychlorinated biphenyls *in utero*. *New Engl. J. Med.* **335**(11):783, 1996.
25. H. Tilson, J.L. Jacobsen, and W. Rogan, Polychlorinated biphenyls and the developing nervous system: Cross-species comparisons. *Neurotoxicol. Teratol.* **12**:239, 1990.
26. M. Huisman, C. Koopman-Esseboom, C.I. Lanting, et al., Neurological condition in 18 month-old children perinatally exposed to polychlorinated biphenyls and dioxins. *Early Hum. Develop.* **43**:165, 1996.
27. O. Hutzinger, S. Safe, and V. Zitko, *The Chemistry of PCBs*. CRC Press, Cleveland, Ohio, 1974.
28. I. Pomerantz, J. Burke, D. Firestone, J. Mckinney, J. Roach, and W. Trotter, Chemistry of PCBs and PBBs. *Environ. Health Perspect.* **24**:133. 1978.
29. U.S. Environmental Protection Agency. *Ecological impacts from climate change: An economic analysis of freshwater recreational fishing*. Environmental Protection Agency, Washington, DC, 1995.
30. P.R. Jutro, Biological diversity, ecology, and global climate change. *Environ. Health Perspect.* **96**:167, 1991.
31. H.Q.P. Crick, C. Dudley, D.E. Glue, and D.L. Thompson, U.K. birds are laying their eggs earlier. *Nature* **388**:526, 1997.
32. M.J. Bean, Waterfowl and climate change: A glimpse into the twenty-first century. *Orion* **8**(2):22, 1989.

33. D.L. Larson, Potential effects of anthropogenic greenhouse gases on avian habitats and populations in the northern Great Plains. *Am. Midland Natural.* **131**(2):330, 1994.
34. C.L., Winek, W.D. Collom, and C.H. Wecht, Death from hydrogen sulfide fumes. *Lancet* **1**:1096, 1968.
35. R.E. Simson, and G.R. Simpson, Fatal hydrogen sulfide poisoning associated with industrial waste exposure. *Med. J. Aust.* **1**:331, 1971.
36. L.S. Jaffe, The biological effects of ozone on man and animals. *Am. Ind. Hyg. Assoc. J.* **28**:267, 1967.
37. U.S. Department of Health, Education and Welfare. *Air quality criteria for carbon monoxide. Nat. Air Poll. Control Adm.* Publication No. AP-62; U.S. Department of Health, Education and Welfare; Washington, DC, January, 1977.

## CHAPTER 2

# Principles of Toxicity and Safety

## INTRODUCTION

Toxicology has played a significant role in the development of human health and civilization. Since the beginning of human activity, efforts have been made to improve quality of life. Because of the expansion of societal activities, humans have linked activities with economic progress. In fact, the primary goal of modern society has been toward the improvement of health, achievement of happiness, and consolidation of prosperity. With these goals in mind, an array of substances have entered the world market. Increased human activities and uncontrolled use of chemicals, whose potential was not well understood, have complicated the problem. The consequences of improper use of chemicals soon became known. Today, the emphasis has been on proper, judicious management of toxic and hazardous substances for the benefit of society. Therefore, the manner in which we handle chemicals and the contribution of toxicology for chemical safety must be well understood by all.

Production of organic chemicals increased after World War II and has continued to do so. It is estimated that more than 300 billion pounds of various chemicals are produced, and more than 80 billion pounds of toxic waste is dumped every year in the United States alone. Further, more than 2 million tons of toxic garbage cross the U.S. borders annually for purposes of disposal and dumping.[1,1a] The total tonnage of toxic garbage around the world is staggering. Some shipments of used chemicals even threaten entire communities. For example, a shipment of several tons of poisonous mercury waste to a British reprocessing plant in South Africa resulted in the contamination of river water with a concentration of mercury 1.5 million times higher than the standard set by the World Health Organization (WHO).[2] In another instance, more than 3 million tons of waste were shipped from the industrialized world to less developed nations during 1986–1988.[3,4]

A large number of chemical disasters around the world have reminded us of the importance of knowing the basic requirements to achieve chemical safety. Of all the countries of the world, the United States is considered the largest

*Industrial Guide to Chemical and Drug Safety*, By T.S.S. Dikshith and Prakash V. Diwan
ISBN 0-471-23698-5 

producer of chemicals and hazardous wastes. During 1990, it has been estimated that the total bulk of waste garbage from the United States is over 500 million tons.[4,5] It has been well established that irregular disposal and/or improper transportation of potentially hazardous chemicals has resulted in human tragedies and environmental contamination. The tragedies at Love Canal near Niagara Falls, New York, during the 1970s are still fresh in our memories. Again, dioxin-contaminated waste oil on the streets of Times Beach, Missouri, has emphasized such disasters.

Today more than 80,000 chemicals are used worldwide. Most of these chemicals are lacking in detailed and requisite safety data. Inadequate safety data have resulted in serious consequences on human health. The vast majority of chemicals have not been adequately tested for their potential toxicity. Too often, regulations become weak and seldom enforced by proper authorities for multiple reasons. This results in both short-term and long-term effects vis-à-vis hazards to humans and the environment. For instance, reckless and unplanned disposal of used car batteries, used pesticide containers, plastics and packaging materials, dry cleaning fluids, and detergents of different kinds have caused untold tragedies in different countries of the world. Further, many countries have suffered industrial disasters involving human and animal life. A classic and tragic example of such disasters is Love Canal. Approximately 20 years ago, hundreds of residents of the Love Canal area fled their homes to escape chemical disasters caused by dioxin in the sites around them. The health-related problems from dioxin and Love Canal are still fresh in our memories.[6] The gas leakage of methyl isocyanate (MIC) from the Union Carbide pesticide manufacturing plant (Bhopal, India) is another disaster that is still taking its toll on the population, due to aftereffects of MIC gas exposure.[7,8] There have been numerous other incidents: Chlorine leakage in a textile factory (Trichur, India, Minimata, Japan, and Swesso, Italy), leakage of ammonia (Cubatao, Brazil), leakage of sulfuric acid vapor (Karlskoga, Sweden), and leakage of toxic gas (West Virginia, United States). All suggest an urgent need for proper handling in the manufacture of toxic chemicals or disposal of hazardous wastes. Complete involvement of qualified workers, a basic knowledge of toxicology, stringent observance and adherence to regulatory measures by all workers will enable us to contain similar recurrences of these past examples and strengthen practices in chemical safety (Table 2-1).

Different kinds of chemicals have been widely used by humans for years. The world has become flooded with hundreds of chemicals used in daily life, for example, washing, household cleaning, automobiles and driving, pest control activities, drugs and chemical industries, and a host of other activities. Around the world, the toxic effects of these chemicals add a new dimension. They have resulted in the spread of parent chemicals as well as by-products around us. It has been reported that more than 60,000 different chemicals of common use are present in United States today, and approximately 1,000 new commercial chemicals are added to the list each year.[9,10] In 1979, a mass poisoning involving 2,000 people in central Taiwan was caused by ingestion of cooking oil

**TABLE 2-1 Chemical Disasters in Different Countries of the World**

| |
|---|
| Parathion poisoning and food contamination (India, 1958) |
| Malathion poisoning (Corpus Christi, Texas, 1966) |
| Parathion poisoning in Columbia, Egypt, Iran, Malaysia, Mexico |
| Organic mercury poisoning in seed grains (Iraq, 1972) |
| Malathion poisoning among workers (Pakistan, 1997) |
| Leakage of 20,000 gallons of nitric acid (Denver, Colorado, 1983) |
| MIC gas leakage (Bhopal, India, 1984) |
| Ethylene oxide (Arkansas, 1984) |
| Pipe breakage of tetraethyl (Callao, Peru, 1984) |
| Leakage of MIC (Middleport, New York, 1984) |
| Leakage of ammonia from a fertilizer plant (Matamoras, Mexico, 1984) |
| Leakage of chlorine gas from a textile plant (Trichur, India, 1985) |
| Leakage of ammonia from a fertilizer plant (Cubatao, Brazil, 1985) |
| Cloud of chlorine gas (Westmalle, Belgium, 1985) |
| Leakage of sulfuric acid vapor (Karlskoga, Sweden, 1985) |
| Toxic gas in a pesticide plant (West Virginia, United States) |
| Dioxin contamination (Italy, Love Canal, Niagara Falls) |

contaminated by polychlorinated biphenyls (PCBs) and polychlorinated dibenzofurans (PCDFs).[11] Sprawling of industries around the world further aggravate this concern. In addition, most chemicals have never been adequately tested and are not properly regulated. In short, today almost all humans are directly or indirectly exposed to different chemicals. It has been shown that children, even before birth, have accumulated toxic chemicals in their system.[8a,8b] Chemicals such as DDT, PCBs, and certain heavy metal compounds have become so ubiquitous as to reach virtually all tissues, breast milk, and tissues of the unborn child. Thus, the indiscriminate manner of production, transportation, use, storage, and disposal methods of toxic chemicals have caused human sufferings such as cancer, miscarriages, birth defects, and other health disorders. After studying the problem and potential alternative solutions, authorities of the world felt an urgent need to identify and address the causes of the thousands of annual chemical accidents. This was in an effort to protect life, property, and the environment from the costly consequences of chemical accidents.

As stated previously, the list of chemicals available worldwide is large. Production of synthetic organic chemicals has exploded in the U.S. markets alone. For instance, the amount of chemicals was about 1 billion pounds in 1935, rose to 30 billion pounds in 1950s, and in 1975 the bulk increased to 300 billion pounds.[12] Again, varieties like heavy metals, agricultural chemicals, drugs and pharmaceuticals, food additives, solvents, dyes and pigments, cosmetics, detergents and soaps, plastic and packaging materials, and many more are included in the list. Toxins released by microorganisms and poisonous plants add another chapter in the management of chemical safety vis-à-vis protection of human health.

The term *toxicity* means the capacity to cause adverse effects or harm to the system from microbe to human. The term *safety* means absence of effects that are adverse to toxicity. Chemical safety, therefore, discusses several aspects such as the list of toxic substances, adverse effects during use and disposal, methods and manners of evaluation required for proper regulations, and involvement of qualified workers to achieve this goal. Only by following these tenets in a strict manner can one regulate the use of chemicals and safely provide their benefits to society. Employees of industry must be well aware of the pros and cons of proper handling of chemicals. Workers associated in the design of product labels should have knowledge of product chemistry, toxicology, product safety, and the compliance of regulatory protocols before a product is sent to market or put to common use. This has been the view of the U.S. National Toxicology Program (NTP), the World Health Organization (Geneva), India's Department of Science and Technology (DST), and similar organizations worldwide.

## HISTORY OF TOXICOLOGY

What is the importance of toxicology to modern society? This question is asked repeatedly in different quarters. It is therefore necessary to know the origin of this important discipline. The science of toxicology has a historical base; it dates back to early times of human history. For instance, *Ayurveda* is an ancient traditional school of medicine and health care originated in India approximately 2500 B.C. This system of human health care links to the ancient books of wisdom — the *Vedas,* meaning knowledge. *Ayurveda* is a word in *Sanskrit* and is made up of two words: *Ayuh* meaning life, and *Veda* meaning knowledge (the knowledge of longevity/life). Based on the principles of *Ayurveda*, the medicinal practices and the doctrines of *Charaka, Sushruta*, and *Vagbhata* made monumental contributions to the identification of poisons and the effects on humans. In fact, these ancients, as in the *Astanga Hrudaya of Vagbhata*, have paved the way to understanding human health. Human ailments, including poisoning, are covered by this system of science. It may be stated that Ayurveda discusses the combination of four essential parts, namely, mind, body, senses, and the soul, and unravels the effect of toxins on the body and elimination by adopting different processes.[13–15]

In fact, the way of life and the association with food and drinks for Asian people was quite different from other populations in occidental regions of the world. Elementary knowledge about the use and restrictions of certain substances/food were the guiding principles for these people. This is evident in the native language of India, Sanskrit. The dictum may be grouped under health and hygiene or the Yoga system of philosophy — a path to lead a life of righteousness. The dictum in *Sanskrit* runs as follows: *Ati Sarvatra Varjayet,* meaning avoid excess of everything and everywhere — even *nectar* (*ambrosia*), the drink of the angels, when consumed in excess can cause adverse effects! Two dictums (in *Sanskrit*) are (1) *Langanam Parmaushadham*, meaning fasting is the best medicine for human health; and (2) *Madyam na Pibeyam*, meaning not to be alcoholic.

Much later, Paracelsus, the father of modern toxicology (1493–1541), pronounced a dictum of his own: *Sola dosis facit veneum*, meaning only the dose makes the poison. There is a glorified commonness to these ancient thinkers, neither knowing the other and from different periods of time.

In fact, all substances are poisonous, and there are none that are safe. The right dose alone differentiates a substance as a remedy or as a poison. In other words, no substance is absolutely safe. Important reference could be made to valuable contributions of Theophra Stus (370–286 B.C.), Hippocrates (400 B.C.), and Ebers Papyrus (1500 B.C.). In fact, the use of hemlock by the Greeks to execute the great philosopher Socrates (470–399 B.C.) is an instance of its own. Recently, Ramazzini (1700) documented preventive measures to control industrial hazards among occupational workers (Fig. 2-1).

Toxicity is the ability of a substance/chemical to induce harmful/adverse effects on the organism. Adverse effects range from mild symptoms, such as headaches, to severe symptoms, like coma, convulsions, or death. Toxic chemicals trigger alterations of normal body functions, eventually leading to death of the organism. However, based on timely intervention and conditions of body

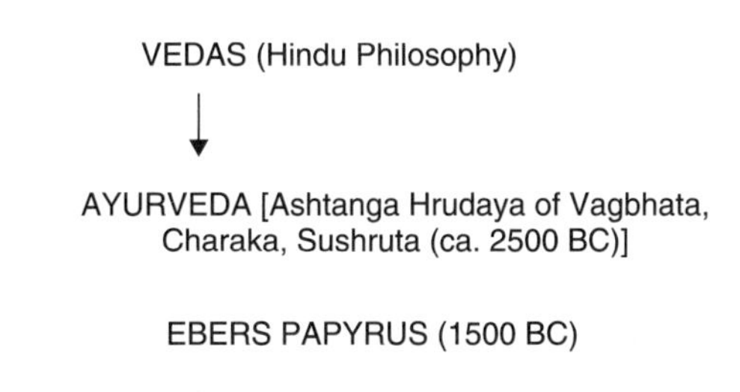


GREEK MEDICINE EGYPTIAN MEDICINE

HIPPOCRATES GREEK MEDICINE (400 BC)
ARISTOTLE (384 – 322 BC)
THEOPHRASTUS (307 – 256 BC)

DIOSCORIDES (ca 40 – 90 AD)
GALEN (131 – 200 AD)

AVICENNA (980 – 1037 AD)
MAIMONIDES (1135 – 1204 AD)

PARACELSUS (1493 – 1541 AD)

ORFILA (1787 – 1853)
CHRISTISO (1797 – 882)
KOBART (1851 – 1918)
LEWIN (1854 – 1929)

**Figure 2-1.** Schematic diagram: pioneers in global toxicology.

function(s), toxicological effects could be reversed to avoid permanent damage to the organism. Presently, each new substance should be evaluated to establish the type of effects and the dose necessary to produce a measurable toxic reaction in humans and animals. Toxicity testing of newer chemicals — particularly drugs, food additives, pesticides, cosmetics, soaps and detergents, and others — involves many phases and will be discussed in subsequent chapters. To control chemical hazards, reduce risks, and increase safety from exposure to chemicals, different toxicity tests are mandated by regulatory authorities of governments.

### Hazard and Risk

The "hazard" and "risk" of chemicals (e.g., pesticides), should be of concern to users. Hazard is the danger of exposure to a chemical; risk is the magnitude of harm resulting from such an exposure and the possibility of its happening. This may be formulated as:

$$\text{Risk} = \text{Toxicity} \times \text{Exposure}$$

A highly toxic chemical (e.g., a pesticide) may be used without causing harmful effect if it is handled with care and caution. Exposure to pesticides can be minimized by wearing protective clothing and equipment and by learning to handle pesticides carefully. If the exposure to the pesticide is low or even nil, the risk is reduced, even when handling highly toxic pesticides.

Toxicology is the discipline of science where evaluations are made regarding the manner of poisoning in animals, humans, and related subjects. The adverse effect of a substance or a chemical on plants, animals, and humans may be sudden, delayed, subtle, or severe. Depending on the manner of exposure, the duration of exposure, and the quantity of the chemical entering the body, the animal develops toxic responses. Essentially, the significance of toxicological investigations is to gain knowledge about the test substance(s) and its evaluation thereof. The following aspects may be listed:

- The nature of toxicity of the chemical under test
- The dose required to develop a response in a test organism
- The type and size of the population normally exposed to chemical(s) in a societal/occupational frame and needs safety and protection
- The quality of toxicological data/information generated in the laboratory and outside (field studies, epidemiology, etc.)
- The responsibility of the concerned organization/government (regulatory bodies) to implement safety measures; and many other aspects as well

Present knowledge regarding the transformation of less harmful chemical(s) into more hazardous ones or vice versa is sketchy. The growing concern about chemical safety and health hazards from innumerable substances present around us in developed and third world countries has been well documented.[1–18] Meaningful conclusions about chemical safety demand the knowledge of the chemical and its use pattern. It is of significance that as the frequency and use of a chemical declines, the adverse effect(s) in the organism also declines. This underlines the importance of correct knowledge, proper use, and the implementation of regulatory measures to contain the chemical hazard. Attempts should be made to obtain quality data/safety factors on the toxicity profile of newer chemicals and to bridge data gaps for existing and economically important chemicals. The value of the safety factor normally depends on: (1) the nature of toxicological effect; (2) the nature of the dose-response curve; (3) the size and type of the population to be safeguarded; and (4) the quality of information on the chemical toxicity.

### Branches of Toxicology

Based on the field of specialization, toxicology is categorized into branches. These branches are often interrelated and could be listed as:

- Analytical toxicology
- Aquatic toxicology
- Biochemical toxicology
- Clinical toxicology
- Ecotoxicology
- Environmental toxicology
- Epidemiological toxicology
- Forest toxicology
- Genetic toxicology
- Immunotoxicology
- Nutritional toxicology
- Mammalian toxicology
- Regulatory toxicology

While analytical and biochemical toxicology discuss chemical reactions, interactions, and the consequences on the organism, ecotoxicology and environmental toxicology focus on the impacts of hazardous chemicals and pollutants on biological systems and humans. Similarly, forensic and regulatory toxicology helps in finding causes of hazards and identifies the legal and regulatory measures to be enforced. In fact, regulatory toxicology mandates for industries the creation and maintenance of product safety data to comply with set national, federal, and international standards.

## THE ELEMENTS IN TOXICOLOGY

### Pathways of Toxicity

***Phase I Metabolic Reactions (Biotransformation)*** All biological species, microorganisms, plants, animals, and humans are continuously exposed to a variety of chemicals directly or indirectly. Based on the manner of entry and physicochemical property of the chemical, the organism/animal either accumulates the chemical in different organs of the body or eliminates it from the body. Before the processes of accumulation or elimination occur in the body, the chemical undergoes certain modifications. In other words, animals have several built-in biochemical mechanisms, equipped with an enzymatic "machine," which helps the conversion of a lipophilic substance into a hydrophilic metabolite. These biochemical reactions help in the elimination and accumulation of chemicals, leading to low/high or severe adverse effects. This phenomenon is called *biotransformation*. In a way, the survival of an organism is dependent on the mechanisms it has developed to remove foreign substances from the body. Many chemicals undergo biotransformation in two phases, namely, phase I and phase II enzyme reactions. Several workers have reported the pattern of disposition of xenobiotics in species of smaller and larger animals.[19–24] While biotransformation occurs in different organs, the mammalian liver plays a significant role in this biological process.

Once a chemical enters the body of an animal (based on the route of entry), the chemical is subjected to metabolism by a variety of mechanisms. The toxicity of many chemicals is dependent on the metabolic rate and pattern of the system. Many tissues are capable of metabolizing substances. However, the maximum activities have been found to occur in the liver, followed by the lung, the intestine, the skin, and the kidneys. The nature of the chemical interactions, biochemical interactions, and metabolic transformations help to determine the toxicity of the chemical. A number of authors have written reviews about the biotransformation of many substances.[19,25–28]

A significant role is played by the mixed function oxidases (MFOs). A number of substances undergo bioalteration reactions in two phases. These are categorized as *phase I metabolism* (biotransformation) and *phase II metabolism* (conjugation). Phase I metabolism involves a number of chemical reactions (e.g., hydrolysis, hydroxylation, isomerization, epoxidation, dehydrohalogenation, dehalogenation, desulfuration, oxidoreduction, and nitroreduction). In contrast, phase II metabolism includes different reactions of endogenous molecule(s) synthesis, which in turn combines with the chemical and/or its metabolite. In phase II metabolism, the formation of conjugates — through glycoside formation, sulfoconjugation, glutathione conjugation, amino acid conjugation, acetylation, and methylation — modulates the toxicological profile of chemicals.[29] The phase I microsomal reactions involve cytochrome p450 monooxygenases, also termed *mixed function oxygenase* (MFO).[30–32] In phase I and phase II reactions, the parent chemical (or more often its metabolite) is transferred to coenzymes of the intermediate metabolism. The enzymes are known to be specific in this process.[33–35] Different activities occur in phase I and phase II metabolism in

**TABLE 2-2 Enzymes in Organs/Tissues of Rat and Metabolic Activities**

| | Name of Organs or Tissues | | | | |
|---|---|---|---|---|---|
| Name of Enzymes | Liver | Kidneys | Lungs | Intestine | Skin |
| Epoxide hydrase* | 138 | 21 | 5 | 5 | 5 |
| Aromatic hydroxylase† | 2.0 | 0.02 | 0.02 | 0.006 | 0.006 |
| UDPG transferase‡ | 36 | 77 | 40 | — | — |

Rate of activity expressed as:
*nmol of styrene glycol/mg microsomal nitrogen/15 min.
†pmol of p-amino/phenol/mg fresh tissue/min.
‡nmol of conjugated 4-methylumbelliferone/mg microsomal proteins/h.

**TABLE 2-3 Subcellular Fractions and Metabolizing Enzymes Associated with Chemicals**

| | Subcellular Fractions | | | |
|---|---|---|---|---|
| Name of Enzymes | Cytosol | Microsomes | Mitochondria | Nucleus + Membrane Residues |
| Epoxide hydrase* | 13 | 392 | 42 | — |
| Aromatic hydroxylase† | 0 | 84 | 8 | 5 |
| UDPG transferase‡ | — | 36 | — | — |
| GSH epoxide transferase§ | 224 | — | — | — |

Rate of activity expressed as:
*nmol of styrene glycol/mg microsomal nitrogen/5 min.
†nmol of p-amino phenol/mg fresh tissue/min.
‡nmol of conjugated 4-methylumbelliferone/mg microsomal proteins/hr.
§nmol of conjugated styrene-oxide/mg microsomal protein/min.

tissues of rat and subcellular fractions (Tables 2-2 and 2-3). More information on this may be found in the literature.[22,23,36–45]

***Phase II Metabolic Reactions*** Reactions involving metabolic conjugations of xenobiotics have been studied by different workers.[24,34,35] The association of conjugation reactions with glucuronic acid, sulfate, glutathione, and amino acids plays a significant role in subcellular fractions. The glucuronic acid conjugation, sulfoconjugation, methylation, acetylation, amino acid conjugation, and glutathione conjugation have all metabolized the chemicals to form more polar molecules for easy, effective excretion. For more information, refer to the literature.[24,30,34–36]

MFO reactions may be shown schematically by the following equation; several factors modulate MFO activity.

$$SH + NAD(P) + H + O_2MFO\ SOH + NAD(P) + H_2O$$

Different enzymatic reactions that occur during the metabolism of chemicals have been well categorized. The biotransformation and conjugation reactions may lead to either the detoxication of the toxicant and the excretion of its metabolites, or to the activation of the toxicant into more reactive intermediates. These may, in turn, react with glutathione, or tissue proteins, or nucleic acids (e.g., RNA, DNA) and undergo different metabolic reactions.

Toxicity of a chemical is the result of several reactions and interactions between a candidate chemical [or its metabolite(s)] and the cellular receptors. These include enzymes, glutathione, nucleic acids, hormone receptors, and the like. The degree of chemical toxicity could be explained as follows:

$$\text{Toxicity C Ar (Chemical) (Receptor)}$$

Ar is the specific affinity of the receptor for the toxic chemical C

Toxicity of a chemical can also be expressed as follows:

$$\text{Toxicity} = k\ (C)\ (R)\ Ac$$

where toxicity is dependent on:

- C = concentration of the candidate chemical in the tissue
- R = concentration of the endogenous receptor of the tissue
- Ac = affinity of the receptor for the chemical

### Factors Influencing Toxicity

Once a chemical enters the body of animal or human, it undergoes metabolic reaction. A host of factors modulate the reaction rate and the induction of toxicological effects. These factors have been termed *intrinsic factors* and include animal species, gender, age, nutritional status, pregnancy, other health status, and circadian rhythms. In addition, there are certain extrinsic factors (e.g., physicochemical properties of chemicals, solvent or vehicle, route of exposure, temperature, and humidity) during exposure to chemicals that also influence the effect of a test chemical. We shall discuss these factors in greater detail.

***Species and Strains*** Many studies have shown that phase I and phase II activities in different animal species differ significantly and influence the detoxification of chemicals. For example, the basal values of hepatic microsomal systems differ to a large extent between species of rodents.[46,47] There are wide variations among animals species such as birds, albino rat, Swiss mouse, guinea pigs, New

**TABLE 2-4 Acute Toxicity of Chemicals in Different Species of Birds**

| Chemical | Species of Birds | Oral $LD_{50}$ (mg/kg) |
|---|---|---|
| Chlorpyrifos | House sparrow | 21 |
| | Coturnix | 17 |
| | Mallard duck | 75 |
| | Pheasant | 13 |
| DDT | Coturnix | 841 |
| | Mallard duck | 2,240 |
| | Pheasant | 1,296 |
| Dieldrin | House sparrow | 47.6 |
| | Coturnix | 69.7 |
| | Mallard duck | 381 |
| Endrin | Mallard duck | 5.6 |
| | Pheasant | 1.8 |
| Malathion | Mallard duck | 1,485 |
| | Bobwhite quail | 3,500 |
| Parathion | House sparrow | 3.4 |
| | Coturnix | 6.0 |
| | Mallard duck | 2.0 |
| | Pheasant | 12.4 |

Zealand white rabbit, and nonhuman primates regarding sensitivity and susceptibility to the chemical action.[48,49] The acute oral toxicity of some insecticides to bird species is a classic example of species sensitivity to toxic action of chemicals (Table 2-4). Different species of mice (e.g., C3H strain and C57 black strain mouse) differ significantly in the induction of neoplasm of the liver.[50] Studies by Oesch and coworkers[51] showed lindane as carcinogenic to CF1 mice but not to B6C3F1 mice. Hayes[52] also observed that rats and humans convert DDT to DDE, whereas rhesus monkeys totally fail to do so, amply substantiating the importance of species in the evasion of toxicity. Thus, differences between species in toxicokinetics and in biotransformation are significant factors for the understanding of chemical safety for animals and humans.[53] Differences in metabolism of animal models, and the extrapolation of toxicity data to humans, therefore, require extra caution. In fact, substances like pesticides vary in their toxicity, from slightly toxic to extremely toxic.

***Animal Gender*** Role of gender and metabolism of xenobiotics has been studied in a variety of experimental animals, particularly rats. The toxicological effects of many chemicals have shown differences between male and female animals. A classic example is the acute toxicity data of different pesticides between male and female rats. This difference was found both when animals were exposed by ingestion or dermal absorption. The ratio between male and female in oral $LD_{50}$ values for of chemicals (e.g., pesticides) ranges from 0.21 to 4.62, which

**TABLE 2-5 Acute Oral Toxicity of Chemicals in Male and Female Rats**

| Pesticides Oral $LD_{50}$ (mg/kg) | Male | Female |
|---|---|---|
| Aldrin | 39 | 60 |
| Chlordane | 335 | 430 |
| Demeton | 6.2 | 2.5 |
| Endosulfan | 18 | 43 |
| Endrin | 17.8 | 7.5 |
| Ethion | 65 | 27 |
| Methyl parathion | 14 | 24 |
| Parathion ethyl | 13 | 3.6 |
| Phorate | 2.3 | 1.1 |

*Source*: Modified from Gaines (1969),[64] Hayes (1991),[52] Dikshith (1991)[82].

indicates the susceptibility of gender (Table 2-5). The activities of drug metabolizing enzymes have shown high dependency on gender—higher in male than in female animals.[61–63]

Chloroform is converted to phosgene, the reactive intermediate made much faster by microsomes from the kidney of male mice than those from female mice. Treatment of male and female rats with mirex and chlordecone showed higher levels of cytochrome p450 in a time- and dose-dependent manner. Exposure of male rats to estrogen also disturbs the MFO system.[64,65] Difference in the profile of cytochrome p450 isozymes in hepatic microsomal system of male and female rats gains much importance. The[39,65] exaggerated gender-dependent metabolism alterations of rats may result from extensive inbreeding and the role of isoforms of cytochrome p450. Further, it has been shown that species-specific gene-duplications and gene-conversion events in the CYP2 and CYP3 families produced different isoforms in rats and humans.[66] Identification of specific forms of cytochrome p450 in male and female animals has gained importance. Epidemiological studies also have revealed that certain types of cancer occur more frequently in males than in females and vice versa.[67]

***Animal Age*** Action of chemicals on different ages of animals requires no emphasis. In other words, the behavior of a newborn or young animal and that of an adult differ significantly, since the former is very sensitive.[54] For example, certain pesticides are more toxic in young rats than in adults.[45,55–57] The difference in chemical metabolism between young and adult animals has been attributed to the MFO system.[58–61] Drug metabolism in young animals (during puberty and senescence) has shown varied cytochrome p450 activities, which again suggests that children are more prone to chemical injury.

***Nutritional and Health Status*** Nutritional status plays an important role in the induction of toxicity in animals. The activity of an MFO system is significantly modulated by the nutrition that animals receive. For instance, a healthy, well-nourished animal with a rich protein diet is able to metabolize a toxic substance more efficiently than an animal given a low-protein diet. Protein restrictions interfere and prevent the induction of the MFO system in animals, thereby making the animal more susceptible to chemical toxicity.[73,74] The nutritional status influences biotransformation; a diet with vitamin deficiency decreases the biotransformation rate of a toxic substance. Different minerals in the animal's diet and fasting/starvation status decrease cytochrome p450-catalyzed oxidation and thus modulate the biotransformation rate (Table 2-6). Evaluation of the chemical(s) toxicity profile in an experimental model or in a weaker section of a population therefore requires understanding of dietary needs. Diet also influences carcinogenic processes through activation/detoxification controlling the enzyme activity.[68] For example, the availability of methyl donors, (e.g., methionine and choline) influence hepatocarcinogenesis in animals and humans.[69] Several authors[70–78] have discussed the role of nutrition for the induction of toxicological effects in animal species.

### *Routes of Exposure*

*Oral Exposure* Animals and humans are exposed to chemicals through different routes. Chemical toxicity differs depending on the route of absorption.[52,64] Entry

**TABLE 2-6 Influence of the Diet and Effect on Biotransformation**

| Diet Composition | Effect on Biotransformation |
|---|---|
| Deficiency in | |
| Calcium | Decrease |
| Copper | Decrease |
| Iron | Decrease |
| Manganese | Decrease |
| Zinc | Decrease |
| Ascorbic acid | Decrease |
| B-complex | Decrease |
| Tocopherol | Decrease |
| Lipid content | Decrease/Increase |
| Natural substances | Decrease/Increase |
| Protein | Decrease |
| Fasting of 12 hours | Decrease/Increase (phase II reactions) |
| Starvation of more than 2 days | Decrease |

of chemicals into the body is often based on the organism's nature of activity and composition of the chemicals it is subjected to. There are three major ports of entry into the body: oral (ingestion), dermal (skin absorption), and respiratory (inhalation). Further, other routes such as intramuscular, intravenous, intradermal, and intratracheal are used for purposes of laboratory studies. Chemicals enter the body orally during food intake (ingestion), accident, or ignorance (e.g., blowing out a plugged nozzle with your mouth, smoking, eating with contaminated hands, eating food with chemical contamination). The seriousness of the exposure depends on the oral toxicity of the material and the amount ingested. Through the oral route (or ingestion), a number of substances enter the body of animals and humans in the form of either food or drink. Eating raw fruits and vegetables can constitute an important source of food poisoning among people, particularly children. Exposure to a substance may be short or long term. Short-term exposure to a large amount of a substance may have significantly different effects than low-dose exposures over a prolonged period, despite an equal total amount of exposure. After entering the body, the toxic chemicals are distributed in organs and tissues in association with body fluids.

*Dermal Exposure* Of the three major routes of exposure, the dermal (skin) route constitutes nearly 90% of chemical exposure, particularly of pesticides. Dermal exposure is common whenever chemicals are mixed or handled. Certain types of dry materials, (e.g., pesticide dusts, wet or dry powders, granules, liquid pesticides) enter the body through quick skin absorption. Many factors influence the rate of dermal exposure of a chemical; these may be as follows:

- The dermal toxicity of the chemical
- Rate of absorption
- The size of the skin area contaminated
- The length of time the chemical is in contact with the skin
- The amount of chemical present on the skin

The skin absorption pattern versus rate of entry through the skin is different for different parts of the human body. Considering absorption through the forearm as an example, absorption is more than 11 times faster in the groin area than on the forearm. (Absorption through the skin in the scrotal area is rapid enough to approximate the effect of injecting toxins directly into the bloodstream.) Absorption continues to take place on all affected skin areas as long as the toxin remains in contact with the skin. The seriousness of the exposure increases if the contaminated area is large or if the material remains on the skin for an extended period of time. Different types of skin structure, as seen under normal conditions, and ill health (e.g., rough skin, smooth skin, dermatitis, broken skin, abrasion) modulates the entry pattern of chemicals into the body. A significant difference also is seen in surface area versus total percent of skin area (Tables 2-7 and 2-8);

**TABLE 2-7 Surface Area of Skin Normally Exposed in Humans After Casual Work Dress**

| Specific Area | Surface Area (in sq. ft.) | Percent Total Area |
|---|---|---|
| Back of neck | 0.12 | 3.80 |
| Face | 0.70 | 22.00 |
| Forearms | 1.30 | 41.30 |
| Front of neck and v of chest | 0.16 | 5.10 |
| Hands | 0.87 | 27.60 |

**TABLE 2-8 Total Surface Area in Different Regions of the Body**

| Body Region | Surface Area (percent of total) |
|---|---|
| Head | 5.60 |
| Neck | 1.20 |
| Upper arms | 9.70 |
| Forearms | 6.70 |
| Hands | 6.90 |
| Back, chest, shoulders | 22.80 |
| Hips | 9.10 |
| Thighs | 18.00 |
| Calves | 13.50 |
| Feet | 6.40 |

as the skin is composed of multiple lipid-containing layers, lipophilic substances easily penetrate the tissue.

There is a wealth of literature showing significant differences in the acute toxicity values of chemicals (particularly pesticides) by oral and dermal routes of exposure. In a majority of instances, chemicals (particularly pesticides) are more toxic by ingestion than by skin absorption. However, it should be remembered that workers handling pesticides invariably can be affected through all three routes of exposure (oral, dermal, inhalation) and must exercise extreme caution (Table 2-9).

*Inhalation Exposure* Inhalation is the absorption of a substance in the form of airborne particles during breathing. This occurs particularly when mixing and spraying a toxic substance, like pesticides. This exposure increases when working in confined or poorly ventilated areas. The chemical must be airborne to cause

**TABLE 2-9 Different Values in Acute Toxicity of Pesticides by Oral and Dermal Routes in Rats**

| Chemical | Oral (mg/kg) | Dermal (mg/kg) |
|---|---|---|
| DDT | 200 | 3,000 |
| DDD (TDE) | 3,400 | >10,000 |
| Dicofol | 640–842 | >4,000 |
| Endosulfan | 18–43 | 78–130 |
| Heptachlor | 100 | 200 |
| Azinphosmethyl | 15 | 225 |
| Diazinon | 66–600 | 379–1,200 |
| Dicapthion | 220–400 | >2,000 |
| Dichlorvos | 25–170 | 59–900 |
| Dicrotophos | 22–75 | 200 |
| Disulfoton | 2.6 | 8.6 |
| Ethion | 27–119 | 62–245 |
| Fonofos | 8–16 | 147 |
| Methyl parathion | 14–24 | 60–67 |
| Monocrotophos | 21 | 354 |
| Parathion | 3.6–15 | 6.8–21.0 |
| Phorate | 1.6–3.7 | 2.5–6.2 |
| Quinolphos | 62–137 | 1,250–1,400 |
| TEPP | 0.2–2.0 | 2–20 |
| Aldicarb | 0.6–1.0 | 2.5–5.0 |
| Methomyl | 17–24 | >1,000 |

*Source*: Modified from Gaines, 1969,[64] Kenaga, and End, 1974[80].

toxicity; this is achieved by the production of small spray particles, gases, or vapors. The nose and throat are very effective barriers, and only very small particles can reach the lungs. The surface of the lung is a very fine membrane, which is a poor barrier against the entry of chemicals. Some chemical reducing its effectiveness also may damage the membrane. Like oral and dermal exposure, inhalation exposure is more serious with some chemicals, for instance, pesticide fumigants and aerosols.

***Dose-Response Relationship*** The most commonly used test method of toxicity for a chemical is $LD_{50}$—meaning the lethal dose for 50% of the animals tested. The poison dose is usually expressed in milligrams of chemical per kilogram of body weight (mg/kg). A chemical with a small $LD_{50}$ (e.g., 5 mg/kg) is highly toxic, and a chemical with a large $LD_{50}$ (e.g., 1,000 to 5,000 mg/kg) is practically nontoxic. The $LD_{50}$, however, suggests nothing about nonlethal toxic effects of the substance. A substance may have a large $LD_{50}$, but may produce illness at small exposure levels. It is, therefore, not correct to term substances with small $LD_{50}$s as more dangerous than substances with large $LD_{50}$s.

They may simply be termed *more toxic*. The danger or risk of adverse effect is mostly determined by how chemicals are used, not by the inherent toxicity of the chemical itself.

Invariably, chemicals are compared by their $LD_{50}$ values and subsequent decisions are made about their safety. This is an oversimplified approach to comparing chemicals, because the $LD_{50}$ is simply one point on the dose-response curve that reflects the potential of the compound to cause death. More important to assessing chemical safety is the threshold dose and the slope of the dose-response curve, which shows how fast the response increases as the dose increases. Although the $LD_{50}$ can provide some useful information, it is of limited value in risk assessment because it only reflects information about the lethal effects of the chemical. It is quite possible that a chemical will produce an undesirable toxic effect (e.g., reproductive toxicity, birth defects) at doses that cause no deaths at all. In reality, an assessment of chemical toxicity involves comparisons of numerous dose-response curves covering different types of adverse effects. In fact, this approach is involved in the determination of restricted use pesticides (RUPs), since some with large $LD_{50}$ values have a strong skin or eye irritation property and require special handling.

The knowledge gained from dose-response studies in animals is used to set standards for human exposure and the amount of chemical residue that is allowed in the environment. As mentioned previously, numerous dose-response relationships must be determined in many different species. Without this information, it is impossible to accurately predict the health risks associated with chemical exposure. With adequate data, decisions may be made about chemical exposure and work needed to minimize the risk to human health. The concept of a dose-response relationship is of fundamental importance for understanding chemical toxicity profiles. To determine the degree of acute toxicity, one assumes that there is a level below which exposures carry minimal risk. This level is termed the *threshold level* of the chemical. Carcinogenic substances are an exception to this concept, where any level is regarded as hazardous. Each chemical has a characteristic dose-response curve. With prolonged exposure to chemicals, the likelihood of an adverse effect is attributed to the chemical. Indirectly, adverse effects from chemicals can be mitigated by reducing the level of exposure. Dose-response curves for different chemicals are drawn by experimentation with animal species and by administering a range of doses under test conditions.

The extensive amount of data developed about a pesticide is often used against it by ignoring the dose-response. For example, some acute toxicity studies, which are designed to include dosage levels high enough to produce death, are cited as proof of the chemical dangers. Chronic effects seen at high doses in lifetime feeding studies are misinterpreted and considered as proof that no exposure to the chemical should be allowed. Major improvements in analytical chemistry permit detection of the presence of chemicals at levels of parts per billion (ppb) or even parts per trillion (ppt). Reports often say that a certain chemical has been found in a food or beverage and the amount found is expressed in ppm or ppb. Often, no information is provided to assist in comprehending the meaning of these numbers. Frequently, this information ignores the issue of dose-response, which simply states "the dose makes the poison." The concentration of a chemical in any product is meaningless unless it is related to the chemical toxicity and the potential for exposure and absorption. Chemicals of low toxicity (e.g., table salt, ethyl alcohol) can be fatal if consumed in large amounts. Conversely, a highly toxic material may pose no hazard when exposure is minimal.

The question of dose-effect relationship is difficult to understand in certain contexts. To observe the effect of a substance in an animal requires that the test animal(s) receive large doses of a suspected toxicant. By effect, we mean to learn and understand the effect of large doses in small animals in a short time and about small doses in humans and large animals over a long time. This means that the high probabilities in small animals must be extrapolated into low probabilities in humans. This requires the theoretical concept of mathematical modeling. In fact, the mathematical model is a relationship between the likelihood of an adverse effect and the magnitude of exposure to the toxic substance. Because of the limited knowledge about this relationship, one may or may not have a basis for conclusions. However, if it has a basis, it is termed an *empirical model*, otherwise it is a *theoretical model*. The most commonly used model links dose-response relationships of a test substance with adverse effects and death in test species, termed a *linear model*. This allows one to link the probability of an adverse effect (e.g., cancer) directly proportional to the total accumulated dose of the test substance. To repeat, the higher the dose, the higher the risk, and the lower the dose, the lower the risk.[79–84]

## Extrinsic Factors

In a majority of instances, particularly with pesticides, combinations of chemicals (e.g., organic solvents, mineral oils) are mixed to enhance the chemical toxicity. Consequently, the toxicity profile is altered. Lipophilicity modulates the absorption rate of the chemical from the site of entry into the system (lung, skin, mucous membrane). Thus, the fat-solubility pattern of a test chemical helps for easy cell membrane transport to reach the active site of intracellular enzymes and trigger possible toxic effects. The toxicity profile of active ingredients of pesticides and those of formulated products of pesticides differ widely.

Since a majority of toxic chemicals trigger multiple adverse effects, it is important to understand the spectrum of effects. Such data on chemical(s) help to evaluate the chemical safety and/or health risks posed to humans. Simultaneous exposure to multiple chemicals makes the problem more complex. All these aspects suggest a plethora of toxic effects of chemical combinations. Further, as seen in the acute toxicity profile of pesticides on bird species, the genetic diversity of animal species and humans underlines the role of individual sensitivities on the induction of toxic effects of chemicals.

The genetic material of the species in question, the habitat, nature, size of the population, and the sensitivity especially susceptible to specific chemical(s) demands an understanding of the molecular mechanisms operating here. Therefore, it is important to be transparent while designing and implementing experiments with these variables. The overall interpretation of the data, generated from species of laboratory animals to humans, is a broad spectrum that requires pragmatism and honesty to interpret. It is here that the chemical safety data requires a touchstone effect. We will discuss this subsequently.

## Types of Toxicological Tests

Different types of tests are carried out to unravel a chemicals toxicological profile. These tests are conducted based on procedures adopted by both national and international laboratories and set regulatory standards. Extensive reports are available regarding the type of and procedures for toxicological studies. However, noting the difficulty in extrapolating data and requiring meaningful conclusions, it is necessary to follow well-established test guidelines. Proper observation of guidelines is also necessary for the generation of quality data to be presented to regulatory bodies. These are well documented in the literature.[85–90]

Generation of quality data based on the proper protocols supported by qualified personnel and laboratories is a must. Human and environmental safety from the large-scale use and disposal of chemicals depends largely on the qualified worker, whether in a laboratory study or a product development or manufacturing industry. Poor investigational techniques and submission of false data to regulatory agencies was identified during the 1970s by governmental authorities. After prolonged investigations, serious lapses were identified in investigation, such as lack of integrity in work and personnel, and inadequate quality data for the product. From then on, the U.S. Food and Drug Administration introduced a system called Good Laboratory Practice (GLP). This system was further made stringent by the inspection of laboratories to assess degree of competence and generation of quality data, per GLP regulations. Toxicological tests and the toxicologists performing them are responsible for the generation of quality data on different products. Because these products (e.g., drugs, cosmetics, soaps and detergents, pesticides, food additives, child care and dairy products, medical devices) are used by humans in daily life and require correct information about toxicity and safety. The importance of these has been discussed in the literature. (For details, refer to Chapter 18). Essential toxicological tests to identify potential chemical toxicity are discussed briefly here, and more details may be found in the literature.[91–95]

***Acute Toxicity*** The acute toxicity of a chemical indicates its ability to do systemic damage as a result of one-time exposure through a specified route, namely, oral (ingestion), dermal (skin), or inhalation (respiratory). A chemical with a high acute toxicity value is extremely toxic and may be fatal to organisms when exposed even to a very small amount (e.g., aldicarb, phorate) (Table 2-5). A chemical toxicity warning is normally present on a container label suggesting appropriate manner of storage, handling, use, and disposal.

*Acute Toxicity Measure* Acute toxicity of a chemical may be measured through different routes and by systemic exposures by injections such as subcutaneous, intravenous, intramuscular, and intratracheal, which are all used in experimental and medical support. Appropriate knowledge about measures and acute toxicity warnings of chemicals are necessary to contain chemical hazards vis-à-vis to achieve chemical safety and human health.

Administration of a test chemical in specific amounts — either as one oral dose or multiples within 24 hours followed by observation for 14 days — for observation of toxicity symptoms is the standard procedure. As previously mentioned, the usual term to describe acute toxicity is $LD_{50}$. This is measured from zero upward; an extremely toxic substance has a very low $LD_{50}$ value, whereas a less toxic substance has a higher $LD_{50}$ value. $LD_{50}$ values are expressed as milligrams per kilogram (mg/kg), which means milligrams of a chemical per kilogram of body weight of the test animal. $LD_{50}$ values are generally expressed on the basis of active ingredient (a.i). In the case of drugs and pesticides, many formulations are made. If a commercial product is formulated to contain 50% a.i., then it would take two parts of the material to make one part of the a.i., namely, drug or pesticide.[95]

$LC_{50}$ is the measure of toxicity through the inhalation route. LC means *lethal concentration*, where the amount of inhaled chemical is indicated. The $LC_{50}$ values are measured in milligrams per liter of air. Again, the lower the $LC_{50}$ value, the more severe the toxicity of the chemical.

### Short-Term Toxicity Studies

*Repeated-Dose Toxicity and Subchronic Toxicity* After the completion of the acute toxicity studies, further studies are required to determine the toxicity profile of the test substance when exposed for an extended period. Normally test animals are treated *daily* for a period ranging from 14 to 90 days, which should not exceed 10% of the life span of the test species. The periods of treatment for all three routes are slightly different (Table 2-10). The information gathered from repeated-dose study and subchronic study provides possible adverse effects likely to arise after repeated doses of the chemical. Further, the studies also provide information on target organ toxicity, residue buildup, accumulation pattern, and estimate of a no-effect level of chemical exposure. In a repeated-dose study and subchronic toxicity study, animals are treated either by the oral route, skin application, or inhalation using more than three test doses of the substance. The data collected become useful to establish chemical safety standards and to design a standard toxicity test for the candidate chemical. For conducting a repeated-dose toxicity or subchronic toxicity study, the following are the standard exposure periods of test animals through different routes.

The parameters of the study include clinical examination (hematology, clinical biochemistry) of the animals in test and control groups, gross necropsy of all animals of the study, and complete histopathology of all organs and tissues.[87,95]

**TABLE 2-10 Exposure Period in a Repeated-Dose and Subchronic Toxicity Study**

| | Route of Exposure | | |
|---|---|---|---|
| Type of Study | Oral | Dermal | Inhalation |
| Repeated-dose toxicity | 14–28* | 21–28 | 14–28 |
| Subchronic toxicity | 90 | 90 | 90 |

*Figures indicate total period of exposure in different routes.

***Chronic Toxicity*** Chronic toxicity indicates harmful effects on animals by the long-term, low-level exposure to chemicals. Information on the chronic toxicity of many chemicals is not well documented because of the complex and subtle nature of the study, which takes into account different parameters as indicators of toxicity after prolonged exposure. Chronic or prolonged exposure to small levels of chemicals occurs in different work conditions (e.g., mixing, loading, applying, and storage of chemicals, especially pesticides). Chronic toxicity includes important areas like carcinogenesis (oncogenesis), teratogenesis, mutagenesis, and reproductive toxicity. Chronic toxicity testing is both lengthy and expensive. EPA and regulatory agencies in other countries require an extensive battery of tests to identify and evaluate the chronic effects of pesticides. These studies, which may last up to 2 years, use several animal species to evaluate toxicity resulting from multiple or long-term exposures.[92]

***Genotoxicity*** The identification of the genotoxic potential of chemicals—particularly those handled by a large group of people around the world—has become increasingly important. Although much of the data are well known with regard to genotoxic potential of radiation, knowledge about chemicals is sketchy. Addition of newer compounds (and their interactions with parent compounds and active metabolites) to the already existing lists has added a new dimension to the problem. Reactive intermediates, which covalently bind to DNA, usually cause DNA damage and are thus called *genotoxins*. The induction of unscheduled DNA synthesis is a common event after genotoxic damage, achieved through the efforts of chemical promoters. Promoters accelerate cancer development by forcing proliferation of damaged DNA cells before the damage could be repaired.

The type of genotoxic effects caused by chemicals on animals and humans is modulated by the location of the genotoxic event. For example, a mutation in somatic tissue may produce a neoplasia or a terata, if a developing embryo is also exposed to the toxic action of the chemical. A germinal mutation induced in germ cells (ova or spermatozoa), on the contrary, may lead to more deleterious effects, since these cells are transmitted to subsequent generations. The genetic changes transmitted can include, for instance, gene mutation, small deletions, and dominant lethals.[81,87,96]

*Mutagenicity* Many physicochemical agents cause an impact in animals and plants. The impact and the spectrum of genotoxic effects in the biological system are termed *mutations*. The genotoxic effects are classified as genetic mutation/point mutation and cytologically identifiable structural changes in cell chromosomes. While point mutations are invisible and involve changes in nucleotide pairs, chromosomal changes are visible (microscopically) as macrolesions and as chromosomal breaks and translocations. A number of substances are known to produce mutagenic effects in animals and plants. Studies have shown that a number of pesticides and environmental chemicals have produced clastogenic effects *in vitro* systems (Figs. 2-2 and 2-3).

Most mutagens and genotoxic carcinogens are able to induce chromosomal alterations in exposed cells. Two important classes of aberrations (namely, structural and numerical) are recognized, and both types are associated with congenital abnormalities and neoplasia in humans. These alterations can be easily detected and quantified in the bone marrow system of mammals *in vivo* (Figs. 2-4 and 2-5), in the *in vitro* cell culture system of mammals (Figs. 2-6 and 2-7), and in the human peripheral blood lymphocytes. Conventional staining techniques can be used to detect these aberrations, and was used to estimate absorbed dose in the case of a radiation accident in Goiania, Brazil.[97,98] A recently introduced fluorescent *in situ* hybridization (FISH) technique using DNA probes has increased the sensitivity and ease of detecting chromosome aberrations. This has helped to identify chromosomal changes, particularly stable chromosome aberrations. This technique allows, to some extent, the estimation of absorbed radiation doses from

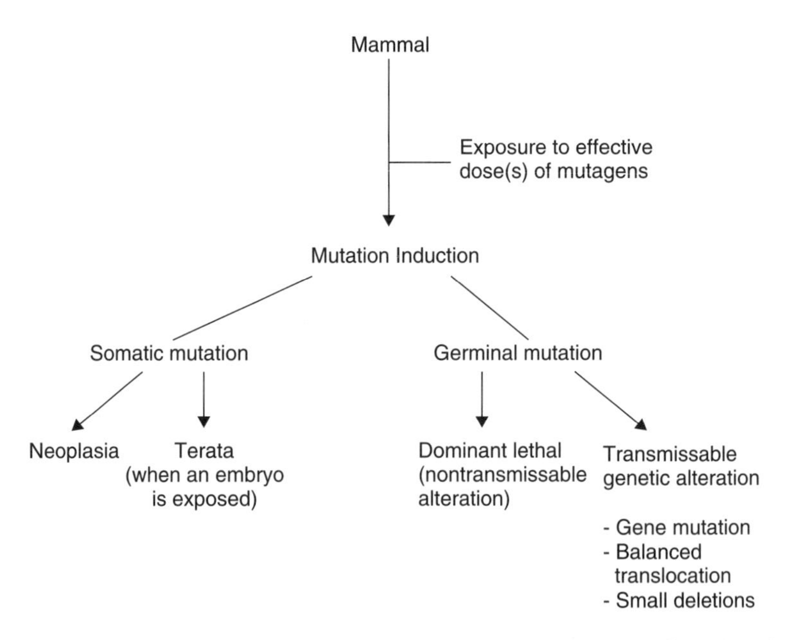


**Figure 2-2.** Exposure to toxic chemicals leading to adverse genotoxic effects in animals.

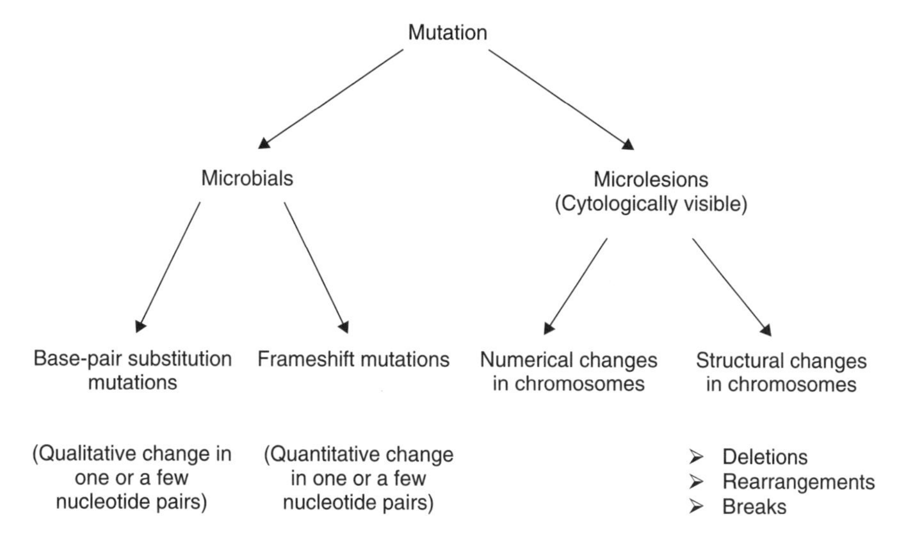


**Figure 2-3.** Classification of molecular changes in DNA potentiated by mutation.

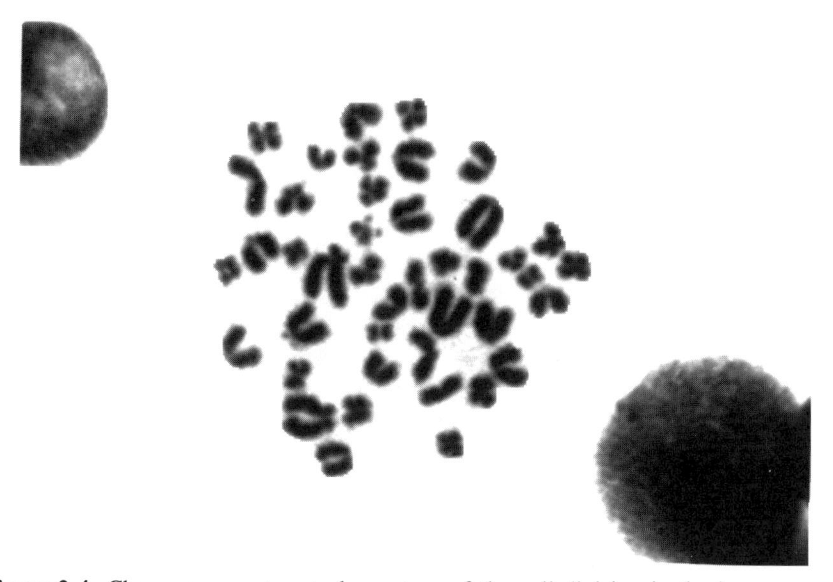

**Figure 2-4.** Chromosomes at metaphase stage of the cell division in the bone marrow cells of albino rats exposed and unexposed to toxic chemicals (*in vivo*).

past exposures. Numerical aberrations can be directly estimated in metaphases by counting the number of FISH-painted chromosomes. Lagging chromosome fragments or whole chromosomes during the anaphase stage of cell division form the micronuclei. The nature of micronuclei, as to whether they possess

**Figure 2-5.** Chromosomes at metaphase stage of the cell division in the bone marrow cells of albino rats exposed and unexposed to toxic chemicals (*in vivo*).

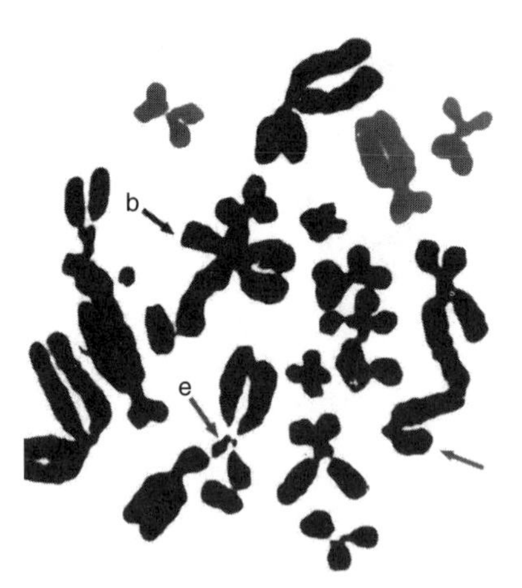


**Figure 2-6.** Chromosomes at metaphase stage of the cell division in the kangaroo cell line exposed to DDT (*in vitro*). Note the aberrations—break, deletion, and exchange.

**Figure 2-7.** Chromosomal changes caused by toxic chemicals on the mammalian cell system in *in vitro*.

a centromere, can be determined either by CREST (calcinosis, Raynaud's phenomenon, esophageal dysmotility, sclerodactyly, telangiectasia) staining or FISH with centromere-specific DNA probes. In several carcinogen-exposed populations (e.g., heavy smokers, people exposed to arsenic) aneuploidy appears to be more common than structural aberrations. In victims of radiation accidents, aneuploidy (hyperploidy) has been common in addition to structural aberrations.[97]

There is an urgent need to extrapolate data to intact animal systems and human situations to develop a more pragmatic approach and suitable remedial measures for human protection. This should not, however, disqualify scientific data but develop more evidence about the adverse effects of chemicals. The testing requirements that identify the mutagenic potential of a substance, based on USEPA, are shown in Tables 2-11 and 2-11a. A number of experimental data suggest that different pesticides, both as technical and formulations, caused clastogenic effects in the bone marrow system of intact rats. Reports have indicated

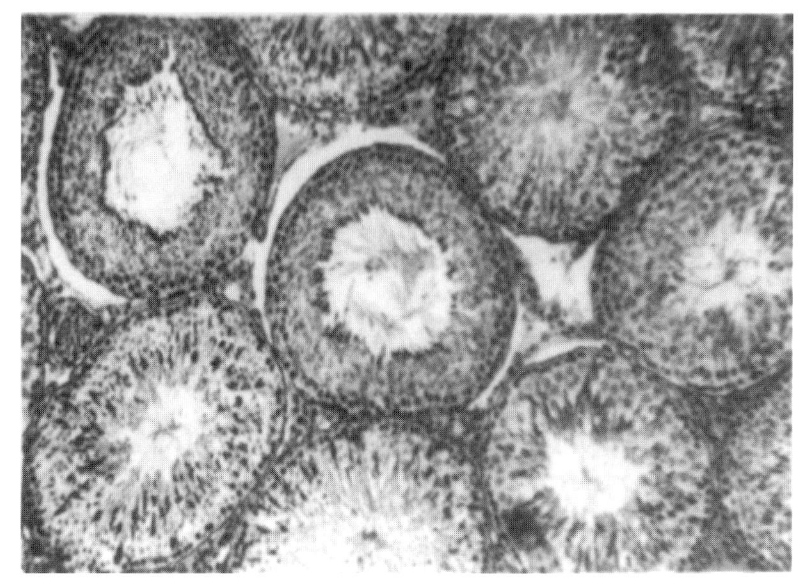

**Figure 2-8.** Section of testis of rat unexposed to chemicals showing normal tubular structure.

that pesticides have caused chromosomal aberrations in mammals.[98,100] Several mutagenic substances do produce tumors, and many oncogenic substances are mutagens. More information may be found in the literature.[101,102]

*Reproductive Toxicity* Many chemicals (e.g., heavy metals, pesticides, PCBs) have been shown to induce effects on the fertility or reproductive rates of animals and humans. Males and females are affected by regular exposure to chemicals. The chemicals have triggered hormonal dysfunction and related effects in mammals as well as terrestrial and aquatic birds.[103–110] Heavy metals, particularly cadmium, and certain pesticides are known to cause adverse effects on the rhythm of spermatogenesis in mammals, indicating tubular damage (Figs. 2-9 and 2-10). Peterson and associates[108] have indicated the toxicity, particularly the developmental and reproductive effects, of dioxin. Much information on malfunctioning of reproductive systems is available in the literature.[103–110]

*Teratogenicity* Teratogenesis studies the formation of birth defects in a pregnant animal. These defects are due to changes triggered by the chemical in the structure or function of the offspring when the embryo or fetus is exposed to chemicals before birth. Studies have shown that many pesticides have caused teratogenic effects in experimental animals.[111–114] The developing fetus has a brief critical period of sensitivity to environmental teratogens. This is particularly true when it is in the process of organogenesis, which ranges from the time of implantation until the end of the embryonic growth period. Gestation periods

**TABLE 2-11 Mutagenicity Testing Requirements**

1. Any chemical or drug need to be tested from each of the following categories of tests to detect the gene mutation

    - *In vitro* bacteria, with and without metabolic activation
    - Eukaryotic microorganisms, with and without metabolic activation
    - Insects (example, sex-linked recessive lethal test)
    - Mammalian somatic cells in culture, with and without metabolic activation
    - Mouse-specific locus test

2. Any chemical or drug need to be tested from each of the following categories of tests to detect the chromosomal aberrations

    - *In vivo* cytogenetic test in mammalians
    - Insect tests for heritable chromosomal changes
    - Dominant-lethal test (DLT test) in rodents to detect heritable translocations

3. Any chemical or drug need to be tested from each of the following categories of tests of detect the primary DNA damage

    - DNA repair in bacteria, with and without metabolic activation
    - Unscheduled DNA repair synthesis in mammalian somatic cells in culture, with and without metabolic activation
    - Mitotic recombination and gene conversion in yeast, with and without metabolic activation
    - Sister-chromatid exchange in mammalian cell culture, with and without metabolic activation

U.S. Environmental Protection Agency proposed guidelines for registering pesticides in U.S.A. Hazard evaluation: Humans and Domestic Animals, FIFRA ACT, Fed, Reg., **43**:37335, August 22, 1978.

vary in species of laboratory animals such as mice, albino rats, rabbits, and hamsters. This sensitive period ranges from day 5 to day 14 in mice and rats, while it differs marginally in other species (Table 2-12).[112]

*Carcinogenicity* The term *carcinogenesis* indicates the formation of tumors (benign or malignant), whereas *oncogenesis* is a generic term. Tumors may be either carcinogenic or noncarcinogenic in nature. In fact, terms like *tumor*, *cancer*, or *neoplasm* are used to suggest an uncontrolled progressive growth of cells and organs of an animal. A number of studies have been conducted on groups of pesticides and documented in the literature.[99,116–119] In certain cases, more confirmation data are needed to establish unequivocal relationships of animal data to humans. In fact, the role of body metabolism in species of experimental animals and the role of microsomal enzyme systems associated therewith have posed

**TABLE 2-12 Period of Gestation (in days) and Characteristic Malformation in Laboratory Animals**

| Animals | Implantation Period | 13 to 20 Somites | End of Embryonic Period | End of Metamorphosis | Fetal Development | Parturition |
|---|---|---|---|---|---|---|
| Mouse | 5 | 9 | 13 | 17 | 18–20 | 19 |
| Rat | 11 | 14 | 17 | 17 | 18–20 | 19 |
| Rabbit | 9 | 10 | 11 | 15 | 16–32 | 32 |
| Hamster | 7 | 9 | 10 | 14 | 15–16 | 15 |
| Major susceptibility | Embryo lethality | Birth defects | Embryo lethality | | Growth retardation, functional defects; fetus death; transplacental carcinogenesis | Growth retardation; other defects |

**Figure 2-9.** Section of the testis in rat unexposed to toxic chemicals. Note the tubular structure with different types of germinal cells.

important questions in data evaluation for humans. In a brief way, one may take the example of PCBs. As already stated, the development of a tumor or cancerous tissue depends on the actions of cytochrome p450. PCBs can be hydroxylated to reactive arene oxide intermediates, which then bind covalently to critical cell components. For example, the hydroxylated 4-chlorobiphenyl metabolite is produced from PCB metabolism, which fastens to DNA, RNA, and proteins in Chinese hamster ovaries. The PCBs are an example of promoters, which follow a specific mechanism of effect. Promoters prefer to target those altered cells for

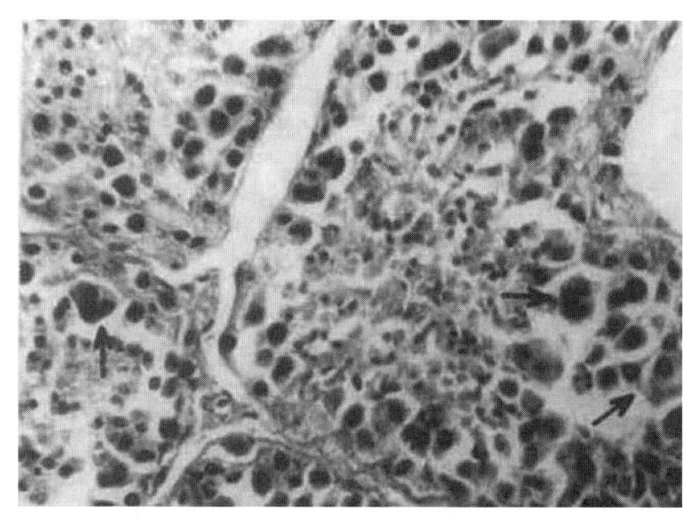

**Figure 2-10.** Sections of the testis in rats exposed to toxic chemicals. Note the tubular structure and degenerative change in different types of germinal cells.

growth that are more responsive, resulting in the differentiation of target organs to form distinct areas: papillomas, polyps, plaques, or nodules. The first indication of carcinogenesis is the presence of phenotypically altered cells. Most promoted cells subsequently become independent of promoters and actively proliferate into "resistant" carcinomas. Subsequent exposure to PCBs only promotes the tumor-producing process. In conclusion, cytochrome p450 has an important role in the studies of toxicity and carcinogenesis. Furthermore, mechanisms that involve certain reactive or unreactive intermediates result in the vast range of observed cellular responses. The tumor induction potential of a number of chemicals, including pesticides, is documented in the literature.[115–119]

***In Vitro Studies*** These studies gain significance in the absence of susceptibility among workers exposed to pesticides (e.g., DDT, dieldrin) with no increase in tumor formation.[96] The cost and time-consuming aspects of carcinogenicity studies have resulted in a paucity of reliable and confirmatory data on a number of older and newer chemicals. The *in vivo* carcinogenicity testing is an expensive and time-consuming process, and as a result, only a relatively small fraction of new and existing chemicals have been tested. Therefore, the development and validation of alternative approaches is desirable. Studies were conducted to develop mammalian *in vitro* assay.[97a,97b] These studies were based on the genotoxic ability of cells to increase their level of the tumor-suppressor protein p53 in response to DNA damage.

Cultured cells are treated with various doses of test substances in varying time periods. After the completion of treatment, they are harvested and lysed. The lysates are analyzed for p53 by Western blot and/or enzyme-linked immunosorbent assay analysis. An increase in cellular p53 following treatment is interpreted as evidence of DNA damage. Studies were conducted with 25 chemicals to determine the ability of the p53-induction assay to predict carcinogenicity in rodents and compare such results with those obtained using alternate approaches. The results indicated citral, cobalt sulfate heptahydrate, D&C Yellow No. 11, oxymetholone, and t-butylhydroquinone as positive in this assay, and emodin, phenolphthalein, and sodium xylenesulfonate as not positive. A comparison of the results of *in vitro* assay with those of *in vivo* protocol using *Salmonella* assay, and the Syrian hamster embryo (SHE) cell assay indicated that the p53-induction assay is an excellent predictor system. These observations are confirmed by other studies as well.[120] To avoid such shortcomings, many researchers have developed *in vitro* systems as alternate approaches. More information on other chemicals that trigger carcinogenic effects in animal species may be found in the literature.[118–123] It is important to add that despite our exposure to environmental chemicals and carcinogens, genetic predisposition undoubtedly plays an important role in cancer development. The organism's genetic predisposition plays an important role in the determination of individuals who will ultimately have or not have cancer. As suggested by the growing literature base on different metals (and more particularly on nickel carcinogenesis), initial events in environmentally induced cancers may be a combination of gene induction and gene silencing by epigenetic DNA methylation, which leads to cancer cell selection. The extent to which this concept applies to other environmental carcinogens requires more challenging investigations. All studies suggest that molecular epidemiology has significant potential in preventing cancer and other diseases caused by environmental exposures that may be due to lifestyles, occupations, or ambient pollution.

Comparative studies, based on laboratory methods and preclinical data, have provided valuable leads in the prevention of cancer. In fact, by incorporating laboratory methods to document the molecular dose and preclinical effects of carcinogens, as well as factors that increase individual susceptibility to carcinogens, newer approaches have become open to this important health problem. Validation studies of biologic markers (e.g., carcinogen-DNA, carcinogen-protein adducts, gene and chromosomal mutations, alterations in target oncogenes or tumor suppressor genes, polymorphisms in putative susceptibility genes, individual p450s, glutathione transferase M1, and serum levels of micronutrients) have received new approaches. These studies have been associated with adults, infants, and children exposed to varying levels of carcinogens, as well as cancer cases and controls. On a group level, dose-response relationships have frequently been seen between various biomarkers and environmental exposures (e.g., polycyclic aromatic hydrocarbons, active and passive cigarette smokers).

However, there is significant interindividual variation in biomarkers, which appears to reflect a modulating effect on biomarkers (hence potential risk) by genetic and acquired susceptibility factors. Ongoing retrospective and nested case control studies of lung and breast cancer examine the association between biomarkers and cancer risk. Results of these studies are encouraging; they suggest that biomarkers, once validated, can be useful in identifying at-risk populations and individuals in time to intervene effectively.

Exposure to particulate matter (or associated air pollutants) early in pregnancy may adversely affect fetal growth. Regardless of which toxicant associated with particulate matter could affect fetal growth, the biologic mechanisms remain to be explained. The active components of these complex mixtures must be inhaled and absorbed into the maternal bloodstream. Highly biologically active compounds (e.g., PAHs) might interfere with some processes in the development or nourishment of the fetus. Analyzing the same mothers from an earlier study, Dejmek et al.[121] observed an increased risk of intrauterine growth rate (IUGR) after exposure to carcinogenic PAHs at more than 15 ng/m$^3$ in the first gestational month.

*Neurotoxicity* A number of chemicals (e.g., organic solvents, pesticides, heavy metals) cause neurological disturbances and other target organ effects.[122–124] Studies have shown that pesticides, more particularly organophosphate insecticides, have the potential to inhibit cholinesterase (ChE) enzyme activity by phosphorylation and accumulation of acetylcholine (ACh) at receptor site areas.[122–124] Acetylcholine, which acts as a neurotransmitter in animals and humans, is present at terminal endings of all postganglionic parasympathetic nerves, neuromuscular junctions, autonomic nervous systems, and the sympathetic and parasympathetic ganglia. The ACh is synthesized by the acetylcholinesterase (AChE) from acetyl-CoA and choline. The ACh contacts cholinergic receptor protein molecules of the postsynaptic membrane and changes its configuration, allowing Na and K cations to penetrate. The inhibition of AChE activity results in the accumulation of endogenous ACh in nerve tissues and the effector organs. This chain reaction eventually ends up in the development of signs and symptoms of organophosphorus (OP)-induced poisoning. The signs and symptoms of OP poisoning are classified as (1) muscarinic; (2) nicotinic; and (3) central nervous system, since they mimic them. Depending on the nature of exposure and the degree of OP poisoning, animals and humans exhibit profuse salivation, lacrimation, tremors, diarrhea, convulsions, and death. In the case of humans, many early warning signs and symptoms become evident, providing an opportunity for the protection of the individual if medical attendance is timely. (refer to Chapter 6, "Organophosphorous Pesticides").[125]

In the case of severe acute neurotoxicity, as more often occurs among agricultural and occupational pesticide spray workers, the cholinergic reaction stimulates the nerve endings, leading to paralysis of the limbs. Regular monitoring of workers and pest control operators goes a long way toward identifying dangers and providing timely care to avoid poisoning and human fatalities. More information may be found in the literature.[52,93]

***Immunotoxicity*** Occupational workers are exposed to a variety of chemicals in work conditions. The assessment of immunotoxicity in humans exposed to potentially immunotoxic chemicals is a complex task. An immune response in fully immunologically competent individuals will provide protection against several infectious agents and environmental hazards. The immune system is a self-restoring system, otherwise known as a homeostatic system, which fosters a quick return to normal function after a period of marked stimulation and response to chemicals and/or drugs.

Immunotoxicity is the study of interactions of chemicals with the immune system. A variety of chemical insults alter the immune function; this interference leads to decreased resistance to infection, certain neoplasia, immune dysregulation, or stimulation such as allergy or autoimmunity. Certain drugs are known to induce antiimmunity; studies also have shown that heavy metals (e.g., lead and mercury) are known to cause immunosuppressive effects, hypersensitivity, and antiimmunity. There is a growing concern about the effects of chemicals and drugs on public health vis-à-vis the immune system.[125a–125d]

***Ecotoxicology*** The term *ecotoxicology* involves the study of adverse effects on the ecological system triggered from substances, chemicals, or pollutants. The list of pollutants ranges from agrochemicals, waste substances, heavy metals, packaging materials, detergents, toxic gases and vapors, and others. The ecosystem includes three major fields, namely, aquatic, terrestrial, and aerial atmosphere. The number of organisms susceptible to toxicant exposure is as large as the natural population, ranging from microorganisms to invertebrates and vertebrate animals. The ecotoxicological effects of pollutants have made a significant impact on the health of plants, animals, and humans. Examples of soil contamination, air pollution, and water pollution and the consequences thereof need no emphasis.

The manner of waste disposal in different countries and the consequences on human health and subsequent human tragedies still persist. The dioxins in Love Canal (United States), the methyl isocyanate (MIC) in Bhopal (India), and the malathion in Pakistan are well-known cases. There are many more such cases caused by other chemicals that need proper documentation. The multiple effects of pollution on surface waters, deep sea, cultivable soil, lawn and farmland, and damage to the biosphere has repeatedly stressed the urgency of the growing problem. Who could be held responsible for these tragedies? The answer is elusive. It has become evident that all tragedies caused by hazardous chemicals have occurred because of worker negligence and related industries. Unless everyone related to the use and management of chemicals is well aware of the elementary rules and responsibilities, human safety will not be accomplished.[126–134]

## CONCLUSION

Safeguarding health and protecting the environment from the adverse effects of chemicals has been a major goal of human activity. Conducting toxicological investigations is a path to reach this goal. The synthesis and production of chemicals is large and ever-increasing, and people from every walk of life are touched by the world of chemicals. Most chemicals that are disposed of — directly or indirectly — ultimately reach the common sink, soil, or surface water systems.

The cultivable soil systems, water bodies, and free air are our most precious natural assets and must not be contaminated further. Soil seepage, wind drift, runoff surface and groundwaters, microbial activity, and interaction have further aggravated the problem.

In fact, a host of factors have complicated the toxicity profile of individual and combination chemicals. In light of the improper use and disposal of toxic chemicals, it is paramount to educate personnel on basic principles of toxicology. By knowing the elements and adopting principles of toxicology, the long-range benefits and goals of chemical safety could be achieved in the workplace. To do so, it is important that all people — workers, managers, students, and others associated with the use of chemicals — are required to understand basic chemical safety and the dangerous implications of lack of such knowledge. A thorough knowledge of chemicals allows the worker to follow healthy practices, develop pragmatic approaches, safeguard the work environment, and eventually help the organization to comply with regulations for common good.

## REFERENCES


1. Environmental Protection Agency. EPA's program to control exports of hazardous waste, March 31. *Report Audit.* EID 37.05-0456-80855; 1988.

1a. U.S. Environmental Protection Agency, *Voluntary U.S. EPA/industry program commits to cut toxic wastes.* U.S. EPA press release, Washington, DC, July 19, 1991.

2. D'Ttri F. Department of Fisheries and Wildlife, Michigan University. *St. Louis Post–Dispatch.* November 26, 1989.

3. Anonymous. *The global poison trade. Newsweek,* Nov. 7, **66**: 1988.

4. Anonymous, *National survey of hazardous waste treatment, storage, disposal and regulating facilities.* Report to Environment Protection Agency by Kate Blow. United States Environmental Protection Agency. July 30, 1990.

5. Bill Moyers, *Global dumping ground. The international traffic in hazardous waste.* Seven Locks Press, Washington, DC, 1990.

6. P.A. Bertazzi, A.C. Pesatori, D. Consonni, et al., Cancer incidence in a population accidentally exposed to 2,3,7,8-tetrachlorodibenzo-paradioxin. *Epidemiology* **4**(5):398, 1993.

7. T.J. Callender, Long-term neurotoxicity at Bhopal. *Internat. Perspect. Pub. Health,* Vols. **11** and **12**:36, 1996.

8. J.P. Cullinan, S.D. Acquilla, and V.R. Dhara, Long term morbidity in survivors of the 1984 Bhopal gas leak. *Nat. Med. J. India.* **9**(1):5, 1996.

8a. W.J. Rogan, et al., Polychlorinated biphenyls (PCBs) and dichlorophenyl dichloroethene (DDE) in human milk: Effect on growth, morbidity and duration of lactation. *Am. J. Public Health.* **77**:1294–1297, 1987.

8b. D.L. Adams, Living in a Healthy Environment. American Medical Women's Association Alexandria, VA, 1999.

9. P.A. Bertazzi, C. Zocchetti, A.C. Psatori, et al., Ten year mortality study of the population involved in the Seveso incident in 1976. *Am. J. Epidemiol.* **129**:1187, 1989.

10. National Research Council, *News Report*, Vol. **XLI**(6): p. 19, June–July, 1991.
11. Y.L. Guo, Mei-Lin Yu, C.C. Chen-Chin Hsu, and W.J. Rogan, Chloracne, goiter, arthritis, and anemia after polychlorinated biphenyl poisoning: 14-year follow-up of the Taiwan Yucheng cohort. *Environ. Health Perspect.* **107**:715, 1999.
12. S. Epstein, and J. Swartz, The fallacies of lifestyle cancer theories. *Nature*, **289**: (January), 1981.
13. Anonymous, *An Introduction to Ayurveda*, Ayurvedic Foundation, India, 1996.
14. Vasant Lad, *The Ayurvedic Institute, India*, Albuquerque, New Mexico, 1998.
15. Sukh Dev, Ancient-modern concordance in ayurvedic plants: Some examples. *Environ. Health Perspect.* **107**:783, 1999.
16. M. Stevenson, *Mexican refinery cleanup could rival Love Canal's*, The Associated Press, May 30, 1999.
17. H.A. Tilson, Developmental neurotoxicity of endocrine disruptors and pesticides. Identification of important gaps and research needs. *Environ. Health Perspect.* **106**:807, 1998.
18. ENDS, Sweedish research spotlights brominated flame retardant risks. *ENDS Report* **276**:6, 1998.
19. G. Sundroik, O. Hutzinger, S. Safe, and N. Platonow, The metabolism of $p,p'$-DDT and $p,p'$-DDE in the pig, in *Fate of Pesticides in Large Animals*, I. Vie, G. and Dorough, H., eds., Academic Press, New York, **175**, 1977.
20. R.D. Brien, *Insecticides Action and Metabolism*, Academic Press, New York, 1967.
21. U.V. Gopalaswamy, and A.S. Aiyar, Biotransformation of lindane in the rat. *Bull. Environ. Contam. Toxicol.* **32**:148, 1984.
22. J.B. Schenkman, and D. Kupfer, eds., *Hepatic Cytochromr p450 Monooxygenase System*, Pergamon Press, Ltd., New York, 1982.
23. T.A. Popov, and B.J. Blaauboer, Biotransformation of pesticides. The role of mixed function oxidase system in pesticide toxicity, in *Toxicology of Pesticides in Animals*, ed., T.S.S. Dikshith, CRC Press, Boca Raton, FL, **41**, 1991.
24. J.N. Smith, The comparative metabolism of xenobiotics. *Adv. Comp. Physiol. Biochem.* **3**:173, 1968.
25. M.C. Antunes-Madeira, and V.M.C. Madeiru, Interaction of insecticides with lipid membranes. *Biochem. Biophys. Acta.* **550**:384, 1979.
26. G.P. Vlasuk, D.E. Ryan, P.E. Thomas, W. Levin, and F.G. Walz Jr., Polypeptide patterns of hepatic microsomes from Long-Evans rats treated with different xenobiotics, *Biochemistry* **21**:6288, 1982.
27. R.C. Baker, L.D. Cooris, R.B. Mailman, and E. Hodgson, Induction of hepatic mixed function oxidase by insecticide mirex. *Environ. Res.* **5**:418, 1972.
28. R.K. Hindere, and R.E. Menzer, Comparative enzyme activities and cytochrome p450 levels of some rat tissue with respect to their metabolism of several pesticides. *Pest. Biochem. Physiol.* **6**:148, 1976.
29. V. Ullrich, A. Roots, A. Hildbrandt, R.W. Estabrook, and A.H. Conney, *Microsomes and Drug Oxidations*, Pergamon Press, Oxford, 1971.
30. D.V. Parke, and R.L. Smith, eds., *Drug Metabolism*, Taylor and Francis, London, 1976.
31. M.W. Anders, ed., *Bioactivation of Foreign Compounds*, Academic Press, New York, 1985.

32. F.P. Guengerich, *Mammalian Cytochromes P450*, CRC Press, Boca Raton, FL, 1987.
33. D.H. Hutson, Glutathione conjugates in *Bound and Conjugated Pesticide Symposium*, Series 29, American Chemical Society, Washington, DC, **103**, 1976.
34. P. Jenner, and B. Testa, Novel pathways in drug metabolism. *Xenobiotica* **8**:1, 1978.
35. H.W. Dorough, Metabolism of insecticides by conjugation mechanisms. *Pharmacol. Ther.* **4**:433, 1979.
36. M.R. Juchau, Mechanisms of drug biotransformation reactions in the placenta. *Fed. Proc.* **31**:48–51, 1972.
37. C.D. Klassen, D.L. Eaton, and S.Z. Cagen, Hepatobiliary disposition of xenobiotics, in J.W. Bridges, and L.F. Chasseaud, eds., *Progress in Drug Metabolism*, John Wiley & Sons, New York, 1981.
38. G.R. Wilkinson, Plasma and tissue binding considerations in drug disposition. *Drug Metab. Rev.* **14**:427–465, 1983.
39. R. Kato, Sex related differences in drug metabolism. *Drug Metab. Rev.* **3**:1, 1974.
40. G. Carrera, and A. Periquet, Metabolism and toxicokinetics of pesticides in animals, in *Toxicology of Pesticides in Animals*, ed., T.S.S. Dikshith, CRC Press, Boca Raton, FL, 1991.
41. J.D. Yarbrough, J.E. Chambers, K.M. Robinson, Alterations in liver structure and functions resulting from chronic insecticide exposure, in *Effects of Chronic Exposure to Pesticides on Animal System*, J.E. Chambers, and J.D. Yarbrough, eds., Raven Press, New York, 1982.
42. L.B. Tee, T. Seddon, A.R. Boobis, and D.S. Davis, Drug metabolizing activity of freshly isolated human hepatocytes. *Br. J. Clin. Pharmacol.* **19**:379, 1985.
43. T.C. Orton, A.E. Sorman, D.N. Crisp, A.P. Sturdie, Dynamics of xenobiotic metabolism by isolated rat hepatocytes using a multichannel perfusion system. *Xenobiotica* **13**:743, 1983.
44. G.J. Dutton, and B. Burchell, Newer aspects of glucuronidation. *Prog. Drug Metab.* **2**:1, 1977.
45. M.A. Kamrin, *Toxicology: A Primer on Toxicology Principles and Application*, CRC Press, Boca Raton, FL, 1988.
46. G.P. Quinn, J. Axelrod, and B.B. Brodie, Species, strain and sex differences in metabolism of hexobarbitone, amidopyrine and aniline. *Biochem. Pharmacol.* **1**:152, 1958.
47. R.L. Cram, M.R. Juchau, and J.R. Fouts, Differences in hepatic drug metabolism in various rabbit strains before and after pretreatment with phenobarbital. *Proc. Soc. Exp. Biol. Med.* **118**:872, 1965.
48. K.L. Davidson, V.J. Feil, and C.H. Lamoureux, Methoxychlor metabolism in goats. *J. Agric. Food Chem.* **30**:130, 1982.
49. C.H. Walker, Species differences in microsomal mono-oxygenase activity and their relationship to biological half-lifes. *Drug Metab. Rev.* **7**:295, 1978.
50. D.E. Stevenson, J.A. Popp, J.M. Ward, R.M. McClain, T.J. Slaga, and H.C. Pitot, eds., Mouse liver carcinogenesis, *Mechanisms and Species Comparisons*, Wiley-Liss, New York, 1990.

51. F. Oesch, T. Frieberg, M. Herbst, W. Paul, N. Wilhelm, and P. Bentley, Effects of lindane treatment on drug metabolizing enzymes and liver weight of CF1 mice in which it evoked hepatomas and in non-susceptible rodents. *Chem. Biol. Interact.* **40**:1, 1982

52. W.J. Hayes Jr., *Toxicology of Pesticides*, Williams & Wilkins, Baltimore, MD, 1975.

53. R.T. Williams, Interspecies variations in the metabolism of xenobiotics. *Biochem. Soc. Trans.* **2**:359, 1974.

54. D.L. Fabacher, and E. Hodgson, Induction of hepatic mixed oxidase enzymes in adult and neonatal mice by kepone and mirex. *Toxicol. Appl. Pharmacol.* **39**:71, 1976.

55. J. Brodeur, and K.P. Du Bois, Comparison of acute toxicity of anticholinesterase insecticides of weanling and adult male rats. *Proc. Soc. Exp. Biol. Med.* **114**:509, 1963.

56. F.C. Lu, D.C. Jessup, and A. Lavellee, Toxicity of pesticides in young rats versus adult rats. *Food Cosmrt. Toxicol.* **3**:591, 1965.

57. R.W. Chadwick, R.S. Linko, J.J. Freal, and A.L. Robbins, The effect of age and long-term low level DDT exposure on the response to enzyme induction in the rat. *Toxicol. Appl. Pharmacol.* **31**:469, 1975.

58. G.J. Dutton, Developmental aspects of drug conjugation. *Ann. Rev. Pharmacol. Toxicol.* **18**:17, 1978.

59. J.R. Gillette, and B. Stripp, Pre and post natal enzyme capacity for drug metabolite production. *Fed. Proc.* **34**:172, 1975.

60. D.P. Richey, and D.A. Bender, Pharmacokinetic consequence of aging. *Annu. Rev. Pharmacol. Toxicol.* **17**:49, 1977.

61. M.G. MacDonald, M.J. Fasco, and L.S. Kainsky, Development of hepatic mixed function oxidase system and its metabolism warfarin in the perinatal rat. *Dev. Pharmacol. Ther.* **3**:1, 1981.

62. A.H. Conney, and J.J. Burns, Factors influencing drug metabolism. *Adv. Pharmacol.* **1**:31, 1962.

63. J.R.E. Gillette, and D.J. Jollow, Drug metabolism in liver, in *The Liver*, Becker, F.F., ed., Marcel Dekker, New York, 1974.

64. T.B. Gaines, The acute toxicity of pesticides to rats. *Toxicol. App. Pharmacol.* **14**:515, 1969.

65. R.E. Ebel, Hepatic microsomal p-nitroanisole O-demethylase, Effects of chlordecone or mirex induction in male and female rats. *Biochem. Pharmacol.* **33**:559, 1984.

66. C.A. Mugford, and G.L. Kedderis, Sex-dependent metabolism of xenobiotics. *Drug Metab. Rev.* **30**(3):441, 1998.

67. American Cancer Society, Facts and Figures, *American Cancer Society, Inc.*, Atlanta, GA, 1990.

68. D.V. Parke, ed., Development of mixed-function oxidases. *Biochem. Soc. Trans.* **18**:7–36, 1990.

69. A.E. Rogers, and M.S. Longnecker, Dietary and nutritional influences on cancer: A review of epidemiologic and experimental data. *Lab. Invest.* **59**:729–759, 1988.

70. D.V. Parke, The effects of nutrition and enzyme induction in toxicology. *World Rev. Nutr. Dietetics* **29**:96, 1978.
71. R. Truhaut, and R. Ferrando, eds., *Toxicology and Nutrition: World Review of Nutrition and Dietetics* S. Karger, Basel, I, 29, 1978.
72. R. Truhaut, and R. Ferrando, and C. Fourlon, Interactions entre la vitamine A et le DDT: Influence due froid et du bruit. *C. R. Acad. Sci. III*, **299**:759, 1984.
73. C.J. Krijnen, and E.M. Boyd, The influence of diet containing 0–81% of protein on tolerated doses of pesticides. *Comp. Gen. Pharmacol.* **22**:373, 1971.
74. E.M. Boyd, *Protein Deficiency and Pesticide Toxicity*, Charles Thomas, Springfield, IL, 1972.
75. L.W. Wattenberg, W.D. Loub, L.K. Lam, and J.L. Speier, Dietary constituents altering the response to chemical carcinogens. *Fed. Proc.* **35**:1327, 1976.
76. G. Debry, Nutrition food and drug interactions in man, *World Review of Nutrition and Dietetics*, S. Karger, Basel, 43, 1984.
77. T. Purshottam, and R.K. Srivatsava, Effects of high-fat and high-protein diets on toxicity of parathion and dichlorvos. *Arch. Environ. Health.* **39**:425, 1984.
78. J.H. Weisburger, and C.L. Horn, Causes of cancer, in A. Holleb, and D. Fink, eds., *American Cancer Society Textbook on Clinical Oncology*, 6th edition, American Cancer Society, Atlanta, GA, Chapter **7**, 1990.
79. W.J. Hayes Jr., ed., *Pesticides Studied in Man*, Williams & Wilkins, Baltimore, MD, 1982.
80. E.E. Kenaga, and C.S. End, *Commercial and experimental organic insecticides*, Entomological Society of America, Special publication 74-1, 77, 1974.
81. R.B. Conolly, B.D. Beck, Goodman J.I., Stimulating research to improve the scientific basis of risk assessment. *Toxicol. Sci.* **49**:1, 1999.
82. T.S.S. Dikshith, *Toxicology of Pesticides in Animals*, CRC Press, Boca Raton, FL, 1991.
83. I. Janku, Problems in the evaluation of the relative biological potency of chemical agents, *in Adverse Effects of Environmental Chemicals and Psychotropic Drugs*, **VI**, M. Horvat, ed., Elsevier, London, 1973.
84. C.S. Weil, Guidelines for experiments to predict the degree of safety of a material for man. *Toxicol. Appl. Pharmacol.* **21**:194, 1972 a.
85. C.S. Weil, Statistics versus safety factors and scientific judgments in the evaluation of safety for man. *Toxicol. Appl. Pharmacol.* **21**:454, 1972 b.
86. World Health Organization. Principles and methods for evaluating the toxicity of chemicals, Part I, *Environ. Health Criterion* **6**: 1978.
87. U.S. Environmental Protection Agency, Office of the Testing and Evaluation. Proposed Health Effects Test Standards for Toxic Substances Control Act Test Rules, 40 CFR, Part 772, Standards for Development of Test Data sub part D- Chronic Health Effects. *Fed. Reg.* **44**(91):27350, 1979.
88. T.S.S. Dikshith, *Safety Evaluation of Environmental Chemicals*, New Age International Publishers, (formerly Wiley Eastern Ltd.,), New Delhi, India, 1996.
89. Department of Human Health Service, Good Laboratory Practice Regulations, FDA 21 CFR Part 58, Final Rule. *Fed. Reg.* **52**(172):33768, 1987.
90. U.S. Environmental Protection Agency, Toxic Substance Control, Good Laboratory Practice Standards, Final Rule. *Fed. Reg.* **48**(230):53937, 1983.

91. T.R. Stiles, Quality assurance in toxicology studies, in *General and Applied Toxicology*, Brayn, B., Timothy, M., and Turner, P., eds., Chapter **16**, Part II, Macmillan Press, UK, 1992.
92. S. Ramamoorthy, *Chlorinated Organic Compounds in the Environment: Regulatory and Monitoring Assessment*, CRC Press, Lewis Publishers, Boca Raton, FL, 1997.
93. C.S. Auletta, Acute, subchronic and chronic toxicology, in *CRC Handbook of Toxicology*, Chapter **2**, Derelanko, M.I., and Hollinger, M.A., eds., CRC Press, Boca Raton, FL, 1995.
94. W.J. Hayes Jr., and E.R. Laws, *Handbook of Pesticide Toxicology*, Volumes **I, II, III**, Academic Press, New York, 1991.
95. J.A. Timbell, *Introduction to Toxicology*, 2nd ed., Taylor & Francis, London, 1995.
96. L.G. Cockerhanu, and B.S. Shane, *Basic Environmental Toxicology*, CRC Press, Boca Raton, FL, 1993.
97. A.P. Li, *Genetic Toxicology*, CRC Press, Boca Raton, FL, 1991.
97a. Report of the International Workshop on *In vitro methods for assessing acute systemic toxicity results of an international workshop.* Organized by the Interagency Coordinating Committee on the Validation of Alternative Methods (ICCVAM), and the National Toxicology Program (NTP) Interagency Center for the Evaluation of Alternative Toxicological Methods (NICEATM). National Institute of Environmental Health Sciences, National Institutes of Health U.S. Public Health Service, Department of Health and Human Services, Washington, DC, NIH Publication No. 01-4500. October 17–20, 2000.
97b. National Institute of Environmental Health Services *In vitro methods for assessing acute systemic toxicity*, Report of the International Workshop on *In Vitro* Methods for Assessing Acute Systemic Toxicity, NIH Publication No. 01-4499. NIEHS, North Carolina, USA, August 2001.
98. A.T. Natarajan, J.J.W.A. Boei, F. Darroudi, P.C.M. Van Diemen, F. Dulout, M.P. Hande, and A.T. Ramalho, Current cytogenetic methods for detecting exposure and effects of mutagens and carcinogens. *Environ. Health Perspect.* **104**:(suppl. 3):445, 1996.
99. W.F. Durham, and C.H. Williams, Mutagenic, teratogenic and carcinogenic properties of pesticides. *Ann. Rev. Entomol.* **17**:123, 1972.
100. H.H. Chen, J.L. Hsuch, S.R. Sirianni, and C.C. Huang, Induction of sister chromatid exchanges and cell cycle delay in cultured mammalian cells treated with organophosphorus pesticides. *Mut. Res.* **88**:307, 1981.
101. H. Tezuka, N. Ando, and R. Susuki, Sister chromatid exchanges and chromosomal aberrations in cultured Chinese hamster cell treated with pesticides positive in microbial reservation assays. *Mut. Res.* **78**:177, 1980.
102. F.N. Dulout, M.C. Pastori, and O.A. Oliver, Malathion induced chromosomal aberration in bone marrow cells of mice: Dose-response relationships. *Mut. Res.* **122**:163, 1983.
103. R.D. Hood, *Handbook of Developmental Toxicology*, CRC Press, Boca Raton, FL, 1996.
104. P.J. Landrigan, L. Claudio, S.B. Markowitz, G.S. Berkowitz, B.L. Brenner, et al., Pesticides and inner-city children: Exposures, risks, and prevention. *Environ. Health Perspect.* **107**(Suppl. 3):431, 1999.

105. T. Colborn, F.S. vom Saal, and A.M. Soto, Developmental effects of endocrine-disrupting chemicals in wildlife and humans. *Environ. Health Perspect.* **101**(5):378, 1993.

106. C.J. Henny, and G.B. Herron, DDE, selenium, mercury, and white-faced ibis reproduction at Carson Lake, Nevada. *J. Wildlife Mgmt.* **53**:1032, 1989.

107. H.T. Jansen, P.S. Cooke, J. Porcelli, T.C. Liu, and L.G. Hansen, Estrogenic and antiestrogenic actions of PCBs in the female rat: In vitro and in vivo studies. *Reprod. Toxicol.* **7**:237, 1993.

108. R.E. Peterson, H.M. Theobald, and G.L. Kimmel, Developmental and reproductive toxicity of dioxins and related compounds: Cross-species comparisons. *Crit. Rev. Toxicol.* **23**(3):283, 1993.

109. A.M. Soto, K.L. Chung, and C. Sonnenschein, The pesticides endodusulfan, toxaphene, and dieldrin have estrogenic effects on human estrogen-sensitive cells. *Environ. Health Perspect.* **102**(4):380, 1994.

110. W.R. Kelce, E. Monosson, P. Gamcsik, S.C. Laws and L.E. Gray Jr., Environmental hormone disruptors: Evidence that vinclozolin developmental toxicity is mediated by antiandrogenic metabolites. *Toxicol. Appl. Pharmacol.* **126**:276, 1994.

111. L.S. Khera, Evaluation of dimethoate (Cygon4E) for teratogenic activity in the cat. *J. Environ. Pathol. Toxicol.* **2**:1283, 1979.

112. R.D. Harbinson, Teratogens, in *Toxicology the Basic Science of Poisons*, J. Dull, C.D. Klaussan, and M.O. Amdur, eds., Macmillan, New York, 1980, 158.

113. D.J. Hoffman, and P.H. Albers, Evaluation of potential embryotoxicity and teratogenicity of 42 herbicides, insecticides, and petroleum contaminants to Mallard egg. *Arch. Environ. Contam. Toxicol.* **13**:15, 1984.

114. R.J. Jager, Kepone chronology. *Science* **193**:95, 1979

115. E.R. Laws Jr., W.D. Maddrey, A. Curley, V.W. Burse, Long term occupational exposure to DDT. *Arch. Environ. Health* **27**:318, 1973.

116. Anonymous, *IARC Monographs on the Evaluation of Carcinogenic Risk of Chemicals to Man, Some Organochlorine Pesticides*, Vol. 5, International Agency for Research on Cancer, Lyon, France, 1974.

117. Anonymous, *IARC Monographs on the Evaluation of Carcinogenic Risk of Chemicals to Humans, Some Halogenated Hydrocarbons*, Vol. 20, International Agency for Research on Cancer, Lyon, France, 1979.

118. Anonymous, IARC *Monographs on the Evaluation of Carcinogenic Risk of Chemicals to Humans, Miscellaneous Pesticides*, Vol. 30, International Agency for Research on Cancer, Lyon, France, 1983.

119. M.D. Ruber, Carcinogenicity of dimethoate. *Environ. Res.* **34**:193, 1984.

120. P.J. Duerksen-Hughes, J. Yang, and O. Ozcan, P. 53 Induction as a genotoxic test for twenty-five chemicals undergoing *in vivo* carcinogenicity testing. *Environ. Health Perspect.* **107**:805, 1999.

121. J. Dejmek, S.G. Selevan, I. Solansky, I. Benes, J. Lenicek, R.J. Sram, Exposure to carcinogenic PAHs in utero and fetal growth. *Epidemiology* **10**:S126, 1999.

122. R.S. Wadia, C. Sadagopan, R.B. Amin, and H.V. Sardesai, Neurological manifestations of organophosphorus pesticide poisoning. *J. Neurol. Neurosurg. Psychiatry* **37**:891, 1974.

123. M. Lotti, M.K. Johnson, Repeated small doses of neurotoxic organophosphate. Monitoring of neurotoxic esterase in brain and spinal cord. *Arch. Toxicol.* **45**:263, 1980.

124. M.B. Abou-Donia, Organophosphorus ester-induced delayed neurotoxicity. *Ann. Rev. Pharmacol. Toxicol.* **21**:511, 1981

125. J.L. Jacobson, S.W., Jacobson and H.E.B. Humphrey, Effects of exposure to PCBs and related compounds on growth and activity of children. *Neurotoxicol. Teratol.* **12**(4):319, 1989.

125a. Food and Drug Administration (1997), International Conference on Harmonization. Guidance for Industry. S6 Preclinical safety evaluation of biotechnology-derived pharmaccuticals. *Fed. Reg.* **62**, 61515–61519, 1997.

125b. S.M. Furst, M. Chen, and A.J. Gandolfi, Use of halothane as a model for investigating chemical-induced autoimmune hepatotoxicity. *Drug Info. J.* **30**, 301–307, 1996.

125c. U.S. Food and Drug Administration, *Guidance for industry immunotoxicology evaluation of investigational new drugs*. U.S. Department of Health and Human Services, Food and Drug Administration Center for Drug Evaluation and Research (CDER), April 2001.

125d. J. Bernier, P. Brousseau, K. Krzystyniak, H. Tryphonas, and M. Fournier, Immunotoxicity of heavy metals in relation to Great Lakes. *Environ. Health Perspect.* **103**(Suppl 9):23–34, 1995.

126. R. Truhaut, Ecotoxicology objectives, principles and perspectives, in *Evaluation of Toxicological Data for the Protection of Public Health*, Hunter, W.J., Smeets, J.G.M., eds., Pergamon Press, Oxford, 1977, 339.

127. G.C. Butler, ed., *Principles of Ecotoxicology*, Scope 12, John Wiley & Sons, New York, 1978.

128. International Union of Pure and Applied Chemistry and International Programme of Chemical Safety, *Chemical Safety Matters*, Cambridge University Press, UK, 1992.

129. M.D. Hawkins, *Safety and Laboratory Practice*, 3rd eds., Cassell Publisher, London, 1988.

130. Croner Publications, *Croner's Hazardous Waste Disposal Guide*, Croner Publishers, New Malden, UK, 1988.

131. L. Bretherick, *Handbook of Reactive Chemical Hazards*, 4th eds., Butterworths, London, 1990.

132. D.C. Malins, and G. Ostrander, *Aquatic Toxicology: Molecular, Biochemical, and Cellular Perspectives*, CRC Press, Boca Raton, FL, 1994.

133. G. Ostrander, *Techniques in Aquatic Toxicology*, CRC Press, Boca Raton, FL, 1996.

134. M.C. Newman, *Ecosystem Ecology*, CRC Press, Boca Raton, FL, 1996.

CHAPTER 3

# Heavy Metals

## INTRODUCTION

Application of metals for different purposes is as old as human history. Metals have been found among the oldest toxic elements known to man, and references are available in the literature about their use by early human civilization. Mercury as mercuric sulfide (cinnabar) has been in existence for the past 2,000 years. During 370 B.C. Hippocrates observed people suffering from abdominal colic and excreting metals. The story of lead as a silent hazard and killer is well documented.[1–5] Robert Grosseteste, Bishop of Lincoln, described the transmutation of metals in *De artibus liberati* and *De generatione stellarum* in 1235. Metals are an integral part of our planet and are found in almost all rocks and soils. However, metals are rarely found in nature in their pure metallic state. Only gold, silver, and copper (so-called native silver and copper) have been found as pure metals. A majority of the metals occur in nature as compounds combined in structures called minerals. Thus, minerals have many metals and occur naturally as inorganic solids with regular chemical compositions and crystal structures.

In general, the metal-bearing minerals have several elements, except the precious metal gold, which is found in its elemental form as a mineral (native gold). The minerals may be classified as (1) oxides (e.g., bauxite, magnetite, hematite, cassiterite, chromite, pyrolusite, laterite, and uraninite); (2) carbonates of sodium, calcium; (3) sulphides (e.g., chalcopyrite, sphalerite, pentlandite, galena, cinnabar, and pyrite); (4) sulfates (e.g., gypsum); (5) phosphates; (6) silicates; and (7) halides. It is from this host of naturally occurring minerals that the mining industry derives a large bulk of structural and functional materials. The metallic elements are often divided into light and heavy metals. Generally, the concept *light metal* refers to beryllium, magnesium, aluminum, titanium, and their alloys, and the term *heavy metal* refers to those with potential toxicity such as, cadmium, mercury, lead, and bismuth.[6–10]

In Australia at the turn of the century (1904), people were grappling with a disease termed *toxicity of habitation*. This was characterized in children with complaints of stomach cramps, paralysis, pain in the limbs, and convulsions. Later, these adverse effects became associated with several metals, namely,

*Industrial Guide to Chemical and Drug Safety*, By T.S.S. Dikshith and Prakash V. Diwan
ISBN 0-471-23698-5 

cobalt, copper, fluorine, iron, manganese, selenium, and zinc. Several researchers discovered that metals caused signs and symptoms of nausea, vomiting, abdominal pain, diarrhea, optic atrophy, deafness, and many kinds of central nervous system (CNS) disturbances in addition to causing bladder cancer.[11–15] In several instances, the enigma of metal poisoning and human fatalities has remained an unsolved problem.

A variety of industrial materials, including metallic compounds (both organic and inorganic), are produced by multiple anthropogenic activities. Because of the manner of use and disposal, these often reach the environment and cause a plethora of potential health hazards.[4] Of the above symptoms, the risk for CNS disturbances in a child's development or in the human adult has gained importance. Information concerning different metals (e.g., arsenic, cadmium, chromium, lead, mercury, many other compounds) and their possible carcinogenicity has been documented as a national priority in the literature (Table 3-1).[6–10,11]

However, the mechanism of heavy metal toxicity used in combination with other chemical agents has remained unexplored. In fact, this subject requires more emphasis because of the occupational complexity and exposure pattern of today's society. The main anthropogenic sources of heavy metals are industrial point sources, mining activities, foundries and smelters, piping, product constituents, combustion by-products, traffic, and, above all, the pattern of waste disposal.[7] Mining, manufacturing, and industrial production of metals have become major

**TABLE 3-1 Carcinogenicity of Metals Under Completed Exposure Pathway Priority Substances**

| Metals | Percent | Category* |
|---|---|---|
| Antimony | 7 | 3 |
| Arsenic | 23 | 1 |
| Barium | 8 | 3 |
| Beryllium | 6 | 2 |
| Cadmium | 17 | 2 |
| Chromium | 17 | 1 |
| Copper | 10 | 3 |
| Lead, inorganic | 34 | 3 |
| Manganese | 11 | 3 |
| Mercury, metallic | 12 | 3 |
| Zinc | 12 | 3 |

*Source*: National Priority List of 530 Completed Exposure Pathway Priority Substances.
*Carcinogenicity categories: 1—Designated a human carcinogen by DHHS, EPA, or IARC; 2—Designated as reasonably anticipated to be a human carcinogen by DHHS, EPA, or IARC; 3—Not classified by DHHS, EPA, or IARC.

commercial activities. Because of this, a number of metals and their combinations have entered the global market in the name of economic improvement and trade development. Because of geologic and environmental changes and related biological transformations (e.g., rain, storm, floods, earth tremors, volcanoes, and geological degradations), different metals have brought the finer elements of rocks and soil into water bodies.

## METALS AND TOXICITY

Workers in industries are constantly exposed to metal processing operations (e.g., ore extraction, grinding, digestion, casting, making alloys, welding, soldering, brazing, wire patenting, painting, plating, printing, tinning, powder mixing, scraping); continuous exposure during these operations makes them susceptible to metal dusts and fumes, leading to health hazards. Biologic changes resulted in bioconcentration in plants and animals. These changes have entered the food cycle and human and animal bodies. Some metallic substances are not easily excreted or chemically transformed by the body. For example, repeated exposures (as in occupational conditions) have caused metals to accumulate in tissues, organs, skeletal tissue, and bones, leading to buildup of the metal(s) within the body and subsequent long-term health effects. Entry of different metals (particularly arsenic and sodium arsenite) into the system — either directly from the work environment or indirectly by intake of contaminated food and water — has caused poisoning and adverse health effects.[12–15]

The impact to health has been mostly dependent on the concentration of the candidate metal. Some metals (e.g., mercury, lead, arsenic, cadmium, iron, copper) ultimately find their way into human systems via soil, minerals, and water. Studies have shown the presence of many metals in daily consumable products (e.g., food, fruits, milk, fabric materials, drinking water). Further, heavy metals associated with particle material can be accumulated in areas suitable for sedimentation or particle concentration (e.g., upstream from sills or dams, in estuary sludge clog, etc.). These accumulation areas are creating possible pollution sources, as particles pooled could be resuspended during punctual hydrologic periods (floods, drains). Bioavailability, and therefore toxicity of heavy metals, is strongly bound to the current chemical form.

The free and labile forms (slightly stable) are known to be more toxic than colloidal and particle forms. An excess intake and major deficiency of certain metals also trigger health effects, which require further study. Studies showed many metals such as lead, tin, mercury, and cadmium compounds have caused adverse effects in the developing and adult rat.[4,6,9,10] Trimethyltin (TMT), for instance, causes neurotoxic disturbances in rats, as evidenced by histological, histochemical, *in situ* hybridization, receptor binding, and biochemical adverse effects. Repeated exposure to metals triggers neurodegenerative changes in the limbic system, including severe transient gliosis and neuronal degeneration in animals.

It is well known today that environmental or ionic and electromagnetically active metals — especially heavy metals — can increase the production of free radicals 1 million-fold.[16,17] Free radicals are considered the root cause of all chronic diseases, from rheumatism to diabetes, cardiovascular diseases to cancer. Free radicals also promote the process of aging through oxidation. This phenomenon is similar to the rusting metal: the body slowly decays until it "dies." The human body has the ability to shift certain oxidation processes through antioxidants, thus limiting the body damage. Because ionic metals are present everywhere in today's environment (in the air we breathe, the water we drink, the food we eat, etc.) and because of other free radical–producing pollutants, diets, and lifestyle, the human body is often fighting a losing battle. The amount of free radicals produced can no longer be offset by available antioxidants. Even ingesting additional antioxidants may not be sufficient to help the body control damage done by excess free radicals.

Estimations of heavy metals in blood and tissues, through hair analysis, atomic absorption spectrophotometric, or electron stripping investigations, can reveal the overall burden to the system.[16,17] Although authorities worldwide have set regulations, guidelines, and standards to protect humans, animals, and the environment from the health effects of exposure to heavy metals, much work must still be done to improve the capabilities of individual workers. In fact, proper knowledge, responsibility, and understanding of the implications are a must for all workers. Strict implementation of regulations is possible only with a qualified, trained worker. Awareness of toxicology elements helps to strengthen the foundations of chemical safety. A proper understanding of the status of metal toxicity vis-à-vis human health and safety should be a goal to achieve chemical safety. The following pages discuss some heavy metals of industrial importance.

## Aluminum

***Uses*** Aluminum (Al) is not a heavy metal but has been found to be appreciably toxic, depending on the manner of handling and disposal. Aluminum is a popular substance used to make cookware, cooking utensils, and foil. Aluminum has many forms, as alumina ($Al_2O_3$), sodium aluminate ($NaAlO_3$), aluminum fluoride ($Al_2F_6$), cryolite ($Na_3Al_6$), and aluminum chloride ($AlCl_3$). Aluminum casting alloys contain silicon, magnesium, copper, zinc, and nickel in different combinations. Because aluminum permeates air, water, and soil, small amounts are present in our food. The average person consumes between 3 and 10 milligrams of aluminum a day. Studies reveal that aluminum is absorbed and accumulated in the body.[18] Excessive use of antacids is the most common cause of aluminum toxicity. Preparations such as Mylanta, Maalox, Gelusil, Amphojel, and many others have a high aluminum hydroxide content. In addition, many drugs used for the relief of inflammation and pain, such as arthritis pain (e.g., Ascriptin, Bufferin, Vanquish) contain aluminum.[18]

***Toxicity*** Aluminum toxicity can lead to a number of ailments, including colic, rickets, gastrointestinal disturbances, poor calcium metabolism, extreme nervousness, anemia, headache, decreased liver and kidney function, forgetfulness, speech disturbances, memory loss, softening and weakness of the bones, and aching muscles. Bone loss and increased intestinal absorption of aluminum and silicon combine to form compounds that accumulate in the cerebral cortex of the brain. These compounds prevent impulses from being carried to or from the brain. An accumulation of aluminum salts in the brain has been connected to seizures and reduced mental faculties. Autopsies performed on Alzheimer's patients revealed that four times the normal amount of aluminum had accumulated in the nerve cells of their brains.[18]

Many symptoms of aluminum toxicity are similar to those of Alzheimer's disease and osteoporosis. This suggests that long-term accumulation of aluminum in the brain may contribute to the development of Alzheimer's disease. In addition, an unidentified protein not found in normal brain tissue has been discovered in the brain tissue of Alzheimer's patients. Because aluminum is excreted by the kidneys, toxic amounts of aluminum also may impair kidney function. People who worked in aluminum smelting plants for long periods have been found to experience dizziness, impaired coordination, and losses of balance and energy. Accumulation of aluminum in the brain was cited as a possible cause for these symptoms as well. This has caused concern, especially as it relates to neurotoxicology (Alzheimer's disease).[19–21]

### Arsenic

***Uses*** Arsenic (As) is a naturally occurring element in the Earth's crust. Pure arsenic is a gray-colored metal and is not common in the environment. In fact, arsenic is often combined with one or more other elements such as oxygen, chlorine, and sulfur. When combined with these elements, arsenic is termed *inorganic arsenic*; arsenic combined with carbon and hydrogen is known as *organic arsenic*. Thus, several substances — inorganic and organic — containing arsenic do occur naturally. The important compounds of arsenic are arsenic trioxide ($As_2O_3$), arsenic trichloride ($AsCl_3$), and arsine ($AsH_3$). The inorganic form of arsenic is more toxic than the organic form. Inorganic arsenic has been recognized as a human poison since ancient times.[8]

Although arsenic is very widely distributed in the environment, humans are exposed to low levels of this element. Although human exposure to arsenic can come through food, water, and air, the largest source of arsenic intake (approximately 25–50 μg/day) is from food. Some edible fish and shellfish do contain elevated levels of arsenic, but this is predominantly in an organic form ("fish arsenic"). Humans are exposed to levels of arsenic in different situations.[22] For instance:

- By drinking water containing large natural mineral deposits in some geographic areas
- By working in areas associated with waste-chemical disposals containing large quantities of arsenic and improper management
- By playing in soil that might have high levels of arsenic

- By working in metal smelting, exposed to inorganic arsenic in the air
- By exposure to releases in the burning of fossil fuels such as oil, coal, gasoline, wood, and cigarette smoke (all of which contain inorganic arsenic)
- By using arsenic-containing pesticides for the control of pests and weeds in orchards and for wood preservations

***Toxicity*** Arsenic enters the human system principally through ingestion of food or drinking water. Most ingested arsenic is quickly absorbed through the stomach and intestine and enters the bloodstream. Inhalation of contaminated air leads to absorption of arsenic through the lungs into the bloodstream. Arsenic can also enter the body through the skin. Most of the arsenic that is absorbed is converted in the liver to a less toxic form and excreted in the urine.

Exposures to low levels of arsenic often cause injury to body tissue and result in systemic effects. Oral exposure causes irritation of the digestive tract, leading to pain, nausea, vomiting, pallor, and diarrhea. Exposure also causes a decrease in the production of red and white blood cells, abnormal heart function, blood vessel and liver damage, kidney damage, coma, and convulsion. The disturbances in nerve function results in a "pins-and-needles" feeling in the feet and hands. Studies in animals exposed to high oral doses of arsenic during pregnancy indicated injury to the fetus, but the studies are still inadequate[8] and need further confirmation. The single most characteristic systemic effect of oral exposure to inorganic arsenic is a pattern of skin abnormalities, which include the appearance of dark and light spots on the skin and small "corns" on the palms, soles, and trunk. While these skin changes are not considered a health concern in their own right, some of these abnormalities may ultimately progress to skin cancer. Repeated and long-term ingestion of arsenic is known to increase the risk of cancer, especially in the liver, bladder, kidney, and lung.

Occupational exposure to arsenic often involves inhalation, as in industrial areas, manufacturing alloys, and during metal smelting. Arsenic is both an animal and a human toxicant causing skin, gastrointestinal, liver problems, and peripheral neuropathy, and is a human carcinogen associated with lung cancer.[23,24] Skin cancer has been associated with both oral and dermal exposures of arsenic. Recent studies have linked exposure to arsenic via drinking water with cancer of the urinary bladder, kidney, and liver.[25,26] Although it is difficult to demonstrate the arsenic-related carcinogenicity in standard animal models, it has caused tumors in hamsters[27] and has been shown to be a "progressor" in *in vitro* experiments.[28–32]

Association of arsenic in drilled wells and water in Finland and the association of arsenic exposure to bladder and kidney cancers has been studied.[33] Between 1981 and 1995, as many as 61 cases of bladder cancers and 49 cases of kidney cancer in 275 subjects have been observed. While statistical studies show no significance with kidney cancer, bladder cancer did show a risk associated with low-level exposure to arsenic through drinking water.[33–35] Further studies are needed to confirm these findings.

### Barium

***Uses*** Barium (Ba) is a silvery-white metal found in nature. The important compounds of barium used in industry are barium sulfate ($BaSO_4$), barium carbonate ($BaCO_3$), barium nitrate (($BaNO_3)_2$), and barium fluoride ($BaF_2$). Barium compounds are used by oil and gas industries to make drilling muds, which make it easier to bore through rock. They are also used to make paint, bricks, tiles, glass, enamel, and rubber. Barium is emitted into the air through the burning of coal and oil. Some barium compounds dissolve easily in water, and some are found in lakes, accumulating in aquatic organisms and fish.

***Toxicity*** Ingestion of high levels of barium compounds causes adverse effects such as, breathing difficulty, increased blood pressure, changes in heart rhythm, stomach irritation, brain swelling, muscle weakness, damage to the liver, kidney, heart, and spleen in humans.[36–38] Barium oxide dust is a dermal and nasal irritant. Information regarding genotoxicity of barium is not available.[34,35]

### Beryllium

***Uses*** Beryllium (Be) (when pure) is a hard, grayish metal. In nature, beryllium can be found in compounds in mineral rocks, coal, soil, and volcanic dust. The important compounds of beryllium used in industry are beryllium oxide (BeO), beryllium hydroxide ($Be(OH)_2$), beryllium sulfate ($BeSO_4$), and beryllium fluoride ($BeF_2$). Beryllium compounds are commercially mined and purified for use in electrical parts, machine parts, ceramics, aircraft parts, nuclear weapons, and mirrors. Beryllium dust is emitted into the air from burning coal and oil.

Beryllium is present in the Earth's crust, in emissions from coal combustion, in surface water and soil, in house dust, food, drinking water, and cigarette smoke. Breathing air in contaminated workplaces (e.g., in areas of mining, ore processing, alloy and chemical manufacturing, machining, recycling plants for metals containing beryllium) are all sources of health hazards. Similarly, breathing tobacco smoke from leaves high in beryllium, breathing contaminated air near industry, and ingestion of food and water at hazardous waste sites also results in adverse effects.[39–41]

***Toxicity*** Beryllium in high levels in air causes lung damage and a disease that resembles pneumonia. Chronic beryllium disease can occur long after exposure to small amounts of beryllium. Both short-term, pneumonia-like disease and the chronic beryllium disease (CBD) can cause death. Toxicity due to chronic inhalation of beryllium is termed *berylliosis*. Beryllium is scientifically recognized as a health problem, primarily because of inhalation exposure. Exposure through the inhalation route, via the air pathway, is the primary route for beryllium to cause health problems. Acute exposure to beryllium by inhalation can cause inflammation of the lungs, chest tightness, coughing, and fatigue. In contrast, long-term exposure by inhalation results in shortness of breath, some scarring of the lung and chest, joint pain, cough, and skin rashes, all leading to CBD.

Epidemiological studies have suggested that beryllium and its compounds could be considered human carcinogens.[39] In a study that covered 15 regions in the United States, Berg and Burbank[43] found a significant correlation between cancers of the breast, bone, and uterus and the detection and concentration of beryllium in drinking water.[41] In laboratory studies, beryllium oxide and beryllium sulfate caused increased incidences of pulmonary tumors (lung anaplastic carcinoma and osteosarcoma) in rats, rabbits, and rhesus monkeys. Based on sufficient evidence for animals and inadequate evidence for humans, beryllium has been placed in the EPA weight-of-evidence classification B2, as a "probable human carcinogen."[42–45]

## Cadmium

***Uses*** Cadmium (Cd) (L. cadmia; Gr. kadmeia, ancient name for calamine, zinc carbonate) was discovered by Stromeyer in 1817 through an impurity in zinc carbonate. Cadmium most often occurs in small quantities associated with zinc ores, such as sphalerite (ZnS). The important compounds used in industry are cadmium oxide (CdO), cadmium chloride ($CdCl_2$), cadmium nitrate ($Cd(NO_3)_2$), cadmium sulfide (CdS), and cadmium sulfate ($CdSO_4$). Greenockite (CdS) is the only mineral of any consequence bearing cadmium. Cadmium is also obtained as a by-product in the treatment of zinc, copper, nonferrous metal industry, and lead ores. Cadmium is a highly toxic heavy metal that forms complex compounds with other metals and elements.

Cadmium is a bluish-white soft metal with no definite taste or odor. Its ingestion causes serious damage to tissues and organs, particularly the kidneys. It is recovered primarily through the smelting of zinc.[46–48] Occupational exposure and its associated hazards are due to the inhalation of dust and fumes of cadmium oxide. The operations in an industrial environment include melting, pouring, welding, and heating of cadmium-plated steel (Table 3-2).

Cadmium is widely used in industry for metal coating, pigments and paints, batteries, and in solder alloys. Its principal use is in the production of nickel-cadmium batteries required for portable electronic and electrical equipments. According to the U.S. Geological Survey, the total world refinery production in 1997 rose to 19,500 tons from 18,900 tons a year.

**TABLE 3-2 Cadmium Processing and Operations in Industries**

| |
|---|
| Cadmium casting |
| Cadmium smelting |
| Cadmium spraying |
| Copper-cadmium melting |
| Heating of cadmium-plated steel |
| Revert annealing |
| Storage battery manufacture |

Breathing contaminated workplace air, battery manufacturing, metal soldering or welding, eating food such as shellfish, liver, and kidney meats, and breathing cadmium in cigarette smoke are easy ways for cadmium to enter the body.[49,50]

***Toxicity*** Cadmium is persistent and accumulates in organs (e.g., liver, kidney, pancreas). The symptoms of toxicity are disorders of the respiratory system and kidneys. Breathing high levels of cadmium severely damages the lungs and can cause death. Eating food or drinking water with very high levels severely irritates the stomach, leading to vomiting and diarrhea, and possibly kidney disease. Other potential long-term effects are lung damage and fragile bones.

Prolonged low-level exposure to cadmium has caused biochemical and morphological changes (e.g., emphysema, lung cancer, kidney damage). Based on studies of animals with increased lung cancer, it has been observed that cadmium and its compounds may act as carcinogens in animals and humans. Cadmium is a probable human carcinogen (IARC) category 2A and, according to USEPA, B1. The data appear to be inadequate and more research is needed.[47–54]

### Chromium

***Uses*** Chromium (Cr) is a naturally occurring element that is found in rocks, soil, plants, animals, volcanic dust, and gases. It is used for making steel and other alloys, bricks in furnaces, dyes and pigments, chrome plating, leather tanning, and wood preserving. Chromium is an essential nutrient required for normal sugar and fat metabolism. It occurs primarily in the trivalent and hexavalent forms, and the form in higher organisms is trivalent. The selected compounds of industrial importance are chromyl chloride ($CrO_2Cl_3$), chromium oxide ($Cr_2O_3$), chromium trioxide ($CrO_3$), chromium chloride ($CrCl_3$), chromous chloride ($CrCl_2$), and potassium dichromate ($K_2Cr_2O_7$).

***Toxicity*** Chromium is present throughout the body, with the highest concentrations in the liver, kidney, spleen, and bone. The hexavalent chromium compounds are irritating and corrosive when absorbed through the digestive tract, skin, or lungs. Because of the slow absorption of chromium, acute exposure does not produce generalized symptoms. Inhalation of dusts or mists of hexavalent chromium is irritating to the upper respiratory passage and can cause sneezing, nasal discharge, and congestion. All forms of chromium are toxic at high levels, but chromium (VI) is more toxic than chromium (III). Breathing high levels of chromium (VI) in air can damage and irritate nose, lungs, stomach, and intestine.

People who are allergic to chromium may also have asthma attacks after breathing high levels of either chromium (VI) or (III). Long-term exposure to high or moderate levels of chromium (VI) causes damage to the nose (e.g., bleeding, itching, sores) and the risk of noncancer lung diseases. Ingestion of high doses of chromium also can cause stomach upsets and ulcers, convulsions, kidney and liver damage, and even death.

Chromium (VI) compounds are known to be potent toxic and carcinogenic agents. Studies have shown that chromium (III) and chromium (VI) induce cellular damage.[55,56] Chromate is an established human carcinogen. A number of hypotheses have been indicated, and one such is the generation of strand breaks by chromate, during its reduction by glutathione involving the generation of DNA lesions by free hydroxyl radicals. Kinetic, spin-trapping, and competition kinetic studies, based on a strand-breaking assay, have been reported in support of this conclusion.[57,58]

## Cobalt

***Uses*** Cobalt (Co) is a compound that occurs in nature in many different chemical forms, as cobalt carbonyl ($Co_2(CO)_4$), cobaltous oxide (CoO), cobaltic oxide ($Co_2O_3$), cobalt hydrocarbonyl ($HCo(CO)_4$), and cobaltic cobaltous oxide ($Co_3O_4$). Pure cobalt is a steel-gray, shiny, hard metal. Cobalt is not currently mined in the United States; all cobalt used in industry is imported or obtained by recycling scrap metal that contains cobalt. It is used in industry to make alloys (mixtures of metals), colored pigments, and driers for paint and porcelain enamel. Occupational workers and the general public who live near hazardous waste sites containing cobalt are exposed to higher levels of the compound.[2,3,5]

***Toxicity*** Exposure to high levels of cobalt causes adverse effects on the lungs, leading to asthma, pneumonia, and wheezing. Cobalt is a possible carcinogen to humans. Animal studies have shown that cobalt causes cancer when placed directly into the muscle or under the skin; studies are inconclusive regarding cobalt and human cancer.[59] Some carcinogenic metal compounds [e.g., cobalt (II), chromate (VI), iron (III) nitrilotriacetate, nickel (II)] have been reported as inducers of various oxygen radical species in the presence of hydrogen peroxide. These oxygen radicals are suggested to cause different site-specific DNA damage. Using pulsed-field gel electrophoresis, nickel sulfide was shown to induce oxidative DNA cleavage in cultured cells. In light of these findings, metal carcinogenesis, like cobalt, and the role of oxygen radicals gains significance.[60] More studies are required for confirmation.

## Lead

***Uses*** Lead (Pb) has been recognized as one of the important metals, used extensively since time immemorial and presently in a large number of industries. The word *plumbum* in Latin means lead. The term *plumbarius* describes a lead worker. The Greek physician Dioscorides (circa 40–90) made reference to the metal's ill effects on the body and mind of humans who had inhaled its fumes. The decline and fall of the Roman Empire has been associated with lead poisoning. Romans, as has been described by historians, drank wine from lead-lined wine jugs and grape juice that was simmered in lead-lined pots, utensils, and wine cups. The important industrial compounds include lead acetate ($Pb(C_2H_3O_2)_3$, lead chloride ($PbCl_2$), lead oxide ($PbO$), lead sulfide (Galena, $PbS$), tetraethyl lead ($Pb(C_2H_5)_4$), and tetramethyl lead ($Pb(CH_3)_4$).

Humans can be exposed to lead in more than 120 occupations. Smelters have been associated as the worst offenders of lead pollution. Heating ore or scrap metal to melt and separate lead and testing soil near smelters produced as high as 100,000 ppm of lead. The plume of lead pollution from smelters has transmigrated long distances from workplaces through wind current or by groundwater.[61–66] Airborne lead contamination and the presence of lead has been identified in the snow peaks and ice layers of Greenland.[67]

There are several major sources and occupations closely associated with lead exposure (Tables 3-3 and 3-4). Approximately 17,000,000 pounds of lead was released into the environment by processing industries, and as much as 88,000 pounds by the manufacturing industries.[63,66] The growing global concern about lead pollution has been traced to improper incineration of municipal waste garbage. Methods of dumping trash, garbage, lead-acid batteries, old TV sets, computer monitors, and glass in cathode-ray-tubes have increased the degree of lead contamination in living areas. The presence of lead in a variety of paints, auto exhausts, and industrial emissions has become an issue of health around the globe.

The automobile and battery industries account for more than 80% of all lead consumed in United States alone, and certainly more in the global context.

**TABLE 3-3 Occupations Associated with Lead**

| |
|---|
| Ammunition (guns and bullets) manufacturers |
| Auto body repairs |
| Auto radiator repair shops |
| Battery workers |
| Brass/copper foundry |
| Bridge and highway construction |
| Cable makers |
| Gas stations |
| Glass manufacturers |
| Industrial machinery and related works |
| Inorganic pigment manufacturers |
| Lead miners, smelters, refiners |
| Plastic manufacturers |
| Painters and printers |
| Plumbers and fitters |
| Pottery and ceramic workers |
| Rubber product manufacturers |
| Shipbuilders and shipyard workers |
| Stained glass makers |
| Steel welders |
| Textile workers |
| Welders and related workers |

**TABLE 3-4 Major Sources of Lead in the Environment**

| |
|---|
| Lead-based paint—Present on many surfaces in homes not recently rebuilt or remodeled |
| Lead pipes—More common in older homes |
| Lead solder—On pipes and water heaters |
| Enameled or ceramic pots and dishware—Improper glazing can leach lead into foods |
| Paper wrappings—Holiday paper and party decorations (10 g/kg) |
| Food packages—Polythene plastic bags, flour bags (20 mg/kg), cardboard boxes with dyes (50 mg/kg) |
| Candy packaging—Candy bar wrappers (7 g/kg), colored sports trading cards packaged with gum (88 mg/kg) |

The application of lead in X-ray and radiation shielding, cable sheathing, acoustical damping, and electronics and computer industries is ever-increasing. Lead enters the body by both inhalation and ingestion. As dust, mist, and fumes of lead enter the system by inhalation. Lead appears in drinking water (via lead pipes) in lead solder used for brass fixtures. Industrial operations, like smelting, scraping, or heating of lead-based paints, makes lead dust settle on floors, walls, and furniture. Lead accumulates in the body once it enters the system—even in small quantities—by ingestion or inhalation. In the occupational atmosphere, lead can enter the system through contaminated food and drinks. Another less common source of lead contamination is the ingestion of spitballs made from newspaper that contains as high as 4,000 ppm of lead. Extensive data are available about the adverse effects of lead.[62,66,68–72]

***Toxicity*** Occupational workers and the general public exposed to high concentrations of lead show symptoms of lead poisoning that can include fatigue, headache, vomiting, ataxia, loss of appetite, leg cramps, numbness, muscle weakness, depression, brain damage, convulsions, coma, and death. Infants are especially susceptible to lead poisoning. In children, a lead blood level of 70 μg/dL or higher lead poisoning can cause bizarre behavior, apathy, vomiting, loss of muscular coordination, loss of recently acquired skills, and brain damage as evidenced by altered state of consciousness, blindness, seizures, and coma (Table 3-5). A report by the American Association of Poison Control Centers cites more than 2,000 cases of lead poisoning reported in 1990.[68]

Chronic exposures to lead by inhalation or the oral route cause adverse effects that include damage to the peripheral and central nervous system, anemia, and chronic kidney damage. Lead accumulates in the soft tissues and bones, with the highest accumulation in the liver and kidneys, and elimination is slow. Lead has shown developmental and reproductive toxicity in both male and female

**TABLE 3-5 Acute and Common Symptoms of Lead Poisoning in Humans**

| |
|---|
| Anorexia and hallucination |
| Altered state of consciousness |
| Apathy |
| Clumsiness |
| Bizarre behavior |
| Constipation |
| Coma |
| Cramps |
| Crankiness |
| Fatigue and lethargy |
| Headaches |
| Loss of acquired skills |
| Loss of appetite |
| Loss of muscular coordination |
| Seizures |
| Sleep disorders |
| Stomach aches, constipation, and nausea |
| Urinary coproporphyrin increase |
| Vomiting |

animals and humans. Lead is listed by IARC in Group 2B ("possible human carcinogen") and by NTP as "reasonably anticipated to be a carcinogen," but is not considered to be a "select carcinogen" under the criteria of the OSHA Laboratory Standard.[69,70,72–77]

The lead (II) ion also is toxic to fishes and plants. In a recent report on lead poisoning, an adult was hospitalized due to severe anemia and other health-related problems. A high lead-blood concentration was found in all the family members. Subsequently, it was discovered that the family was eating flour contaminated with a high level of lead (38.7 mg/g).

Exposure to high concentrations of lead is known to cause acute encephalopathy (brain disease), which may eventually result in seizure, coma, and death from cardiovascular arrest. Any conditions that would elevate lead blood levels to 100 $\mu$g/dL or breathing an atmosphere contaminated with 100 mg lead/$m^3$ are extremely dangerous to life and health. Chronic overexposure to lead, depending on the blood lead levels, induces different adverse effects (Table 3-6). The lead can cause deleterious effects to cardiovascular, nervous, urinary, and reproductive systems. The poisoned individual exhibits difficulty in speech and learning, loss of hearing, loss of appetite, weight loss, headache, joint and body pain, limp wrist, high blood pressure, reproduction problems, miscarriages, infertility, retarded fetal development, low birth weight, tiredness, hypertension, constipation, and kidney disease. It is also well known that lead has caused a significant increase in mortality rates among battery workers as a result of increased stomach

**TABLE 3-6 Blood-Lead Levels and Human Health Effects**

| Blood-Lead Levels (μg/dl) | Effects |
|---|---|
| 10–20 | Initial biochemical changes decreased metabolism of vitamin D |
| 20–30 | Hearing impairment, central nervous system damage |
| 40–50 | Slowing of red blood cell production, lower sperm production |
| 50–100 | Anemia, colic, seizure, brain damage, decreased longevity |
| Over 100 | Convulsions, permanent brain damage, death |

*Source*: Compiled from different reports, 1997.

**TABLE 3-7 Concentration of Lead in Different Systems and Related Health Effects in Humans**

| | | Concentration (μg/DL) | |
|---|---|---|---|
| System | Effect | Adults | Children |
| Nervous system | Encephalopathy | 80–100 | 100–120 |
| Subclinical | Encephalopathy | — | 50 |
| | IQ deficit | <30 | — |
| *In utero* effects | | <15 | — |
| Peripheral | Neuropathy | 40 | 40 |
| Renal system | Acute nephropathy | 80–100 | — |
| Chronic | Neuropathy | — | 60 |
| | Vitamin D metabolism | 30 | — |
| | Blood pressure | — | 30 |
| | Blood Anemia | 80–100 | 80–100 |
| | U-aminolevulinic acid | 40 | 40 |
| | B-EPP | 15 | 15 |
| | Aminolevulinic acid inhibition | 10 | <10 |
| | Py-5-N-inhibition | <10 | — |

*Source*: U.S. EPA Air Quality Criteria for Lead, EPA 600/8–83/02aF, Washington, DC, 1986.

and lung cancer.[5,9,74,75] Different levels of lead induced severe adverse health effects in humans (Table 3-7).

***Lead Exposure in Children*** Lead poisoning is a silent epidemic and affects all ages, including infants and children. Children are more susceptible to lead-induced toxicity. Children absorb as much as 50% of lead, compared with approximately only 10% in adults when the metal enters the system. More important, this accumulation occurs during the formative years of the child and when the nervous system (including the brain) has yet to unfold. A child needs exposure to only a small amount of lead (e.g., a small granule of sugar) daily to accumulate to a level of 35 μg/dL over a period of time; the threshold for lead poisoning in children is as low as 10 μg/dL.

Lead has been shown to cause a variety of neurologic disturbances in the growing child such as reduced intelligence as evidenced by reading, learning, and memory tests. Lead has produced irritability, memory lapses, sleep disturbances, hyperactivity, and other behavioral problems in children. Further, lead also induces physical disabilities (e.g., loss of hearing, loss in body weight and height, altered gait, loss of words in speech, other general development skills). The greatest impact of lead on the young child is on his or her developing brain and CNS. The fetus and child's developing nervous system is most sensitive to the lead toxicity. In fact, movement of lead through the transplacental barrier has resulted in spontaneous abortion. In certain cases, lead has caused stunting of fetal growth and childhood stature by disturbing the secretion of growth hormone. Lead has severe effects on pregnant women as well.[73–82] Millions of children under 6 years of age have shown high blood-lead levels from different exposures (Table 3-8). Human poisoning caused by lead has been classified (Appendix 3-1) based on blood-lead levels. This should guide occupational workers during pre- and postemployment, help to monitor children's health status, and initiate appropriate safety measures for everyone[75a] (Table 3-9).

**TABLE 3-8 Blood-Lead Levels in Children (less than 6 years of age)**

| Exposure | Total Number of Children | Blood-Lead Level > 10.0 μg/DL | |
|---|---|---|---|
| | | Percent | Number |
| Drinking water | 30 million | 3 | 1 million |
| Paint | 12 million | 17 | 2 million |
| Soil and dust | 12 million | — | — |

*Source*: U.S. EPA Strategy for reducing lead exposures, U.S. EPA Report, February 21, 1991.

## Manganese

***Uses*** Manganese (Mn) is a reddish-gray soft metal. The principal ore is pyrolusite and ferrous scrap. The most commonly occurring compounds are manganese dioxide ($MnO_2$), manganese tetraoxide ($Mn_3O_4$), manganous carbonate ($MnCO_3$), manganese sulfate ($MnSO_4$), manganese chloride ($MnCl_2$), and methylcyclopenta-dienyl manganese tricarbonyl (MMT). Humans are most commonly exposed to manganese during mining, smelting, metal refining, welding operations, and the production of different materials. Manganese is a naturally occurring substance found in many types of rock; it is ubiquitous in the

**TABLE 3-9 Classification of Lead Poisoning Based on Blood-Lead Level**

| Class | Concentration (μg/DL) |
|---|---|
| I | Less than 10 |
| II A | 10–14 |
| II B | 15–19 |
| III | 20–44 |
| IV | 45–69 |
| V | More than 69 |

*Source*: U.S. Department of Health and Human Services, Public Health Services, Centres for Disease Control, October, 1991 preventing lead poisoning in young children.

environment and is found in low levels in water air, soil, and food.[83] Workers handling ores and disposing of waste can suffer from manganese poisoning.

***Toxicity*** Manganese causes adverse effects in humans. In Chile, where much manganese is mined, workers have shown a strange syndrome called *locura manganica*, or "manganese madness." The first symptoms of this manganese poisoning are anorexia, weakness, and apathy. However, there may be an initial "manic" phase, characterized by inappropriate or uncontrollable laughter, increased sexuality, insomnia, delusions or hallucinations, violence, and other mental changes. The earlier mania may shift to depression, impotence, and excessive sleeping. Parkinsonian symptoms (e.g., tremors and muscle rigidity) may also appear in later stages of manganese poisoning. These symptoms may appear (as in Parkinson's disease) from a loss of dopamine in the brain cells. L-dopa, which converts to dopamine in the brain, is used to reduce symptoms in the treatment of manganese toxicity.

Humans with chronic exposure to high levels of manganese via inhalation primarily experience effects on the CNS. These effects are termed *manganism* and typically begin with feelings of weakness and lethargy, and progress to speech disturbances, a mask-like face, tremors, and psychological disturbances. Other chronic effects from inhalation exposure to manganese are respiratory effects such as an increased incidence of cough and bronchitis and an increased susceptibility to infectious lung disease.[83–85] In children, severe manganese deficiency may lead to convulsions, paralysis, or blindness. In adults, dizziness, weakness, and hearing problems such as strange ear noises are associated with manganese deficiency. Decreased strength and ataxia (unstable gait) also have been reported in addition to weight loss, irregular heartbeat, and skin problems.

Manganese is an essential nutrient, but also one that can be extremely toxic. Deficiency of manganese causes problems in domestic animals, and manganese toxicity is an important problem in many human miners and ore smelters. The homeostasis mechanism maintains the natural manganese level in the body system. However, the iron level in the diet interferes with absorption of manganese. Information is inadequate regarding storage of iron in the body and its effect on the absorption and retention of manganese. Studies have shown that healthy young women differ between iron storage and pattern of absorption and retention of manganese.[82] Low levels of iron increase absorption of manganese and its excretion, suggesting the role of iron in the diet for the induction of manganese toxicity.[83]

### Mercury and Methyl Mercury

***Uses*** Mercury (Hg) occurs naturally as a free metal or as cinnabar (HgS) and is produced from ore by roasting or reduction. One common form of mercury is used in thermometers. This form is called "metallic mercury." Mercury is also used in barometers and other common consumer products. Mercury combines with other chemicals (e.g., chlorine, carbon, oxygen) to form either "inorganic" or "organic" mercury compounds. The organic form of mercury, called methylmercury, builds up in certain fish, indicating that low levels of mercury in water bodies can contaminate these fish. Selected compounds of mercury are mercuric chloride ($HgCl_2$), mercurous chloride ($Hg_2Cl_2$), methyl mercuric chloride ($CH_3HgCl$), and phenyl mercury acetate ($C_8H_8HgO_2$).

Mercury released into the environment is persistent and can change between organic and inorganic forms. For example, organic mercury slowly breaks into inorganic mercury and some inorganic mercury slowly changes into organic mercury, particularly into methylmercury. The interactions of soil and water with microorganisms work to participate in various chemical processes. The wastes and by-products from paper manufacturing, chloralkali, and burning fossil fuels can contaminate the environment.[86–89] However, it has been reported that mercury levels in air and water have not increased much over the past 20 years.

The opportunities for occupational and environmental exposures to mercury are varied. Although industrial workers associated with chloralkali and electrical equipment manufacture have shown greater incidence of exposures, other human activities shown relatively less (Tables 3-10 and 3-11). Mercury can easily enter the body if its vapor is breathed in, if it is eaten in organic forms in contaminated fish or other foods, or if food or water contaminated with inorganic mercury is ingested. Mercury in all forms also may enter the body directly through the skin. Exposure above normal levels at National Priorities List (NPL) sites may occur through drinking water contaminated with salts of inorganic mercury. Some sites may have such high amounts of mercury in the soil or in containers that breathing mercury metallic vapor may be a health problem.

Mercury leaves the body mostly through the urine and feces. One of the worst-known cases of mercury poisoning occurred at Minamata, Japan, when methylmercury compounds formed during the manufacture of a paint solvent were discharged into Minamata Bay. Local people who ate a large amount of fish began

**TABLE 3-10 Environmental and Occupational Exposure to Mercury in Humans**

| Industry | Exposure (%) |
|---|---|
| Electrical equipments | 20 |
| Paints | 15 |
| Laboratory | 12 |
| Dental | 13 |
| Chloralkali (e.g., bleach) | 25 |
| Thermometers | 10 |

**TABLE 3.11 Average Daily Retention of Total Mercury and Mercury Compounds in the General Population**

| | Hg Retention Mercury/Day | | |
|---|---|---|---|
| Exposure | Hg Vapor | Inorganic Hg Salt | Methylmercury |
| Air | 0.024 | 0.001 | 0.0064 |
| Dental amalgams | 3–17 | 0.00 | 0.0 |
| Drinking water | 0.00 | 0.0035 | 0.0 |
| Food | 0.00 | 0.29 | 0.0 |
| Total | 3–17 | 0.3 | 2.31 |

to suffer from the above symptoms. Eventually, nearly 1,000 people died. Human fatalities associated with mercury and methylmercury are well documented in the literature.[90–104]

***Toxicity*** Exposure to mercury produces multiple adverse effects in vital systems. For instance, mercury induces irritability, tremor in the extremities, anxiety or nervousness, and often difficulty breathing. Symptoms of poisoning include restlessness; fits of anger with violent, irrational behavior; loss of memory; inability to concentrate; lethargy; drowsiness; insomnia; mental depression; numbness and tingling in hands, feet, fingers, toes, or lips; muscle weakness progressing to paralysis; ataxia; tremors/trembling of hands, feet, lips, eyelids, or tongue; incoordination; and other CNS disturbances.[89,90]

The adverse effects caused by mercury also include bleeding gums, loosening of teeth, excessive salivation, foul breath, metallic taste, burning sensation, tingling of lips and face, tissue pigmentation (amalgam tattoo of gums), leukoplakia, stomatitus, or ulceration of the gingiva, palate, and tongue. Dizziness, acute or chronic vertigo, ringing in ears, and difficulty with hearing and speech also have been observed.

Grandjean and associates[90] indicated that gold-mining activities in the Amazon basin used mercury to capture the gold particles and subsequently contaminated the freshwater system and aquatic organisms with methylmercury. Further, a number of children between 7 and 12 years of age were found to have high concentrations (more than 10 μg/g) of mercury and neurologic disturbances, including adverse effects on brain development.

Long-term exposure to either organic or inorganic mercury can permanently damage the brain, kidneys, and developing fetuses. The specific form of mercury and the manner of exposure determine which of these health effects will be more severe. For instance, certain organic mercury compounds (e.g., methylmercury, iodide, dimethylmercury) cause greater harm than soluble mercury salts. Metallic mercury vapors also cause greater harm to the brain. In contrast, inorganic mercury salts cause greater harm to the kidneys. Thus, maternal exposure to organic mercury may lead to brain damage in fetuses, whereas exposed adults may develop shakiness (tremors), memory loss, and kidney disease. Short-term exposure to high levels of inorganic and organic mercury also have similar health effects, but full recovery is more likely after short-term exposures, once the body clears itself of the contamination. Mercury has not been shown to cause cancer in humans because of lack of data.[90a,90b]

The gastrointestinal effects of mercury poisoning consist of food sensitivities (especially to milk and eggs) abdominal cramps, colitis, and chronic diarrhea or constipation. Cardiovascular effects include abnormal heart rhythm, characteristic findings on EKG, and unexplained elevated serum triglyceride. The unexplained elevated cholesterol and abnormal blood pressure can be either high or low. Some of the immunologic effects include repeated infections, autoimmune disorders, arthritis, multiple sclerosis (MS), scleroderma, and hypothyroidism. The systemic effects also consist of chronic headaches, allergies, severe dermatitis, thyroid disturbance, subnormal body temperature, cold, clammy skin (especially hands and feet), excessive perspiration with frequent night sweats, unexplained sensory symptoms including pain, unexplained numbness or burning sensations, and unexplained anemia. People who eat large amounts of fish, such as tuna and swordfish, may be exposed because these fish can contain high levels of organic mercury compared with other foods.

***Children at Risk*** Very young children are more sensitive to mercury than adults are. Mercury in the mother's body passes to the fetus and can pass to a nursing infant through breast milk. However, the benefits of breast-feeding may be greater than the possible adverse effects of mercury in breast milk.[90]

Mercury's harmful effects to the developing fetus include brain damage, mental retardation, incoordination, blindness, seizures, and an inability to speak. Children poisoned by mercury may develop problems of their nervous and digestive systems, as well as kidney damage. The CNS is very sensitive to all forms of mercury. Methylmercury and metal vapors are more harmful than other forms, because more mercury reaches the brain in these forms. Exposure to high levels of metallic, inorganic, or organic mercury can permanently damage the brain, kidneys, and developing fetus. Effects on brain function may result in irritability, shyness, tremors, changes in vision or hearing, and memory problems.

The incidence of chronic toxicity of methylmercury is well documented in the cases of food contamination. Epidemic poisonings in Iraq and Minamata and Niigata, Japan, took a large human toll. More than 6,000 individuals were hospitalized in Iraq, and nearly 500 people died after eating bread prepared with flour, which was made from wheat and barley treated with fungicide.[96] Subsequent estimations indicated that the level of methylmercury in the wheat flour ranged from 4.8 to 14.6 μg/g. The clinical symptoms of victims included paresthesia, visual disorders, dysarthria, and deafness. The most severe cases resulted in coma and death due to CNS failure. The USEPA is moving toward the development of integrated assessments involving parameters such as developmental toxicity, neurotoxicity, immunotoxicity, reproductive effects, and germ cell mutagenicity. It has been reported that methylmercury has a high potential as a germ cell mutagen, and more information may be found in the literature.[86]

## Nickel

***Uses*** Nickel (Ni) is a hard, silvery white metal. The industrially important nickel compounds are nickel oxide (NiO), nickel acetate ($Ni(C_2H_3O_2)$, nickel carbonate ($NiCO_3$), nickel carbonyl ($Ni(CO)_4$), nickel subsulfide ($NiS_2$), nickelocene $(C_5H_5)_2Ni$, and nickel sulfate hexahydrate ($NiSO_4$. $6H_2O$). Nickel compounds are known human carcinogens. Investigations into the molecular mechanisms of nickel carcinogenesis have revealed that not all nickel compounds are equally carcinogenic. Certain water-insoluble nickel compounds exhibit potent carcinogenic activity, whereas highly water-soluble nickel compounds exhibit less potency. The high carcinogenic activity of certain water-insoluble nickel compounds relates to their bioavailability and the ability of the nickel ions to enter cells and reach chromatin. The water-insoluble nickel compounds enter cells quite efficiently via phagocytic processes and subsequent intracellular dissolution. Nickel is classified as a borderline metal ion because it has both soft and hard metal properties, and it can bind to sulfur, nitrogen, and oxygen groups. Nickel ions are very similar in structure and coordination properties to magnesium. Not only have nickel compounds been shown to be responsible for a number of human cancers in occupationally exposed workers, but also carcinogenic nickel compounds have been shown to induce many adverse effects.[102,103] As early as 1907 and 1934, reports have shown that occupational exposure to nickel (in copper nickel refineries) caused respiratory tract cancer in workers.[105]

***Toxicity*** Nickel has a carcinogenic property and may be associated with hypersensitivity reactions. Acute poisoning causes headache, dizziness, nausea and vomiting, chest pain and tightness, dry cough with shortness of breath, rapid respiration, cyanosis, and extreme weakness. Lesions resulting from acute exposure appear mainly in the lungs and brain. Soluble nickel compounds cause dermatitis or *nickel itch*, particularly among electroplate workers. Nickel carbonyl is highly toxic. Nickel poisoning has induced edema and hemorrhages in lungs, brain injury, and death.

Studies have indicated that carcinogenic nickel compounds induce chromosomal aberrations, including those that are specific to heterochromatic chromosome regions found to increase the extent of DNA methylation.[105,106] Although the mechanisms of DNA hypermethylation by nickel are not well documented, nickel may trigger a *de novo* methylation of the genome region. In addition to producing mutations, chromosome aberrations, and gene silencing by DNA methylation, exposure of cells to nickel also induces a variety of gene expression changes that yield cells with a spectra of expressed genes similar to cancer cells. Nickel has been found to induce the hypoxia-inducible factor (HIF), which is responsible for increasing the expression of glycolytic enzymes and other genes that allow cells to survive under low oxygen tension. As suggested by the growing literature on nickel carcinogenesis, initial events in environmentally induced cancers may be a combination of gene induction and gene silencing by epigenetic DNA methylation that leads to cancer cell selection. Determining the extent to which this concept applies to other environmental carcinogens is the challenge of the future.[104–108]

## Zinc

***Uses*** Zinc (Zn) is one of the most common elements in the Earth's crust. Metal zinc was first produced in India and China during the Middle Ages. Industrially important compounds of zinc are zinc chloride ($ZnCl_2$), zinc oxide (ZnO), zinc stearate ($Zn\ (C_{16}H_{35}O_2)_2$), and zinc sulfide (sphalerite, ZnS). It is found in air, soil, and water, and is present in all foods. Pure zinc is a bluish-white shiny metal. Zinc has many commercial uses, as coatings to prevent rust, in dry cell batteries, and mixed with other metals to make alloys like brass and bronze. Zinc combines with other elements to form zinc compounds. Zinc compounds are widely used in industry to make paint, rubber, dye, wood preservatives, and ointments.[109–112]

***Toxicity*** Zinc and its compounds are relatively nontoxic, but very large doses can produce acute gastroenteritis characterized by nausea, vomiting, and diarrhea. The recommended dietary allowance (RDA) for zinc is 15 mg/day for men, 12 mg/day for women, 10 mg/day for children, and 5 mg/day for infants. Not enough zinc in the diet can result in a loss of appetite, a decreased sense of taste and smell, skin sores and slow wound healing, or a damaged immune system.

Pregnant women with low zinc intake may have babies with growth retardation. Exposure to zinc in excess, however, also can be damaging. Harmful health effects generally begin at levels from 10 to 15 times the RDA (in the 100 to 250 mg/day range). Ingesting large amounts of zinc, even for a short time, can cause stomach cramps, nausea, and vomiting.[109] Chronic exposure to zinc chloride fumes causes irritation, pulmonary edema, bronchopneumonia, pulmonary fibrosis, and cyanosis.[110] It also causes anemia, pancreas damage, and lower levels of high-density lipoprotein cholesterol. Breathing large amounts of zinc (as dust or fumes) can cause a specific short-term disease called *metal fume fever*, which includes disturbances in adrenal secretion. There is little information on the toxicological effects followed by prolonged exposure to high concentrations of zinc.[109–115]

## CONCLUSION

Education can effectively minimize exposure to hazardous substances. Information and training has been provided to health care providers in Palmerton, Pennsylvania, concerning biological monitoring and potential adverse health effects in people exposed to lead and cadmium. The general public also has been advised about potential adverse health effects associated with exposure to lead and cadmium.

Heavy metals, like lead and mercury, have been recognized as toxic poisons for centuries. Further, toxic concentrations of mercury, for example, can trigger several effects like autoimmune diseases, infections, unexplained chronic fatigue, depression, nerve impairment, memory problems, decreased mental clarity, and bowel disorders. For several decades, mercury vapor exposure has caused severe health problems among chloralkali workers. This is only an example. It may be repeated that education can effectively minimize exposure to hazardous metals. Basic information and training for proper handling of toxic chemicals will reduce potential adverse health effects.

Exposure to heavy metals in the industrial environment has caused global concern, because several substances have been shown to cause severe adverse effects among workers and children. This situation requires proper training and monitoring by associated agencies worldwide.

## REFERENCES


1. H.A. Schroeder, A sensitive look at air pollution of metals. *Arch. Environ. Health* **21**:798, 1970.
2. C.D. Klaassan, Heavy metals and heavy metal—Antagonists. In Hardman, J.G., Molinoff, P.B., Ruddon, R.W., and Gilman, A.S., eds. *Goodman & Gilman's The Pharmacological Basis of Therapeutics*, 9th ed., McGraw-Hill, New York, 1996.
3. R. Albert, Accumulation of toxic metals with special reference to their absorption, excretion and biological half times. *Environ. Physiol. Biochem.* **3**:65, 1978.
4. R.J. Lewis, *Sax, Dangerous Properties of Industrial Materials*, 9th ed. Vol. I, II, III. A Wiley Publication, 1995.
5. C. Winder, Toxicology of metals, in *Occupational Toxicology*, Stacey, N.H., ed., Taylor & Francis, London, 165, 1993.

6. I.T. Brakhnova, *Environmental Hazards of Metals: Toxicity of Powdered Metals and Metal Compounds*, Plenum Press, New York, 1975.
7. U.S. Department of Commerce, *Waste Disposal Practices: A Threat to Health and Nation's Water Supply*, Washington, DC, June 1978.
8. Agency for Toxic Substances and Disease Registry, *Toxicological profile for arsenic*, U.S. Department of Health and Human Services, Public Health Service, Atlanta, GA, 2000, PB/2000/108021.
9. R.L. Singhal, and J.A. Thomas, eds., *Lead Toxicity*, Urban & Schwarzenberg, Baltimore, MD, 1980.
10. R.A. Goyer, Toxic effects of metals, Chapter 19, In, *Casarett and Doull's Toxicology: The Basic Science of Poisons*, 4th ed., Amdur, M.O., Doull, J., and Klaassen, C.D., eds., Pergamon Press, New York, 1991.
11. E.J. Massaro, S.J. Yaffe, and C.C. Thomas, Jr., Mercury levels in human brain, skeletal muscle and body fluids. *Life Sci.* **14**:1939, 1974.
12. O.E. Roses, Gaarcia, J.C. Fernandez, J.C. Villaamil, et al., Mass poisoning by sodium arsenite. *Clin. Toxicol.* **29**:20, 1991.
13. M.M. Wu, T.L. Kuo, Y.H. Hwang, and C.J. Chen, Dose response relation between arsenic concentration in well water and mortality from cancers and vascular diseases. *Am. J. Epidemiol.* **130**:1123, 1989.
14. J.B. McKinney, Metabolism and disposition of inorganic arsenic in laboratory animals and humans. *Environ. Geochem. Health* **14**:43, 1992.
15. J. Cuzick, P. Sasieni, and S. Evans, Ingested arsenic, keratoses and bladder cancer. *Am. J. Epidemiol.* **136**:417, 1992.
16. H.V. Aposhian, Biochemical toxicology of arsenic. *Rev. Biochem. Toxicol.* **10**:265, 1989.
17. U.S. Environmental Protection Agency, *Health assessment document of inorganic arsenic*, Office of Health and Environmental Assessment, Washington, DC, 1984, EPA/600/8-83/021F.
18. Agency for Toxic Substances and Disease Registry, *Toxicological profile for aluminum*, U.S. Department of Health and Human Services, Public Health Service, Atlanta, GA, 1999, PB/99/166613.
19. R.A. Robert, A. Yokel, and M.S. Golub, eds. *Research Issues in Aluminum Toxicity*, Taylor & Francis, 1997.
20. P.O. Ganrot, Metabolism and possible health effects of aluminum. *Environ. Health Perspect.* **65**:363–441, 1986.
21. R.A. Yokel, and P.J. McNamara, Aluminum bioavailability and disposition in adult and immature rabbits. *Toxicol. Appl. Pharmacol.* **7**:344–352, 1985.
22. S.S. Pinto, and K.W. Nelson, Arsenic Toxicology and industrial exposure. *Ann. Rev. Pharmacol. Toxicol.* **16**:95, 1976.
23. International Agency for Research on Cancer, Arsenic and arsenic compounds. In: *IARC Monograph on the evaluation of carcinogenic risks to humans — Overall evaluations of carcinogenicity: An update of IARC Monographs 1 to 42*, IARC, Lyon, France (suppl. 7), 100, 1987.
24. K. Harrington-Brock, T.W. Smith, C.L. Doerr, and M.M. Moore, Mutagenicity of the human carcinogen arsenic and its methylated metabolites, monomethylarsenic

and dimethylarsenic acids in L5178Y TK +/− mouse lymphoma cells. *Environ. Mol. Mutagen.* **21**(suppl. 22): 1993.

25. M.F. Hughes, M. Menache, and D.J. Thompson, Dose dependent disposition of sodium arsenate in mice following acute oral exposure. *Fundam. Appl. Toxicol.* **22**:80, 1994.
26. A.H. Smith, C. Hopenhayn-Rich, M.N. Bates, H.M. Goeden, I. Hertz-Picciotto, and H.M. Duggan, Cancer risks from arsenic in drinking water. *Environ. Health Perspect.* **97**:259, 1992.
27. M.J. Mass, Human carcinogenesis by arsenic. *Environ. Geochem. Health* **14**:49, 1992.
28. N. Scott, K.M. Hatlelid, N.E. MacKenzie, and D.E. Carter, Reactions of arsenic (III) and arsenic (V) species with glutathione. *Chem. Res. Toxicol.* **6**:102, 1993.
29. K.S. Squibb, and B.A. Fowler, The toxicity of arsenic and its compounds, in Fowler, B.A., ed., *Biological and Environmental Effects of Arsenic*, Amsterdam: Elsevier Science, 233, 1983.
30. G. Stohrer, Arsenic: Opportunity for risk assessment. *Arch. Toxicol.* **65**:525, 1991.
31. G. Pershagen, and N.E. Bjorklund, On the pulmonary tumorigenicity of arsenic trisulfide and calcium arsenate in hamsters. *Cancer Lett.* **27**:99, 1985.
32. D.J. Thompson, A chemical hypothesis for arsenic methylation in mammals. *Chem. Biol. Interact.* **88**:89, 1993.
33. P. Kurttio, E. Pukkala, H. Kahelin, A. Auvinen, and J. Pekkanen, Arsenic concentrations in well water and risk of bladder and kidney cancer in Finland. *Environ. Health Perspect.* **107**(9): 1999.
34. Agency for Toxic Substances and Disease Registry, *Toxicological profile for barium.* U.S. Department of Health and Human Services, Public Health Service, Atlanta, GA, 1993, PB/93/110658/AS.
35. G.R. Brenniman, and P.S. Levy, High barium levels in public drinking water and its association with elevated blood pressure, in *Advances in Modern Toxicology IX*, E.J. Calabrese, ed., Princeton Scientific Publications, Princeton, NJ, P. 231, 1984.
36. R.P. Beliles, The metals, in *Patty's Industrial Hygiene and Toxicology*, 4th ed., G.D. Clayton, and F.E. Clayton, eds. John Wiley & Sons, New York, p. 1925, 1994.
37. C.H. Johnson, and V.J. VanTassell, Acute barium poisoning with respiratory failure and rhabdomyolysis. *Ann. Emer. Med.* **20**:1138, 1991.
38. W. Zschiesche, K.H. Schaller, and D. Weltle, Exposure to soluble barium compounds: An interventional study in arc welders. *Int. Arch. Occup. Environ. Health* **64**:13, 1992.
39. S. Budavari, M.J. O'Neil, A. Smith, P.E. Heckelman, and J.F. Kinneary, eds., *Beryllium*, in *The Merck Index, An Encyclopedia of Chemicals and Biologicals*, 12th ed., Merck Research Laboratories, Merck & Co., Inc., Whitehouse Station, NJ., pp. 195–197, 1997.
40. Agency for Toxic Substances and Disease Registry, *Case studies in environmental medicine: Beryllium toxicity*, U.S. Department of Health and Human Services, Public Health Service, Atlanta, GA, 1993.
41. Environmental Protection Agency, *Beryllium Integrated Risk Information System (IRIS)*, Environmental Criteria and Assessment Office, Office of Health and Environmental Assessment, Cincinnati, Ohio, 1991.

42. Agency for Toxic Substances and Disease Registry, *Toxicological profile for beryllium*, U.S. Department of Health and Human Services, Public Health Service, Atlanta, GA, 1993, PB/93/182392/AS.
43. J.W. Berg, and F. Burbank, Correlations between carcinogenic trace metals in water supply and cancer mortality. *Ann. N.Y. Acad. Sci.* **199**:249, 1972.
44. H.C. Williams, *Beryllium workers sarcoidosis or chronic beryllium disease sarcoidosis*, **6**(supp.):34–35, 1989.
45. International Agency for Research on Cancer, *Beryllium, Cadmium, Mercury, and Exposures in the Glass Industry, IARC Monograph on the Evaluation of Carcinogenic risk of chemicals to man*, Vol. 58, International Agency for Research on Cancer, pp. 88–89, Lyon, France, 1993.
46. Agency for Toxic Substances and Disease Registry, *Toxicological profile for cadmium*, U.S. Department of Health and Human Services, United States, Public Health Service, Atlanta, GA, ATSDR/TP-88/08, 1989.
47. A. Bernard, and R. Lauwerys, Cadmium in human populations. *Experimentia* **40**:143, 1986.
48. Agency for Toxic Substances and Disease Registry, *Toxicological profile for cadmium*, U.S. Department of Health and Human Services, Public Health Service, Atlanta, GA, 1999, PB/99/166621.
49. L. Friberg, C.G. Elinder, T. Ljellstrom, and G. Nordberg, Ed., Cadmium and Health, A Toxicological and Epidemiological Appraisal, Vol. I, General Aspects, Vol. II, *Effects and Responses*, CRC Press, Inc., Boca Raton, FL, 1986.
50. National Toxicology Program, *Sixth Annual Report on Carcinogens*, Research Triangle Park, NC, 1991.
51. T. Kido, R. Honda, I. Tsuritani, H. Yamaya, M. Ishizaki, Y. Yamada, and K. NogWA, Progress of renal dysfunction in inhabitants environmentally exposed to cadmium. *Arch. Environ. Health* **43**:213, 1988.
52. T. Kjellstrom, L. Friberg, and B. Rahnster, Mortality and cancer morbidity among cadmium-exposed workers. *Environ. Health Perspect.* **28**:199, 1979.
53. S. Takenaka, H. Oldiges, H. Konig, D. Hochrainer, and G. Oberdorster, Carcinogenicity of cadmium chloride aerosols in Wistar rats. *J. Natl. Cancer Inst.* **70**:367, 1983.
54. M.J. Thun, T.M. Schnorr, A.B. Smith, W.E. Halperin, and R.A. Lemen, Mortality among cohort of U.S. cadmium production workers — an update. *J. Natl. Cancer Inst.* **74**:325–333, 1985.
55. Agency for Toxic Substances and Disease Registry, *Toxicological profile for chromium*, U.S. Department of Health and Human Services, Public Health Service, Atlanta, GA, 2000, PB/2000/108022.
56. Agency for Toxic Substances and Disease Registry, *Case studies in environmental medicine: Chromium toxicity*, U.S. Department of Health and Human Services, Public Health Service, Atlanta, GA, 1993.
57. P. O'Brien, and A. Kortenkamp, Chemical models important in understanding the ways in which chromate can damage DNA. *Environ. Health Perspect.* **102**(suppl. 3):3, 1994.
58. M. Sugiyama, Role of paramagnetic chromium in chromium(VI)-induced damage in cultured mammalian cells. *Environ. Health Perspect.* **102**(suppl 3):31, 1994.

59. Agency for Toxic Substances and Disease Registry, *Toxicological profile for cobalt*, U.S. Department of Health and Human Services, Public Health Service, Atlanta, GA, 1993, PB/93/110724/AS.
60. S. Kawanishi, S. Inoue, and K. Yamamoto, Active oxygen species in DNA damage induced by carcinogenic metal compounds. *Environ. Health Perspect.* **102**(suppl. 3):17, 1994.
61. I. Bremner, Heavy metal toxicities. *Quart. Rev. Biophysiol.* **7**:75, 1974.
62. World Health Organization, *Lead, Environ. Health Crite.*, Document 3, WHO Task Group on Environ. Health, Geneva, Switzerland, 1977.
63. National Research Council, Lead, Airborne Lead in Perspective. *National Acad. Sci.* Washington, DC, 1972.
64. L.S. Moore, A.I. Fleischman, Subclinical lead toxicity. *J. Orthomolec. Psychiat.* **4**:61, 1975.
65. L.F. Durback, G.P. Wedin, and D.E. Sedler, Management of lead foreign body ingestion. *Clin. Toxicol.* **27**:173, 1989.
66. R.M. Stapleton, *Lead Is a Silent Hazard.* Walker and Company, New York, 1994.
67. Anonymous, *Guidelines for the Detection and Management of Lead Poisoning for Physicians and Health Care*, Report of Illinois Department of Public Health, Illinois, 1992.
68. Agency for Toxic Substances and Diseases Registry, *Report on nature and extent of lead poisoning in children in United States*, U.S. Public Health Services, Atlanta, GA, 1988.
69. J.J. Jr. Chisolm, *Increased lead absorption and lead poisoning (plumbism), in Nelson Textbook of Pediatrics*, Behrman, ed., 14th ed., W.B. Saunders, Philadelphia, 1992.
70. R.J. Levine, R.M. Moore, G.D. Mc Laren, W.F. Barthel, and P.J. Landrigan, Occupational lead poisoning, animal deaths, and environmental contamination at a scrap smelter. *Am. J. Pub. Health* **66**:548, 1976.
71. K.R. Mahaffey, P.E. Corneliussen, C.F. Jelinek, and J.A. Florino, Heavy metal exposure from foods. *Environ. Health Perspect.* **12**:63, 1975.
72. K.R. Mahaffey, Ed., *Dietary and Environmental Lead: Human Health Effects*, Elsevier Scientific, New York, 1985.
73. M. Cook, W. Chappell, R. Hoffman, and E. Mangione, Assessment of blood-lead levels in children living in a historic mining and smelting community. *Am. J. Epidemiol.* **137**(4):447, 1994.
74. Agency for Toxic Substances and Disease Registry, *Toxicological Profile for Lead.* Update. Prepared by Clement International Corporation under contract No. 205-88-0608 for ATSDR, U.S. Public Health Service, Atlanta, GA, 1993.
75. W.C. Cooper, O. Wong, L. Kheifets, Mortality among employees of lead battery plants and lead-producing plants, 1947–1980. *Scand. J. Work Environ. Health*, **11**:331–345, 1985.
75a. Agency for Toxic Substances and Disease Registry, *Preventing lead poisoning in young children*, U.S. Department of Health and HumanKindly check the corrections marked in the proof are not clear. Services, Public Health Service, Atlanta, GA, 1991.

76. P.A. Baghurst, A.J. McMichael, N.R. Wigg, et al., Environmental exposure to lead and children's intelligence at the age of seven years, The Port Pirie Cohort Study. *New Engl. J. Med.* **327**:1279, 1992.
77. H. Muñoz, I. Romieu, E. Palazuelos, et al., Blood lead level and neurobehavioral development among children living in Mexico City. *Arch. Environ. Health* **48**:132, 1993.
78. U.S. Environmental Protection Agency, *Evaluation of the Potential Carcinogenicity of Lead and Lead Compounds*, Office of Health and Environmental Assessment, EPA/600/8-89/045A., Washington, DC, 1989.
79. U.S. Environmental Protection Agency, *Guidance Manual for the Integrated Exposure Uptake Biokinetic Model for Lead in Children*, Office of Solid Waste and Emergency Response, Washington, DC, EPA/540/R-93/081, PB93-963510, 1994.
80. R.F. White, R. Diamond, S. Proctor, C. Morey, and H. Hu, Residual cognitive deficits 50 years after lead poisoning during childhood. *Br. J. Ind. Med.* **50**:613, 1993.
81. H.L. Needleman, The persistent threat of lead, A singular opportunity. *Am. J. Public Health* **79**:643, 1989.
82. H.L. Needleman, A. Schell, D. Bellinger, A. Leviton, and E. Allred, Long-term effects of childhood exposure to lead at low dose, An eleven-year follow-up report. *New Engl. J. Med.* **322**:83, 1990.
83. Agency for Toxic Substances and Disease Registry, *Toxicological Profile for Manganese (Draft)*. U.S. Public Health Service, U.S. Department of Health and Human Services, Altanta, GA, 2000, PB/2000/108025.
84. F.W. John, *Manganese absorption and retention by young women is dependent on serum ferritin concentration*, USDA, Agricultural Services, Washington, DC, 1998.
85. U.S. Environmental Protection Agency, *Integrated Risk Information System (IRIS) on Manganese*, Environmental Criteria and Assessment Office, Office of Health and Environmental Assessment, Office of Research and Development, Cincinnati, Ohio, 1993.
86. P.A. Krenkel, *Mercury, Environmental Considerations*, Part I, CRC Press, Cincinnati, Ohio, 303, 1973.
87. World Health Organization, Mercury, *Environ. Health Criteria*, 1, World Health Organization, Geneva, Switzerland, 1976.
88. R. Schoeny, Use of genetic toxicology data in U.S. EPA risk assessment: The Mercury Study Report as an example. *Environ. Health Perspect.* **104**(Suppl 3):663, 1996.
89. Agency for Toxic Substances and Disease Registry, *Toxicological profile for mercury*, U.S. Department of Health and Human Services, Public Health Service, Atlanta, GA, 1999, PB/99/142416.
90. P. Grandjean, R.F. White, A. Nielsen, D. Cleary, and E.C. de Oliveira Santos, Methylmercury neurotoxicity in Amazonian children downstream from gold mining. *Environ. Health Perspect.* **107**(7):587, 1999.

90a. Agency for Toxic Substances and Disease Registry (ATSDR), *Toxicological profile for mercury*. U.S. Department of Health and Human Services, Public Health Service, Atlanta, GA, Update: September 1, 1995.

90b. Agency for Toxic Substances and Disease Registry (ATSDR), *Toxicological profile for mercury*. U.S. Department of Health and Human Services, Public Health Service, Atlanta, GA, 1999.

91. N.K. Mottet, C.M. Shaw, and T.M. Burbacher, Health risks from increases in methylmercury exposure. *Environ, Health Perspect.* **63**:133–140, 1985.
92. C. Cox, T.W. Clarkson, D.O. Marsh, L. Amin-Zaki, S. Tikriti, and G.G. Myers, Dose-response analysis of infants prenatally exposed to methyl mercury: An application of a single compartment model to single strand hair analysis. *Environ. Res.* **49**:318, 1989.
93. H. Roels, J.P. Gennart, R. Lauwerys, J.P. Bucher, J. Malchaire, and S. Bernard, Surveillance of workers exposed to mercury vapor: Validation of a previously proposed biological threshold limit value for mercury concentration in urine. *Am. J. Ind. Med.* **7**:45, 1985.
94. P.A. D'Itri, and F.M. D'Itri, *Mercury Contamination: A Human Tragedy*, John Wiley & Sons, New York, 1977.
95. M.M. Agocs, R.A. Etzel, R.G. Parrish, et al., Mercury exposure from interior latex paint. *New Engl. J. Med.* **323**:1090, 1990.
96. C.A. Glomski, H. Brody, and S.K.K. Pillay, Distribution and concentration of mercury in autopsy specimens of human brain. *Nature (London)* **232**:200, 1971.
97. E.J. Massaro, S.J. Yaffe, and C.C. Thomas, Jr., Mercury levels in human brain, skeletal muscle and body fluids. *Life Sci.* **14**:1939–1974.
98. F. Kakir, S.F. Kamluji, L. Amin-Zaki, et al., Methylmercury poisoning in Iraq. *Science* **181**:230, 1973.
99. L. Bårregard, G. Lindstedt, A. Schütz, and G. Sällsten, Endocrine function in mercury exposed chloralkali workers. *Occup. Environ. Med.* **51**:536, 1994.
100. N. Ishihara, and K. Urushiyama, Longitudinal study of workers exposed to mercury vapor at low concentrations: Time course of inorganic and organic mercury concentrations in urine, blood, and hair. *Occup. Environ. Med.* **51**:660, 1994.
101. R. Kishi, R. Doi, Y. Fukuchi, H. Satoh, T. Satoh, A. Ono, et al., Subjective symptoms and neurobehavioral performances of ex-mercury miners at an average of 18 years after the cessation of chronic exposure to mercury vapor. *Environ. Res.* **62**:289, 1993.
102. M. Sakamoto, A. Nakano, Y. Kajiwara, I. Naruse, and T. Fujisaka, Effects of methyl mercury in postnatal developing rats. *Environ. Res.* **61**:43, 1993.
103. World Health Organization, *Methylmercury*, Environ. Health Crite., 101, World Health Organization, Geneva, Switzerland, 1990.
104. World Health Organization, *Inorganic mercury*, Environ. Health Crite., 118, World Health Organization, Geneva, Switzerland, 1991.
105. Agency for Toxic Substances and Disease Registry, *Toxicological profile for nickel (update)*, U.S. Department of Health and Human Services, Public Health Service, Atlanta, GA, 1998, PB/98/101199/AS.
106. F.W. Jr. Sunderman, Mechanisms of nickel carcinogenesis. *Scand. J. Work Environ. Health* **15**:1–2, 1989.
107. K. Salnikow, S. Cosentino, C. Klein, M. Costa, The loss of thrombospondin transcriptional activity in nickel-transformed cells. *Mol. Cell Biol.* **14**:851, 1994.
108. K. Salnikow, S. Wang, and M. Costa, Induction of activating transcription factor 1 by nickel and its role as a negative regulator of thrombospondin I gene expression. *Cancer Res.* **57**(22):5060, 1997.

109. R.L. Bertholf, Zinc, in *Handbook on Toxicity of Inorganic Compounds*, H. Sigel, and H.G. Seiler, eds., Marcel Dekker, New York, 787, 1988.
110. National Institute of Occupational Safety and Health, *Pocket Guide to Chemical Hazards*, Department of Health and Human Services, Bethesda, MD, NIOSH—90–117, pp. 226–228, 1990.
111. J.M. Llobet, J.L. Domingo, M.T. Colomina, et al., Subchronic oral toxicity of zinc in rats. *Bull. Environ. Contam. Toxicol.* **41**:36, 1988.
112. Agency for Toxic Substances and Disease Registry, *Toxicological Profile for Zinc*, Agency for Toxic Substances and Disease Registry, U.S. Public Health Service, Atlanta, GA, 121 pp. ATSDR/TP-89-25, 1989.
113. J. Brandao-Neto, B.B. de Mendonca, T. Shuhama, et al., Zinc acutely and temporarily inhibits adrenal cortisol secretion in humans. *Biol. Trace Element Res.* **24**:83, 1990.
114. Agency for Toxic Substances and Disease Registry, *Toxicological Profile for Zinc*, U.S. Public Health Service, Agency for Toxic Substances and Disease Registry, Atlanta, GA, October 1994.
115. Agency for Toxic Substances and Disease Registry, *Toxicological profile for zinc*, U.S. Department of Health and Human Services, Public Health Service, Atlanta, GA, 1995, PB/95/100236/AS.

## APPENDIX 3-1 LEAD LEVELS ARE CONSIDERED ELEVATED IN HUMANS

- At levels above 80 μg/dL, serious, permanent health damage may occur
- Between 50 and 80 μg/dL, serious health damage may occur
- At lead levels between 30 and 50 μg/dL, health damage may be occurring, even if there are no symptoms
- From 20 to 30 μg/dL, regular exposure is occurring. There is some evidence of potential physiologic problems
- From 1 to 20 μg/dL, lead is building up in the body and some exposure is occurring

All health care providers should be aware of the blood lead levels to monitor the health status.

Source: 1. Occupational Safety and Health Administration (OSHA) Lead Standard (29CFR1910.1025 and 29CFR1926.62.); K.L. Hipkins, B.L. Materna, M.J. Kosnett, J.W. Rogge, and J.E. Cone, Medical surveillance of the lead exposed worker, *AAOHN Journal*, **46**(7):330–339, 1998; 2. D. Rempel, The lead-exposed worker, *JAMA* **262**(4):532–534, 1989; 3. U.S. Department of Labor, OSHA, *Lead in Construction, OSHA* 3142, 1993.

## CHAPTER 4

# Organochlorine Pesticides

## INTRODUCTION

The manufacture and use of pesticides have been increasing over the decades around the world. In view of their importance as pest control chemicals, many pesticides have entered the world market.[1] The first synthetic organic pesticides, which attracted attention in the 1930s, were dinitro compounds and thiocyanates. As has been documented, the appearance of DDT, an important pesticide to global market, became more prevalent after its discovery in 1939, although it was first synthesized by Zeidler in 1874. During subsequent years (1945 to late 1950s) other compounds entered the market and found use as pest control chemicals. From 1983 through 1993, significant changes occurred with the use of millions of pounds of fungicides.[2–4] Organophosphate insecticides entered the global market much earlier, in 1945. The pioneering work of Gerhard Schrader and the growth of the German industry initiated the use of many chemical warfare agents as pest control chemicals.[5] Over the years — and after learning the consequences of several organochlorinate pesticides that showed persistent effects on human health — attention was diverted to newer and safer compounds for combating crop pests, unwanted plants (weeds), and vectors of animal and human diseases. This shift resulted in the introduction of many compounds, such as synthetic pyrethroids and herbicides and weedicides.

Organochlorine insecticide can be considered pesticide of the greatest historical significance, because of its effect on the environment, agriculture, and human health. First synthesized by a German graduate student in 1873, DDT was rediscovered by Dr. Paul Mueller, a Swiss entomologist, in 1939 while searching for a long-lasting insecticide for the clothes moth. Application of DDT subsequently proved extremely effective against flies and mosquitoes, ultimately leading to the award of the Nobel Prize in medicine for Dr. Mueller in 1948. On January 1, 1973, the United States Environmental Protection Agency (USEPA) officially canceled all uses of DDT, but not before more than 1 billion kilograms of DDT had been introduced into the United States. Like endosulfan, DDT disrupts the delicate balance of sodium and potassium within neurons. The pesticide

*Industrial Guide to Chemical and Drug Safety*, By T.S.S. Dikshith and Prakash V. Diwan
ISBN 0-471-23698-5 

is effective against a wide spectrum of insects in the agricultural arena as well as mosquitoes that transmit malaria, yellow fever, body lice, and typhus.[6]

## GROWTH AND PRODUCTION

In addition to the economic cost of the pesticides, the amount of pesticides produced is also very large. In 1992, over 1 billion pounds of pesticides were sold in the United States. Approximately 58.7% (647 million pounds) were herbicides; 23.2% (255 million pounds) were insecticides; 10.9% (120 million pounds) were fungicides; and 7.2% (80 million pounds) were other pesticides. Production of pesticides from 1983 to 1993 showed significant changes. Growth was found to be slow, as is evident from the numbers: 975 million pounds in 1983 to 1,152 million pounds of pesticides in 1993. Thus, a growth rate of approximately 18.2% could be documented for the industry, which corresponds to an average annual growth rate of 1.8% per year during the 10-year period. A moderately steady growth in herbicide production — from 587 million pounds in 1983 to 754 million pounds in 1993 — was recorded during this period. The growth rate of 28.4% is nearly 50% more than the rate of overall pesticide production, corresponding to an average annual growth rate of 2.8% per year. The annual overall production of insecticides, however, remained constant at approximately 200 million pounds per year from 1983 to 1993. In contrast, the production of fungicides almost doubled; for instance, from 43 million pounds in 1983 to 78 million pounds in 1993. This represents an overall growth rate of 81.4% and an average annual growth rate of 8.4% per year.[1–3]

In 1994, companies in the United States exported more than 3 million pounds of pesticides designated by the United Nations (UN) Environment Program as "likely to cause problems under conditions of use in developing countries." These chemicals are methamidofos, methyl parathion, and parathion. According to the report, parathion accounts for 80% of pesticide poisoning cases in Central America.[7] The third world countries that received the greatest quantities of pesticides between 1992 and 1994 included Argentina (14.3 million pounds), Brazil (10.9 million pounds), Colombia (11.0 million pounds), Costa Rica (17.6 million pounds), Guatemala (11.7 million pounds), and the Philippines (16.1. million pounds). A total of more than 108.5 million pounds of hazardous pesticides were shipped to Latin American countries during this period. Thus, export of hazardous pesticides to many countries has continued unabated, leading to untold human suffering resulting from improper management by authorities. This calls for an immediate halt to such activities if society and government regulatory bodies are serious about achieving global chemical safety in the present century.[7]

## USE PATTERN OF PESTICIDES

Approximately 85% of American homes maintain an average inventory of three to four pesticide products (e.g., pest strips, bait boxes, bug bombs, flea collars, pesticide pet shampoos, aerosols, granules, liquids, dusts). In fact, nearly 70 million household people use more than 4 billion pesticides every year. In other words, there is an average of 57 applications of pesticides in each household each year, which amounts to one application every month. This is certainly not a healthy practice to keep the family and the environment healthy. Studies conducted by the National Home and Garden Pesticide Use Survey of USEPA have indicated that approximately 39% of pesticides are used for the control of insect pests irrespective of chemical use. It is indeed a matter of improper use of pesticides that has resulted in nearly 140,000 pesticide exposures during 1993. Of this number, approximately 93% involved improper use in and around residential areas. About 25% of these exposures involved pesticide poisoning symptoms. Over half of all reported exposures involved children younger than 6 years of age.

By law, inert ingredients are not listed on pesticide product labels; only "active" ingredients are listed. Furthermore, government officials are forbidden by law from revealing the inert ingredients in pesticide products. U.S. government evaluation of pesticides has focused rather narrowly on cancer, and the evidence shows that pesticide exposures can cause other health effects aside from cancer. Specifically, damage to the immune system (including, but not limited to, allergic reactions) and the central nervous system (CNS) is known to result from pesticide exposure. In addition to the potential effects on the immune system, much is still not known about the subtle effects on the nervous system regarding learning, memory, and potential psychological effects. Studies using animal models have limitations. It has been strongly advised that pregnant women should avoid contact with pesticides at home and in workplaces. The rapidly growing fetus may be particularly susceptible to genetic damage because pesticides may cause mutagenic, teratogenic and carcinogenic effects.[2,3]

## Use of Pesticides in Tanzania

The amount of pesticides used in Tanzania increased from 330 g capita-1 in 1977 to 500 g capita-1 in 1988. With population growth in the country at 2.7%, this implies a significant increase (140%) in pesticide use. The control of importation, formulation, and use of pesticides is inefficient. Imported amounts can exceed the authorized amounts because of donations and the specific projects that import pesticides without official authority from the licensing agent (Tropical Pesticides Research Institute).

The types of pesticides used in Zanzibar have changed significantly over the past 10 years. More toxic pesticides have been phased out and replaced by less toxic ones. Pesticide use decreased from 13 tons in 1990 to approximately 2 tons in 1995. Among the reasons for the decrease are the efforts by the Plant Protection Department and the Ministry of Agriculture to support a more organic form of agriculture, and the removal of government subsidies for agrochemicals. The pesticides discussed here are those ordered by the Plant Protection Department (90% of all the pesticides used in Zanzibar). Pesticides also are ordered by public health sectors and individuals. However, the malaria project—the main source in the public health sector—was stopped in 1989.

Recent reports released by Department of Pesticide Regulation (DPR) in California indicated that in 1997 pesticide use showed a 3.5% increase in pounds applied from 1996 and a slight decrease from 1995.[1a,7a,8] Reported pesticide use in California totaled 204,779,717 pounds in 1997, compared with 197,828,481 pounds in 1996 and 205,133,950 pounds in 1995. Reported use of pesticides includes production agriculture and postharvest fumigation of crops, structural pest control, landscape maintenance, and other uses. Exempt from reporting requirements (and therefore not included in these totals) are home and garden use of pesticides and most industrial and institutional uses.[8] All these developments have indicated that the use of organophosphate and carbamate pesticides increased substantially. In California alone, for instance, the use pattern of pesticides ranged from 13.8 million pounds in 1991 to 16.1 million pounds in 1997, meaning an annual average of 500,000 pounds.[9] Recent observations of Dewailly and associates[10] revealed the body burden of organochlorine (OC) compounds as analyzed in tissues of Greenlanders and in comparison to Canadians in Quebec. Their growth and use in other developed countries of the world has added a new dimension to this industry.

### Benefits from Pesticides

Because of the productivity demanded in modern agriculture to feed the millions of increasing world population, the use of different pesticides has become a necessity to contain a host of crop and household pests. A major benefit of pesticides is their ability to contain hosts of pests on commercial crops, food crops, lawn, and turf in and around household surroundings. Hop destruction by locust infestation in Africa is an example. Insecticides also are used for control of vector-borne diseases such as typhus and malaria.

### Adverse Effects from Pesticides

Pesticides are toxic chemicals that are used for the control of crop pests and household infestation. Any kind of improper use is, therefore, known to cause adverse effects on man, animals, and the environment. Children are at high risk for exposure to pesticides that are used extensively in urban areas such as schools, homes, and daycare centers for pest control. Approximately 6 million children who live in American inner cities live in poverty and are most often exposed to different pesticides. In view of the activity pattern of children, namely, playing close to the ground, hand-to-mouth behavior, and dietary pattern, they are naturally more vulnerable to pesticides. In fact, inadequate data are available that would indicate that pesticides, like chlorpyrifos and certain pyrethroids, trigger neurodevelopmental hazards in children.[11] Children, particularly those in agricultural areas, are exposed to much higher concentrations of pesticides and have shown health effects. These effects are attributed to pesticide drift in the air, children playing on ground soil in nearby crop fields, and children feeding on breast milk from their farmworker mother.[12]

Food grains, fruits, and vegetables are important as nutrients. The sources of nutrients are always attacked by rodents, insects, fungi, molds, and other microorganisms. Therefore, humans have synthesized a number of organic chemicals (pesticides) to fight these competitors of food source or pests. Based on their mode of action, the chemicals is classified as acaricides, fumigants, fungicides, herbicides (weedicides), insecticides, nematicides, or rodenticides. Each commonly known pesticide belongs to different classes or groups (e.g., organochlorine, organophosphorus, carbamates, herbicides, fungicides, rodenticides, acaricides). Each pesticide also has a common name and one or more trade names (Table 4-1). Studies have shown that pesticides are toxic chemicals; improper use is known to cause adverse effects to animals and humans.[11–13] Laboratory studies and epidemiological surveys have established pesticides with the potential to trigger neurotoxicity, cardiotoxicity, genotoxicity, behavioral changes, and disturbances in hormone function in animals and humans (Tables 4-2 and 4-3).[12,13] Names of pesticides, the classes to which they belong, and the acute oral dermal and inhalation toxicity values are indicated in Table 4-4. These toxicity values were generated after conducting studies on laboratory animals, such as albino rats and rabbits. This study provides the basic information necessary for any worker in a laboratory or field as well as a layman. In fact, this basic information helps during the use and disposal of toxic chemicals, like pesticides.

As discussed earlier, OC pesticides have played an important part in the management of crop pests. The following pages discuss, in brief, the use and toxicity of selected OC pesticides.

## Cyclodienes

Cyclodienes are an important group of chlorinated pesticides. The group heptachlor includes insecticides such as chlordane, aldrin, dieldrin, endosulfan, and heptachlor and its epoxide. These are used for the control of a variety of plant pests in agriculture and household environments. The entry of cyclodienes to the global market has created easy management for the control of crop pests. They appeared after World War II as tools to protect food crops and control diseases from pests. In fact, humans were protected from malaria, typhus, and loss of food crops by pesticides.

The improper use of and lack of management capabilities for toxic chemicals made them outcast and disastrous. Chlordane came into the market in 1945; aldrin and dieldrin in 1948; heptachlor in 1949; endrin in 1951; mirex in 1954; endosulfan in 1956; and chlordecone (Kepone) in 1958. Synthetic chemistry continued adding newer products to reduce hazardous chemicals from further use. However, lack of knowledge of basic toxicology, proper management, and pragmatic approach among occupational workers did not receive due emphasis and resulted in human fatalities, pest resistance, and contamination. There is no easy way to have benefits from chemicals yet ignore the basic tenets of proper handling, waste disposal, and management of toxic chemicals.

**TABLE 4-1 Pesticides of Different Classes* with Common Names and Trade Names**

| Common Name | Trade Name† |
|---|---|
| Alachlor (H) | Alazine, Alanex, Crop Star |
| Aldicarb (A, N, I) | Temik, UC 21149 |
| Aldrin (I) | Aldrex |
| Allethrin (I) | Pynamir |
| Amitrole (H) | Amerol, Amitrol-T, Amizol, Weedazol |
| Amitraz (A, I) | Taktic, Mitac |
| Atrazine (H) | Aatrex, Atratol |
| Azinfos-methyl (A, I) | Azimil, Bay, Guthion |
| Bendiocarb (I) | Dycarb, Ficam, Garvox, Rotate, Sexox, Seedoxin |
| Benomyl (F) | Agrocit Benlate, Benosan, Du Pont, 1991, Fundazol |
| Binapacryl (A) | Acricid, Endosan, Morocide |
| Bromacil (H) | Borea, Bromax, Croptex Onyx |
| Bromophos (I) | Nexion |
| Butachlor (H) | Machetc |
| Captan (F) | Captaf, Capanex, Captazel, Captec, Captol |
| Carbaryl (I) | Arylam Bug Master, Carbamec, Carbamine, Denapon, Devicarb, Dicarbam, Hexavin, Sevin, Tornado |
| Carbendazim (F) | Bavistin, Delsene |
| Carbendazim (F) | Bavistin, Delsene |
| Carbofuran (I) | Bay 70143, Chinufur, Curaterr, Furacarb, Furadan, Yaltox |
| Carbophenothion (A) | Garrathion, Trithion |
| Carboxin (F) | Vitavax |
| Chlordane (I) | Belt, Carbodan, Carbosip, Chlor Kil, Chlortox, Corodane, Octa-Klor, Octachlor, Topiclor, Velsicol 1068 |
| Chlordimeform (A) | Fundal, Spanone, Galecron |
| Chlorobenzillate (I) | Acaraben, Akar, Benzilan |
| Chlorothalonil (F) | Bombardier, Bravo, Daconil 2787 |
| Chlorpyrifos (I) | Dursban, Lorsban |
| Chlorpyrofos methyl (I) | Dowco 214, Reldan |
| Cyanazine (H) | Bladex, Fortrol |
| Cypermethrin (I) | Cymbush, Imperator, Ripcord, Barricade |
| 2,4-D (H) | 2,4-D |
| Dalapon (H) | Dalapon 85, Radapon |
| DDT (I) | Gesarol, Neocid |
| Demethol-S-methyl (A, I) | Metasystox |
| Diazinon (A, I) | Basudin, Diazitol |
| Dichlorvos (I) | DDVP, Vapona, Nuvon |
| Dicofol (A, I) | Kelthane |
| Dicrotopos (A, I) | Ektaphos |
| Dieldrin (I) | Dieldrex |
| Dimethoate (A, I) | Rogor, Fostion |
| Dinocap (F) | Karathane |
| Diquat (H) | Ortho Diquat, Aquacide, Weedol |
| Disulfoton (I) | Disyston |
| Diuron (H) | Karmex |
| Dimethoate (I) | Cygon |
| Edifenfos (F) | Hinosan |

**TABLE 4-1** (*continued*)

| Common Name | Trade Name[†] |
|---|---|
| Endosulfan (I) | Thiodan, Cyclodan, Thimul |
| Endrin (I) | Endrex |
| Esfenvalerate (I) | Asana Sumi-Alpha |
| Ethion (A, I) | Ethiol |
| Ethylene dibromide (Fm) | EDB, Bromofume |
| Fenamiphos (N) | Nemacur |
| Fenitrothion (I) | Sumithion, Cytel |
| Fensulfothion (I, N) | Dasanit, Terracur |
| Fenthion (I) | Baytex, Entex, Lebaycid |
| Fenvalerate (I) | Pydrin, Belmark |
| Fonofos (I) | Dyfonate |
| Gamma BHC (I) (Lindane) | Gammexane |
| Glyphosate (H) | Roundup |
| Heptachlor (I) | Heptamul |
| Hexachlorobenzene (F) | HCB |
| Iodofenphos (A, I) | Nuvanol-N |
| Isoproturon (H) | Arelon, Graminon |
| Malathion (A, I) | Cythion |
| Mancozeb (F) | Dithane M-45 |
| Maneb (F) | Dithane M-22 |
| Methamidofos (A, I) | Tamaron |
| Methomyl (I) | Lannate |
| Methoxychlor (I) | Methoxychlor, Marlate |
| Methyl bromide (Fm, A) | Brom-O-Gas |
| Methyl parathion (I) | Metacid, Folidol-M |
| Mevinphos (A, I) | Phosdrin |
| Monocrotofos (A, I) | Azodrin, Nuvacron |
| Parathion, (ethyl) (I) | Folidol, Niran |
| Paraquat (H) | Dexuron, Gramuron |
| Pentachlorophenol (H) | PCP, Santobrite |
| Phorate (I) | Thimet |
| Phosalone (A, I) | Zolone |
| Phosphamidon (I) | Dimecron |
| Picloram (H) | Tordon |
| Propoxur (I) | Baygon, Blattanex |
| Pyrethrum (I) | Pyrenone |
| Quinalfos (A, I) | Ekalux |
| Rotenone (I) | Rotacide |
| Turbufos (I) | Counter |
| Warfarin (R) | Rosex |
| Zinc phosphide (R, Fm) | |
| Zineb (F) | Dithane Z-78 |
| Ziram—(F) | Zirlate |

*A—Acaricide; Fm—Fumigant; F—Fungicide; H—Herbicide; I—Insecticide; N—Nematicide; R—Rodenticide.

[†]The citation of trade names is not an endorsement or an approval for product safety.

**TABLE 4-2 Pesticides and Their Potential Toxicity to Mammals**

| Group and Name of the Pesticide | Nature of Toxicity |
|---|---|
| **Organochlorine Pesticides** | |
| Aldrin | General malaise, anxiety, irritability vomiting, convulsions |
| Benzene hexachloride | Hyperexcitability, neurologic disorders, myoclonic jerks, aplastic anemia, hepatotoxicity, neurotoxicity, cerebral seizures |
| Chlordane | Generalized convulsions, reproductive toxicity, birth defects, loss of consciousness, change in EEG pattern, hepatic disorders, neurologic disturbances, mutagenicity, carcinogenicity |
| DDT | Loss of weight, anorexia, tremors paresthesia, hepatotoxicity, reproductive toxicity, cancer |
| Dicofol | Nausea, vomiting, muscular weakness |
| DDD | Ataxia, confusion, abnormal gait, mild anemia |
| Dieldrin | Violent headache, muscular pain, reproductive toxicity, birth defects, cancer |
| Dimethoate | Cancer, mutagenicity, reproductive toxicity, birth defects |
| Endrin | Nausea, dizziness, headache, hyperexcitability, abdominal discomfort |
| Endosulfan | Agitation, diarrhea, foaming, vomiting, hyperplexia, muscle twitching, cyanosis, chronic toxicity |
| Heptachlor | Myoclonic jerking, psychological disorders, irritability, anxiety, carcinogenicity |
| Isodrin | Motor hyperexcitability, intermittent, muscle twitching |
| Lindane (Gamma HCH) | Neurotoxicity, hepatotoxicity |
| Methoxychlor | Neurotoxicity, hepatotoxicity |
| Telodrin | Nausea, vomiting, hyperexcitability |
| Toxaphene | Loss of consciousness, epileptiform convulsions, carcinogenicity |
| **Organophosphorus Pesticides** | |
| Azinphos-methyl | Neurotoxicity, depression, slurred speech |
| Bromophos ethyl | Neurotoxicity, depression, slurred speech |
| Chlorpyrifos | Neurotoxicity, depression, slurred speech |
| Crotoxyphos | Neurotoxicity, depression, slurred speech |
| Demeton | Mutagenicity, birth defects |
| Diazinon | Neurotoxicity, neurobehavioral disturbances |
| Dichlorvos | Neurotoxicity, depression, slurred speech |
| Dimethoate | Muscle weakness, respiratory distress |

**TABLE 4-2** (*continued*)

| Group and Name of the Pesticide | Nature of Toxicity |
|---|---|
| Edifenfos | Dizziness, vomiting, nausea |
| Ethion | Discomfort, vomiting, muscular twitching, nausea, nervousness convulsions |
| Fenitrithion | Tremors, fatigue, memory loss, lethargy |
| Fensulfothin | Vomiting, diarrhea, muscular twitching, pulmonary edema, convulsions, coma |
| Fenthion | Muscle weakness, respiratory distress, neurotoxicity |
| Methamidofos | Muscle weakness, respiratory distress |
| Mevinphos | Mutagenicity |
| Monocrotofos | Muscle weakness, respiratory distress |
| Parathion (ethyl) | Headache, miosis, nervousness, salivation, diarrhea, respiratory distress, convulsions, coma, cancer, mutagenicity |
| Parathion methyl | Diarrhea, salivation, nervousness, respiratory distress, convulsions chronic toxicity, mutagenicity |
| Phosmet | Cancer, mutagenicity |
| Phosphomidon | Respiratory distress, nervousness, diarrhea, convulsions, salivation, paralysis, coma |
| Quinolfos | Respiratory distress, nervousness, diarrhea, convulsions, salivation, paralysis, coma |
| **Carbamate Pesticides** | |
| Aldicarb | Extremely toxic, even in very small concentrations |
| Carbaryl | Mutagenicity, nephronotoxicity |
| Chlorpropham | Mutagenicity |
| Fenvalerate | Cancer |
| Methomyl | Chronic toxicity, mutagenicity |
| Chlorophenoxy compounds | |
| 2,4-D and 2,4,5-T | Nausea, dizziness, vomiting, limited information |
| **Synthetic Pyrethroids** | |
| Cypermethrin | Burrowing, sinous writhing |
| Deltamethrin | Clonic seizures |
| Fenpropanthrin | Dermal tingling |
| Fenvalerate | Profuse salivation, enhanced startle response |
| Permethrin | Prostration |
| Phenorthrin | Whole body tremor |
| Resmethrin | Enhanced startle response |
| Tetramethrin | Aggressiveness |

**TABLE 4-3 Pesticides Known to Cause Hormone Disturbances in Mammals**

**Organochlorine Compounds**

Alachlor, Benzene hexachloride, Cyclodienes: Aldrin, Chlordane, Dieldrin, Endrin, Endosulfan, Heptachlor, Isodrin, Telodrin and Toxaphene, DDT and metabolites, Dicofol, Dimethoate, Lindane, Methoxychlor, Mirex, Pentachlorophenol, Perthane

**Organophosphorus Compounds**

Azinphosmethyl, Bromophos ethyl, Chlorpyrifos, Crotoxyphos, Demeton, Diazinon, Dichlorvos, Ethion, Fenitrothion, Fensulfothin, Fenthion, Flusulfothion, Methamidofos, Mevinphos Monocrotophos and Dicrotofos, Oxamyl, Phorate, Parathion ethyl, Parathion methyl, Phosphomidon, Quinalphos, Temefos

**Carbamate Compounds**

Aldicarb, Benomyl, Carbaryl, Chlorpropham, Fenvalerate, Methomyl, Chlorophenoxy compounds 2,4-D, 2,4,5-T

**TABLE 4-4 Acute Oral, Dermal, and Inhalation Toxicity of Pesticides in Rats and Rabbits**

| Pesticides | Group | Acute Oral $LD_{50}$ (mg/kg) Rat | Acute Dermal $LD_{50}$ (mg/kg) Rat*/Rabbit† | Acute Inhalation $LD_{50}$ (mg/L air) Rat |
|---|---|---|---|---|
| Acephate | I | 866–945 | >200 | >15 mg/L |
| Acrolein | H | 29 | 231 | 8.3 mg/L (4 hours) |
| Alachlor | H | 930–1200 | 13,300 | 1.04 mg/L (4 hours) |
| Aldicarb | A, I, N | 0.93 | 20.0 | 0.2 mg/L (5 min.) |
| Allethrin | I | 585–110 | >2,500 | >2,000 $mg/m^3$ |
| Amitraz | A, I | 600–800 | >200 | 65 mg/L (6 hours) |
| Amitrole | H | 1,100–24,600 | >10,000* | — |
| Atrazine | H | 1,839–3,080 | 7,500 | >5.8 mg/L (4 hours) |
| Azadirachtin | I | >5,000 | >2,000 | — |
| Azinphos-ethyl | A, I | 12.0 | 500 Rat | 0.15 mg/L (4 hours) |
| Azinphos-methyl | I | 9.0 | 150–200* | 0.15 mg/L (4 hours) |
| Bendiocarb | I | 40–156 | 566–800* | 0.55 m/L (4 hours) |
| Bemonyl | F | >10,000 | >10,000 | >2.0 m/L (4 hours) |
| Bromacil | H | 5,200 | >5,000 | >4.8 mg/L (4 hours) |
| Butachlor | H | 2,000 | >13,000 | >4.7 mg/L (4 hours) |
| Capatafol | F | 5,000–6,200 | >15,400 | — |
| Captan | F | 9,000 | >4,500 | 4.0 mg/L (4 hours) |
| Carbaryl | I | 850 | >4,000*<br>>2000† | 206 mg/L |
| Carbendiazim | F | >15,000 | >10,000 | 10 g/L (4 hours) |

TABLE 4-4 (*continued*)

| Pesticides | Group | Acute Oral $LD_{50}$ (mg/kg) Rat | Acute Dermal $LD_{50}$ (mg/kg) Rat*/Rabbit† | Acute Inhalation $LD_{50}$ (mg/L air) Rat |
|---|---|---|---|---|
| Carbofuran | I, N | 8.0 | >3,000* | 0.075 mg/L (4 hours) |
| Carbosulfan | I | 185–250 | >2,000* | 0.63–1.53 mg/L |
| Carboxin | F | 3,820 | >8,000 | >20 mg/L (1 hours) |
| Chlordane | I | 457–590 | 200–2,000 | >200 mg/L (4 hours) |
| Chlorfenvinfos | A, I | 24–39 | 31–108*<br>400–4,700† | 0.05 mg/L (4 hours) |
| Chloropicrin | I, N | 250 | — | 0.08–0.12 mg/L (1 hours) |
| Chlorothalonil | F | >10,000 | >10,000 | >4.7 mg/L (1 hours) |
| Chlorpropham | H | 5,000–7,500 | — | — |
| Chlorpyrifos | I | 135–163 | 2,000 | >0.2 mg/L (4–6 hours) |
| Chlorpyrifosmethyl | A, I | >3,000 | >2,000 | >0.67 mg/L (4 hours) |
| Chlorsulfuron | H | 5,545–6,293 | >3,400 | >5.9 mg/L (4 hours) |
| Copper oxychloride | F | 700–800 | >2,000 | >30 mg/L (4 hours) |
| Coumafos | I | 15.5–41.0 | 860* | >341–1,081 $mg/m^3$ (1 hours) |
| Cyanazine | H | 182–334 | >2,000 | >2,460 $mg/m^3$ |
| Cyanofos | I | 710–730 | >2,000* | >1,500 $mg/m^3$ (4 hours) |
| Cypermethrin | I | 250–4150 | >2,460 | 2.5 mg/L (4 hours) |
| 2,4, D | H | 639–764 | >2,400 | >1.79 mg/L (24 hours) |
| Dalapon | H | 7,570–9,330 | >2,000 | >20 mg/L (8 hours) |
| DDT | I | 113–118 | 2,510* | — |
| Deltamethrin | I | 135–5,000 | >2,000 | 2.2 mg/L (4 hours) |
| Demeton-S-methyl | A, I | 30 | 30* | 0.13 mg/L (4 hours) |
| Diazinon | A, I | 300–400 | 540–650 | 3.5 mg/L (4 hours) |
| Dicamba | H | 1,707 | >2,000 | 9.6 mg/L (4 hours) |
| Dichlorvos | A, I | 50 | 300* | >0.1 mg/L (4 hours) |
| Dicofol | A | 587–595 | 2,500†<br>7,500* | >0.1 mg/L (4 hours)<br>>5.0 mg/L (4 hours) |
| Dicrotofos | A, I | 17–22 | 224 | 0.09 mg/L (4 hours) |
| Dimethoate | A, I | 290–325 | >800* | 0.2 mg/L (4 hours) |
| Diquat | H | 231 | >750†<br>>2,000* | — |
| Diuron | H | 3,400 | >2,000 | >5 mg/L (4 hours) |
| Dodine | F | 1,000 | 1,500 | — |
| Edifenfos | F | 100–260 | 700–800* | 0.32–0.36 mg/l (4 hours) |

(*continued overleaf*)

**TABLE 4-4** (*continued*)

| Pesticides | Group | Acute Oral $LD_{50}$ (mg/kg) Rat | Acute Dermal $LD_{50}$ (mg/kg) Rat*/Rabbit† | Acute Inhalation $LD_{50}$ (mg/L air) Rat |
|---|---|---|---|---|
| Endosulfan | A, I | 18–160 | 359† >500–4,000* | >21.0 mg/L (1 hours) |
| EPN | A, I | 24–36 | 538–2,850* | — |
| Esfenvalerate | I | 75–458 | >2,000† >5,000* | 480–570 mg/m³ (4 hours) |
| Ethion | A, I | 208 | 915 | 0.45 mg/L (4 hours) |
| EDB | I, N | 146–420 | — | 200 mg/L |
| Fenamifos | N | 6.0 | 80 | 0.12 mgl/L (4 hours) |
| Fenitrothion | I | 1,700–1,720 | 810–840 | >2,210 mg/m³ (4 hours) |
| Fenothiocarb | A | 1,150–1,200 | — | 1.79 mg/L (4 hours) |
| Fenthion | I | 250 | 700* | 0.5 mg/L (4 hours) |
| Fentin | A, F | 140–298 | 450* | 0.044 mg/L (4 hours) |
| Fenuron | H | 6,400 | — | — |
| Fenvalerate | | 451 | 1,000–3,200† | >101 mg/m³ |
| Folpet | F | >10,000 | >22,600† | 1.89 mg/L (4 hours) |
| Fonofos | I | 5.5–11.5 | 32–261† | 17–511 μg/L (4 hours) |
| Gamma HCH | I | 88–270 | 147* | 511 μg/L (4 hours) |
| Glyphosate | H | 5,600 | >5,000† | 12.2 mg/L (4 hours) |
| Heptachlor | I | 147–220 | >2,000† 119–250* | >2.0–200 mg/L (4 hours) |
| Hydrogen Cyamide | R, I | 10–15 | — | 0.36 mg/L (30 min) |
| Isofenfos | I | 20.0 | 70* | 0.3 mg/L (4 hours) |
| Isoproturon | H | 1,826–5,000 | >2,000* | >1.9 mg/L (4 hours) |
| Linuron | H | 1,500–400 | >2,000* | >0.849 mg/L (4 hours) |
| Malathion | A, I | 1,375–2,800 | 4,100† | — |
| Mancozeb | F | >5,000 | >10,000* >5,000† | — |
| Maneb | F | >5,000 | >5,000 | >3.8 mg/L (4 hours) |
| Methamidofos | A, I | 20 | 130* | 0.2 mg/L (4 hours) |
| Methidathion | A, I | 25–54 | 200† 1,546* | 3.6 mg/L (4 hours) |
| Methiocarb | A, I, M | 20 | >5,000* | 0.3 mg/L (4 hours) |
| Methomyl | A, I | 17–24 | >5,000† | 0.3 mg/L (4 hours) |
| Methoxychlor | I | 6,000 | >2,000† | — |
| Methyl isothiocyanate | H, F, N, I | 72–220 | 2,780* 263† | 1.9 mg/L (1 hours) |

**TABLE 4-4** (*continued*)

| Pesticides | Group | Acute Oral $LD_{50}$ (mg/kg) Rat | Acute Dermal $LD_{50}$ (mg/kg) Rat*/Rabbit† | Acute Inhalation $LD_{50}$ (mg/L air) Rat |
|---|---|---|---|---|
| Metylpertin | I | 6 | 45 | 0.17 mg/L (4 hours) |
| Mevinfos | A, I | 3–12 | 4–90*<br>16–33† | 0.125 mg/L (1 hours) |
| Monocrotophos | A, I | 18–20 | 130–250†<br>112–126* | 0.08 mg/L (4 hours) |
| Nabam | A, F | 395 | — | — |
| Omethoate | A, I | 25 | 200* | 0.3 mg/L (4 hours) |
| Oxamyl | A, I, N | 5.4 | 2,000–5,027† | 0.12–0.17 (1 hours) |
| Oxydemethionmethyl | I | 50 | 130* | 0.25 mg/L (4 hours) |
| Paraquat | H | 157 | 236–500† | — |
| Parathion | A, I | 2.0 | 50* | 0.05 mg/L (4 hours) |
| Pentachlorophenyl | F | 210 | — | — |
| Permethrin | I | 430–6,000 | >4,000*<br>>2000† | >685 mg/L (4 hours) |
| Phenthoate | I | 300–400 | 2,100 | 0.8 mg/L (4 hours) |
| Phorate | I | 1.6–3.7 | 2.5–6.2*<br>2.9–5.6† | 0.01 mg/L (1 hours) |
| Phosalone | A, I | 120–175 | 1,500*<br>>1,000† | — |
| Phosmet | A, I | 113–160 | >5,000† | 2.76 mg/L (1 hours) |
| Phosfamidon | A, I | 17.9–30.0 | 374–539*<br>267† | 0.18 mg/L (4 hours) |
| Phosfine (Zinc Phosfate) | I | 45.7 | 2,000–5,000† | 10 $mg/m^3$ (6 hours) |
| Picloram | H | 8,200 | >4,000† | — |
| Propoxur | I | 50 | >5,000 | >0.5 mg/L (4 hours) |
| Quinalfos | A, I | 71 | 1,750* | 0.45 mg/L (4 hours) |
| Simazine | H | >5,000 | 3,100*<br>10,200† | >2.0 mg/L (4 hours) |
| Temefos | I | 4,204–10,000 | >4,000*<br>2,181† | — |
| Tetradifon | A | 14,700 | >10,000† | >3.0 mg/L (4 hours) |
| Thiram | F | 1,800 | >1,000* | 4.4 mg/L (4 hours) |
| Trichlorfon | I | 250 | >5,000* | 0.5 mg/L (4 hours) |
| Warfanin | R | 186 | — | — |
| Zineb | F | >5,200 | >6,000* | — |
| Ziram | F | 320 | >6,000* | — |

*Rat.
†Rabbit.
A—Acaricide; Fm—Fumigant; F—Fungicide; H—Herbicide; I—Insecticide; N—Nematicide; R—Rodenticide.

### **Chlordane** (Belt)

IUPAC name: 1,2,4,5,6,7,8,8-octachloro-2,3,3a,4,7,7a-tetrahydro-4,7-methano-indane
Molecular formula: $C_{10}H_6Cl_8$
Toxicity class: USEPA: II; WHO: II

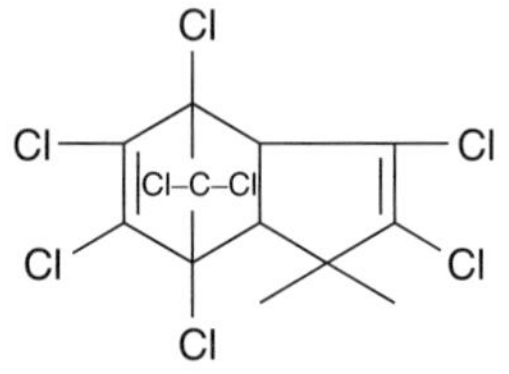

**Figure 4-1.** Structural formula of chlordane.

***Uses*** Chlordane is a viscous, amber-colored liquid. Technical-grade chlordane is a mixture of many structurally related compounds including *trans*-chlordane, *cis*-chlordane, chlordane, heptachlor, and *trans*-nonachlor.[14,15] Chlordane was used as a broad-spectrum pesticide in the United States from 1948 to 1988. Its uses included termite control in homes and pest control on agricultural crops (e.g., corn, citrus, home lawns, gardens, turf, ornamental plants).

***Toxicity*** The acute oral $LD_{50}$ values of technical-grade chlordane for the rat range from 137 to 590 mg/kg, and acute dermal $LD_{50}$ for the rabbit is 1,720 mg/kg. Signs of acute chlordane poisoning include ataxia, convulsions, and cyanosis, followed by death due to respiratory failure. Rats treated by gavage with 100 mg/kg once a day for 4 days had increased absolute liver weights, fatty infiltration of the liver, increased serum triglycerides, creatine phosphokinase, and lactic acid dehydrogenase. Sheep treated by stomach tube with 500 mg/kg showed signs of poisoning but recovered fully within 5 to 6 days; a dose of 1,000 mg/kg resulted in death after 48 hours.[16–18]

Male and female rats and mice given chlordane (1,600 and 800 ppm) for 6 weeks demonstrated adverse effects and high mortality.[18a] All treated animals showed generally poor physical condition at the end of the study, with significantly reduced weight gain measured in both genders of rats. Laboratory animals fed for prolonged periods with high doses experience liver hypertrophy, hepatocellular swelling, and necrosis.[19–21]

Ingestion of chlordane induces vomiting, dry cough, agitation and restlessness, hemorrhagic gastritis, bronchopneumonia, muscle twitching, convulsions, and death among humans. Nonlethal, but accidental, poisoning of children has resulted in convulsions, excitability, loss of coordination, dyspnea, and tachycardia. Recovery, however, was complete. When a municipal water supply was contaminated with chlordane in concentrations of up to 1.2 g/L, 13 persons had symptoms of gastrointestinal and neurologic disorders. Signs of toxicity from chronic inhalation exposure in chlordane-treated homes include sinusitis, bronchitis, dermatitis, neuritis, migraine, gastrointestinal distress, fatigue, memory deficits, personality changes, decreased attention span, numbness or paresthesia, disorientation, loss of coordination, dry eyes, and seizures. Blood dyscrasias, including production defects and thrombocytopenic purpura, have been described for both professional applicators and homeowners and their families following home termite treatment. Chlordane can promote cancer and is an endocrine disrupter.[22,23]

### Heptachlor and Heptachlor Epoxide (Heptamul)

IUPAC name: 1,4,5,6,7,8,8-heptachloro-3a,4,7,7a-tetrahydro-4,7-methanoindene
Molecular formula: $C_{10}H_5Cl_7$
Toxicity class: USEPA: II; WHO: II

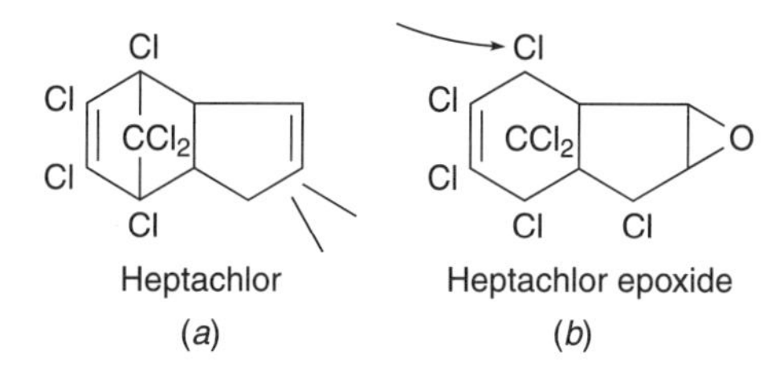


**Figure 4-2.** Structural formula of heptachlor.

***Uses*** Heptachlor was first isolated from technical chlordane in 1946. Its extensive use from 1960 to 1970 was primarily for the control of termites, ants, and soil insects. Different formulations such as dusts, wettable powders, emulsifiable concentrates, and oil solutions were in use for pest management before the imposition of its use. Heptachlor has both nonsystemic stomach action and contact action. Heptachlor epoxide is the principal metabolite (oxidation product) of heptachlor and is formed by different plants and animals.

Heptachlor was in extensive use until the 1970s for the control of certain soil-inhabiting insects that attack corn and other field crops, cotton insects, grasshoppers, and for the treatment of crop seeds. Along with other cyclodiene insecticides, heptachlor is uniquely suited for termite control. Since 1988, it has not been used for the control of termites, but it is permitted for commercial use in the United States for fire ant control in power transformers.[24] Nonachlor is a by-product created during the manufacture of chlordane and heptachlor.

***Toxicity*** The acute oral and dermal $LD_{50}$ values of heptachlor range from moderate to severe toxicity in animal species.[25,26] Accordingly, acute oral $LD_{50}$ for mice ranges from 30 to 68 mg/kg, and for rats ranges from 110 to 220 mg/kg. Similarly, the acute dermal $LD_{50}$ for rats ranges from 119 to 320, but rabbits have greater than 2,000 mg/kg.[14,25,26]

Heptachlor induces tremors, convulsions, paralysis, and hypothermia in rats. Young calves fed with different doses (2.5, 5, or 10 mg/kg/day) of heptachlor formulation for 3 to 15 days showed muscle spasms in the head and neck region, convulsive seizures, elevated body temperatures, and engorged brain blood vessels. The photoisomer of heptachlor (photoheptachlor) and the major metabolite of heptachlor (heptachlor epoxide) are considered more toxic than the parent compound.

Heptachlor is known to cause hepatic effects in rats following acute exposures, leading to increased liver weight, necrosis, cellular vacuolization, steatosis, and elevated liver enzymes (serum glutamic pyruvate transaminase and alkaline phosphatase), cholesterol, and bilirubin in animals.[25] Dietary administration of 10 mg/kg of heptachlor (96%) given to rats for 5 or 7 days resulted in alterations in liver function, as evidenced by increased blood glucose, decreased liver glycogen content, and decreased glutamine oxalacetic transaminase and increased acid and alkaline phosphatase levels. Animal species, such as Osborne-Mendel rats, pigs, mink, and sheep, exposed to heptachlor for prolonged periods developed adverse effects (e.g., liver necrosis, hyperexcitability, incoordination, paralysis of hind legs).[26–29] Humans exposed to heptachlor in homes during termite control operations showed signs of neurotoxicity, (e.g., irritability, salivation, lethargy, dizziness, labored respiration, muscle tremors, convulsions, and death due to respiratory failure). Heptachlor interfered with nerve transmission and caused hyperexcitation of the CNS, lethargy, incoordination, tremors, convulsions, and stomach cramps or pain, leading to coma and death.

### Heptachlor Epoxide

***Uses*** Heptachlor epoxide (HCE), also known as epoxyheptachlor, is a white crystalline solid. Heptachlor is converted to heptachlor epoxide and other degradation products in the environment. HCE degrades more slowly and, as a result, is more persistent than heptachlor. Both compounds adsorb strongly to sediments and are bioconcentrated in terrestrial and aquatic organisms; biomagnification of both is significant.[14]

***Toxicity*** The acute oral $LD_{50}$ of HCE for rats, mice, and rabbits ranges from 39 to 144 mg/kg. Single oral doses of a heptachlor:heptachlor epoxide mixture (25:75:) induce moderate hypoactivity, ruffled fur, and mortality of animals. Calves exposed to different doses of heptachlor epoxide died within 3 hours to 3 days. Young calves had muscle spasms in the head and neck region, convulsive seizures, elevated body temperatures, and engorged brain blood vessels.[26,30] More information regarding the adverse effects (including carcinogenicity) is available in the literature.[24,30–32]

### Aldrin, Dieldrin, and Endrin (Aldrex, Dieldrex, Endrex)

IUPAC name: Aldrin: not less than 95% of 1,2,3,4,10,10-hexachloro-1,4,4a,5,8,8a-hexahydro-1,4-endo,exo-5-,8-dimethanonaphthalene. Dieldrin:

not less than 85% of 1,2,3,4,10,10-hexachloro-6,7-epoxy-1,4,4a,5,6,7,8,8a-octahydro-1,4-endo,exo-5-8-dimethanonaphthalene. Endrin: (1,2,3,4,10,10-hexachloro-1,4,4a,5,6,7,8,8a-octahydro-6,7-epoxy-1,4:5:,8-dimethanonaphthalene

Molecular formula:$C_{12}H_8Cl_6O$

Molecular formula: $C_{12}H_8Cl_6$

Toxicity class: USEPA: II; WHO: Ib

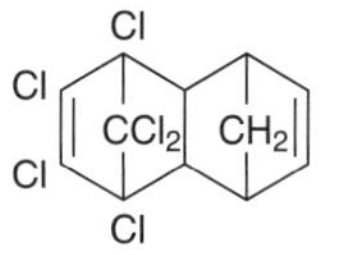

**Figure 4-3.** Structural formula of aldrin.

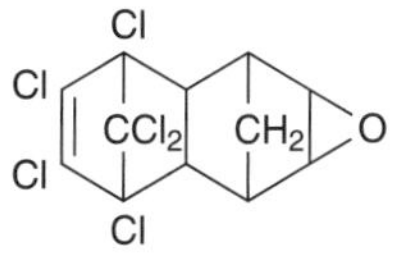

**Figure 4-4.** Structural formula of dieldrin.

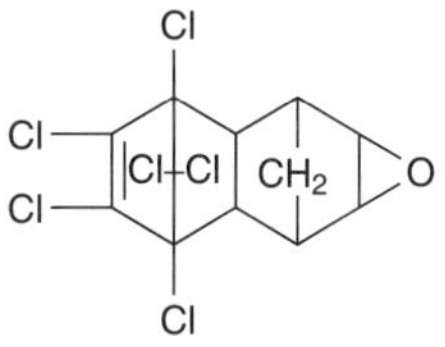

**Figure 4-5.** Structural formula of endrin.

***Uses*** Aldrin, dieldrin, and endrin are the common names of three insecticides that are powerful and closely related chemically. Technical aldrin is a light tan to brown solid or powder. Aldrin is readily converted to dieldrin in the environment, and both are considered as closely related compounds by regulatory bodies. Their toxicities do not differ significantly. Aldrin and dieldrin were widely used from the 1950s to the early 1970s.[14]

Aldrin has been used as a soil insecticide to control root worms, beetles, and termites. Dieldrin has been used in agriculture for soil and seed treatment and in public health to control disease vectors such as mosquitoes and tsetse flies. Dieldrin also has had veterinary use as a sheep dip and has been used in the treatment of wood and mothproofing woolen products. Workers could be occupationally exposed to aldrin or dieldrin from inhalation and absorption through the skin. Most uses for aldrin and dieldrin were banned in 1975; since 1986, these compounds have not been produced or imported into the United States.

Endrin is a solid, white, and odorless insecticide. It has been found in groundwater and surface water. Endrin is known to cling to the bottom sediments of rivers, lakes, and other bodies of water. Endrin was in extensive use as a pesticide to control insects, rodents, and birds. Endrin generally is not found in the air, except when it is applied to fields during agricultural applications. Endrin has a long persistence in soil, lasting over 10 years.[14]

***Toxicity*** Aldrin, dieldrin, and endrin are highly toxic to animals and humans. The acute oral and dermal toxicity of aldrin and dieldrin in rats ranges from 39 to 98 mg/kg; in contrast, endrin has a very low value of oral and dermal $LD_{50}$ — 7.5 to 15 mg/kg in rats — and is highly toxic. Although dieldrin is persistent in soil, environmental background levels are decreasing slowly. Residual contamination may be present at waste sites from the disposal of used stocks. Where this residual amount is appreciable, a potential exists for exposure to cleanup workers. Air and water appear to be sources of minor importance to the general population, with regard to aldrin and dieldrin exposure. In the past, food products grown in soil treated with aldrin or dieldrin have probably been the primary source of dieldrin residues in fatty tissues of the general population; however, since 1970, dietary intake has shown a significant decrease. Because neither aldrin nor dieldrin is currently produced in or imported into the United States, their use is believed to be minimal. Possible new releases may come from individually owned stockpiles of aldrin for the underground control of termites, although because importation of aldrin ceased more than 3 years ago, it is believed that there is very little, if any, termicide stock left in this country. Higher exposure rates can be expected for persons residing in homes treated with aldrin for termite control. Improper application practices may result in unnecessarily high exposure to occupants of treated structures.[14,32] It is possible to breathe air containing aldrin or dieldrin in homes that have been treated with these compounds for termite control. There also is potential for exposure to applicators and nearby residents if improper procedures are used. Aldrin and dieldrin may enter the body by penetrating the skin. However, since these compounds are no longer available to any extent, this route of exposure is no longer likely, except as a potential threat to cleanup workers at hazardous waste sites. The main effects from short-term exposure to high levels or doses of aldrin and dieldrin are headache, dizziness, irritability, loss of appetite, nausea, muscle twitching, convulsions, and loss of consciousness; death may occur at extremely high exposures or doses. All symptoms disappear with time after removal from a nonlethal exposure. The use of protective clothing and respirators is necessary under conditions when high exposures may occur.

Long-term occupational exposure to fairly low levels of aldrin and dieldrin has not been documented to show any demonstrable adverse effects. Studies with animals fed dieldrin have shown that the liver can be damaged, and the immune system can be suppressed. Oral doses of aldrin and dieldrin have caused liver cancer in mice but not in rats. Data are inadequate to judge the carcinogenicity of aldrin and dieldrin in humans, but the USEPA considers them probable carcinogens based on sufficient evidence in animals.

Endrin is highly toxic to animals by both ingestion and dermal absorption. The acute oral $LD_{50}$ for rats ranges from 7.5 to 17.5 mg/kg, and acute dermal $LD_{50}$ for female rats is 15 mg/kg. Endrin exposure can cause various harmful effects, including death and severe CNS injury. Swallowing large amounts of endrin may cause convulsions and death within a few minutes or hours. Workers exposed to high doses of endrin showed symptoms of headache, dizziness, nervousness, confusion, nausea, vomiting, and convulsions. Endrin toxicity may be the result of an oxidative stress associated with increased lipid peroxidation, decreased glutathione content, and inhibition of glutathione peroxidase activity. Animal species exposed to endrin showed degeneration and necrotic changes, with inflammatory cell infiltration in the liver and kidneys.[33,34]

## DDT

IUPAC name: 1,1,1′-trichloro-2,2-bis(4-chlorophenyl) ethane
Molecular formula: $C_{14}H_9Cl_5$
Toxicity class: USEPA: II; WHO: II

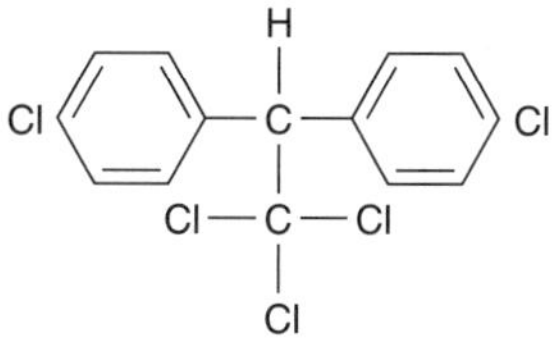

**Figure 4-6.** Structural formula of DDT.

***Uses*** The technical *p,p*′-DDT is a waxy solid, but in its pure form, it appears as colorless crystals. It is a mixture of three isomers: *p,p*′-DDT isomer (approximately 85%), *o,p*′-DDT, and *o,o*′-DDT (in smaller levels). DDT is soluble in solvents such as cyclohexanone, dioxane, benzene, xylene, trichloroethylene, dichloromethane, acetone, chloroform, and diethyl ether. It also is soluble in ethanol and methanol. The USEPA has grouped DDT under restricted use pesticide (RUP). Two similar chemicals that sometimes contaminate DDT products are DDE (1,1-dichloro-2,2-bis(chlorophenyl) ethylene) and DDD (1,1-dichloro-2,2-bis(p-chlorophenyl) ethane). DDD also was used to kill pests, but its use has been banned. One form of DDD has been used medically to treat cancer of the adrenal gland, but DDE has no commercial use.[14,35]

DDT is available and used in different formulations (e.g., aerosols, dustable powders, emulsifiable concentrates, granules, wettable powders). It is used mainly to control mosquito-borne malaria. Its use on crops has decreased because of its persistent residues. DDT was extensively used during World War II among Allied troops and certain civilian populations to control insect typhus and malaria vectors, and was extensively used as an agricultural insecticide after 1945. DDT was banned for use in Sweden in 1970 and in the United States in 1972. In view of its large-scale use over the decades, many insect pests may have developed resistance to DDT. It is no longer registered for use in the United States barring public health emergency (e.g., outbreak of malaria).

The latest group meeting by 110 countries on DDT met in Geneva (September 17, 1999) to phase out production and impose a total ban on its use, even for public health purposes. However, the conference delegates could not arrive at the conclusion for a global ban on DDT. Absence of a suitable substitute of DDT for the control of malaria and the absence of an antimalaria vaccine emphasized the continuation of DDT use for malaria control.[36]

***Toxicity*** The acute oral toxicity $LD_{50}$ of DDT showed differences between species and ranged from moderate to slight toxicity in mammals. For instance, 113 to 800 mg/kg for rats, 150 to 300 mg/kg for mice, 300 mg/kg for guinea pigs, 400 mg/kg for rabbits, 500 to 750 mg/kg for dogs, and greater than 1,000 mg/kg for sheep and goats.

DDT is slightly nontoxic to test animals via the dermal route, with reported dermal $LD_{50}$ of 2,500 to 3,000 mg/kg in female rats, 1000 mg/kg in guinea pigs, and 300 mg/kg in rabbits. It is not readily absorbed through the skin unless it is in solution form. It is believed that inhalation exposure to DDT will not result in significant absorption through the lung alveoli (gas exchanging sacs) but due to its probable trapping in mucous secretions and later by swallowing. DDT has caused chronic effects on the CNS, liver, kidneys, and immune systems in experimental animals. In rats and mice, doses of 16 to 32 mg/kg and 6.5 to 13 mg/kg for approximately 26 weeks and 80 to 140 weeks, respectively, caused tremors. Similarly, DDT caused changes in cellular chemistry in the CNS of monkeys at doses of 10 mg/kg/day over 100 days. Higher doses of DDT (50 mg/kg/day) caused loss of equilibrium in monkeys. It has been reported that prolonged exposure of animal species to DDT caused adverse effects; rats, mice, hamsters, dogs, and monkeys showed effects on the liver, kidney, CNS, and adrenal glands.[37,38]

It is now well known that most of the OC insecticides have neurotoxic properties and cause adverse effects on the CNS. Poisoned animals and humans show symptoms like tremor, eye jerking, changes in gait, convulsions, paralysis, loss of memory, and death.

Humans exposed to DDT have shown many adverse effects (e.g., nausea, diarrhea, increased liver enzyme activity, eye irritation, nose and throat irritation, disturbed gait, malaise, excitability); at higher doses, tremors and convulsions are evident. Despite extensive use of DDT, evidence seems inadequate to establish the reproductive toxicity effect on the kidney, liver, and immune systems in humans. The blood cell culture studies of humans occupationally exposed to DDT showed an increase in chromosomal damage. In a separate study,[38a] significant increases in chromosomal damage were reported in workers who had direct and indirect occupational exposure to DDT. Thus, it appears that DDT may have the potential to cause genotoxic effects in humans, but it does not appear to be strongly mutagenic. Further studies are required to confirm whether these effects may occur at exposure levels likely to be encountered by most people. More information may be found in the literature.[5,14,39,40]

### **Dicofol** (Kelthane)

IUPAC name: 2,2,2-trichloro-1,1-bis(4-chlorophenyl)ethanol
Molecular formula: $C_{14}H_9Cl_{15}O$
Toxicity class: USEPA: II or III (depending on the formulation); WHO: III

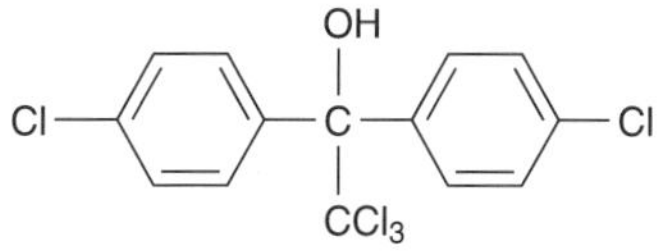

**Figure 4-7.** Structural formula of dicofol.

***Uses*** Pure dicofol is a miticide and a white crystalline solid. Technical dicofol is a red-brown or amber viscous liquid with an odor like fresh-cut hay. Dicofol is manufactured from DDT. In 1986, the use of dicofol was temporarily banned by the USEPA because of concerns raised by high levels of DDT contamination. However, it was reinstated when it was shown that modern manufacturing processes can produce technical-grade dicofol that contains less DDT (0.1%).[14] It is used for the control of infestation on a wide variety of fruit, vegetable, ornamental, and field crops.[41]

***Toxicity*** The acute oral $LD_{50}$ in rats is 575 to 960 mg/kg; in rabbits and guinea pigs it is 1810 mg/kg; and in mice it is 420 to 675 mg/kg. The dermal $LD_{50}$ in rats is 1,000 to 5,000 mg/kg, and in rabbits it is between 2,000 and 5,000 mg/kg. The inhalation $LC_{50}$ (4-hour) in rats is greater than 5 mg/L. Exposure of animal species (e.g., mice, rats, dogs) to dicofol caused liver enlargement/damage, damage to the kidney and adrenals, and eventually death.[42–45]

Dicofol is absorbed through ingestion, inhalation, or skin contact. Symptoms of exposure include nausea, dizziness, weakness, vomiting from ingestion or respiratory exposure, skin irritation or rash from dermal exposure, and conjunctivitis from eye contact. Poisoning may affect the liver, kidneys, or CNS. Overexposure by any route may cause nervousness and hyperactivity, headache, nausea, vomiting, unusual sensations, and fatigue. Very severe cases may result in convulsions, coma, or death from respiratory failure. Dicofol is a moderate skin and eye irritant. Since dicofol is stored in fatty tissues, intense activity or starvation may mobilize the pesticide, resulting in the reappearance of toxic symptoms long after actual exposure.

### **Endosulfan** (Thiodan)

IUPAC name: 1,4,5,6,7,7-hexachloro-8,9,10-trinorborn-5-en-2,3-ylenebismethylene
Molecular formula: $C_9H_6Cl_6O_3S$
Toxicity class: USEPA: I; WHO: II

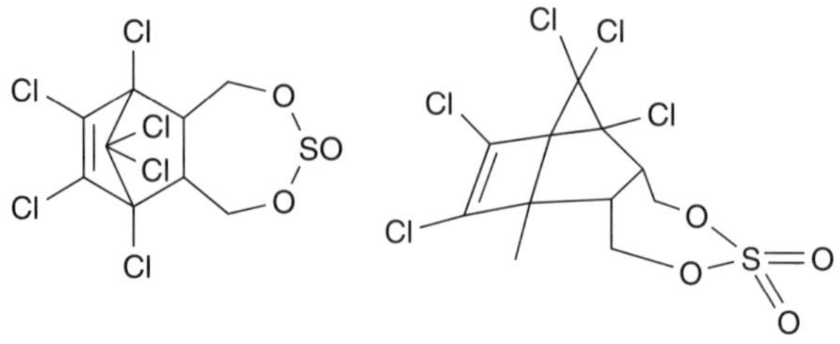

Endosulfan cyclic sulfate
(Endosulfan sulfate)

**Figure 4-8.** Structural formula of endosulfan.

***Uses*** Pure endosulfan is a colorless crystal. The USEPA has grouped it under RUP. The technical-grade endosulfan is made up of a mixture of two molecular forms (isomers) of endosulfan, the alpha- and beta-isomers and appears brown to yellow in color. The formulations of endosulfan include emulsifiable concentrate, wettable powder, ultra-low volume (ULV) liquid, and smoke tablets.[14]

Endosulfan is used as an insecticide and as a contact poison. It is used primarily on food crops, cereals, tea, coffee, fruits, and vegetables. It is also used as a wood preservative. It is compatible with many other pesticides and may be found in formulations with dimethoate, malathion, methomyl, monocrotofos, pirimicarb, triazophos, fenoprop, parathion, petroleum oils, and oxine copper. It is not compatible with alkaline materials.[14]

***Toxicity*** The acute oral $LD_{50}$ for rats ranges from 18 to 160 mg/kg, for mice 7.36 mg/kg, and for dogs 77 mg/kg. The acute dermal $LD_{50}$ values in rats ranges from 78 to 359 mg/kg. Endosulfan is considered slightly toxic via inhalation with an $LC_{50}$ 21 mg/L for 1 hour and 8 mg/L for 4 hours. Stimulation of the CNS is the major characteristic of endosulfan poisoning. Endosulfan earned a bad reputation for an accidental fish kill in the Rhine River in 1969. As a consequence of extreme toxicity and persistence in the environment, organochlorines are gradually being phased out. Prolonged exposure (2 years) to endosulfan caused reduced growth and survival, changes in the kidney structure, and changes in the blood chemistry of rats.

Humans exposed to endosulfan showed incoordination, imbalance, difficulty breathing, gagging, vomiting, diarrhea, agitation, convulsions, and loss of consciousness. Reversible blindness has been documented for cows that grazed in a field sprayed with the compound. The animals, however, completely recovered after 1 month following exposure. In an accidental exposure, sheep and pigs grazing on a sprayed field suffered a lack of muscle coordination and blindness.[46,47]

### **Hexachlorocyclohexane** (HCH)

IUPAC name: 1,2,3,4,5,6-hexachloro-cyclohexane-mixed isomer
Molecular formula: $C_6H_6Cl_6$

Toxicity class: USEPA: II; WHO: II

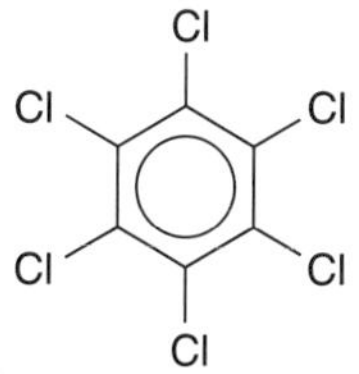

**Figure 4-9.** Structural formula of hexachlorocyclohexane.

***Uses*** Hexachlorocyclohexane (HCH) exists in eight chemical forms (called isomers). One of these forms, gamma-HCH (also known as lindane), is a white solid substance that may evaporate into the air as a colorless vapor with a slightly musty odor. Lindane was used as an insecticide on fruit and vegetable crops (including greenhouse vegetables and tobacco) and forest crops (including Christmas trees).[14]

HCH is still used in ointments, lotions, and shampoos to treat head and body lice and scabies. Lindane has not been produced in the United States since 1977. It is still imported and formulated in the United States, although its use is restricted by the USEPA and it can be applied only by a certified applicator. Technical-grade HCH is a mixture of several isomers of HCH. It also was used as an insecticide in the United States, but production stopped in 1983.[12,14,48]

***Toxicity*** The toxicity of technical HCH to animals shows differences depending on the percent content of the gamma isomer, lindane. The acute oral and dermal $LD_{50}$ values for rats have been reported as 1,752.8 mg/kg and 8,000 mg/kg, respectively. In comparison to rats, rabbits are more sensitive to technical HCH. The acute oral and dermal $LD_{50}$ vaues are 1,362.5 mg/kg and 1,786.3 mg/kg, respectively. Breathing HCH-contaminated workplace air causes blood disorders, dizziness, headaches, and changes in sex hormone levels. Animals fed high levels of HCH had convulsions and some became comatose. Liver and kidney effects occurred at moderate levels. A reduced ability to fight infection was reported in animals given moderate levels of HCH. Animal studies indicated a decreased ability to reproduce when fed with moderate to high levels of HCH. No significant chromosomal changes were seen in the bone marrow cells of male rats given low oral doses of HCH.[49–51]

### Lindane (gamma-HCH)

IUPAC name: $1\alpha,2\alpha,3\beta,4\alpha,5\alpha,6\beta$-hexachlorocyclohexane
Molecular formula: $C_6H_6Cl_6$
Toxicity class: USEPA: II; WHO: II

***Uses*** Lindane has been used as a fumigant on a range of soil-dwelling and plant-eating insects. It is commonly used on a variety of crops, in warehouses, in public health to control insect-borne diseases, and (with fungicides) as a seed treatment. Lindane is also presently used in lotions, creams, and shampoos for the control of lice and mites (scabies) in humans. It is available as a suspension, emulsifiable concentrate, fumigant, seed treatment, wettable and dustable powder, and ultra-low volume (ULV) liquid.

Because of the high toxicity of lindane, formulations are classified by the USEPA under RUP. Therefore, it is recommended that any purchase and use of lindane should be by trained, qualified, and certified pesticide applicators. Lindane is no longer manufactured in the United States, and most agricultural and dairy uses have been cancelled by the USEPA because of concerns about the compound's carcinogenic potential.[12,14]

***Toxicity*** The acute oral $LD_{50}$ in animal species differs greatly (e.g., 88 to 190 mg/kg in rats, 59 to 562 mg/kg in mice, 100 to 127 mg/kg in guinea pigs, and 200 mg/kg in rabbits).[14] Lindane is moderately toxic via the dermal route, and again shows wide difference in animal species (500 to 1,000 mg/kg in rats, 300 mg/kg in mice, 400 mg/kg in guinea pigs, and 300 mg/kg in rabbits). Younger animals are found to be more susceptible to lindane toxicity.

Acute exposure to high doses of lindane is known to cause CNS stimulation (usually developing within 1 hour), mental/motor impairment, excitation, clonic (intermittent) and tonic (continuous) convulsions, increased respiratory rate and/or failure, pulmonary edema, and dermatitis. Toxic symptoms in humans are more behavioral in nature (e.g., loss of balance, teeth grinding, and hyperirritability. Most acute effects in humans have been the result of accidental or intentional ingestion, although inhalation toxicity occurred (especially among children) when lindane was used in vaporizers.

Very high and repeated doses of lindane caused kidney, pancreas, testes, and nasal mucous membrane damage in animals. There have been reported links of lindane to immune system damage; however, these results have not been amply demonstrated in test animals or in humans in a long-term study.[48] Evaluation of several studies has indicated that the evidences on carcinogenicity and mutagenicity are contradictory and inadequate.[14,48,50]

### **Methoxychlor** (Marlate)

IUPAC name: 1,1,1-trichloro-2,2-bis(4-methoxyphenyl) ethane
Molecular formula: $C_{16}H_{15}Cl_3O_2$
Toxicity class: USEPA: IV; WHO: III (Table 4-5)

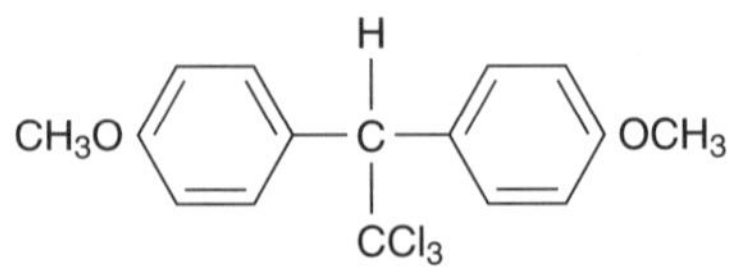

**Figure 4-10.** Structural formula of methoxychlor.

The pure form of methoxychlor is a gray to pale yellow powder with a fruity or musty odor, or a crystalline solid. Technical methoxychlor (88% to 90% pure) is a gray powder. Methoxychlor is a practically nontoxic compound, and the USEPA has grouped it under general use pesticide (GUP).[14]

***Uses*** Methoxychlor is used as an insecticide against a variety of pests (e.g., flies, mosquitoes, cockroaches, chiggers, insects of household and ornamental plants). It is used on agricultural crops, livestock, animal feed, grain storage, home gardens, and pets. It is registered for use on fruits, vegetables, forage crops, and in forestry. Methoxychlor also is registered for veterinary use to kill parasites on dairy and beef cattle. Methoxychlor is one of a few OC pesticides that have seen an increase in use since the 1972 ban on DDT.

Methoxychlor is available in wettable and dustable powders, emulsifiable concentrates, granules, and an aerosol. It may be found in formulations with malathion, parathion, piperonyl butoxide, and pyrethrins.[14,48]

***Toxicity*** Methoxychlor is practically nontoxic via the oral route, with different acute oral $LD_{50}$ values of 5,000 to 6,000 mg/kg in rats, 1,850 mg/kg in mice, and 2,000 mg/kg in hamsters. The lowest oral dose that can cause lethal effects in humans is estimated to be approximately 6,400 mg/kg, and the lowest dose through the skin that produces toxic effects in humans is 2,400 mg/kg based on behavioral symptoms. It reportedly is slightly to practically nontoxic dermally, with an acute dermal $LD_{50}$ in rabbits of greater than 2,000 mg/kg.

High doses of methoxychlor induce adverse changes in animals and humans (e.g., depression, progressive weakness, diarrhea, and in severe cases, death within 36 to 48 hours of exposure). Reports have shown that rats given methoxychlor (500 mg/kg/day) for 2 years showed no significant weight gain (which has been attributed to low food intake rather than chemical toxicity).[52] However, much higher doses (1,500 mg/kg/day) caused weight loss and mortality in experimental rats. More information on methoxychlor may be found in the literature.[52,53]

### **Mirex and Chlordecone** (Kepone)

IUPAC name: Dodecachloropentacyclodecane
Molecular formula: $C_{10}Cl_{12}$

***Uses*** Mirex is used in North America and in the southeastern United States for the control of fire and ants. Mirex is also effective against the hamster ant, the yellow-jacket, the Texas leaf-cutting ant, and the Hawaiian mealy bug. A ban on the use of mirex for pest control with exemptions was brought in on June 30, 1978. It has its use as a fire retardant in plastics, rubber, paint, paper, and electrical goods from 1959 to 1972 due to its high melting point and high chemical stability.[14,48]

Chlordecone was used as an insecticide on tobacco, ornamental shrubs, bananas, and citrus trees, and in ant and roach traps. Mirex was sold as a flame

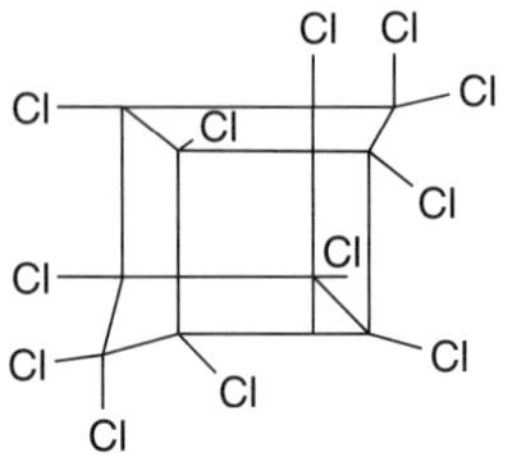

Figure 4-11. Structural formula of mirex (Kepone).

retardant under the trade name Dechlorane, and chlordecone was also known as Kepone.

The greatest concentration of mirex in the environment was found in Lake Ontario, Canada, where several factories dumped effluent containing mirex. Almost all of the mirex in North America was produced by the Hooker Chemical and Plastics Company of Niagara Falls, New York. A Canadian study found that although effluent runoff was isolated to Lake Ontario, mirex was found in sediment and fish samples from each of the Great Lakes.[48a,48b] This raises the question of how mirex is being transported upstream from Lake Ontario.[54–56]

***Toxicity*** The acute oral $LD_{50}$ of mirex to rats has been reported as 600 mg/kg, and that of chlordane was 132 mg/kg and the acute dermal $LD_{50}$ for a rabbit was 410 mg/kg. Both mirex and chlordecone are known to cause severe hepatotoxicity, as demonstrated by liver enlargement and hepatobiliary dysfunction. Studies in rats have shown that mirex does not undergo metabolic changes.[57] Conversion of mirex into chlordecone occurs in soil medium. Chlordecone failed to demonstrate biotransformation in animal species like rats, guinea pigs, and hamsters.[54–58]

Mirex has been found to be both carcinogenic and teratogenic to test animals. Effects of mirex in short-term studies are characterized by a decrease in body weight, hepatomegaly, induction of mixed-function oxidases, morphological changes in liver cells, and sometimes death.[55,57] Mirex is also known to decrease the number of offspring, decrease offspring survival, and increase fetal mortality rates. More information is available in the literature.[55,57]

### Pentachlorophenol (PCP)

IUPAC name: Pentachlorophenol
Molecular formula: $C_6HCl_5O$
Toxicity class: USEPA: II; WHO: Ib

***Uses*** Pentachlorophenol is covered by the USEPA under RUP for preservation of wood material and as a GUP for other purposes.[14,48] PCP was one of the most heavily used pesticides in the United States. Now, only certified applicators can purchase and use PCP. It is still used in industry as a wood preservative for

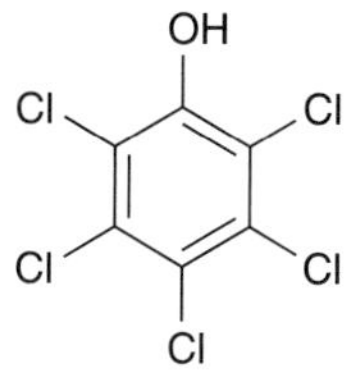

Figure 4-12. Structural formula of pentachlorophenol.

power line poles, railroad ties, cross arms, and fence posts. It is no longer found in wood-preserving solutions or in formulations of insecticides and herbicides that are for home and garden use. It generally sticks to soil particles.[14,15,48]

***Toxicity*** Exposure to PCP may occur in different ways: breathing contaminated air while working at wood treatment facilities and lumber mills; by touching treated lumber in wood-treatment facilities and lumber mills, in construction, or farming; by breathing contaminated air from log homes made from pentachlorophenol-treated logs; by breathing contaminated air near waste sites, sites of accidental spills, and work sites; by touching contaminated soil at waste sites, and landfills; by drinking contaminated water near waste sites, areas of accidental spills, and work sites; by eating contaminated food, such as fish, or drinking contaminated water, but these exposures are low and are not very common.[59]

Studies have shown that PCP causes adverse effects to the liver, kidneys, blood, lungs, CNS, immune system, and gastrointestinal tract both after a short- or long-term exposure.[60,61] Direct contact with PCP causes irritation to the skin, eyes, and mouth, particularly when it is a hot vapor. In view of its importance during earlier years, extensive studies of Braun and associates[61–63] demonstrated the metabolism and pharmacokinetic disposition of PCP in rats, nonhuman primates, and humans. All these studies indicated the adverse effects of PCP to animals and humans.[59–67]

### **Toxaphene** (Camphechlor)

IUPAC name: A reaction mixture of chlorinated camphenes containing 67–69% chlorine
Molecular formula: $C_{10}H_{10}Cl_6$

***Uses*** Toxaphene is a yellow to amber, waxy solid that smells like turpentine. It does not burn and readily changes to vapor when in solution. Toxaphene in soil vaporizes to air or stick or soil particles. Toxaphene is a mixture containing more than 670 chemicals.[12,14]

Toxaphene was one of the most heavily used insecticides in the United States until most uses were banned in 1982 by the USEPA. Toxaphene was used primarily in the southern United States to control insect pests on cotton and other agricultural crops. It also was used to control insect pests on livestock and to kill unwanted fish species in lakes. Since 1982, toxaphene is allowed to be used only on livestock and for emergency use (determined by the USEPA on a case-by-case basis). Toxaphene is specifically used for the control of insects on cotton, soybeans, sorghum, peanuts, banana, pineapple; to control insect pests on livestock; for the control of household insects; and to kill unwanted fish species in lakes. Toxaphene is banned in the United States because of its high toxicity.[68]

Toxaphene is a very stable, persistent, solid compound that requires decades to break down into different components. In fact, nearly 45% of the toxaphene applied to the soil in 1951 was detected 20 years later. Toxaphene has very low solubility in water, but in organic solvents, it is miscible. Toxaphene is very volatile and quickly changes to vapor, which transmigrates in the atmosphere to very long distances. Reports indicate that toxaphene bioaccumulates, rapidly breaks down in the intact body of animals, and eventually discharges out.[68]

***Toxicity*** The acute oral $LD_{50}$ and dermal $LD_{50}$ toxicity of toxaphene in rats are 40 and 600 mg/kg, respectively. Toxaphene is an active nerve poison and interferes with fluxes of cations across nerve cell membranes, which increases neuronal irritability and results in convulsions and seizures. Toxaphene also has been found to damage the lungs, liver, and kidney of animals and humans. Although the dermal adsorption efficiency of toxaphene is less than that of other organochlorines, its absorption is enhanced by fat and fat solvents. Toxaphene has been shown to cause cancer in pregnant animals and to induce birth defects.[68]

## CONCLUSION

Pesticides are toxic chemicals. Organochlorinated pesticides, because of their high persistence, have been identified in tissues of animals and humans. Insecticides (e.g., DDT and heptachlor) have been found in the milk and body fluids of humans and have caused global concern. Proper care and disposal of organochlorinated pesticides by qualified and trained workers under proper management is the answer to achieve human safety and to contain chemical toxicity.

## REFERENCES

1. A.L. Aspelin, Pesticides Industry Sales and Usage. 1992 and 1993 Market Estimates. Biological and Economic Analysis Division; Office of Pesticide Programs; Office of Prevention, Pesticides, and Toxic Substances, United States Environmental Protection Agency, Washington, DC, 1994.

1a. A.L. Aspelin, and A.H. Grube, *Pesticides industry sales and usage: 1996 and 1997 market estimates, biological and economic analysis division.* Office of Pesticide Programs Office of Prevention, Pesticides and Toxic Substances, U.S. Environmental Protection Agency, Washington, DC, (733-R-99-001), November 1999.

2. American Crop Protection Association (ACPA). Pesticides production: Total gained in 1993, despite fungicides decline. *Chem. Engin. News* **73**(66):44, 1995.

3. L.P. Gianessi, and J.E. Anderson, Pesticide use in U.S. crop production: National data report. National Center for Food and Agricultural Policy, Washington, DC, 1995.
4. Grossman, J., Dangers of household pesticides. *Environ. Health Perspect.* **103**(6):550, 1995.
5. W. Lorenz, and K. Sasse, Gerhard Schrader and the development of organophosphorus compounds for crop protection. *Pflanzenchutz-Nachr.* **21**:5, 1968.
6. P. Muller, *DDT, The Insecticide Dichlorodiphenyl Trichloroethane and Its Significance*, Vol. 2, Birkhauser, Verlag, Basel, Switzerland, 1949.
7. Report, *Exporting Risk: Pesticide Exports from U.S. Ports, 1992–1994, FASE Research Report, Spring 1996*, Foundation for Advancements in Science and Education (FASE), 1996.
7a. L. Wilhoit, D. Supkoff, J. Steggall, A. Braun, C. Goodman, B. Hobza, B. Todd, and M. Lee, *An Analysis of Pesticide Use in California, 1991–1995*, Department of Pesticide Regulation, Environmental Monitoring and Pest Management Branch Pest Management Analysis and Planning Program, Environmental Protection Agency, Sacramento, California, December 1998.
8. Report of Department of Pesticide Regulation, USEPA, Sacramento, California, September, 1999.
9. PANUPS, Pesticide Action Network North America, July 16, 1999.
10. E. Dewailly, G. Mulvad, H.S. Pedersen, P. Ayotte, A. Demers, J.P. Weber, and J.C. Hansen, Concentration of organochlorines in human brain, liver, and adipose tissue autopsy samples from Greenland. *Environ. Health Perspect.* **107**:823, 1999.
11. P.J. Landrigan, L. Claudio, S.B. Markowitz, G.S. Berkowitz, B. Brenner, et al., Pesticides and inter-city children: Exposure, risks, and prevention. *Environ. Health Perspect.* **107**(suppl. 3):431, 1999.
12. B. Eskenazi, A. Bradman, and R. Castorina, Exposure of children to organophosphate pesticides and their potential adverse health effects. *Environ. Health Perspect.* **107**(suppl. 3):409, 1999.
13. W.W. Au, C.H. Sierra-Torres, N. Cajas-Salazar, B.K. Shipp, M.S. Legator, Cytogenetic effects from exposure to mixed pesticides and the influence from genetic susceptibility. *Environ. Health Perspect.* **107**(6):501, 1999.
14. H. Kidd, and D.R. James, eds., *The Agrochemicals Handbook*, 3rd ed., Royal Society of Chemistry Information Services, Cambridge, UK, 1991.
15. S. Budavari, M.J. O'Neil, A. Smith, and P.E. Heckelman, and J.F. Kinneary, eds., *The Merck Index: An Encyclopedia of Chemicals, Drugs, and Biologicals*, 12th ed., Merck and Co., Inc., Whitehouse Station, NJ, 321, 1996.
16. U.S. Agency for Toxic Substances and Disease Registry, *Toxicological Profile for Chlordane (ATSDR/TP-89/06)*, Atlanta, GA, 6–21, 1995, (Update) PB/95/100111/AS.
17. U.S. Environmental Protection Agency, *Health Advisory Summary: Chlordane*, Office of Drinking Water, Washington, DC, 6–22, 1989.
18. F.D. Aldrich, and J.H. Holmes, Acute chlordane intoxication in a child: Case report with toxicological data. *Arch. Environ. Health* **19**:29, 1969.
18a. U.S. National Library of Medicine, *Hazardous Substances DataBank*, Bethesda, MD, 1995.

19. International Agency for Research on Cancer, *Monograph on the Evaluation of Carcinogenic Risk of Chemicals in Man*, Vol. 5, *Some Organochlorine Pesticides*, International Agency for Research on Cancer, Lyon, France, 1974.
20. World Health Organization, Chlordane, *Environ. Health Crit.*, Vol. 34, WHO, Geneva, Switzerland, 1982, 984.
21. S. Tashiro, and F. Matsumara, Metabolism of *trans*-nonachlor and related chlordane components in rat and man. *Arch. Environ. Contam. Toxicol.* **7**:113, 1978.
22. Agency for Toxic Substances and Disease Registry, *Toxicological Profile for Chlordane (Update)*. Prepared by Clement International Corporation, under Contract No. 205-88-0608 for ATSDR, Public Health Service, U.S. Department of Health and Human Services, Atlanta, GA, 1994.
23. J.B. Barnett, B.L. Blaylock, J. Gandy, J.H. Menna, R. Denton, and L.S.F. Soderberg, Alteration of fetal liver colony formation by prenatal chlordane exposure. *Fundam. Appl. Toxicol.* **15**:820, 1990.
24. Agency for Toxic Substances and Disease Registry, *Toxicological Profile for Heptachlor/Heptachlor Epoxide*. Prepared by Clement International Corporation, under Contract No. 205-88-0608, U.S. Public Health Service, TP-92/11, 1993, (Update) PB/93/182467/AS.
25. A.P. Leber, and T.J. Benya, Chlorinated hydrocarbon pesticides, in *Patty's Industrial Hygiene and Toxicology*, 4th ed., Volume II, Part E. G.D. Clayton and F.E. Clayton, eds., New York, John Wiley & Sons, p. 1540, 1994.
26. International Agency for Research on Cancer, Heptachlor and heptachlor epoxide. In *IARC Monographs on the Evaluation of the Carcinogenic Risk of Chemical to Humans. Some Halogenated Hydrocarbons*, Vol. 20. World Health Organization, Lyon, France, 129, 1979.
27. M.T. Akay, and U. Alp, The effects of BHC and heptachlor on mice. *Hacette Bull. Nat. Sci. Eng.* **10**:11, 1981.
28. J. Formanek, M. Vanickova, J. Plevova, et al., The effect of some industrial toxic agents on EEG frequency spectra in rats. *Adv. Effects Environ. Chem. Psychotropic Drugs* **2**:257, 1976.
29. R.J. Aulerich, G.J. Bursian, and A.C. Napolitano, Subacute toxicity of dietary heptachlor to mink (*Mustela vison*). *Arch. Environ. Contam. Toxicol.* **19**:913, 1990.
30. D.W. Arnold, G.L. Kennedy Jr., M.L. Keplinger, and J.C. Calandra, Dominant lethal studies with technical chlordane, HCS-3260, and heptachlor: Heptachlor epoxide. *J. Toxicol. Environ. Health* **2**:547, 1977.
31. Environmental Protection Agency, *Carcinogenicity Assessment of Chlordane and Heptachlor Epoxide*. EPA/600-/6-87/004. Carcinogen Assessment Group, Office of Health and Environmental Assessment, Washington, DC, 1986.
32. W.J. Hayes Jr., Heptachlor, in *Pesticides Studied in Man*, Williams & Wilkins, Baltimore, MD, 1982.
33. M.Q. Hassan, I.T. Numan, N. al-Nasiri, and S.J. Stohs, Endrin-induced histopathological changes and lipid peroxidation in livers and kidneys of rats, mice, guinea pigs and hamsters. *Toxicol. Pathol.* **19**(2):108, 1991.
34. Agency for Toxic Substances and Disease Registry, *Profile for Endrin*. U.S. Department of Health and Human Services, Public Health Service in Atlanta, GA, 1997, (Update) PB/97/121040/AS.

35. Agency for Toxic Substances and Diseases Registry, U.S. Public Health Service, *Toxicological Profile for 4,4′-DDT, 4,4′-DDE, 4,4′-DDD* (Update), U.S. Public Health Service, Atlanta, GA, 1995, (Update) PB/95/100137/AS.
36. Ravi Kanth, *Anti-malaria vaccine must be developed first, India, China, Mexico oppose ban on toxic DDT, DH News Service, Geneva*, September 17, 1999, (Deccan Herald News Service, Bangalore, India).
37. World Health Organization, DDT and its derivatives, *Environ. Health Crite.*, 9, WHO, Geneva, Switzerland, 1979.
38. Sax, N. Irving., *Dangerous Properties of Industrial Materials*, 6th ed., Van Nostrand Reinhold, New York, 1984.
38a. Agency for Toxic Substances and Diseases Registry (ATSDR), U.S. Public Health Service, *Toxicological Profile for 4,4′-DDT, 4,4′-DDE, 4,4′-DDD* (Update), ATSDR, Atlanta, GA, 1994.
39. M. Wasserman, M. Ron, B. Bercovici, D. Wasserman, S. Cucos, and A. Pines, Premature delivery and organochlorine compounds: Polychlorinated biphenyls and some organochlorine insecticides. *Environ. Res.* **28**:106, 1982.
40. D.H. Garabrant, J. Held, B. Langholz, J.M. Peters, and T.M. Mack, DDT and related compounds and risk of pancreatic cancer. *J. Natl. Cancer Inst.* **84**:764–771, 1992.
41. I.R. Edwards, D.G. Ferry, and W.A. Temple, Fungicides and related compounds, in *Handbook of Pesticide Toxicology*, W.J. Hayes Jr., and E.R. Laws Jr., eds., Academic Press, New York, 1991.
42. U.S. Environmental Protection Agency, *Integrated Risk Information System*, Washington, DC, 1995.
43. S.S. Hurt, *Dicofol: Toxicological Evaluation of Dicofol Prepared for the WHO Expert Group on Pesticide Residues* (*Report No. 91R-1017*). Toxicology Department, Rohm and Haas Company, Spring House, PA, 1991.
44. Rohm and Haas Company, *Material Safety Data Sheet for Kelthane Technical B Miticide*, Philadelphia, PA, 1991.
45. A. Tillman, Residues, Environmental Fate and Metabolism Evaluation of Dicofol Prepared for the FAO Expert Group on Pesticide Residues. (*Report No. AMT 92-76*), Rohm and Haas Company, Philadelphia, PA, 1992.
46. U.S. Agency for Toxic Substances and Disease Registry, *Toxicological Profile for Endosulfan*, Draft Report. Atlanta, GA, 2000, (Update) PB/2000/108023.
47. National Cancer Institute, *Bioassay of Endosulfan for Possible Carcinogenicity*, *Technical Report Series No. 62*, National Institutes of Health, Bethesda, MD, 6, 53, 1978.
48. A.G. Smith, Chlorinated hydrocarbon insecticides, in *Handbook of Pesticide Toxicology*, W.J. Hayes Jr., and E.R. Laws Jr., eds. Academic Press Inc., New York, 1991.
48a. K. Michael, Great Lakes water clear, but contamination high, Chicago Tribune, April 2, 2002.
48b. Government of Canada, *Toxic chemicals in the Great Lakes and associated effects*. Environment Canada, Dept. of Fisheries and Oceans, Health and Welfare Ottawa, Canada, 1991.
49. T.S.S. Dikshith, R.B. Raizada, R.P. Singh, S.N. Kumar, K.P. Gupta, R.A. Kaushal, Acute toxicity of hexachlorocyclohexane in mice, rats, rabbits, pigeons, and fresh water fish. *Vet. Hum. Toxicol.* **31**:113, 1989.

50. T.S.S. Dikshith, Cytogenetic effects of pesticides in the bone marrow system of male rats, (unpublished data 1982, Scientific report ITRC, Lucknow, India 1981–1984), in *Toxicology of Pesticides in Animals*, Ed., T.S.S. Dikshith, CRC Press, Boca Raton, FL, 194, 1999 (Update) PB/99/166662.
51. Agency for Toxic Substances and Disease Registry, *Toxicological profile for alpha-, beta-, gamma-, and delta-hexachlorocyclohexane* (update), U.S. Department of Health and Human Services, Public Health Service, Atlanta, GA, 1995, PB/95/100202/AS.
52. Agency for Toxic Substances and Disease Registry, *Toxicological profile for methoxychlor*, U.S. Department of Health and Human Services, Public Health Service, Atlanta, GA, 1994.
53. U.S. Environmental Protection Agency, *Health Advisory Summaries: Methoxychlor, Office of Drinking Water*, Washington, DC, 6–66, 1989.
54. W.W. Johnson, and M.T. Finley, *Handbook of Acute Toxicity of Chemicals to Fish and Aquatic Invertebrates, Resource Publication* 137, U.S. Department of Interior, Fish and Wildlife Service, Washington, DC, 6–56, 1980.
55. Agency for Toxic Substances and Disease Registry, *Toxicological profile for mirex and chlordecone*, U.S. Department of Health and Human Services, Public Health Service, Atlanta, GA, 1995, PB/95/264354.
56. World Health Organization, Mirex, *Environ. Health Crite.*, 44, Geneva, Switzerland, 1984.
57. World Health Organization, *Mirex Health and Safety Guide. Health and safety* guide No. 39, Geneva, Switzerland, 1990.
58. D.B. Sergeant, et al., Mirex in the North American Great Lakes: New detections and their confirmation. *J. Great Lakes Res.* **19**(1):145, 1993.
59. Agency for Toxic Substances and Disease Registry, *Toxicologic Profile for Pentachlorophenol* (Update Draft), U.S. Dept. of Health and Human Services, Public Health Service, Atlanta, GA, 2001 (Update) PB/2001/109106/AS.
60. U.S. Agency for Toxic Substance and Disease Registry, *Toxicological Profile for Pentachlorophenol*, Draft Report. Atlanta, GA, 1992.
61. W.H. Braun, and M.W. Sauerhoff, The pharmacokinetic profile of pentachlorophenol in monkeys. *Toxicol. Appl. Pharmacol.* **38**:525, 1976.
62. W.H. Braun, J.D. Young, G.E. Blau, et al., The pharmacokinetics of Pentachlorophenol in rats. *Toxicol. Appl. Pharmacol.* **41**:395, 1997.
63. W.H. Braun, G.E. Blau, and M.B. Chenoweth, The metabolism/pharmacokinetics of pentachlorophenol in man, and a comparison with the rat and monkey, in W.E. Deichmann, ed., *Toxicology and Occupational Medicine*, New York, Amsterdam, Oxford, Elsevier/North-Holland, 289, 1992.
64. G. Renner, Urinary excretion of pentachlorophenol (PCP) and its metabolite tetrachlorohydroquinone (TCH) in rats. *Toxicol. Environ. Chem.* **25**:29, 1989.
65. G. Renner, C. Hopfer, and J.M. Gokel, Acute toxicities of pentachlorophenol, pentachloroanisole, tetrachlorohydroquinone, tetrachlorocatechol, tetrachlororesorcinol, tetrachlorodimethoxybenzenes, and tetrachlorodibenzenediol diacetates administered to mice. *Toxicol. Environ. Chem.* **11**:37, 1986.
66. G. Renner, and C. Hopfer, Metabolic studies on pentachlorophenol (PCP) in rats. *Xenobiotica* **20**:573, 1990.

67. P.L. Williams, Pentachlorophenol, an assessment of the occupational hazard. *J. Am. Ind. Hyg. Assoc.* **43**:799, 1982.

68. Agency for Toxic Substances and Disease Registry, *Toxicological Profile for Toxaphene produced by the Public Health Service*, U.S. Department of Health and Human Services, Public Health Service in Atlanta, GA, 1997, (Update) PB/97/121057/AS.

## APPENDIX 4-1 ORGANOCHLORINATED PESTICIDES AND CARCINOGENICITY

| Pesticides | IARC | NTP | USEPA | Type of Changes |
|---|---|---|---|---|
| Aldrin | 3 | — | B2 | Mouse liver tumors |
| Chlordane | 3 | — | B2 | Mouse liver tumors |
| Heptachlor | | | | |
| Kepone (Chlordecone) | 2B | e | — | Rat, mouse liver tumors |
| DDT | 2B | e | B2 | Mouse liver, lung tumors, lymphomas; rat liver tumors; no tumors in three hamster studies |
| Dieldrin | 3 | e | B2 | Mouse liver tumors |
| Endrin | 3 | — | — | No evidence of tumor |
| Lindane | — | e | B2/C | Mouse liver tumors |
| Mirex | 2B | e | B2 | Mouse, rat liver tumors and thyroid tumors |
| Toxaphene | 2B | e | B2 | Mouse, rat liver tumors |

B2—Probable human carcinogen (no human evidence).
2B—Possibly carcinogenic to humans.
C—Possible human carcinogen.
3—Not classifiable as to carcinogenicity in humans.
e—Reasonably anticipated to be carcinogenic to humans.

CHAPTER 5

# Organophosphorus Pesticides

## INTRODUCTION

Many organophosphorus pesticides (OPs) are used for the control of crop pests. In modern agriculture and pest management activities, a host of chemicals including OPs are put to use. Modern agriculture requires the distribution of large volumes of pesticides. Although known since 1854, OP compounds were not in extensive use until 1930. As a substitute for the botanical insecticide nicotine, the Germans developed the OP compound tetraethyl pyrophosphate (TEPP).[1,2]

During World War II, studies carried out in Germany unraveled the potency and insecticide qualities of nerve gases such as sarin and soman. The OPs generally are more toxic to vertebrates than other classes of insecticides; they are also chemically unstable and less persistent. In fact, the "no residue" property of OPs gave them a boost for more use in agriculture and pest control programs. OPs differ from one another in many important respects, although chemically they are the derivatives of phosphoric or thiophosphoric acids.

Essentially all OPs are nerve poisons that cause severe toxicity to the infected organism. They differ widely in inherent toxicity and in their ability to penetrate the skin. Some OPs act directly, whereas others require activation by enzymes within the body. Some are destroyed and eliminated more rapidly than others. They also differ with respect to their reaction with the cholinesterase enzyme. An important property of OP compounds is their inhibition of the cholinesterase enzyme by forming covalent chemical bonds through a process called *phosphorylation*. A spontaneous enzymatic regeneration half-life may last from days to months. The nature of the compound, the dose received, and the duration of exposure all affect the regeneration period. Because of the prolonged regeneration half-life (i.e., the time required for half of the cholinesterase to reactivate), intoxication by OPs usually is considered more serious. The OPs cause the inhibition in the activity of both red blood cells (RBC) and plasma cholinesterase. The important role of OPs and other nerve gases has gained significance both in crop protection management and during global wars and human tragedies.[3,6]

*Industrial Guide to Chemical and Drug Safety*, By T.S.S. Dikshith and Prakash V. Diwan
ISBN 0-471-23698-5 

Different OP compounds have structural similarities within classes. The phosphorus compounds have the characteristic "phosphoryl" bond, P=O. Most OP compounds have a phosphoryl bond or a thiophosphoryl bond (P=S). All OP compounds are esters of phosphorus with varying combinations of oxygen, carbon, sulfur, and nitrogen attached. These are classified as (1) phosphates; (2) phosphonates; (3) phosphorothioates; (4) phosphorodithioates; (5) phosphorothiolates; and (6) phosphoramidates. Further, the OP compounds are categorized as (1) aliphatic; (2) phenyl; and (3) heterocyclic derivatives. The aliphatic are carbon chain–like in structure. TEPP, which was used in agriculture for the first time in 1946, is a member of this group. Others include malathion, trichlorfon, monocrotophos, dimethoate, oxydemetonmethyl, dimethoate, dicrotophos, disulfoton, dichlorvos, mevinphos, methamidophos, and acephate.

The chemistry and metabolism of OPs involve many reactions. More important, the degradation of OPs is associated with oxidative and hydrolytic systems. Cholinesterase (ChE) is one of many important enzymes required for proper functioning of nervous systems in humans, other vertebrates, and insects. The OPs and carbamates are the chemicals used to control different crop pests by causing disturbances in ChE activity. Since these are toxic chemicals specifically designed against pests, they are, of course, poisonous. Improper use and careless handling results in human poisoning and death.[6–8] Exposure to workers can occur by inhalation, ingestion, and eye or skin contact during manufacture, mixing, and field application.

Acetylcholinesterase (AChE) deesterifies the neurotransmitter acetylcholine (ACh). AChE belongs to a group of enzymes considered serine esterases and has a mechanism similar to that of chymotrypsin. AChE has an anionic binding site that attracts the positively charged quaternary ammonium group of ACh. The serine then attacks and cleaves the ester.[9,10]

## NEUROTOXICITY

All organisms, including vertebrates, have electrical switching centers, termed *synapses*. The muscles, glands, and nerve fibers called *neurons* are stimulated or inhibited by the constant firing of signals across these synapses. Stimulating signals are usually carried by a specific chemical, namely, ACh, and the stimulating signals are discontinued by another specific enzyme, AChE. AChE breaks down the ACh. These important chemical reactions occur continuously and very fast. In other words, ACh causes stimulation, and AChE ends the signal. When ChE-affecting chemicals, like OPs, are present in the synapses, the harmonized situation is disturbed. The presence of OPs (otherwise ChE-inhibiting chemicals) prevents the breakdown of ACh and the buildup of ACh, causing a jam in the nervous system. In this manner, when a person or occupational worker is exposed to OPs or other nerve-affecting poisons, the body is unable to break down the ACh.

Typically in the nervous system and in a synapse, the movement of muscle is controlled by a nerve. An electrical signal or nerve impulse is controlled by ACh across the junctions between the nerve and the muscle, stimulating the muscle to move. The electrical impulses can fire away continuously, unless the number of messages being sent through the synapse is limited by ChE action. Repeated and unchecked firing of electrical signals can cause uncontrolled, rapid twitching of muscles, paralyzed breathing, convulsions, and, in extreme cases, death. The AChE present at the neurosynaptic junctions breaks ACh into acetyl and choline fragments. It has been observed that AChE helps increase the precision of nerve firing, enabling nerve cells to fire as rapidly as 1,000 times per second without the overlap of the neural impulses.[11,12]

## CHOLINERGIC RECEPTOR SITES

Organs with cholinergic receptor sites include smooth muscles, skeletal muscles, the central nervous system (CNS), and most exocrine glands. In addition, cranial efferents and ganglionic afferents are cholinergic nerves. Muscarine stimulates some cholinergic sites, and these are known as muscarinic sites. Organs with these sites include the smooth muscles and glands. Nicotine will stimulate other cholinergic sites, known as nicotinic sites, which are found in skeletal muscle and ganglia. The CNS contains both types of receptors. Atropine and similar compounds block the effects of excess ACh more effectively at muscarinic sites than at nicotinic sites.[13,17]

### Signs and Symptoms of Poisoning

OPs cause adverse effects in poisoned animals and humans, which are based on the dose and duration of exposure. Examples of the effects include nausea, coma, convulsions, respiratory failure, and death. Workers in pesticide manufacturing industries and farming industries may contract OP poisoning via skin absorption. The inhibition of AChE results in accumulation of ACh, causing the wide variety of Henry's symptoms.

These compounds differ from one another in many important respects, although chemically all can be considered derivatives of phosphoric or thiophosphoric acids. They differ widely in inherent toxicity and ability to penetrate the skin. Some OPs act directly, whereas others require activation by enzymes within the body. Some are destroyed and eliminated more rapidly than others. Also, they differ with respect to their reaction with the ChE enzyme. As already described, OPs inhibit ChE by forming covalent chemical bonds through a process called *phosphorylation*. Since spontaneous enzymatic regeneration half-life takes a long time, OP poisoning has been considered more serious and often fatal. OPs affect both RBC and plasma ChE activities; in contrast, the carbamate group of pesticides normally affects only the plasma fraction.[17,18] The signs and symptoms of poisoning linked with different body systems is an important aspect in the identification of OP poisoning for taking appropriate therapeutic measures (Fig. 5-1) (Table 5-1).[19]

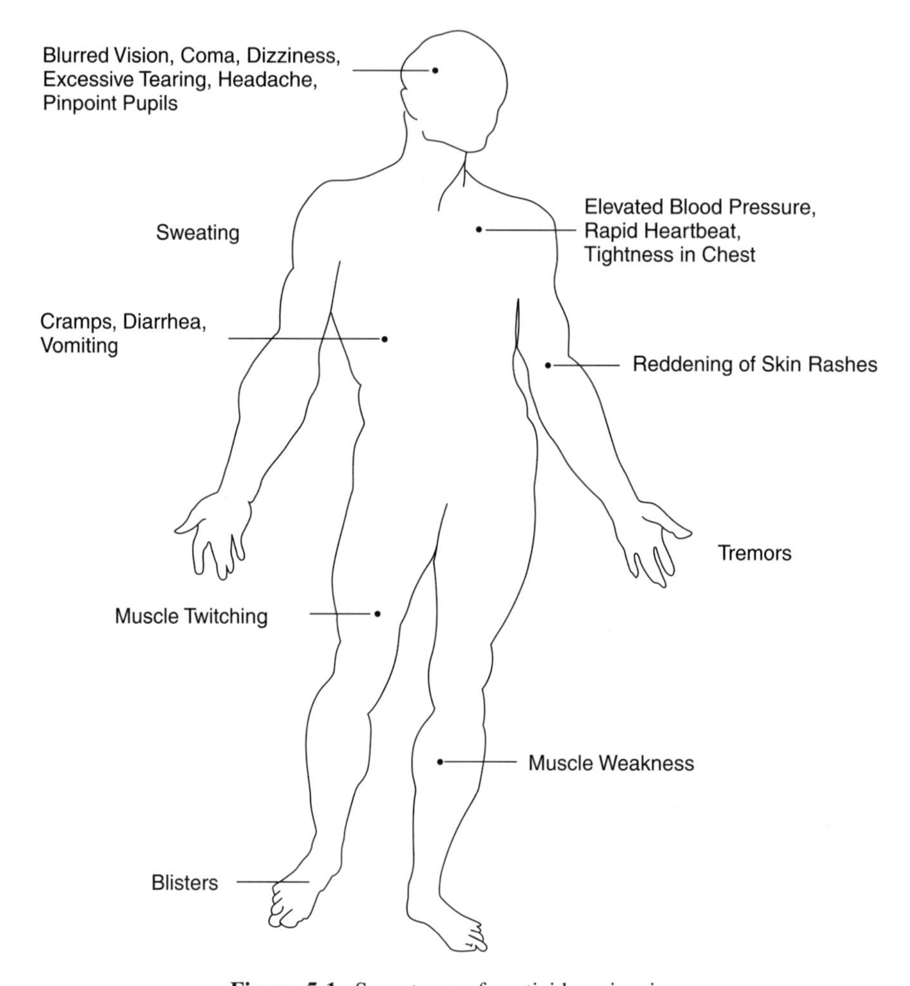


**Figure 5-1.** Symptoms of pesticide poisoning.

Skin contact with OPs may cause localized sweating and involuntary muscle contractions. Eye contact will cause pain, bleeding, tears, pupil constriction, and blurred vision. Following exposure by any route, systemic effects may begin within a few minutes or be delayed for up to 12 hours. These may include pallor, nausea, vomiting, diarrhea, abdominal cramps, headache, dizziness, eye pain, blurred vision, constriction or dilation of the eye pupils, tears, salivation, sweating, and confusion. Severe poisoning will affect the CNS, producing incoordination, slurred speech, loss of reflexes, weakness, fatigue, involuntary muscle contractions, twitching, tremors of the tongue or eyelids, and eventual paralysis of the body extremities and the respiratory muscles. In severe cases, there also may be involuntary defecation, urination, psychosis, irregular heartbeat, unconsciousness, convulsions, and coma. Death may be caused by respiratory failure

**TABLE 5-1 Signs and Symptoms of Organophosphorus Pesticide Poisoning**

| | |
|---|---|
| I. EXOCRINE GLANDS | |
| | Excessive salivation, lacrimation, perspiration |
| II. GASTROINTESTINAL | |
| | Nausea, vomiting, abdominal tightness, swelling, cramps, diarrhea, involuntary defecation |
| III. RESPIRATORY | |
| | Cough, dyspnea, wheezing, edema, bronchospasms, bronchoconstrictions, tightness, excessive bronchial secretions |
| IV. CARDIOVASCULAR | |
| | Altered blood pressure (increase/decrease), tachycardia, pallor |
| V. NERVOUS SYSTEM | |
| | Fatigue, drowsiness, lethargy, confusion, headache, generalized weakness, tremors, convulsions, depression of respiratory centers, coma with absence of reflexes, and cyanosis |

or cardiac arrest.[20–22] Some OPs may cause delayed symptoms, beginning 1 to 4 weeks after an acute exposure that may or may not have produced immediate symptoms. Whenever OPs cause delayed toxicity, numbness, tingling, weakness, and cramping may appear in the lower limbs and progress to incoordination and paralysis. Improvement may occur over months or years, but some residual impairment will remain.[21,22]

The ChE activity of blood components is inhibited by OPs, and any assay of the activity is useful for poison detection. Namba and coworkers[23,24] observed a good correlation between the maximum inhibition of AChE activity and severity of toxic symptoms. Recent research[22] with a genetic approach has been found to be advantageous and offering more protection. However, these studies are inadequate and require more confirmatory data. As mentioned earlier, OPs cause inhibition of AChE because of phosphorylation of the esteric site. The inhibition of this enzyme leads to ACh accumulation. Laboratory animals exposed to OPs have demonstrated morphological changes in the cerebellum, specifically in the Purkinje cells, which lie between the granular layer and the molecular layer of the cerebellum. OPs have caused severe symptoms in movement/gait, AChE inhibition, and damage of Purkinje cells.

### Neurotoxicity

The nature of neurotoxicity induced by different OPs invariably follows the same biochemical reactions. The lethal effects of OPs are the result of an irreversible AChE inhibition and increased ACh levels in the nervous system, causing diverse neurologic effects. The neurotoxic effects of OPs often become so strong that the animal or individual—despite of pre- or post-treatment—often fails to respond completely. This is due to a strong potential in the continuation of toxicity termed *delayed neurotoxicity*.

It should be noted that, while acute toxicity can be treated with atropine, oxime, and diazepam, no treatment is available for delayed neurotoxicity. The phenomenon of delayed neurotoxicity was first reported during the 1950s. It can occur even 5 to 10 years after OP exposure. The individual demonstrates symptoms of impaired concentration and memory, depression, fatigue, and irritability. Recent studies have shown that OPs induce delayed neurotoxicity (OPIDN) due to inhibition of neurotoxic esterase enzyme in the nervous system and axonal degeneration. This produces muscular weakness, ataxia, and paralysis. The syndrome is characterized by a delay period of 4 to 21 days following exposure to nerve gas and occurs before the manifestation of clinical symptoms. The primary molecular target for OPIDN initiation is the inhibition of membrane-bound enzyme neuropathy target esterase (NTE) in the nervous system. A minimum of 70% NTE inhibition after a single exposure to OPs, 45% NTE inhibition after multiple exposures, and subsequent aging of NTE are biochemical prerequisites.[23,24]

Application of *in vitro* test methods have become advantageous in specific cases, such as structurally defined compounds and delayed neuropathy, since target cell data and biochemical processes associated in delayed neuropathy are known. Microscopic studies reveal that cases of OPIDN have degeneration of axons followed by demyelination of the nervous system.[25,26] Epidemiologic studies have indicated mild impairment of the brainstem, spinal cord, and peripheral nerve functions in Gulf War veterans.[27] Such studies are consistent with the spectrum of OPIDN syndrome. The main nerve agents have been shown to inhibit NTE *in vitro* as well as *in vivo*. Sarin has been shown to produce delayed neurotoxicity when administered at higher doses in protected hens.[25–27]

## Synergism, Antagonism, and Additive Effects

In the preparation of commercial pesticide formulations, the biochemical and toxic properties undergo phenomenal modifications. These reactions (and the effects thereof) are called *synergism*, *antagonism*, and *additive effects*. In pest control management, different OPs are mixed to achieve quick knock-down effects, and absence of residues effects, for the better killing of crop pests (Table 5-2).

OPs are chemicals used in agriculture as acaricides, herbicides, and insecticides (Appendix 5-A-1). Because of their toxicity, several of these chemicals are being phased out from use; parathion (ethyl) is an example. Many have been now classified by the United States Environmental Protection Agency (USEPA) as a restricted use pesticide (RUP) or a general use pesticide (GUP). Pesticide chemistry has taken a turn for the synthesis, manufacture, and use of still safer compounds. A list of OPs considered for banning has been identified by the USEPA (Table 5-3). The following pages will briefly discuss the uses and toxicity of different OPs.

**TABLE 5-2 Synergistic Effects of Organophosphorus Insecticides in Rats**

| Nature of Effect | Combination of Insecticides | Percent Mortality |
|---|---|---|
| Synergistic | Azinphosmethyl and Trichlorfon | 100 |
| | Malathion and Trichlorfon | 100 |
| | Malathion and EPN | 100 |
| Antagonistic | Azinphosmethyl and Demeton | 5 |
| | Azinphosmethyl and Malathion | 10 |
| | Azinphosmethyl and Parathion | 10 |
| | Demeton and Trichlorfon | 10 |
| | Malathion and Parathion | 10 |
| | Trichlorfon and EPN | 30 |
| Additive | Azinphosmethyl and EPN | 60 |
| | Demeton and EPN | 45 |
| | Demeton and Malathion | 60 |
| | Parathion and Trichlorfon | 55 |
| | Parathion and Demeton | 50 |
| | Parathion and EPN | 45 |

**TABLE 5-3 Organophosphorus Pesticides Cancelled or Proposed for Cancellation***

| |
|---|
| Cadusafos† |
| Chlorthiophos‡ |
| Dialiflor‡ |
| Dioxathion‡ |
| Fonofos‡ |
| Isazophos |
| Isophenphos‡ |
| methyl‡ |
| Mevinphos† |
| Monocrotophos‡ |
| Phosalone† |
| Phosphamidon‡ |
| Sulprofos‡ |

*Source*: U.S. Environmental Protection Agency, March 1999.
†Import tolerances only; no U.S. registrations for use.
‡These chemicals have been cancelled or proposed for cancellation. They will be included in the organophosphate risk assessment if import tolerances remain after other tolerances are revoked.

### **Azinphos-Methyl** (Gusathion, Guthion)

IUPAC name: S-(3,4-dihydro-4-oxobenzo[d]-(1,2,3)-triazin-3-ylmethyl) O,O-dimethyl phosphorodithioate
Molecular formula: $C_{10}H_{12}N_3O_3PS_2$
Toxicity class: USEPA: I; WHO: Ib

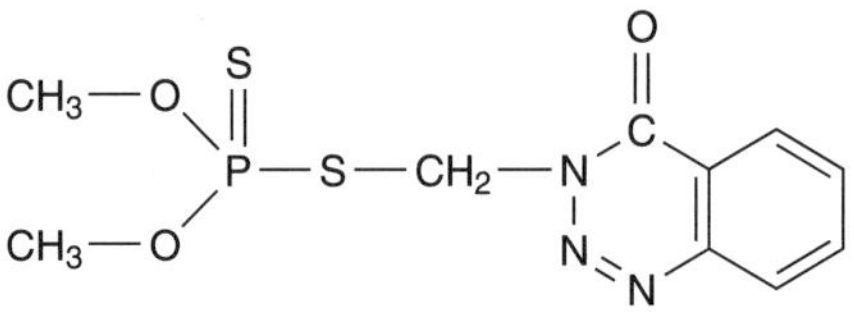

**Figure 5-2.** Structural formula of azinphos-methyl.

***Uses*** Pure azinphos-methyl is a white crystalline solid; technical azinphos-methyl is a brown waxy solid. Based on toxicity, the USEPA has grouped it under RUP. Azinphos-methyl is a highly persistent, broad-spectrum insecticide. It is used for the control of mites and ticks, and it is poisonous to snails and slugs. It also is used in the control of many insect pests on a wide variety of fruit, vegetable, nut, and field crops, as well as on ornamentals, tobacco, and forest and shade trees. Outside the United States, azinphos-methyl is used in lowland rice production. Azinphos-methyl is available in emulsifiable liquid, liquid flowable, ultra-low volume (ULV) liquid, and wettable powder formulations.[28,29]

***Toxicity*** The acute oral $LD_{50}$ of azinphos-methyl for rats is 4.4 to 16 mg/kg, for mice it is 8 to 20 mg/kg, and for guinea pigs it is 80 mg/kg. The acute dermal $LD_{50}$ for rats is 88 to 220 mg/kg, and for mice is 65 mg/kg. The 1-hour $LC_{50}$ for rats is 0.15 mg/L. Azinphos-methyl is highly toxic by inhalation, dermal absorption, ingestion, and eye contact. Workers with a history of reduced lung function, convulsive disorders, or recent exposure to other cholinesterase inhibitors are at increased risk from exposure to azinphos-methyl. Ingestion of azinphos-methyl by humans, even as low as 1.5 mg/day, can cause severe poisoning with symptoms such as dimmed vision, salivation, excessive sweating, stomach pain, vomiting, diarrhea, unconsciousness, and death.

Inhalation of the dust or aerosol preparation of azinphos-methyl may cause wheezing, tightness in the chest, blurred vision, and tearing of the eyes. Eye contact with concentrated solutions of azinphos-methyl can be life-threatening. Exposed animals and humans demonstrate respiratory difficulties, gastrointestinal problems, and CNS disturbances.

Long-term exposure to azinphos-methyl, above the average 8-hour standard set by the Occupational Safety and Health Administration (OSHA), can impair concentration and memory, and cause headache, irritability, nausea, vomiting, muscle cramps, and dizziness. Cholinesterase inhibition from exposure to azinphos-methyl may persist for 2 to 6 weeks. Repeated exposure to small amounts may result in an unexpected inhibition of cholinesterase, causing symptoms that resemble flu-like illnesses, including general discomfort, weakness, and lack of appetite. The effects of azinphos-methyl exposure may be greater in a previously exposed person than in an individual with no previous exposure. Studies of mice and rats also showed adverse effects.[30–33]

### **Carbophenothion** (Garrathion)

IUPAC name: S-(4-chlorophenylthio)methylO,O-diethyl phosphorodithioate
Molecular formula: $C_{11}H_{16}ClO_2PS_3$
Toxicity class: USEPA: I; WHO: Ib

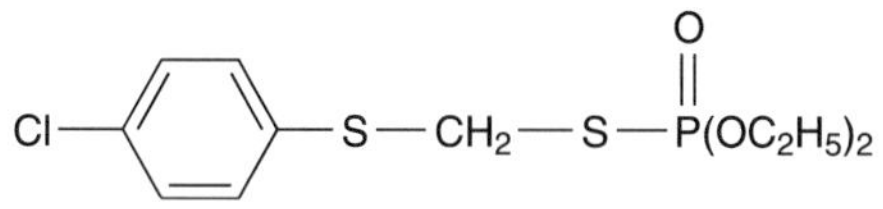

Cl—C6H4—S—CH2—S—P(O)(OC2H5)2

**Figure 5-3.** Structural formula of carbophenothion.

***Uses*** Carbophenothion in its pure form is a yellow-brown liquid with a mild mercaptan-like odor. The USEPA has grouped it under RUP. It is used as both an insecticide and an acaricide to control pests on citrus fruit and to control aphids and spider mites on cotton. When combined with petroleum oil, it controls overwintering mites, aphids, and scale bugs. It also is used to control aphids, mites, suckers, and other pests on fruit, nuts, citrus seeds, vegetables, sorghum, maize, and others. It is used to control parasites on animals.

***Toxicity*** The acute oral $LD_{50}$ of carbophenothion for rats ranges from 10 to 30 mg/kg. The acute dermal $LD_{50}$ for rats ranges from 27 to 54 mg/kg and for rabbits 1,270 to 1,850 mg/kg. Symptoms of poisoning include headache, blurred vision, weakness, nausea, discomfort in the chest, abdominal cramps, vomiting, diarrhea, salivation, sweating, and pinpoint pupils. It is highly toxic when eaten and nearly as toxic when absorbed through the skin. Carbophenothion affects the nervous system by inhibiting ChE activity.[34]

### **Chlorpyrifos** (Dursban)

IUPAC name: 0,0-diethyl 0-(3,5,6-trichloro-2-pyridyl) phosphothorothioate
Molecular formula: $C_9H_{11}Cl_3NO_3PS$
Toxicity class: USEPA: II; WHO: II

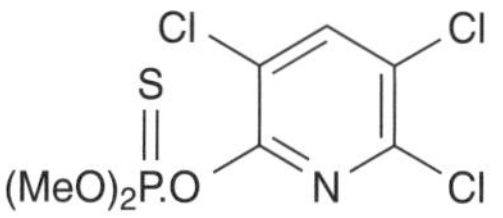

Cl, Cl, Cl, S, (MeO)2P.O, N

**Figure 5-4.** Structural formula of chlorpyrifos.

***Uses*** The synthesis of chlorpyrifos was first described by Rigterink and Kenaga. Chlorpyrifos has been used for the control of agricultural and domestic pests since the 1970s. It is used on the farm to control pests in turfgrass and ornamentals and is used in the home to control cockroaches, fleas, and termites. Formulations of chlorpyrifos include emulsifiable concentrates, dusts, granular wettable powders, microcapsules, pellets, and sprays. Chlorpyrifos is widely used as an active ingredient in many commercial insecticides, such as Dursban and Lorsban, to control household pests, mosquitoes, and pests in animal houses.[28] The estimated use of chlorpyrifos in the state of California is approximately 7,023,190 pounds of active ingredient per year.

***Toxicity*** The acute oral $LD_{50}$ for rats is 135 to 163 mg/kg, and the acute dermal $LD_{50}$ for rabbits is more than 2,000 mg/kg. Exposure to chlorpyrifos causes fatigue, headache, dizziness, loss of memory, respiratory problems, muscular and joint pains, and gastrointestinal disturbances both in experimental animals and in occupational workers. Recent studies have indicated that chlorpyrifos causes several types of autoantibody, automicrosomal, and tissue injury.[35] Chlorpyrifos caused acute inhibition of DNA synthesis in newborn test animals. Although low exposure (0.5 μg/mL) could trigger a variety of cellular changes, repeated low dose caused persistent inhibition of DNA synthesis, leading to deficits in cell numbers and suppression of macromolecular constituents. In view of these findings, chlorpyrifos has been reported to cause disturbances in the brain development process.[36]

### **Coumaphos** (Asuntol)

IUPAC name: O-(3-chloro-4methyl-2-oxo-2H-Chromen-7-ylO,O-diethyl phosphorothioate)
Molecular formula: $C_{14}H_{16}ClO_5PS$
Toxicity class: USEPA: II; WHO: Ia

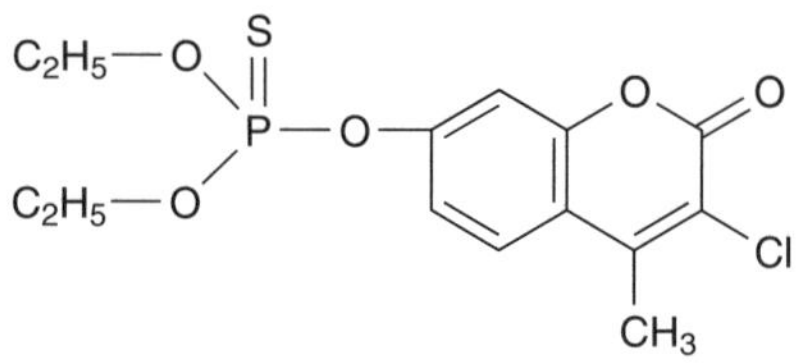

**Figure 5-5.** Structural formula of coumaphos.

***Uses*** Technical coumaphos is a tan crystalline solid with a slight sulfur odor. The USEPA has grouped it under GUP. Coumaphos is extensively used for the control of a wide variety of livestock insects, including cattle grubs, screw worms, lice, scabies, flies, and ticks. It is used against ectoparasites, which are insects that live on the outside of host animals (e.g., sheep, goats, horses, pigs, poultry).

It is added to the feed of cattle and poultry to control the development of fly larvae that breed in manure. It also is used as a dust, dip, or spray to control mange, horn flies, and face flies of cattle.[28,37]

***Toxicity*** The acute oral $LD_{50}$ of coumaphos for rats ranges from 13 to 41 mg/kg, for mice 28 to 55 mg/kg, for guinea pigs 58 mg/kg, and for rabbits 80 mg/kg. The acute dermal $LD_{50}$ is 860 mg/kg for rats, and 500 to 2,400 mg/kg for rabbits. The 1-hour inhalation $LC_{50}$ for rats ranges from 0.34 to 1.1 mg/L. Signs of poisoning include diarrhea, difficulty breathing, and stiffness of the legs and neck. The acute symptoms of inhalation are headache, dizziness, and incoordination. Moderate poisoning is characterized by muscle twitch and vomiting. Severe poisoning is indicated by diarrhea, fever, toxic psychosis, fluid retention (edema) in the lungs, and high blood pressure.

Occupational workers repeatedly exposed to coumaphos have shown impaired memory and concentration, disorientation, severe depression, irritability, confusion, headache, speech difficulties, delayed reaction times, nightmares, sleepwalking, and drowsiness or insomnia. An influenza-like condition with headache, nausea, weakness, loss of appetite, and malaise also has been reported. Data on reproductive effect and genotoxicity are inadequate.[37–39]

### **Demeton-S-Methyl** (Metasystox, Metaphor)

IUPAC name: S-2-ethylthioethyl-O,O-dimethyl phosphorothioate
Molecular formula: $C_6H_{15}O_3PS_2$
Toxicity class: USEPA: I; WHO: Ib

$$CH_3CH_2SCH_2CH_2S\overset{\overset{\displaystyle O}{\|}}{P}(OCH_3)_2$$

**Figure 5-6.** Structural formula of demeton-s-methyl.

***Uses*** Demeton-s-methyl is a pale yellow oil that has a sulfur-like odor. It is a systemic and contact insecticide and acaricide. It kills insects that feed on plants by sucking juices. It is used to control aphids, sawflies, and spider mites in fruits, vegetables, potatoes, cereals, ornamentals, and forestry. Demeton-s-methyl replaces methyl demeton, a mixture of demeton-s-methyl and demeton-o-methyl sold as systox meta. Demeton-s-methyl is more toxic to insects than demeton-o-methyl. It is available as an emulsifiable concentrate.[28,40]

***Toxicity*** The acute oral $LD_{50}$ of demeton-s-methyl for rats is 37.5 mg/kg, and the acute dermal $LD_{50}$ for rats is 85 mg/kg. The acute $LC_{50}$ for rats (4 hours) is 500 $mg/m^3$ of air. Workers repeatedly exposed to demeton-s-methyl showed symptoms of headache, nausea, vomiting, diarrhea, sweating, and dizziness. In addition, symptoms in some cases of poisoning include tremors, lack of coordination, and hiccough. With time, anorexia and loss of concentration may occur. Demeton-s-methyl is nonirritating to the skin and eyes. Daily oral doses of oxydemeton caused decreased cholinesterase in pregnant rats, as well as decreased body weight. The females also developed tumors, but there was no evident effect on fetuses. Test results indicate that demeton-s-methyl is mutagenic.[20,41]

## **Diazinon** (Basudin, Knox-out)

IUPAC name: O, O-diethyl 0-2-isopropyl-6-methyl (pyrimidin-4-yl) phosphorothioate
Molecular formula: $C_{12}H_{21}N_2O_3PS$
Toxicity class: USEPA: II or III; WHO: II

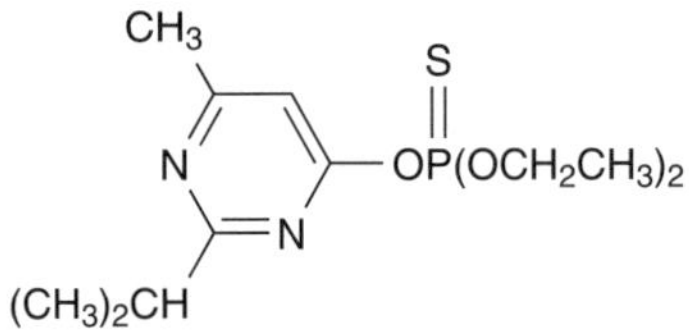

$CH_3$
S
N
$OP(OCH_2CH_3)_2$
N
$(CH_3)_2CH$

**Figure 5-7.** Structural formula of diazinon.

***Uses*** Diazinon is a colorless to dark brown liquid. Based on the Toxic Release Inventory (TRI) database for 1995, approximately 107,510 pounds of production-related waste was generated by the United States. Further, releases from facilities to the environment totaled 11,498 pounds, most of which was in the form of stack emissions. Diazinon is used for the control of pest insects in soil, on ornamental plants, and on fruit and vegetable field crops. It also is used to control household pests such as flies, fleas, and cockroaches.[28,42]

Diazinon has been used since 1956 for the control of soil insects and pests of fruit trees, vegetables, grasslands, and other crops. Now it is used to control flies around animal facilities, greenhouses, fairgrounds, and other businesses and public places where food or animal wastes might accumulate.[51] In addition, diazinon is the active ingredient added to most pet collars and ear tags to control biting insects or skin parasites on pets and livestock. For home and garden applications (i.e., to control crickets and cockroaches), diazinon is applied in the form of strips placed near entryways or sprayed near certain areas as wettable powders, emulsifiable concentrates, and pressurized sprays.[43]

***Toxicity*** The acute oral $LD_{50}$ of diazinon for rats ranges from 300 to 800 mg/kg, and the acute dermal $LD_{50}$ for rats is greater than 2,150 mg/kg. Exposures to diazinon are most significant in people who work in the manufacture and professional application of this insecticide. Some mild symptoms of exposure include headache, dizziness, weakness, anxiety, constriction of the pupils of the eye, and inability to see clearly. Most cases of unintentional diazinon poisoning in people have resulted from short exposures to very high concentrations of the material. Very high levels of exposure to diazinon have resulted in death in people accidentally exposed and in those who have swallowed large amounts of the chemical to commit suicide.

### **Dichlorvos** (DDVP, Vapona, Phosvit, Vantaf, Uniphos, Swing, Nuvon)

IUPAC name: 2,2-dichlorovinyl dimethyl phosphate.
Molecular formula: $C_4H_7Cl_2O_4P$
Toxicity class: USEPA: I; WHO: Ib

$$(CH_3-O)_2P(=O)-O-CH=CCl_2$$

**Figure 5-8.** Structural formula of dichlorvos.

***Uses*** Dichlorvos is a colorless to amber liquid with a mild chemical odor. The USEPA has classified it under RUP, and hence it should be purchased and used only by certified applicators. Dichlorvos is used primarily to control household insects such as flies, aphids, spider mites, caterpillars, and thrips. The chemical also is used therapeutically against parasitic worm infections in dogs, livestock, and humans. The mechanism of action appears to be a contact poison and stomach poison in insects caused by interference with cholinesterase, a neurologic enzyme that is important to nerve transmissions. It is used as a fumigant and has been used to make pet collars and pest strips. It is available as an aerosol and a soluble concentrate.[28,44]

***Toxicity*** The acute oral $LD_{50}$ for rats is 25 to 80 mg/kg, for mice 61 to 175, for rabbits 11 to 12 mg/kg, for dogs 100 to 1,090 mg/kg, for pigs 157 mg/kg, and for chickens 15 mg/kg. The acute dermal $LD_{50}$ for rats is 70.4 to 250 mg/kg, for mice 206 mg/kg, and for rabbits 107 mg/kg. The acute $LC_{50}$ (4 hours) for rats is greater than 0.2 mg/L. Repeated or prolonged exposure is known to cause delayed effects of toxicity. Exposed workers have shown impaired memory and concentration, disorientation, severe depression, irritability, confusion, headache, speech difficulties, delayed reaction times, nightmares, sleepwalking, and drowsiness or insomnia. Persons with reduced lung function, convulsive disorders, liver disorders, or recent exposure to cholinesterase inhibitors will be at increased risk. Alcoholic beverages may enhance the toxic effects of dichlorvos. High environmental temperatures or exposure to light may enhance its toxicity. All poisoned workers have shown numbness, tingling sensations, incoordination, headache, dizziness, tremor, nausea, abdominal cramps, sweating, blurred vision, difficulty breathing or respiratory depression, and slow heartbeat.

Dichlorvos has been classified as a possible human carcinogen because it caused tumors in rats and mice in some studies. When dichlorvos was administered by gavage to mice 5 days per week for 103 weeks at doses of 20 mg/kg/day in males and 40 mg/kg/day in females, there was an increased incidence of benign tumors in the lining of the stomach in both sexes. When rats were given doses of 4 or 8 mg/kg/day for 5 days per week for 103 weeks, there was an increased incidence of benign tumors of the pancreas and of leukemia in male rats at both doses. At the highest dose, there also was an increased incidence of benign lung tumors in males. In female rats, there was an increase in the incidence of benign tumors of the mammary gland.[44,45]

### **Dimethoate** (Rogor)

IUPAC name: O,O-dimethyl S-methylcarbamoylmethyl phosphorodithioate
Molecular formula: $C_5H_{12}NO_3PS_2$
Toxicity class: USEPA: II; WHO: II

$$(CH_3O)_2P(=O)—S—CH—C(=O)—N(H)—CH$$

**Figure 5-9.** Structural formula of dimethoate.

***Uses*** Dimethoate is a gray-white crystalline solid at room temperature and the USEPA has grouped it under GUP. Dimethoate is used for the control of mites and insects systemically and on contact. It is used against a wide range of insects, including aphids, thrips, planthoppers, and whiteflies on ornamental plants, alfalfa, apples, corn, cotton, grapefruit, grapes, lemons, melons, oranges, pears, pecans, safflower, sorghum, soybeans, tangerines, tobacco, tomatoes, watermelons, wheat, and other vegetables. It also is used as a residual wall spray in farm buildings for houseflies. Dimethoate has been administered to livestock for control of botflies. Formulations of dimethoate include aerosol spray, dust, emulsifiable concentrate, and ultra-low volume concentrate.[28]

***Toxicity*** The acute oral $LD_{50}$ for rats ranges from 180 to 330 mg/kg, for mice 160 mg/kg, for rabbits 400 to 500 mg/kg, and for guinea pigs 550 to 600 mg/kg. The acute dermal $LD_{50}$ for rats ranges from 100 to 600 mg/kg. The acute $LC_{50}$ (4 hours) is greater than 2.0 mg/L. Dimethoate is reportedly not irritating to the skin and eyes of laboratory animals.[28] Severe eye irritation has occurred in workers manufacturing dimethoate, although this may be due to impurities in the compound. Workers repeatedly exposed to dimethoate have shown symptoms of numbness, tingling sensations, incoordination, headache, dizziness, tremor, nausea, abdominal cramps, sweating, blurred vision, difficulty breathing or respiratory depression, slow heartbeat, and speech difficulties. Very high doses have caused unconsciousness, convulsions, and death. A work environment with high temperature, invisible or ultraviolet light enhances the toxicity. In fact, the manner of dimethoate biotransformation in cell culture systems and the nature of degradation in human liver has offered clues to the overall management of the toxic compound.[46–49]

Studies have shown that dimethoate is teratogenic to cats and rats at a dose of 12 mg/kg/day.[49a,49b] The adverse effects are in the form of extra toes on kittens, and the same dose in pregnant rats produced certain birth defects in bone or skeleton formation, runting, and malfunction of the bladder. However, administration of dimethoate 9.5 to 10.5 mg/kg day in drinking water caused no teratogenic effects in mice. Mice exposed to a single high dose experienced mutagenic effects. Although prolonged exposure to dimethoate produced an increase in malignant tumors, the results are inconclusive.[50,51]

### **Disulfoton** (Disyston)

IUPAC name: O, O-diethyl S-2-ethylthioethyl phosphorodithioate
Molecular formula: $C_8H_{19}O_2PS_3$
Toxicity class: USEPA: I; WHO: Ia

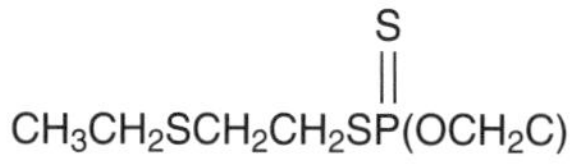

$$CH_3CH_2SCH_2CH_2S\overset{\overset{\displaystyle S}{\|}}{P}(OCH_2C)$$

**Figure 5-10.** Structural formula of disulfoton.

***Uses*** Disulfoton is a yellowish oil. It is grouped by the USEPA under RUP and therefore must be handled by qualified and trained applicators. Disulfoton is a selective systemic insecticide and acaricide. It is specifically effective against sucking insects and is used to control aphids, leafhoppers, thrips, beet flies, spider mites, and coffee leaf-miners. Disulfoton products are used on cotton, tobacco, sugar beets, cole crops, corn, peanuts, wheat, ornamentals, cereal grains, and potatoes.[28]

***Toxicity*** The acute oral $LD_{50}$ for rats ranges from 1.9 to 12.5 mg/kg. The acute dermal $LD_{50}$ for rats is to 3.6 mg/L, and inhalation $LC_{50}$ for rats (1 hour) is 0.3 mg/L. Disulfoton is highly toxic to all mammals by all routes of exposure, whether absorbed through the skin, ingested, or inhaled; early symptoms in humans may include blurred vision, fatigue, headache, dizziness, sweating, tearing of the eyes, and salivation. It inhibits cholinesterase and affects nervous system function. Symptoms occurring at high doses include defecation, urination, fluid accumulation in the lungs, convulsions, or coma. Death can occur if high doses stop respiratory muscles or constrict the windpipe. Workers chronically exposed to OP-like disulfoton have developed irritability, delayed reaction times, anxiety, slowness of thinking, and memory defects. Chronic exposure also may lead to cataracts.[52–54]

### **Ethion** (Cethion, Dhanumit, Ethiol, Royethion, Rodocide, Tafethion)

IUPAC name: O,O,O′,O′-tetraethyl S,S′-methylene bis (phosphorodithioate)
Molecular formula: $C_9H_{22}O_4P_2S_4$
Toxicity class: USEPA: II; WHO: II

```
                S          S
                ||         ||
(CH3CH2O)2PSCH2SP(OCH2CH3)2
```

**Figure 5-11.** Structural formula of ethion.

***Uses*** Technical ethion is an odorless amber liquid. Ethion is used to kill aphids, mites, scales, thrips, leafhoppers, maggots, and foliar feeding larvae. It may be used on a wide variety of foods, fibers, and ornamental crops, including greenhouse crops, lawns, and turf. Ethion often is used on citrus and apples. It is mixed with oil and sprayed on dormant trees to kill eggs and scales. Ethion also may be used on cattle. It is available in dust, emulsifiable concentrate, emulsifiable solution, granular, and wettable powder formulations.[28]

***Toxicity*** Ethion is highly to moderately toxic to animals and humans by the oral route. The acute oral $LD_{50}$ of pure ethion for rats is 208 mg/kg, and that of technical ethion is 21 mg/kg. The acute dermal $LD_{50}$ for mice is 40 mg/kg. Ethion is moderately toxic via inhalation, with a reported 4-hour $LC_{50}$ in rats of 0.864 mg/L. It is highly to moderately toxic via the dermal route, with a reported dermal $LD_{50}$ of 62 mg/kg in rats, 915 mg/kg in guinea pigs, and 890 mg/kg in rabbits. Adverse effects in humans include nausea, cramps, diarrhea, excessive salivation, blurred vision, headache, fatigue, tightness in the chest, abnormal heartbeat and breathing, loss of coordination, convulsions, coma, and death. Skin exposure may cause contact burns. Persons with respiratory ailments and workers repeatedly exposed have shown impaired memory and concentration, disorientation, severe depression, irritability, confusion, headache, speech difficulties, delayed reaction times, nightmares, sleepwalking, and drowsiness or insomnia. An influenza-like condition with headache, nausea, weakness, loss of appetite, and malaise also has been reported.[55,56]

### **Fenamiphos** (Nemacur)

IUPAC name: ethyl-4-methylthio-m-tolyl isopropylphosphoramidate
Molecular formula: $C_{13}H_{22}NO_3PS$
Toxicity class: USEPA: I; WHO: Ia

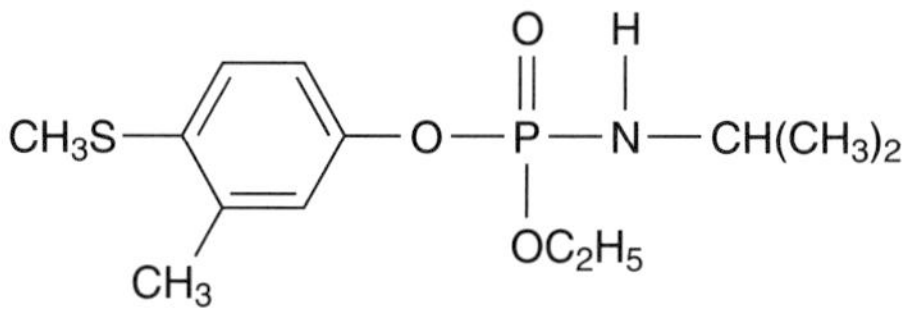

**Figure 5-12.** Structural formula of fenamiphos.

***Uses*** Fenamiphos is used as a systemic and contact insecticide to control several genera of nematodes. Pest infestation of fruit crops, vegetables, and turfgrasses are extensive. For example, crops such as bananas, pineapples, citrus fruit, pome fruit, stone fruit, vines, cotton, tobacco, cocoa, coffee, okra, groundnuts, soybeans, and vegetables (e.g., curcurbits, tomatoes, potatoes, sugar beets) have been sprayed with fenamiphos for the control of pests. The compound also has secondary activity against sucking insects and spider mites. Fenamiphos is used in the form of emulsifiable concentrate, granules, or oil-in-water emulsion by itself or can be mixed with isofenphos, carbofuran, or disulfoton.[28]

***Toxicity*** The acute oral $LD_{50}$ for rats is 15.3 to 19.4 mg/kg, for dogs 10 mg/kg, and for guinea pigs 75 to 100 mg/kg. The acute dermal $LD_{50}$ for rats is 500 mg/kg. The acute $LC_{50}$ (4-hour) for rats is approximately 0.12 mg/L. Fenamiphos is known to cause severe toxicity such as impaired memory and concentration, disorientation, severe depression, irritability, confusion, headache, speech difficulties, delayed reaction times, nightmares, sleepwalking, and drowsiness or insomnia. An influenza-like condition with headache, nausea, weakness, loss of appetite, and malaise also has been reported.

### **Fenitrothion** (Sumithion)

IUPAC name: O,O-dimethyl O-4-nitro-m-tolyl phosphorothioate
Molecular formula: $C_9H_{12}NO_5PS$
Toxicity class: USEPA: II; WHO: II

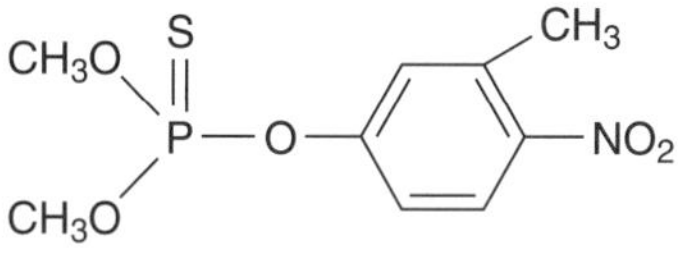

**Figure 5-13.** Structural formula of fenitrothion.

***Uses*** The formulations of fenitrothion include dusts, emulsifiable concentrate, flowable, fogging concentrate, granules, ultra-low volume, oil-based liquid spray, and wettable powder formulations (Novathion 500-E, EC as a 95% concentrate,

50% emulsifiable concentrate, 40% and 50% wettable powder and 2%, 2.5%, 3%, and 5% dusts).[28]

Fenitrothion is a contact insecticide and selective acaricide of low ovicidal properties. Fenitrothion is effective against a range of pests, namely, penetrating, chewing, and sucking insect pests (e.g., coffee leaf-miners, locusts, rice stem borers, wheat bugs, flour beetles, grain beetles, grain weevils) on cereals, cotton, orchard fruits, rice, vegetables, and forests. It also may be used as a fly, mosquito, and cockroach residual contact spray for farms and public health programs. Fenitrothion also is effective against household and nuisance insects. The World Health Organization (WHO) has confirmed its effectiveness as a vector control agent for malaria. It is extensively used in other countries, including Japan, where parathion has been banned.

***Toxicity*** The acute oral $LD_{50}$ for rats ranges between 250 and 800 mg/kg, for mice between 715 and 870 mg/kg, and for guinea pigs 500 mg/kg. The acute dermal $LD_{50}$ for rats and mice is more than 890 and 3,000 mg/kg, respectively. The acute inhalation $LC_{50}$ in rats is reported as 5 mg/L. Sumithion 50-EC has been shown to cause delayed neurotoxicity in adult rats as well as in humans.[57–60]

### **Fenthion** (Lebaycid)

IUPAC name: O,O-dimethyl O-4-methylthio-m-tolyl phosphorothioate
Molecular formula: $C_{10}H_{15}O_3PS_2$
Toxicity class: USEPA: II; WHO: II

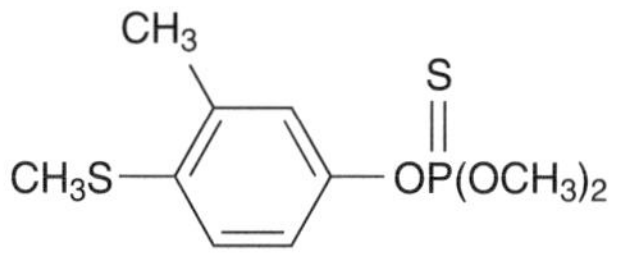

$CH_3$
S
$CH_3S$ — — $OP(OCH_3)_2$

**Figure 5-14.** Structural formula of fenthion.

***Uses*** Pure fenthion is a colorless liquid. Technical fenthion is a yellow or brown oily liquid with a weak garlic odor. It is grouped by the USEPA under RUP and requires handling by qualified, certified, and trained workers. Fenthion is used for the control of sucking and biting pests (e.g., fruitflies, stem borers, mosquitoes, cereal bugs). In mosquitoes, it is toxic to both the adult and the immature forms (larvae). The formulations of fenthion include dust, emulsifiable concentrate, granular, liquid concentrate, spray concentrate, ultra-low volume, and wettable powder.[28]

***Toxicity*** Fenthion is moderately toxic to mammals and highly toxic to birds. The acute oral $LD_{50}$ of fention for rats is 250 mg/kg and the acute dermal $LD_{50}$ for rats is 700 mg/kg. It is slightly toxic via inhalation for rats, with an acute $LC_{50}$

(1-hour) that ranges from 2.4 to 3.0 mg/L. Fenthion poisoning has caused numbness, tingling sensations, incoordination, headache, dizziness, tremor, nausea, abdominal cramps, sweating, blurred vision, difficulty breathing or respiratory depression, and slow heartbeat. Very high doses may result in unconsciousness, incontinence, convulsions, or death.

In Nigeria, sprayers without protective clothing experienced a decrease in whole-blood cholinesterase activity. Veterinary clinic workers who did not use skin protection when applying a 20% topical application to dogs experienced symptoms ranging from tingling and numbness of the hands and feet to generalized weakness and shooting pains. Animal tests and human use experience indicated that fenthion induces adverse effects on the central and peripheral nervous systems, as well as on the heart.[61]

### **Fonophos** (Dyfonate)

IUPAC name: O-ethyl S-phenyl (RS)-ethylphosphorodithioate
Molecular formula: $C_{10}H_{15}OPS_2$
Toxicity class: USEPA: I or II; WHO: Ia

S
||
SPOCH$_2$CH$_3$
|
CH$_2$CH$_3$

**Figure 5-15.** Structural formula of fonophos.

***Uses*** Fonophos is a highly toxic compound. It has been grouped by the USEPA as RUP and hence requires special handling by qualified, certified, and trained workers. Fonophos is a soil insecticide primarily used on corn crops. It also is used on sugarcane, peanuts, tobacco, turf, and some vegetable crops. It controls aphids, corn borers, corn root-worm, corn wireworm, cutworms, white grubs, and some maggots. Formulations of fonophos includes granular, microgranular, emusifiable concentrate, suspension concentrate, microcapsule, and suspension.[28]

***Toxicity*** Fonophos is highly toxic via the oral route. The acute oral $LD_{50}$ for rats ranges from 5 to 11.5 mg/kg. The acute dermal $LD_{50}$ for rats is 147 mg/kg, for rabbits, it ranges from 30 to 261 mg/kg, and for guinea pigs, 278 mg/kg.

Symptoms of fonophos poisoning are often delayed but can occur within a few minutes to 12 hours after exposure. Early symptoms of poisoning include blurred vision, headache, and dizziness. Skin contact often brings about sweating and muscle twitching. Eye contact causes tearing, pain, and blurring. Ingestion may cause nausea, abdominal cramps, and diarrhea. Deaths resulting from high exposures are often due to respiratory arrest. A number of human poisonings by fonophos have been recorded.[4] One woman exposed orally to a large amount of fonophos developed nausea, sweating, and respiratory arrest in addition to muscle twitching, low blood pressure and pulse rate, and pinpoint pupils. A pancreatic cyst (also attributed to fonophos exposure) was located and drained externally in the course of her treatment. She recovered after 2 months of hospitalization.[4,62–64]

### **Isofenphos** (Oftanol)

IUPA name: O-ethyl O-2-isopropoxycarbonylphenyl N-isopropylphosphoramidothioate
Molecular formula: $C_{15}H_{24}NO_4PS$
Toxicity class: USEPA: I; WHO: Ib

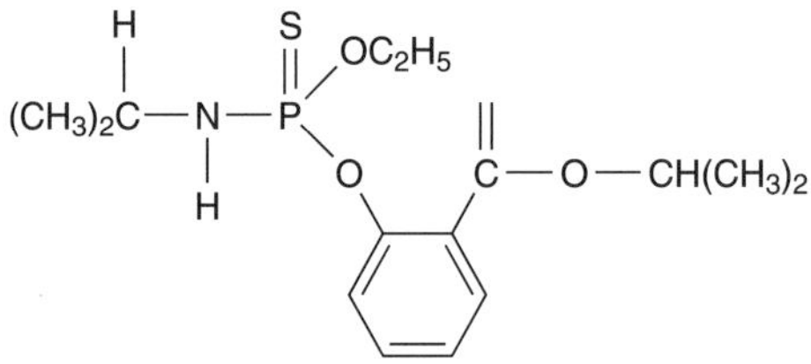

**Figure 5-16.** Structural formula of isofenphos.

***Uses*** Isofenphos is a colorless oil at room temperature, and the USEPA has grouped it as RUP, requiring special handling by qualified, certified, and trained workers. Isofenphos is used to control soil-dwelling insects such as white grubs, cabbage root flies, corn roundworms, and wireworms. The product is used on vegetables including corn and carrots, on soil insects in fruit crops like bananas, and on soils with turfgrass. It is a selective contact and stomach poison in insects. It is applied as a preplant or preemergence soil treatment. It also is used to control termites in and around building structures.[28] Isofenphos is transported (to a limited extent) from the roots to plant leaves and stems. In the United States, it is mostly used in the corn-growing areas of the Ohio Valley.

***Toxicity*** The acute oral $LD_{50}$ of isofenphos for rats ranges from 20 to 125 mg/kg, and the acute dermal $LD_{50}$ for rats is 70 mg/kg. The acute $LC_{50}$ for rats is 0.144 mg/L and 1.3 mg/L. The acute inhalation $LC_{50}$ (4 hours) for hamsters is 0.23 mg/L. These values indicate that the compound is highly toxic to organisms and humans who inhale even small amounts of isofenphos. In combination with malathion, isofenphos becomes extremely toxic.[65,67]

### **Malathion** (Celthion, Cythion)

IUPAC name: Diethyl(dimethoxythiophosphorylthio) succinate
Molecular formula: $C_{10}H_{19}O_6PS_2$
Toxicity class: USEPA: III; WHO: III

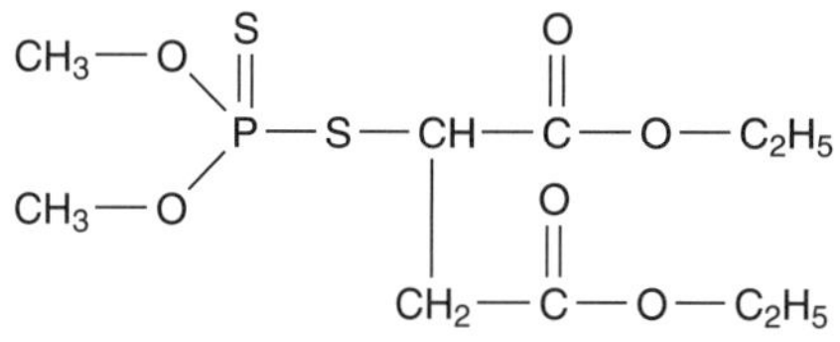

```
            S                O
CH3—O       ||               ||
      \     ||               ||
       P—S—CH——————C—O—C2H5
      /       |      O
CH3—O         |      ||
              CH2——C—O—C2H5
```

**Figure 5-17.** Structural formula of malathion.

***Uses*** Malathion is a clear amber liquid, and the USEPA has grouped it under GUP. Malathion is nonsystemic and has been in use since the 1950s in different formulations in association with other pesticides. It has wide use such as the control of sucking and chewing insects on fruits and vegetables, and the control of mosquitoes, flies, household insects, animal parasites (ectoparasites), and head and body lice.

***Toxicity*** Toxicity of malathion is modulated by different factors such as its purity, the solvent, animal species, diet, route, and duration of exposure. The acute oral $LD_{50}$ for rats ranges from 1,375 to 2,800 mg/kg and for mice from 775 to 3,320 mg/kg. The acute dermal $LD_{50}$ for rabbits is 4,100 mg/kg. Acute symptoms in humans include nausea, headache, tightness in the chest, and other symptoms typical of AChE inhibition. Unconsciousness, convulsions, and a "prolonged worsening illness" also are typical of malathion poisoning at high doses. Dermal absorption of malathion has caused severe poisoning and human fatalities.[68]

Several fatalities due to skin absorption of malathion are found among occupational workers handling malathion for malaria control operations in Pakistan.[69,70] Malathion has been shown to convert rapidly into malaoxon — the oxygen analog known for its severe toxicity. It has been reported that malathion (0.5%) caused poisoning of an infant as a result of aerosol inhalation.[70a] The child exhibited severe signs of ChE inhibition. Numerous malathion intoxication incidents have occurred among pesticide workers and with small children through accidental exposure. Human exposure can occur through ingestion, inhalation, and skin absorption.[69]

Malathion produced detectable mutations in three types of cultured human cells, including white blood cells and lymph cells. It is possible that malathion could pose a mutagenic risk when humans are chronically exposed. Studies by Flessel[71] indicated that technical-grade malathion (or formulations other than pure malathion) has the potential to produce genotoxic effects in mammalian systems. Thus, in test animals, technical-grade malathion has triggered chromosomal changes, including chromosomal aberrations and micronuclei. Studies on human and animal cells in culture suggested that both technical grade and purified malathion do produce cytogenetic damage, including chromosomal aberrations and sister chromatid exchanges. *In vivo* studies with humans, however, failed to confirm these observations. Malathion has not produced any point mutations in standard gene mutation assays in bacteria. The metabolite malaoxon (94% pure) was found to be positive in mammalian cell mutation tests. In fact, the interaction of malathion with DNA has not been adequately studied.[71]

### **Methamidophos** (Monitor, Tamaron)

IUPAC name: O,S-dimethylphosphoramidothioate
Molecular formula: $C_2H_8NO_2PS$
Toxicity class: USEPA: I; WHO: Ib

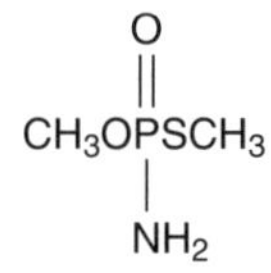


**Figure 5-18.** Structural formula of methamidophos.

***Uses*** Methamidophos is a highly active, systemic, residual insecticide/acaricide/avicide with contact and stomach action. It has been grouped by the USEPA under RUP and requires special handling by qualified, certified, and trained workers. Methamidophos is effective against chewing and sucking insects and is used to control aphids, flea beetles, worms, whiteflies, thrips, cabbage loopers, Colorado potato beetles, potato tube worms, armyworms, mites, leafhoppers, and many others. Crop uses include broccoli, Brussels sprouts, cauliflower, grapes, celery, sugar beets, cotton, tobacco, and potatoes. It is used abroad for many vegetables, hops, corn, peaches, and other crops. Commercially available formulations include soluble concentrate, emulsifiable concentrate, wettable powder, granules, ULV spray, and water miscible spray concentrate.[28]

***Toxicity*** Methamidophos is slightly corrosive to mild steel and copper alloys. This compound is highly toxic to mammals, birds, and bees. One should not permit grazing in treated areas and should wear protective clothing, including a respirator, chemical goggles, rubber gloves, and impervious protective clothing. The acute oral $LD_{50}$ for rats ranges from 21 to 30 mg/kg, for guinea pigs from 30 to 50 mg/kg, and for rabbits from 10 to 30 mg/kg. The acute dermal $LD_{50}$ for rats ranges from 50 to 110 mg/kg, and acute $LC_{50}$ for rats is 9 mg/$m^3$. Inhalation of methamidophos causes weakness, tightness in the chest, wheezing, headache, blurred vision, pinpoint pupils, tearing of the eyes, and runny nose. If it is ingested, nausea, vomiting, diarrhea, and cramps are the most common early signs of poisoning. Sweating and twitching in the area of absorption are seen with skin exposure. Weakness, shakiness, blurred vision, tightness in the chest, sweating, confusion, changes in heartbeat rate, convulsions, and coma.

An intermediate syndrome has been described in cases of poisonings in Sri Lanka, where patients experienced paralysis of limb, neck, and respiratory muscles 24 to 96 hours after exposure.[71a,71b] Delayed neurologic problems have been observed 2 to 4 weeks after large exposures. Workers indicate symptoms of needle-type pain in the feet, legs, and hands. It has been observed that workers with high blood pressure, gastrointestinal disorders, heart, liver, lung, or nervous system problems may be more sensitive to methamidophos.[72–74]

### **Methyl Parathion** (Folidol-M, Metacide, Dhanuman, Sweeper)

IUPAC name: O,O-dimethyl-O-4-nitrophenylphosphorothioate
Molecular formula: $C_8H_{10}NO_5PS$
Toxicity class: USEPA: I; WHO: I or Ia

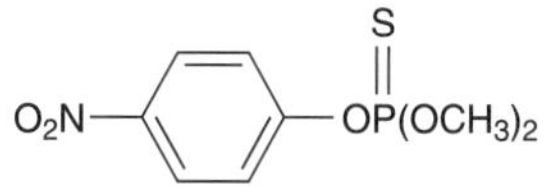

$O_2N-C_6H_4-OP(S)(OCH_3)_2$

**Figure 5-19.** Structural formula of methyl parathion.

***Uses*** Methyl parathion is a contact insecticide and acaricide used for the control of boll weevils and many biting or sucking insect pests of agricultural crops. It kills insects by contact, stomach, and respiratory action. The formulations include dusts, emulsifiable concentrate, ULV liquid, microencapsules, and wettable powders.

***Toxicity*** The acute oral $LD_{50}$ for methyl parathion in rats is 3 mg/kg, for mice 30 mg/kg, and for rabbits 19 mg/kg. The acute dermal $LD_{50}$ for rats is 45 mg/kg. The acute inhalation $LC_{50}$ (4 hours) for rats is 34 mg/m$^3$. Methyl parathion is highly toxic by inhalation and ingestion, and moderately toxic by dermal adsorption. It readily enters through the skin. Skin that has come in contact with this material should be washed immediately with soap and water, and all contaminated clothing should be removed. Accidental skin and inhalation exposures have caused human fatalities. It also can cause contact burns to the skin or eyes. Exposure may occur during mixing, spraying, or application of methyl parathion; during cleaning and repair of equipment; or during early reentry into fields. Persons with respiratory ailments, recent exposure to cholinesterase inhibitors, cholinesterase impairment, or liver malfunction are at increased risk from exposure to methyl parathion. High environmental temperatures or exposure of the chemical to visible or ultraviolet light may increase its toxicity.[75,76]

### **Mevinphos** (Duraphos, Mevindrin, Phosdrin)

IUPAC name: 2-methoxycarbonyl-1-methylvinyl dimethyl phosphate
Molecular formula: $C_7H_{13}O_6P$
Toxicity class: USEPA: I; WHO: Ia

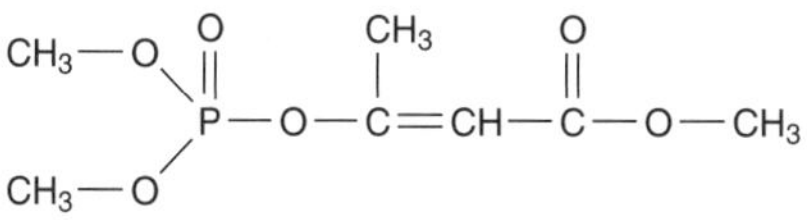

$(CH_3-O)_2P(=O)-O-C(CH_3){=}CH-C(=O)-O-CH_3$

**Figure 5-20.** Structural formula of mevinphos.

***Uses*** Pure mevinphos is a colorless liquid, whereas technical-grade mevinphos is a pale yellow liquid with a very mild odor. It has been grouped by the USEPA under RUP, and hence it should be purchased and used only by certified and trained applicators. Mevinphos is used for the control of a broad spectrum of insects, including aphids, grasshoppers, leafhoppers, cutworms, caterpillars, and many other insects on a wide range of field, forage, vegetable, and fruit crops. It also is an acaricide that kills or controls mites and ticks. It acts quickly both as a contact insecticide, acting through direct contact with target pests, and as a systemic insecticide that becomes absorbed by plants on which insects feed.[28]

***Toxicity*** Mevinphos is highly toxic via the oral route. The acute oral $LD_{50}$ for rats ranges from 3 to 12 mg/kg and for mice from 4 to 18 mg/kg. Dermal absorption is rapid and highly toxic. The acute dermal $LD_{50}$ for rats is 4.2 mg/kg and for rabbits is 16 to 34 mg/kg. The acute $LC_{50}$ (1-hour) for rats is 0.125 mg/L. Acute pulmonary edema (the filling up of lungs with fluid), and changes in the structure or function of salivary glands were seen in rats exposed to this concentration for 1 hour. Effects of acute exposure to mevinphos are similar to those of exposure to other OPs, except that they may occur at much lower doses than OPs.

Symptoms of acute poisoning include numbness, tingling sensations, incoordination, headache, dizziness, tremor, nausea, abdominal cramps, sweating, blurred vision, difficulty breathing or respiratory depression, and slow heartbeat. Very high doses may result in unconsciousness, incontinence, convulsions, or death. In humans, symptoms of poisoning have appeared within as little as 15 minutes to 2 hours after exposure to mevinphos, but onset of symptoms has been delayed for as long as 2 days. Several children were taken ill by unknowingly wearing clothing that had been contaminated with mevinphos.[77]

### **Parathion** (E605, Fighter, Folidol, Fostox)

IUPAC name: O, O-diethyl O-4-nitrophenyl phosphorothioate
Molecular formula: $C_{10}H_{14}NO_5PS$
Toxicity class: USEPA: I; WHO: Ia

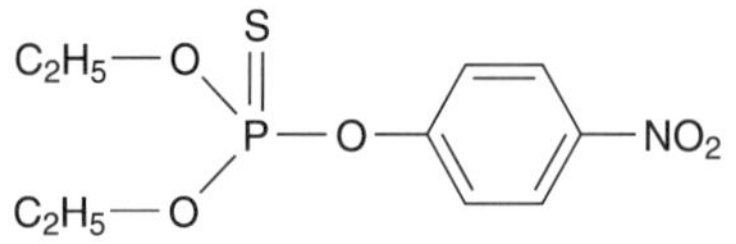

S
$C_2H_5$—O
P—O—
—$NO_2$
$C_2H_5$—O

**Figure 5-21.** Structural formula of parathion.

***Uses*** Pure parathion is a pale yellow liquid with a faint odor of garlic at temperatures above 6°C. Technical-grade parathion is a deep brown to yellow liquid. Parathion is one of the most acutely toxic pesticides registered by the USEPA, which has classified parathion under RUP; hence it should be handled, by qualified, trained, and certified workers in pest control operations. In January

1992, the USEPA announced the cancellation of parathion for all uses on fruit, nut, and vegetable crops.

Parathion was in use for the control of pests of fruits, nuts, and vegetable crops. The only uses retained are those on alfalfa, barley, corn, cotton, sorghum, soybeans, sunflowers, and wheat. Further, to reduce exposure of agricultural workers, parathion may be applied to these crops only by commercially certified aerial applicators, and treated crops may not be harvested by hand. Parathion is a broad-spectrum, organophosphate pesticide used to control many insects and mites. It has nonsystemic, contact, stomach, and fumigant actions. It has a wide range of applications on many crops against numerous insect species. Parathion is available in dust, emulsion concentrate, granular, ULV liquid, and wettable powder formulations.[28]

***Toxicity*** Parathion is highly toxic by all routes of exposure. Occurrences of human fatalities are due to ingestion, dermal adsorption, and inhalation of parathion. The acute oral $LD_{50}$ for rats is 2.6 mg/kg, for mice 12.0 mg/kg, and for guinea pigs 10 mg/kg. The acute dermal $LD_{50}$ for rats is 71 to 76 mg/kg.[1,3,9,78,79]

Persons with cardiovascular, liver, or kidney diseases; glaucoma; or CNS abnormalities may be at increased risk from exposure to parathion. High environmental temperatures or exposure of the chemical to visible or ultraviolet light may increase its toxicity. Parathion may cause thickening and roughening of the skin (hyperkeratinization). It does not cause sensitization (allergies). Parathion is not irritating to the eyes. Splashing parathion into an eye may cause constriction of the pupil, making it difficult to determine the path of moving objects. Parathion primarily affects the nervous system through inhibition of cholinesterase, an enzyme required for proper nerve functioning. In humans poisoned with parathion, an increase in brain weight occurs.

Parathion caused severe poisoning in children. Reports have shown that as many as 17 school-going children who moved through a farm sprayed with parathion suffered severe poisoning and had severe decrease in AChE activity. During shipment, food was contaminated with parathion and the consumption of this food resulted in high human fatalities in a state of India (Kerala).[7,8]

## **Phosphamidon** (Aimphon, Dimecron, Kinadon, Phosron, Rilan, Rimdon)

IUPAC name: 2-chloro-2-diethyl-carbamoyl-1-methylvinyl-dimethylphosphate
Molecular formula: $C_{10}H_{19}Cl\ NO_5P$
Toxicity class: USEPA: I; WHO: Ia

$$(CH_3O)_2P(=O)-O-C(CH_3)=C(Cl)-C(=O)-N(C_2H_5)_2$$

**Figure 5-22.** Structural formula of phosphamidon.

***Uses*** Phosphamidon is used as a broad-spectrum insecticide in agriculture. It is toxic both systemically and by contact, and acts through the inhibition of ChE. Phosphamidon currently is registered for use by both ground and aerial applications on vegetables, fruits, and field crops.[28]

***Toxicity*** The acute oral $LD_{50}$ for rats is 17 to 30 mg/kg. The acute dermal $LD_{50}$ for rats is 374 to 530 mg/kg. It is highly toxic to birds and honeybees. Prolonged treatment of mice and rats ranging for different periods (60 to 110 weeks) and further observation demonstrated hyperexcitability and tremors. The treated animals also demonstrated the development of late appearance of tumors, hemangiomas, and hemangiosarcomas in the spleen. However, these studies are inadequate to provide meaningful conclusions about phosphamidon carcinogenicity.[20,80]

### **Propetamphos** (Safrotin)

IUPAC name: (E)-O-2-isopropoxycarbonyl-1-methylvinyl O-methyl ethyl-phosphoramidothioate
Molecular formula: $C_{10}H_{20}NO_4PS$
Toxicity class: USEPA: II; WHO: Ib

```
             S         CH3       O        H
             ||         |        ||       |
  CH3O—P—O—C=C—C—O—C(CH3)2
             |               |
           NC2H5             H
             |
             H
```

**Figure 5-23.** Structural formula of propetamphos.

***Uses*** Propetamphos is a moderately toxic insecticide. The USEPA has classified it under both GUP and RUP, which suggests the use by certified and qualified applicators. Propetamphos is used for the control of cockroaches, flies, ants, ticks, moths, fleas, and mosquitoes in households and where vector eradication is necessary to protect public health. It also is used in veterinary applications to combat parasites such as ticks, lice, and mites in livestock. The commercial formulations of propetamphos include aerosols, emulsified concentrates, liquids, and powders.[28]

***Toxicity*** Propetamphos is moderately toxic via the oral route. The acute oral $LD_{50}$ for rats is 59 to 119 mg/kg. The acute dermal $LD_{50}$ for rats is 2,260 to 2,825 mg/kg. The acute inhalation $LC_{50}$ (4 hour) is greater than 2.04 mg/L for rabbits. Toxic effects of propetamphos poisoning among occupational workers include numbness, tingling sensations, incoordination, headache, dizziness, tremor, nausea, abdominal cramps, sweating, blurred vision, difficulty breathing or respiratory depression, and slow heartbeat. Very high doses may result in unconsciousness, incontinence, convulsions, or death.[81,82]

### **Temephos** (Abate, Temeguard)

IUPAC name: O,O,O,O′-tetramethyl-O,O′-thiodi-p-phenylene-bis (phosphorothioate)
Molecular formula: $C_{16}H_{20}O_6P_2S_3$
Toxicity class: USEPA: III; WHO: III

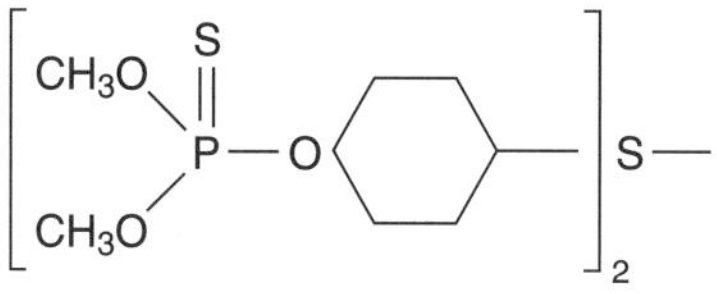

$$\left[ (CH_3O)_2P(=S)-O-C_6H_4- \right]_2 S$$

**Figure 5-24.** Structural formula of temephos.

***Uses*** Temephos is a solid at room temperature and is composed of colorless crystals. As a liquid, it is brown and viscous. It has been grouped by the USEPA under GUP. Temephos is used for the control of mosquito, midge, and blackfly larvae. It is used in lakes, ponds, and wetlands. It also may be used to control fleas on dogs and cats and to control lice on humans. Temephos is available in different formulations, such as emulsifiable concentrates (50%), wettable powder (50%), and granular forms (5%).[28]

***Toxicity*** The acute oral $LD_{50}$ for rats ranges from 1,226 to 13,000 mg/kg, and the acute dermal $LD_{50}$ for rabbits is 1,930 mg/kg. Temephos may potentiate the observed toxicity of malathion when used in combination with it at very high doses. Symptoms of temephos toxicity include nausea, salivation, headache, loss of muscle coordination, and difficulty breathing.[83–85]

### **Trichlorfon** (Dipterex, Neguvon)

IUPAC name: Dimethyl 2,2,2-trichloro-1-hydroxyethylphosphonate
Molecular formula: $C_4H_8Cl_3O_4P$
Toxicity class: USEPA: II; WHO: II

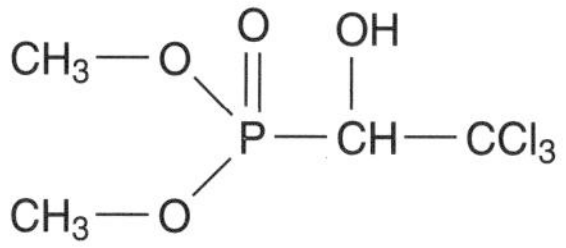

$$(CH_3-O)_2P(=O)-CH(OH)-CCl_3$$

**Figure 5-25.** Structural formula of trichlorfon.

***Uses*** Trichlorfon is a pale, clear, white, or yellow crystalline solid with an ethyl ether odor. It is a solid at room temperature. Trichlorfon is grouped by the USEPA under GUP. Trichlorfon is used for the control of a variety of insect pests, (e.g., cockroaches, crickets, silverfish, bedbugs, fleas, cattle grubs, flies, ticks, leafminers, leafhoppers). It has extensive use in agriculture, applied to vegetable, fruit, and field crops.

It also is used for the control of internal parasites affecting domestic animals and livestock. Trichlorfon is available in dust, emulsifiable concentrate, granular, fly bait, and soluble powder formulations with the percentage active of ingredient ranging from 40% (soluble powder) to 98% (technical). Trichlorfon is a selective insecticide, meaning that it kills selected insects.[28]

***Toxicity*** The acute oral $LD_{50}$ for rats ranges from 250 to 650 mg/kg and for mice from 300 to 860 mg/kg. Other reported oral $LD_{50}$ values are 94 mg/kg in cats, 400 mg/kg in dogs, 420 mg/kg in dogs, and 160 mg/kg in rabbits. The dermal $LD_{50}$ is 2,000 to 5,000 mg/kg in rats and 1,500 to greater than 2,100 mg/kg in rabbits. The acute $LC_{50}$ (4 hour) for trichlorfon in rats is greater than 0.5 mg/L.[86]

The symptoms of acute exposure include headache, giddiness, nervousness, blurred vision, weakness, nausea, cramps, loss of muscle control or reflexes, convulsion, or coma. It has been suggested that impurities or additives may be associated with some cases of delayed polyneuropathy (damage to nerve cells) attributed to ingestion of large amounts of trichlorfon.[87–90]

## CONCLUSION

OPs have been in use for several decades as important chemicals for the control of crop pests. With their chemical and biochemical reactions, OPs have been well established as extremely poisonous chemicals. This classification is due to the inhibition of the marker enzyme ChE, which is produced in the liver. Blood enzymes provide an estimate of tissue enzyme activity. After acute exposure to OPs or a nerve agent, the erythrocyte enzyme activity most closely reflects the activity of the tissue enzyme. Once the OPs inhibit the tissue enzyme, it cannot hydrolyze ACh, and the accumulation stimulates the affected organ. Based on the manner of exposure (dose and duration) to different OPs, a series of toxicity signs and symptoms set in the organism, leading to death. These are important aspects to be closely monitored among pest control operators and occupational workers exposed to OPs.

Selection of a comprehensive testing battery to record the rate of AChE inhibition is a crucial point. In addition to the estimation of AChE activity, other functions (e.g., sensory systems, reflexes, neuromotor development, locomotion, gait, reactivity to medication) require close monitoring to achieve chemical safety to OPs. In the case of children and workers exposed to very low doses of OPs, other tests should indirectly include parameters like learning, memory loss, social behavior, and reproductive behavior. To integrate behavioral data into the context of other manifestations of developmental toxicity, data on physical development of the offspring must have further emphasis. To continue meaningful therapeutic measures, other observations — particularly of brain biochemistry — must be properly recorded and evaluated. Therefore, proper management, guidance, and training are necessary for occupational workers, specifically in industrial and crop protection management environments, during the use, storage, and disposal of OPs for the safety and protection of human health.

## REFERENCES


1. C. Frest, and K.J. Schmidt, *The Chemistry of Organophosphorus Pesticides*, 2nd ed., Springer-Verlag, New York, 1982.
2. W. Lorenz, and K. Sasse, Gerhard Schrader and the development of organophosphorus compounds for crop protection. *Pflanzenschutz-Nachr.* **21**:5, 1968.
3. J. Brodeur, and K.P. Du Bois, Comparison of acute toxicity of anticholinesterase insecticides of weanling and adult male rats. *Proc. Soc. Exp. Biol. Med.* **114**:509, 1963.
4. M.A. Gallo, and N.J. Lawryk, Organic phosphorus pesticides, in *Handbook of Pesticide Toxicology*, Hayes, W.J., Jr. and Laws, E.R., Jr., eds., Academic Press, New York, 1991.
5. M.B. Abou-Donia, Biochemical toxicology of organophosphorous compounds, in: Blum K., and Manzo L., eds., *Neurotoxicology*, Marcel Dekker, New York, 1985.
6. H.V. Smith, and J.M.K. Spalding, Outbreak of paralysis in Morocco due to orthocresyl phosphate poisoning. *Lancet* **2**:1019, 1956.
7. C.O. Karunakaran, The Kerala food poisoning. *J. Indian Med. Assoc.* **31**:204, 1958.
8. K. Kanagaratnam, W.H. Boon, and T.K. Hoh, Parathion poisoning from contaminated barley. *Lancet* **1**:538, 1960
9. R.D. O'Brien, *Toxic Phosphorus Esters, Chemistry, Metabolism and Biological Effects*, Academic Press, New York, 1960.
10. W.T. Thomson, *Agricultural Chemicals, Book III, Miscellaneous Agricultural Chemicals*, Thomson Publications, Fresno, CA, 209, 1995.
11. S.L. Wagner, The acute health hazards of pesticides, in *Chemistry, Biochemistry, and Toxicology of Pesticides*, J.M. Witt, ed., Oregon State University Cooperative Extension Service, Corvallis, OR, 1989.
12. M.B. Abou-Donia, The cytoskeleton as a target for organophosphorus ester-induced delayed neurotoxicity (OPIDN). *Chem. Biol. Interact.* **87**(1–3):383, 1993.
13. *Farm Chemicals Handbook*, Meister Publishing Co., Willoughby, Ohio, Vol. 82, 847, 1996.
14. H.P. Rang, M.M. Dale, and J.M. Ritter, *Pharmacology cholinergic transmission*, Chapter 6, 117–147, Churchill Livingstone, 3rd ed., 1996.
15. D.J. Ecobichon, *The Basis of Toxicity Testing*, CRC Press, Boca Raton, FL, 1992.
16. K.W. Jager, D.V. Roberts, and A. Wilson, Neuromuscular function in pesticide workers. *Br. J. Ind. Med.* **27**:273, 1978.
17. H.W. Drenth, I.E.G. Ensberg, D.V. Roberts, and A. Wilson, Neuromuscular function in agricultural workers using pesticides. *Arch. Environ. Health* **25**:395, 1972.

18. J.E. Davies, and V.H. Freed, ed., *An agromedical approach to pesticide management: Some health and environmental considerations*, Consortium for International Crop Protection, Berkeley, CA, 370, 1981.
19. World Health Organization, Organophosphorus insecticides: A general introduction, *Environ. Health Criteria*, 63, 1–181, WHO, Geneva, Switzerland, 1986.
20. W.J. Hayes Jr., *Toxicology of Pesticides*, Williams & Wilkins, Baltimore, MD, 1975.
21. T. Namba, C.T. Nolte, J. Jackrel, and D. Grob, Poisoning due to organophosphate insecticides, acute and chronic manifestations. *Am. J. Med.* **50**:475, 1971.
22. H.P.M. Van Helden, B. Groen, E. Moor, B.H.C. Westerink, P.L.B. Bruihjnzeel, Newer generic approach to the treatment of organophosphate poisoning: Adenosine receptor mediated inhibition of ACh release. *Drug Chem. Toxicol.* **21**(1):171, 1998.
23. D.S. Barrett, F.W. Oehme, S.M. Kruckenberg, A review of organophosphorus ester-induced delayed neurotoxicity. *Vet. Hum. Toxicol.* **27**(1):22, 1985.
24. M.G. Cherniack, Toxicological screening for organophosphorus-induced delayed neurotoxicity: Complications in toxicity testing. *Neurotoxicology* **9**(2):249, 1988.
25. M. Lotti, Biological monitoring for organophosphate-induced delayed polyneuropathy. *Toxicol. Lett.* **33**(1–3):167–172, 1986.
26. M. Lotti, The pathogenesis of organophosphate polyneuropathy. *Crit. Rev. Toxicol.* **21**(6):465, 1991.
27. N. Aldridge, Postscript to the symposium on organophosphorus compound-induced delayed neuropathy. *Chem. Biol. Interact.* **87**(1–3):463, 1993.
28. C.D.S. Tomlin, Ed., *The Pesticide Manual—A World Compendium*, 11th ed., British Crop Protection Council, Farnham, Surrey, GU9 7PH, UK, 1997.
29. C.A. Anderson, J.C. Cavagnol, and C.J. Cohen, Guthion (azinphosmethyl): Organophosphorus insecticide. *Residue Rev.* **51**:23, 1974.
30. New York State Department of Health, *Chemical Fact Sheet: Guthion. Bureau of Toxic Substances Management*, Albany, NY, 5–24, 1984.
31. U.S. Environmental Protection Agency, *Pesticide Fact Sheet Number 100: Azinphos-methyl, Office of Pesticides and Toxic Substances*, Washington, DC, 5–18, 1986.
32. U.S. Environmental Protection Agency, Pesticide tolerance for O,O-dimethyl S-[(4-OXO-1,2,3-benzotriazin-3(4H)-YL)methyl] phosphorodithioate (azinphos-methyl). *Fed. Reg.* **54**:46082–84, 1989.
33. Occupational Health Services, Inc., *MSDS for Azinphos Methyl*, OHS Inc., Secaucus, NJ, February, 1991.
34. U.S. Public Health Service, *Hazardous Substance Data Bank*, Washington, DC, 1995.
35. J.D. Thrasher, Immune system problems after indoor Dursban applications. *Arch. Environ. Health* **48**(2):89, 1993.
36. T.A. Slotkin, Brain developmental damage occurs from common pesticide Dursban (chlorpyrifos). *Environ. Health Perspect.* **107**(suppl. 1): 1999.
37. U.S. Environmental Protection Agency, *Pesticide Fact Sheet Number 207: Coumaphos. Office of Pesticides and Toxic Substances*, Washington, DC, pp. 5–19, 1989.
38. U.S. Environmental Protection Agency, *Registration Standard for Pesticide Products Containing Coumaphos as the Active Ingredient*, Office of Pesticide Programs, Washington, DC, 1989.

39. National Cancer Institute, *Bioassay of Coumaphos for Possible Carcinogenicity.* (Technical Report No. 96. NCI-CG-TR-96.), National Institutes of Health, Bethesda, MD, 1979.

40. World Health Organization, *Dimethioate*, Environ. Health Criteria, 90, Geneva, Switzerland, 1989.

41. D.F. Heath, and M. Vandekar, Some spontaneous reaction of O,O-dimethyl S-ethylthioethyl phosphorothiolate and related compounds in water and on shortage and their effect on the toxicological properties of the compounds. *Bio. Chem J.* **67**:187, 1957.

42. U.S. Department of Health and Human Services, *Toxicological Profile for Diazinon produced by the Agency for Toxic Substances and Disease Registry*, Public Health Service, Public Health Service in Atlanta, GA, 1996.

43. R. Eisler, *Diazinon Hazards to Fish, Wildlife and Invertebrates: A Synoptic Review (Contaminant Hazard Review No. 9)*, U.S. Department of the Interior, Fish and Wildlife Service, Washington, DC, 1986.

44. U.S. Environmental Protection Agency, *Dichlorvos: Initiation of special review. Fed. Reg.* **53**:5542–49, 1988.

45. U.S. Environmental Protection Agency, Dichlorvos: Revocation of tolerance and food additive regulation. *Fed Reg.* **56**:5788–89, 1991.

46. H.H. North, and R.E. Menzer, Biotransformation of dimethoate by cell culture systems. *Pestic. Biochem. Physiol.* **2**:278, 1972.

47. T. Uchia, and R.D. O'Brien, Dimethoate degradation by human liver and its significance for acute toxicity. *Toxicol. Appl. Pharmacol.* **10**:89, 1967.

48. T. Uchida, W.C. Dauterman, and R.D. O'Brien, The metabolism of dimethoate by vertebrate tissues. *J. Agric. Food Chem.* **12**:48, 1964.

49. P.R.S. Chen, and W.C. Dauterman, Studies on the toxicity of dimethoate analogs and their hydrolysis by sheep liver amidase. *Pestic. Biochem. Physiol.* **1**:340, 1971.

49a. K.S. Khera, Evaluation of dimethoate (Cygon 4E) for teratogenic activity in the cat. *J. Environ. Pathol. Toxicol.* **2**:1283–1288, 1979.

49b. K.S. Khera, C. Whalen, G. Trivett, and Angers, Teratogenicity studies on pesticidal formulations of dimethoate, diuron, and lindane in rats. *Bull. Environ. Contam. Toxicol.* **22**:522–529, 1979.

50. P.H. Howard, Ed., *Handbook of Environmental Fate and Exposure Data for Organic Chemicals*, Vol. 3 *Pesticides*, Lewis Publishers, Chelsea, MI, 1991.

51. H. Kidd, and D.R. James, Eds., *The Agrochemicals Handbook*, 3rd ed., Royal Society of Chemistry Information Services, Cambridge, UK, 1991 (as updated).

52. U.S. Environmental Protection Agency, *Health Advisory Draft: Disulfoton*, Office of Drinking Water, Washington, DC, 1987.

53. U.S. Environmental Protection Agency, *Pesticide Fact Sheet Number 43: Disulfoton*, Office of Pesticides and Toxic Substances, Washington, DC, 1984.

54. Agency for Toxic Substances and Disease Registry, *Toxicological profile for disulfoton (Draft for public comment).* Atlanta, GA: U.S. Department of Health and Human Services, Public Health Service, 1995.

55. U.S. Environmental Protection Agency, *Registration Standards for Pesticide Products Containing Ethion as the Active Ingredient*, Office of Pesticides and Toxic Substances, Washington, DC, 1989.

56. U.S. Environmental Protection Agency, *Pesticide Fact Sheet Number 209: Ethion*, Office of Pesticides and Toxic Substances, Washington, DC, 1989.
57. B.M. Francis, *Toxic Substances in the Environment*, John Wiley & Sons, New York, 1984.
58. U.S. Environmental Protection Agency, *Guidance for the Reregistration of Pesticide Products Containing Fenitrothion*, USEPA, Office of Pesticide Programs, Registration Div., Washington, DC, 132, July 1987.
59. U.S. Environmental Protection Agency, *Pesticide Fact Sheet Number 142*, USEPA, Office of Pesticide Programs, Registration Div., Washington, DC, July 30, 1987.
60. National Research Council, *Proceedings of a symposium on fenitrothion: The long-term effects of its use in forest ecosystems*, Ottawa, Canada, 307, 1975.
61. J.I. Francis, and J.M. Branes, Studies on the mammalian toxicity of Fenthion. *Bull. World Health. Org.* **29**:205, 1963.
62. U.S. Environmental Protection Agency, *Draft Health Advisory Summary: Fonofos*, Office of Drinking Water, Washington, DC, 5–70, 1987.
63. U.S. Environmental Protection Agency, *Pesticide Fact Sheet Number 36: Fonofos*, Office of Pesticides and Toxic Substances, Washington, DC, 5–71, 1984.
64. L.J. Hoffman, I.M. Ford, and J.J. Menn, Dyfonate metabolism studies, I — Absorption, distribution, and excretion of dyfonate in rats. *Pest. Biochem. Physiol.* **1**:349, 1971.
65. R.K. Broadberg, *Estimation of Exposure of Persons in California to Pesticide Products Containing Isofenphos*, California Department of Food and Agriculture, Division of Pest Management, Sacramento, CA, 5–75, 1990.
66. B.W. Wilson, M. Hooper, E. Chow, R.J. Higgins, and J.B. Knaack, Antidotes and neuropathic potential of isofenphos. *Bull. Environ. Contam Toxicol.* **33**:386, 1984.
67. U.S. Environmental Protection Agency, *Pesticide Environmental Fate On-Line Summary: Isofenphos, Environmental Fate and Effects Division*, Washington, DC, 1990.
68. U.S. Environmental Protection Agency, *Guidance for the Reregistration of Pesticide Products Containing Methidathion as the Active Ingredient*, Washington, DC, 1988.
69. National Cancer Institute, *Bioassay of Malathion for Possible Carcinogenicity, Technical Reports 192*, National Institutes of Health, Bethesda, MD, 1979.
70. E.L. Baker, Epidemic malathion poisoning in Pakistan malaria workers. *Lancet* **1**:31, 1956.
70a. R.E. Gosselin, R.P. Smith, and H.C. Hodge, *Clinical Toxicology of Commercial Products*, 5th ed. Williams and Wilkins, Baltimore, MD, 1984.
71. Peter Flessel, Genetic toxicity of malathion: A review. *Environ. Molec. Mutagen.* **22**:717, 1993.
71a. N. Senanayake, and M.K. Johnson, Acute polyneuropathy after poisoning by a new organophosphate insecticide. *N. Engl. J. Med.* **306**:155–157, 1982.
71b. N. Senanayake, and L. Karalliedde, Neurotoxic effects of organophosphorus insecticides. *N. Engl. J. Med.* **316**:761–763, 1987.
72. M.A. Hussain, Anticholinesterase properties of methamidophos and acephate in insects and mammals. *Bull. Environ. Contam. Toxicol.* **38**:131, 1987.
73. L.M. Juarez, and J. Sanchez, Toxicity of the organophosphorus insecticide metamidophos (O,S-dimethyl phosphoramidothioate) to larvae of the freshwater prawn

*Macrobrachium rosenbergii* (De Man) and the blue shrimp *Penaeus stylirostris Stimpson. Bull. Environ. Contam. Toxicol.* **43**:302, 1989.

74. U.S. Environmental Protection Agency, *Pesticide Environmental Fate One Line Summary: Methamidophos*, U.S. EPA Environmental Fate and Effects Division. Washington, DC, 1989.
75. Agency for Toxic Substances and Disease Registry, *Toxicological Profile for Methyl Parathion*. Public Health Service, U.S. Dept. of Health and Human Services, Washington, DC, October 1990.
76. World Health Organization, *Environmental Health Criteria 145, Methyl parathion*, World Health Organization, Geneva, Switzerland, 1993.
77. Shell Chemical Company, *Summary of Basic Data for Mevinphos Insecticide, Technical data bulletin.*, San Ramon, CA, 1972.
78. T.S.S. Dikshith, Mutagenicity and teratogenicity of pesticides, in *Toxicology of Pesticides*, ed., T.S.S. Dikshith, Chapter 8, 190, CRC Press, Boca Raton, FL, 1991.
79. U.S. Environmental Protection Agency, *Ethyl Parathion, Correction to the Amended Cancellation Order*, OPP, USEPA, Washington, DC, February 4, 1992.
80. F. Matsumura, *Toxicology of Insecticides*, 2nd ed., Plenum Press, New York, 1982.
81. U.S. Environmental Protection Agency, Tolerance for pesticides in food, propetamphos. *Fed Reg.* **46**:43964–5, 1981.
82. J. Kumari, and N.B. Krishnamurthy, Mutagenicity studies with Safrotin in *Drosophila melanogaster* and mice. *Environ. Res.* **41**:44, 1986.
83. J.C. Franson, and J.W. Spann, Effects of dietary ABATE on reproductive success, duckling survival, behavior, and clinical pathology in game-farm mallards: Temephos. *Arch. Environ. Contam. Toxicol.* **12**:529, 1983.
84. U.S. Environmental Protection Agency, *Toxicology On-Line Summary: Temephos*, Environmental Fate and Effects Division, Washington, DC, 1985.
85. R.H. Pierce, R.B. Brown, K.R. Hardman, M.S. Henry, C.L. Palmer, T.W. Miller, and G. Witcherman, Fate and toxicity of Temephos applied to an intertidal mangrove community. *J. Am. Mosq. Control Assoc.* **4**:569, 1989.
86. U.S. Environmental Protection Agency, *Pesticide Fact Sheet Number 30: Trichlorfon*, Office of Pesticides and Toxic Substances, Washington, DC, 1984.
87. U.S. Environmental Protection Agency, *Guidance for Registration of Pesticide Products Containing Trichlorfon as the Active Ingredient*, Washington, DC, 1984.
88. U.S. Environmental Protection Agency, *Chemical Profile: Trichlorfon*, Washington, DC, 1985.
89. G.N. Berge, and I. Nafstad, Distribution and placental transfer of trichlorfon in guinea pigs. *Arch. Toxicol.* **59**:26, 1986.
90. U.S. Public Health Service, *Hazardous Substance Data Bank*, Washington, DC, 1995.

## APPENDIX 5-1 SELECTED LIST OF ORGANOPHOSPHORUS PESTICIDES

| | |
|---|---|
| Acephate | Isoxathion |
| Anilofos | Malathion |
| Azamethiphos | Mecarbam |
| Azinphos-ethyl | Methacrifos |
| Azinphos-methyl | Methamidophos |
| Bromophos ethyl | Methidathion |
| Butamifos | Methyl-parathion |
| Cadusafos | Mevinphos |
| Chlorethoxyfos | Monocrotophos |
| Chlorfenvinphos | Naled |
| Chlormephos | Omethoate |
| Chlorpyrifos | Oxydemeton-methyl |
| Chlorpyrifos-methyl | Parathion |
| Coumaphos | Parathion (ethyl) |
| Crotoxyphos | Phenthoate |
| Cyanophos | Phorate |
| Demeton-S-methyl | Phosalone |
| Diazinon | Phosmet |
| Dichlorvos | Phosphamidon |
| Dicrotophos | Phoxim |
| Dimethoate | Piperophos |
| Dimethylvinphos | Pirimiphos-ethyl |
| Disulfoton | Pirimiphos-methyl |
| Ediphenphos | Profenofos |
| EPN | Propaphos |
| Ethion | Propetamphos |
| Ethoprophos | Prothiofos |
| Etrimfos | Pyraclofos |
| Famphur | Pyrazophos |
| Fenamiphos | Pyridaphention |
| Fenitrithion | Quinalphos |
| Fenthion | Sulfotep |
| Fensulfothin | Sulprofos |
| Fenthion | Tebupirimfos |
| Fonofos | Temephos |
| Formothion | Terbufos |
| Fosamine | Tetrachlorvinphos |
| Fosthiazate | Thiometon |
| Heptenophos | Tolclofos-methyl |
| Iprobenfos | Triazophos |
| Isazofos | Trichlorfon |
| Isofenphos | Vamidothion |

# CHAPTER 6

# Herbicides and Fungicides

## INTRODUCTION

Herbicides, or weed killers, may be classified as pesticide chemicals. They can kill plants on contact, or they can be translocated (i.e., absorbed by one part of the plant and carried to other parts where they exert their primary toxic effect). Most commonly used herbicides have a low toxicity and have caused few adverse effects in users. Some herbicides pose more serious problems to the central nervous system (CNS) and can cause depression. The skin absorption of herbicides also may cause skin irritation, dermatitis, and photosensitization in addition to peripheral motor neuropathies.

The term *fungicide* is used for those chemicals that destroy or inhibit the growth of fungi. The history of fungicides is quite old. The use of sulfur compounds and copper sulfate dates back to between 1882 and 1934. A common practice to control fungal infection was the occasional sprinkling of lime and ash. Application of a Bordeaux mixture — a unique practice for infection control in grapevines of France — is well known. Later dithiocarbamates, such as thiram, ziram, zineb, and ferbam, entered the global market. The following pages describe in brief selected herbicides and fungicides vis-à-vis their chemical safety for humans.

Along with improved crop varieties, insecticides and herbicides have increased crop yields, decreased food costs, and enhanced the appearance of food. Without proper controls, however, the residue of some pesticides that remains on foods can create potential health risks. Before 1910, no legislation existed to ensure the safety of food and feed crops that were sprayed and dusted with pesticides. In 1910, the first pesticide legislation was designed to protect consumers from impure or improperly labeled products. During the 1950s and 1960s, pesticide regulation evolved to require maximum allowable residue levels of pesticides on foods and to deny registrations for unsafe or ineffective products. During the 1970s, acting under these strengthened laws, the newly formed U.S. Environmental Protection Agency (USEPA) removed DDT and other highly persistent pesticides from the marketplace. In 1996, the Food Quality Protection Act set

*Industrial Guide to Chemical and Drug Safety*, By T.S.S. Dikshith and Prakash V. Diwan
ISBN 0-471-23698-5 

stricter safety standards and required the review of older allowable residue levels to determine their safety. In 1999, federal and state laws required that pesticides meet specific safety standards. Accordingly, the USEPA regularly reviews and registers each product before it can be used, and sets safety levels and restrictions on each product intended for food or feed crops.

Like other pesticides, herbicides are rated based on their toxicity to humans. These ratings are used to determine the warnings and requirements (e.g., protective clothing and equipment, reentry periods into treated areas, other warnings) that are printed on the pesticide label. Toxicity ratings determine if the pesticide is classified as a restricted use product, thereby requiring a pesticide application license. As synthetic chemicals, herbicides and fungicides found extensive use in the control of plant diseases and eradication of unwanted plants. In fact, these chemicals have significantly reduced the strong competition of weeds with important and essential food crops. The biochemical mechanisms of herbicides and fungicides are complex and diverse. The phytotoxic and physiologic processes in species of plants and mammals, respectively, demonstrate further complexity. Consequently, the evaluation of product safety requires a proper pragmatic approach supported with quality data. Studies on the toxicologic effects and disposition of individual herbicides and fungicides have been reported by several workers.[1–6]

Once an herbicide or fungicide enters the water, even from overspray, its concentration quickly declines because of turbulence associated with mixing and dilution, volatilization, and degradation by sunlight and secondarily by microorganisms. None of the recommended herbicides are known to bioaccumulate and store in fatty tissue of animals and humans, and all are easily biodegraded, both in organisms and in the environment. Based on their chemistry and mechanism of action, herbicides have been classified into many groups, such as, chlorophenoxy compounds, phenols, benzoics, triazines, substituted ureas, dinitroanilines, chloroactamides, thiocarbamates, bipyridyliums, and dicarboxylic acids (Tables 6-1 and 6-2). Different herbicides (approximately 200 products) entered the global market more aggressively from 1945 onward. Some herbicides (e.g., phenoxy acids and benzoic acids) influence protein synthesis and DNA-RNA transcription chemistry. These chemicals have played a significant role in the management of unwanted plants on crop lands, lawns, and surface waters.

Several studies have reported on the synthesis, mode of action, and grouping of herbicides.[1–6] The acute oral and dermal toxicity profiles of different herbicidal compounds are covered under categories III or IV, slightly to moderately toxic compounds (Table 6-2). Herbicides for the most part are comparatively less toxic than normal chemicals to which humans are exposed. None of the herbicides currently listed by the USEPA have caused death, been found to be a carcinogen, or induced adverse effects to the endocrine systems in test animals. However, the manner of contamination of a variety of chemicals has opened Pandora's box, because different adverse effects in animals and humans (even with compounds having a large margin of $LD_{50}$ values) have been reported.[6a,6b,6c]

**TABLE 6-1 Different Groups of Herbicides**

| Group | Chemical |
|---|---|
| Benzoic acids | Chloramben, Dicamba, 2,3,6-TBA |
| Bipyridyliums | Diquat, Paraquat |
| Chloroacetamides | Alachlor, CDAA, Propachlor |
| Dicarboxylic acids | Endothal, Amitrole, Glyphosate |
| Dinitroanilines | Benefin, Butralin, Dinitramine, |
| Fluchloralin | Isopropalin, Nitralin, Oryzalin, Penoxylin, Profluralin, Trifluralin |
| Phenols | Dinoseb, DNOC, |
| Phenoxy aliphatic acids | 2,4-D, 2,4-DB, 2,4,5-T, Dichlorprop, MCPA, MCPB, Mecoprop, Silvex |
| Substituted ureas | Chloroxuron, Diuron, Fenuron, Linuron Metobromuron, Monuron, Norea |
| Thiocarbamates | Butylate, Cycloate, Diallate, EPTC, Molinate, Pebulate, Trillate, Vernolate |
| Triazines | Ametryne, Atrazine, Chlorazine, Cyanazine Prometone, Prometryne, Propazine, Simazine, Terbutryne |

**TABLE 6-2 Acute Toxicity of Herbicides**

| | Acute Toxicity $LD_{50}$ (mg/kg) | | | |
|---|---|---|---|---|
| Herbicides | Oral | Dermal | NOEL* | RfD* |
| Chlorsulfuron | 5545 | 3400 | 5 | 0.05 |
| Clopyralid | 4300 | >2000 | 15 | 0.5 |
| Dicamba | 1707 | >2000 | 125 | 0.03 |
| Glyphosate | >5000 | 7940 | 400 | 0.1 |
| MCPA (many) | 1160 | >4000 | 0.15 | 0.0005 |
| Metsulfuron | >5000 | >2000 | 250 | 0.25 |
| Triclopyr | 713 | >2000 | 3 | 0.05 |
| 2,4-D (acid) | 764 | >2000 | 1 | 0.01 |

*Source*: A.S. Felsot, (1998).[10]

*NOEL is the no observable effect level based on a 2-year chronic feeding study. The RfD is the USEPA's reference dose, expressed as mg/kg/d. It is based on the NOEL divided by an uncertainty factor of 100. The RfD is used to estimate the daily dose that an organism can be exposed to over its lifetime with reasonable certainty of no harmful effects.

For example, glyphosate (Roundup®)—the largest selling herbicide, particularly for genetically engineered plants and food crops—has been implicated in human health cases.

Report on a population-based case control study indicated a significant risk increase in non-Hodgkin's lymphoma (NHL) among certain workers exposed to 2,4-D, 2,4,5-T, and glyphosate. It has been observed that NHL has increased rapidly in certain countries of the West. The American Cancer Society has reported an increase (approximately 80%) in the incidence of NHL since 1970. The data seem to be inadequate and demand more research on animal species and epidemiological observations. Another instance is of simazine with an oral $LD_{50}$ value of 1,000 mg/kg in rats. Studies showed that people exposed to Dursban for a long period (approximately 1 to 5 years) experienced immune disorders. In fact, in the past five decades, information on the mode of action and type of adverse effects on animals and humans exposed to different herbicides is appreciably extensive and often contradictory.[7–10] This has been due to the interaction of many contaminants and related substances with the pure parent compound. The importance of toxicologic disasters caused by TCDD-contaminated herbicides and related chemicals and the importance of quality data to arrive at pragmatic conclusions (after the extrapolation of animal and human data) requires no emphasis.

## HERBICIDES

### Alachlor (Lasso)

IUPAC name: 2-chloro-2′,6′-diethyl-N-(methoxymethyl) acetanilide
Molecular formula: $C_{14}H_{20}ClNO_2$
Toxicity class: USEPA: III; WHO: III

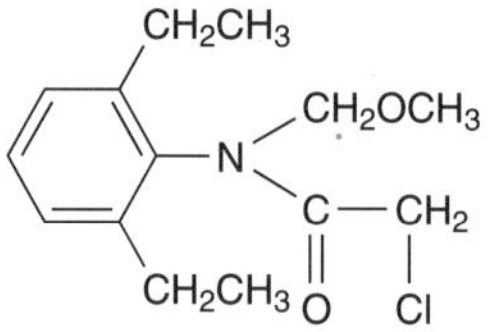

$CH_2CH_3$
$CH_2OCH_3$
N
$C—CH_2$
$CH_2CH_3$ O Cl

**Figure 6-1.** Structural formula of alachlor.

***Uses*** Alachlor is a colorless to yellow crystal compound.[1,11–13] Alachlor is an aniline herbicide used to control annual grasses and broadleaf weeds in field corn, soybeans, and peanuts. It is a selective systemic herbicide, absorbed by germinating shoots and roots. It works by interfering with a plant's ability to produce protein and by interfering with root elongation. This compound is one of the most extensively used herbicides. The USEPA categorizes alachlor under restricted use pesticides (RUP), and it should be purchased and used only by certified applicators.

***Toxicity*** Alachlor is slightly toxic to animals. The acute oral $LD_{50}$ of alachlor in rats ranges from 930 kg to 1,350 mg/kg. In the mouse, the acute oral $LD_{50}$ ranges between 1,910 and 2,310 mg/kg. The acute dermal $LD_{50}$ in rabbits is 13,300 mg/kg, but some formulated materials can be more toxic, with acute dermal $LD_{50}$ values ranging from 7,800 to 16,000 mg/kg. Skin irritation is slight to moderate. The inhalation $LC_{50}$ in rats reportedly is greater than 23.4 mg/L for 6 hours of exposure. High doses of alachlor caused adverse effects in the stomach and thyroid, and produced nasal turbinate tumors. Prolonged exposure to alachlor resulted in toxicity to the liver, spleen, and kidney in rats and dogs.[14–16]

### Amitrole (Weedazol, Azolon)

IUPAC name: H-1,2,4-triazole-3-ylamine
Molecular formula: $C_2H_4N_4$
Toxicity class: USEPA: III; WHO: III

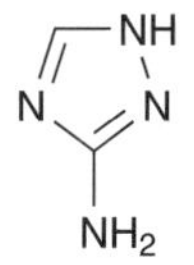

**Figure 6-2.** Structural formula of amitrole.

***Uses*** Amitrole is a white to off-white, odorless crystalline powder with a bitter taste. It belongs to chemical class triazole. Amitrole is an RUP and should be used only by certified applicators. The USEPA canceled application of amitrole on food crops in 1971 because it caused cancer in experimental animals.[2,12]

Amitrole is a nonselective systemic triazole herbicide. It is used for control of annual grasses and perennial and annual broadleaf weeds, for poison ivy control, and for control of aquatic weeds in marshes and drainage ditches. This compound is compatible with many other herbicides. It is available as wettable powders, soluble concentrates, and water-dispersible granules. Amitrole was involved in the Delaney Clause's first enforcement.[12,13]

***Toxicity*** Amitrole has a very low acute toxicity to humans and animals. The acute oral and dermal $LD_{50}$ to rats are greater than 5,000 mg/kg. The acute dermal $LD_{50}$ in rabbits is greater than 200 mg/kg. Birth defects have occurred in the pups of pregnant rabbits, rats, and mice exposed to amitrole, but only at doses high enough to produce signs of toxicity in the mothers. In high doses (5 and 25 mg/kg/day), amitrole caused atrophy of the thymus and spleen in animals.[17–21] Associated symptoms in humans include skin rash, vomiting, diarrhea, and nosebleeds. Poisoning by amitrole is characterized by increased intestinal peristalsis. Amitrole is a mild skin and eye irritant.

## TRIAZINES

The triazines are a group of chemically similar herbicides. The group includes atrazine, cyanazine, propazine, and simazine, which are primarily used to control broadleaf weeds. Atrazine currently is one of the two most widely used agricultural pesticides in the United States. About three-fourths of field corn and sorghum acres are treated with atrazine annually for weed control, which accounts for most of the 75 to 85 million pounds used per year. Other uses are the control of weeds on turf and lawns, pineapples, sugarcane, wheat, and macadamia nuts. In 1994, approximately 21 to 34 million pounds of cyanazine were applied annually, with 85% to 90% used to control weeds on field corn. Presently, cyanazine use is being phased out. With gradual phase-out, estimated annual use has fallen to approximately 12 to 15 million pounds. After December 2002, the use of cyanazine will be banned. In November 1994, the USEPA began a special review of atrazine, cyanazine, and simazine based on concerns of cancer risks to consumers exposed through consumption of food, drinking water, and exposure to treated lawns.

### Atrazine

IUPAC name: 2-chloro-4-ethylamino-6-isopropylamino-1,3,5-triazine
Molecular formula: $C_8H_{14}ClN_5$
Toxicity class: USEPA: III; WHO: III

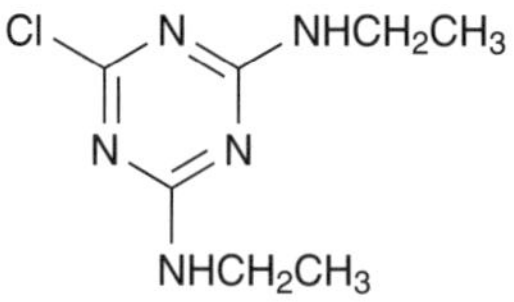

**Figure 6-3.** Structural formula of atrazine.

***Uses*** Atrazine is a symmetrical triazine used for broadleaf control on corn. It forms major metabolite(s) by dealkylation of the ethyl or isopropyl side chains. Approximately 120 million pounds of triazines are used in the United States alone on corn-growing fields.

***Toxicity*** The acute oral toxicity of atrazine in rats is 2,850 mg/kg, and the acute dermal toxicity in rabbits is 7,550 mg/kg. The acute inhalation $LC_{50}$ (1 hour) in rats is greater 167 mg/L. Atrazine did not cause any primary irritation in rabbits, although it caused eye irritation in rabbits.[17] A carcinogenicity study of mice exposed to atrazine through diet (82 ppm) for 18 months is sketchy and requires more confirmatory data.[23,24]

### Bromacil (Hyvarx)

IUPAC name: 5-bromo-3-sec-butyl-6-methyluracil
Molecular formula: $C_9H_{13}BrN_2O_2$

Toxicity class: USEPA: IV (dry), II (liquid); WHO: III

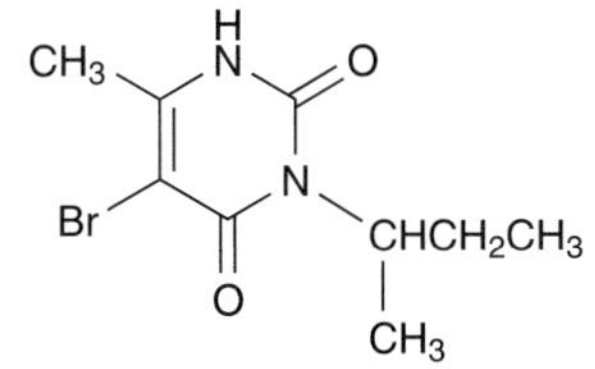

**Figure 6-4.** Structural formula of bromacil.

***Uses*** Bromacil is a colorless crystalline solid. It is used for the control of annual and perennial grasses, broadleaf weeds, and woody plants.[12,13] Bromacil is a herbicide used for bush weed control on non-cropland areas. It is especially useful against perennial grasses. It is also used for selective weed control in pineapple and citrus crops. It interferes with photosynthesis of plants. It is available in granular, liquid, water-soluble liquid, and wettable powder formulations.

***Toxicity*** The acute oral $LD_{50}$ for rats is 5,200 mg/kg, whereas the acute dermal $LD_{50}$ to rabbits is greater than 5,000 mg/kg. The primary skin irritation and eye irritation to rabbits is moderate. The acute inhalation $LC_{50}$ to rats is greater than 4.8 mg/L. Studies on carcinogenicity are inadequate for a meaningful evaluation and require more data.[25]

### **Butachlor** (Machete)

IUPAC name: N-(butoxymethyD-2-Chloro-2–6-diethylacetanilide
Molecular formula: $C_{17}H_{26}ClNO_2$
Toxicity class: USEPA: III; WHO: III

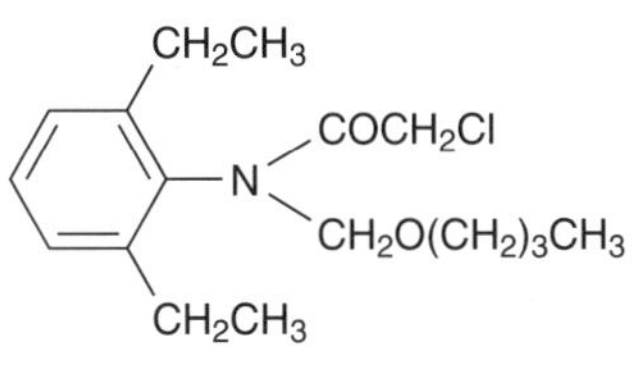

**Figure 6-5.** Structural formula of butachlor.

***Uses*** Butachlor is used as a preemergence herbicide for the control of annual grasses and broadleaf weeds in rice, wheat, barley, cotton, sugar beet, and peanut crops.

***Toxicity*** The acute oral $LD_{50}$ to rats is 3,300 mg/kg, whereas the acute dermal $LD_{50}$ to rabbits is 4,000 mg/kg. Butachlor is a mild irritant to the skin and eyes of rabbits.

## Dicamba (Velsicol 59-58-cs-11)

IUPAC name: 3,6-dichloro-2-methoxybenzoic acid
Molecular formula: $C_8H_6Cl_2O_3$
Toxicity class: USEPA: IV; WHO: III

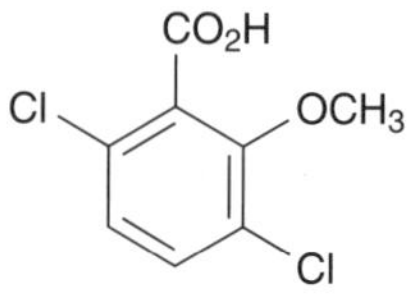

**Figure 6-6.** Structural formula of dicamba.

***Uses*** In pure form, dicamba is a white crystalline solid. The technical acid is a pale buff crystalline solid. Dicamba is a benzoic acid herbicide that is used for the control of infestation on leaves and soil. Dicamba controls annual and perennial broadleaf weeds in grain crops and grasslands, and is used to control brush and bracken in pastures. It kills broadleaf weeds before and after they sprout. Legume weeds are killed by dicamba. In combination with a phenoxyalkanoic acid or other herbicides, dicamba is used in pastures, rangeland, and noncrop areas such as fencerows and roadways to control weeds.[2,17]

***Toxicity*** The acute oral $LD_{50}$ to rats is 2.74 g/kg and the acute dermal $LD_{50}$ to rats is more than 2,000 mg/kg. The acute inhalation $LC_{50}$ to rats is greater than 200 mg/L. Dicamba has slight primary irritation, and primary eye irritation in rabbits has shown corrosive effects. The carcinogenicity of dicamba in mice, rats, and dogs demonstrated negative effects.[26,27]

## Glyphosate (Roundup)

IUPAC name: N-(phosphonomethyl)glycine
Molecular formula: $C_3H_8NO_5P$
Toxicity class: USEPA: III; WHO: III

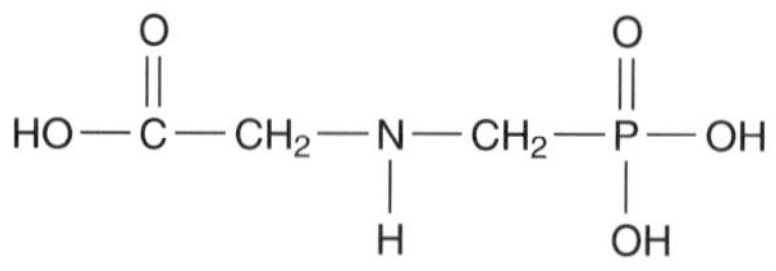

$$HO-C(=O)-CH_2-NH-CH_2-P(=O)(OH)-OH$$

**Figure 6-7.** Structural formula of glyphosate.

***Uses*** Glyphosate is a phosphanoglycine and does not inhibit cholinesterase activity.[12,13] Glyphosate is a broad-spectrum, nonselective systemic herbicide with moderate toxicity. Essentially, it is useful on all annual and perennial plants including grasses, sedges, broadleaf weeds, and woody plants. It is used on non-cropland and on a variety of crops.[17,28–34]

***Toxicity*** The acute oral $LD_{50}$ for rats ranges from 4,320 to 5,600 mg/kg. The acute dermal $LD_{50}$ for rabbits ranges from 794 mg/kg to 5,010 mg/kg. Glyphosate did not demonstrate a primary irritation effect in rabbits but was found to be a mild eye irritant. Human study volunteers showed no visible skin changes, sensitization, or other adverse effects.[32,33]

### **Simazine** (Gesatop)

IUPAC name: 2-chloro-4,6-bis(ethylamine)-1,3,5-triazine
Molecular formula: $C_7H_{12}ClN_5$
Toxicity class: USEPA: IV; WHO: III

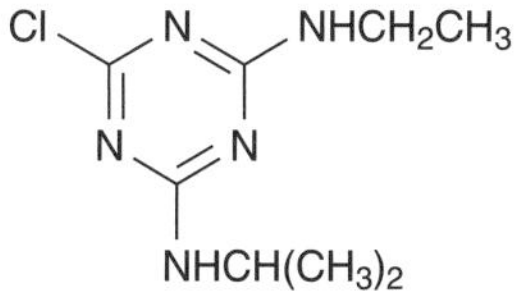

**Figure 6-8.** Structural formula of simazine.

***Uses*** Simazine is a selective triazine herbicide. It is used to control broadleaf weeds and annual grasses in field, berry fruit, nuts, vegetable, and ornamental crops, turfgrass, orchards, and vineyards. At higher rates, it is used for nonselective weed control in industrial areas. Before 1992, simazine was used to control submerged weeds and algae in large aquariums, farm ponds, fish hatcheries, swimming pools, ornamental ponds, and cooling towers. Simazine is available in wettable powder, water-dispersible granule, liquid, and granular formulations. It also is used for soil treatment.[2,12,17]

***Toxicity*** The acute oral toxicity of simazine for rats is more than 15,380 mg/kg, and the acute dermal $LD_{50}$ in rabbits is 10,200 mg/kg. It demonstrated moderate primary skin and eye irritation effects in rabbits. The acute inhalation $LC_{50}$ to rats is greater than 5 mg/L. Reports have indicated that simazine caused no serious effects in humans except minor skin rashes in some cases. Similarly, it caused no long-term effects in humans.[5,36]

### **Picloram** (Tordon)

IUPAC name: 4-amino-3,5,6-trichloropyridime-2-carboxylic acid
Molecular formula: $C_6H_3$ $CL_3N_2O_2$

Toxicity class: USEPA: IV; WHO: III

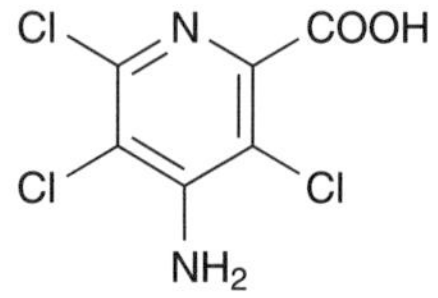

Figure 6-9. Structural formula of picloram.

***Uses*** Picloram belongs to the pyridine family of compounds. It is a systemic herbicide used for the control of woody plants and a range of broadleaf weeds. Most grasses are resistant to picloram. It is classified by the USEPA as RUP because of its mobility in water, combined with extreme sensitivity to many important crop plants.[2,12,13,37]

***Toxicity*** Picloram is slightly to practically nontoxic to animals via ingestion. The acute oral $LD_{50}$ for rats ranges from 5,000 to 8,200 mg/kg. The acute dermal $LD_{50}$ in rabbits is greater than 4,000 mg/kg. The technical grade is moderately toxic by inhalation, with $LC_{50}$ of greater than 0.35 mg/L. Available reports suggest the absence of carcinogenicity in animals, but require more confirmatory data.[37,38]

### **2,4-D** (AgricornD, Capri, Kay-D, Dioweed, Desormone, Weedtox)

IUPAC name: 2,4-dichlorophenoxy acetic acid
Molecular formula: $C_8H_6Cl_2O_3$
Toxicity class: USEPA: II; WHO: II

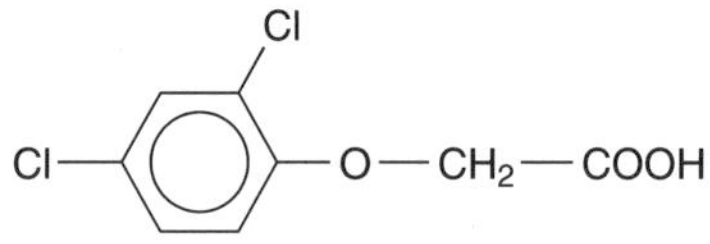

Figure 6-10. Structural formula of 2,4-D.

***Uses*** 2,4-D is a colorless powder with a mild phenylic odor. There are many derivatives of 2,4-D, including esters, amines, and salts. As a systemic herbicide, it is used to control many types of broadleaf weeds. It is used in cultivated agriculture, in pasture and rangeland applications, forest management, home, garden, and to control aquatic vegetation. It may be found in emulsion form, in aqueous solutions (salts), and as a dry compound.

Mixtures with other herbicides also are used for weed control. The product Agent Orange, used extensively throughout Vietnam, was about 50% 2,4-D. However, the controversy about the use of Agent Orange was associated with a contaminant (dioxin) in 2,4,5-trichlorophenoxy acetic acid (2,4,5-T) ($C_8H_5Cl_3O_3$), not with 2,4-D. It should be well understood that 2,4,5-T is different from 2,4-D but similar to 2,4-D as a herbicide component of the defoliant.[39]

***Toxicity*** The acid form is of slight to moderate toxicity to animals. The acute oral $LD_{50}$ of 2,4-D ranges from 375 to 764 mg/kg in the rat, 138 mg/kg in mice, and from less than 320 to 1,000 mg/kg in guinea pigs. The acute dermal $LD_{50}$ values are 1,500 mg/kg in rats and 1,400 to less than 2,400 mg/kg in rabbits, respectively. In humans, prolonged inhalation of 2,4-D causes coughing, burning, dizziness, and temporary loss of muscle coordination. Symptoms of poisoning include fatigue and weakness with possible nausea. Prolonged exposure to 2,4-D (50 mg/kg/day) in the diet for 2 years showed no adverse effects. High doses of 2,4-D, however, are known to cause birth defects with increased skeletal abnormalities, such as delayed bone development and wavy ribs.[39]

Rats fed with 2,4-D for 2 years showed an increase in malignant tumors.[2] Female mice given a single injection of 2,4-D developed cancer (reticulum cell sarcomas). Another study of rodents shows a low incidence of brain tumors at moderate exposure levels (45 mg/kg/day) over a lifetime.[19] However, a number of questions have been raised about the validity of this evidence and thus about the carcinogenic potential of 2,4-D. Several studies suggest a connection with prolonged 2,4-D exposure and cancer in humans.[40,41,41a] An increased occurrence of non-Hodgkin's lymphoma was found among Kansas and Nebraska farm populations associated with the spraying of 2,4-D. In contrast, 2,4-D studies done in New Zealand, Washington, New York, Australia, and on Vietnam veterans from the United States were all negative.[41b] Thus, there remains considerable controversy about the methods used in various studies and their results, and the carcinogenic status of 2,4-D to humans is not clear.[17,40–43] In view of the conflicting reports, a pragmatic approach is necessary to identify the carcinogenic potential versus the dose-response relationship of 2,4-D.

### **Fluazifop-p-Butyl** (Fusilade Zeneca, Venture, Winner)

IUPAC name: 2-(4-(5-trifluoromethyl-2-pyridyloxy) propionic acid
Molecular formula: $C_{19}H_{20}F_3NO_4$
Toxicity class: Not classified

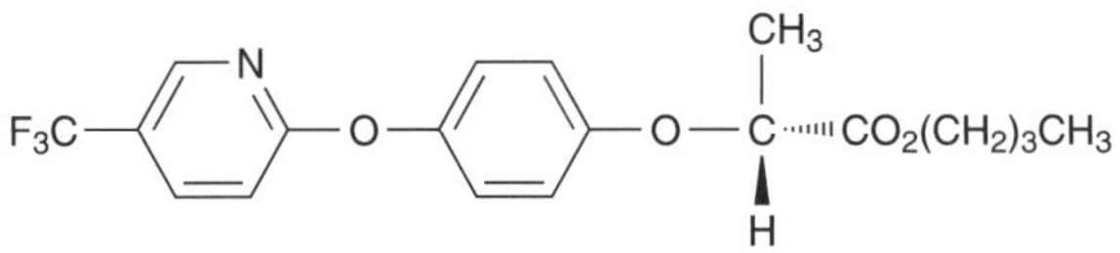

**Figure 6-11.** Structural formula of fluazifop-p-butyl.

***Uses*** Fluazifop-p-butyl is a selective postemergence phenoxy herbicide. It is used for control of the annual and perennial grass weeds in cotton, soybeans, stone fruits, asparagus, coffee, and others. The formulation is an emulsifiable concentrate. Fluazifop-p-butyl formulations may contain some fluazifop-butyl.[13,17]

***Toxicity*** The acute oral $LD_{50}$ of technical-grade fluazifop-p-butyl to rats ranges from 2,451 to 4,096 mgt/kg. The acute dermal $LD_{50}$ in rabbits is greater than 2,400 mg/kg and causes mild skin and eye irritation in animals. The acute inhalation $LC_{50}$ of the formulation Fusilade DX in rats ranges from 0.54 and 0.77 mg/L. Administration of a single large oral dose of Fusilade 2000 caused severe stomach, intestine, and CNS disturbance. The symptoms and signs of toxicity include drowsiness, dizziness, loss of coordination, and fatigue. Fluazifop-p-butyl is slightly toxic via the dermal route as well. It is reported to cause only slight skin and mild eye irritation in rabbits, but no skin sensitization in guinea pigs.[44]

### **Napropamide** (Devrinol Zeneca)

IUPAC name: N-N-diethyl-2-(1-naphthyloxy) propionamide
Molecular formula: $C_{17}H_{21}NO_2$
Toxicity class: USEPA: III; WHO: III

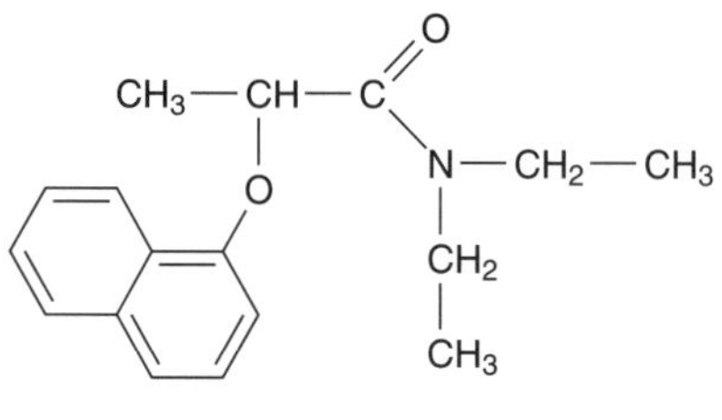

**Figure 6-12.** Structural formula of napropamide.

***Uses*** Napropamide is a slightly toxic amide herbicide. It is grouped by the USEPA under GUP.[13] Napropamide is a selective systemic amide herbicide used to control a number of annual grasses and broadleaf weeds. It also is applied to soil growing vegetables, fruit trees and bushes, vines, strawberries, sunflowers, tobacco, olives, and other crops. The formulations include emulsifiable concentrate, wettable powder, granules, and suspension concentrates.[12]

***Toxicity*** The acute oral $LD_{50}$ of napropamide to rats ranges from 4,680 to 5,000 mg/kg. It is considered practically nontoxic to animals by ingestion. Similarly, the acute dermal $LD_{50}$ to rabbits is as high as 4,640 mg/kg. The acute inhalation $LC_{50}$ for rats is 0.2 mg/L. A formulation of 43.2% napropamide (Devrinol 4F) has an acute $LC_{50}$ of 0.2 mg/L. The toxic effects after an acute exposure in rats include diarrhea, excessive tearing and urination, depression, salivation, rapid weight loss, respiratory changes, decreased blood pressure, and accumulation of fluid in body cavities.

Rats fed napropamide at doses of up to 30 mg/kg/day for 13 weeks showed no significant effects, but at doses of 40 mg/kg/day for the same length of time, female rats experienced a reduction in uterine weight. In a feeding study in dogs over 13 weeks, males experienced decreased liver and body weight, as well as some changes in blood chemistry at the highest dose tested. But there were no tissue changes in either rats or dogs at 100 mg/kg/day.[45–47] One study conducted over three successive generations of rats showed a decrease in body weight gain in fetal pups.[47a]

### **Dalapon** (Dalacide)

IUPAC name: 2,2-dichloropropionic acid
Molecular formula: $C_3H_4\ CL_2O_2$
Toxicity class: USEPA: II; WHO: III

Cl
|
$H_3C$–C(–COOH)(–Cl)

**Figure 6-13.** Structural formula of dalapon.

***Uses*** The pure acid form of dalapon is a colorless liquid with an acrid odor. Dalapon is a type of acid that usually is formulated with sodium and magnesium salts.[44] Commercial products usually contain 85% sodium salt or mixed sodium and magnesium salts of dalapon. As a sodium-magnesium salt, it is a white to off-white powder. Dalapon is used as a plant growth regulator and for the control of a variety of annual and perennial grasses, (e.g., quackgrass, Bermuda grass, Johnson grass, cattails, rushes). Dalapon is selective in action and kills only certain plants, while sparing nontarget vegetation. Dalapon is used on sugarcane, sugar beets, various fruits, potatoes, carrots, asparagus, alfalfa, and flax. It is used in public and domestic sites, forestry, home gardening, and in or near water to control reed and sedge growth. As a pre- and postemergence herbicide, dalapon in water-soluble powder form is applied extensively in agriculture.[2] The USEPA has grouped dalapon under RUP, and it must be handled by trained, experienced persons.

***Toxicity*** Dalapon is moderately toxic to humans. Skin and inhalation exposure could be of significance to dalapon production workers, pesticide applicators, and some agricultural workers. Effects of acute exposure include absence of appetite, slowed heartbeat, skin irritation, eye irritation (e.g., conjunctivitis or corneal damage), gastrointestinal disturbances (e.g., vomiting or diarrhea), tiredness, pain, and irritation of the respiratory tract. Dalapon is an acid that may cause corrosive injury to body tissues. Eye exposure to this material can cause permanent eye damage. Skin burns may occur from dermal exposure to dalapon, especially when skin is moist.[17]

The acute oral $LD_{50}$ of dalapon for male rats is as high as 9,330 mg/kg, and for female rabbits is 3,860 mg/kg, suggesting that rabbits are more susceptible to the herbicide. The sodium salt of dalapon (in a dry powder formulation) to rabbit eyes produced pain and irritation, followed by severe conjunctivitis and corneal injury, which healed after several days. A 10% solution produced slight pain and conjunctivitis. Repeated or prolonged exposure to dalapon caused irritation to the mucous membrane linings of the mouth, nose, throat, lungs, and eyes.

Chronic skin contact with the herbicide can lead to moderate irritation or even mild burns, although occasional contact is not likely to produce irritation. Dalapon is not absorbed through the skin in toxic amounts. Long-term dalapon feeding studies in dogs and rats did not produce lesions, but did show increased kidney weights in animals fed very high daily doses. Rats fed 50 mg/kg/day for 2 years showed a slight average increase in kidney weight. No adverse effects were seen in this study in rats fed 15 mg/kg/day. In a 1-year study of dogs fed 100 mg/kg/day, there was a slight average increase in kidney weight; no adverse effects were seen at 50 mg/kg/day. These mild effects on the kidneys are consistent with data that show that dalapon is rapidly excreted in the urine. Dalapon produced no adverse effects on fertility or reproduction, except at extremely high doses. Doses of 1,500 and 2,000 mg/kg in female rats and 1,000 and 1,500 mg/kg in rat pups resulted in reduced weight or lower rates of weight gain.[45,46]

### **Diquat** (Reglex)

IUPAC name: 1,1′-ethylene-2,2′-bipyridyldiylium dibromide salt
Molecular formula: $C_{12}H_{12}\ Br_2N_2$
Toxicity class: USEPA: II; WHO: II

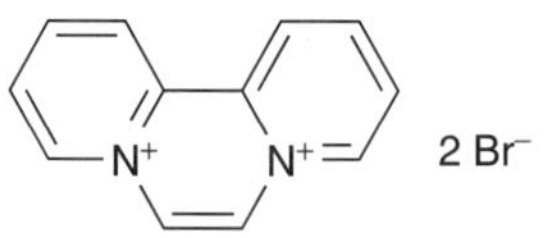

**Figure 6-14.** Structural formula of diquat.

***Uses*** Technical diquat dibromide is more than 95% pure and forms white to yellow crystals. It is used to desiccate potato vines and seed crops, to control flowering of sugarcane, and for industrial and aquatic weed control. It is not residual (i.e., it does not leave any trace of herbicide on or in plants, soil, or water). Diquat dibromide is a nonselective, quick-acting herbicide and plant growth regulator, causing injury only to the parts of the plant to which it is applied. Diquat dibromide is referred to as a desiccant because it causes a leaf or entire plant to dry out quickly.

***Toxicity*** Diquat dibromide is a moderately toxic herbicide grouped under GUP. The acute oral toxicity differs in species of mammals. For instance, in rats 120 mg/kg, in mice 233 mg/kg, in rabbits 188 mg/kg, and in guinea pigs and dogs 187 mg/kg. Cows are sensitive and have much lower oral $LD_{50}$: 30 to 56 mg/kg. The acute dermal $LD_{50}$ for diquat dibromide is approximately 400 to 500 mg/kg in rabbits, indicating moderate toxicity by this route. A single dose of diquat dibromide was not irritating to the skin of rabbits, but repeated dermal dosing did cause mild redness, thickening, and scabbing. Moderate to severe eye membrane irritation occurred when diquat dibromide was administered to rabbits.

Ingestion of sufficient doses is known to cause severe irritation of the mouth, throat, esophagus, and stomach, followed by nausea, vomiting, diarrhea, severe dehydration, and alterations in body fluid balances, gastrointestinal discomfort, chest pain, diarrhea, kidney failure, and toxic liver damage. Skin absorption of high doses may cause symptoms similar to those that occur following ingestion. Very large doses of the herbicide do result in convulsions and tremors in animals.[7,17,47]

### **Paraquat** (Gramoxone)

IUPAC name: 1,1′-dimethyl-4,4′-bipyridinium
Molecular formula: $C_{12}H_{14}N_2$
Toxicity class: USEPA II; WHO: II

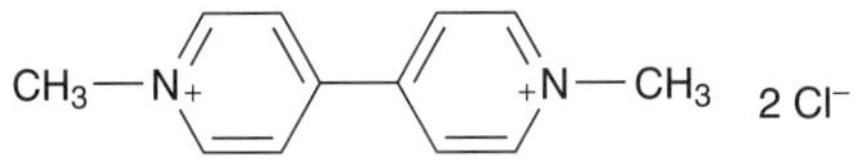

$CH_3-N^+$ (pyridinium)–(pyridinium) $^+N-CH_3$ $2\ Cl^-$

**Figure 6-15.** Structural formula of paraquat.

***Uses*** Paraquat is a colorless, odorless, white or pale yellow crystalline solid that is hygroscopic. Paraquat is a quaternary nitrogen herbicide that is widely used for broadleaf weed control. It is a quick-acting, nonselective compound that destroys green plant tissue on contact and by translocation within the plant. It has been employed for killing marijuana in the United States and Mexico. It also is used as a crop desiccant and defoliant, as well as an aquatic herbicide. Paraquat is highly persistent in the soil environment, with a reported field half-life of greater than 1,000 days.[5–7,17]

***Toxicity*** Paraquat is highly toxic via ingestion, with reported acute oral $LD_{50}$ values of 110 to 150 mg/kg in rats, 50 mg/kg in monkeys, 48 mg/kg in cats, and 50 to 70 mg/kg in cows. The toxic effects of paraquat are due to the cation; halogen anions have little toxic effect. The dermal $LD_{50}$ in rabbits is 236 to 325 mg/kg, indicating moderate toxicity to skin. The acute inhalation $LC_{50}$ (4-hour) is greater than 20 mg/L for the technical grade of the compound. It causes skin and eye irritation in rabbits (severe for some of the formulated products) and in some formulations has caused skin sensitization in guinea pigs. Effects due to high acute exposure to paraquat may include excitability and lung congestion, which in some cases leads to convulsions, incoordination, and death by respiratory failure.

When swallowed, paraquat causes burning of the mouth and throat followed by gastrointestinal tract irritation, resulting in abdominal pain, loss of appetite, nausea, vomiting, and diarrhea. Other toxic effects include thirst, shortness of breath, rapid heart rate, kidney failure, lung sores, and liver injury. Persons with lung problems may be at increased risk from paraquat exposure. In general, paraquat affects the lungs, heart, liver, kidneys, cornea, adrenal glands, skin, and digestive system. Mice fed paraquat dichloride for 99 weeks at high levels did not show cancerous growths. Rats fed high doses for 113 (male) or 124 (female) weeks developed lung, thyroid, skin, and adrenal tumors; thus, the evidence regarding the carcinogenic effects of paraquat is inconclusive.[2,17,51–53]

## FUNGICIDES

For the control of fungal infection on fruits and vegetables, crop workers often spray fungicides. Some of the fungicides convert to ethylene thiourea, a known carcinogen. Steenland et al.[2] measured chemical toxicity in these workers and in 14 lightly exposed landowners and 31 nonexposed controls. There was an increase in thyroid-stimulating hormone and an increase in genotoxicity, suggesting that the fungicide ethylenebis (dithiocarbamate) affects the thyroid gland and the lymphocyte genome in heavily exposed backpack sprayers.[2]

### **Benomyl** (Benlate, Fundazol, Pilarben, Romyl)

IUPAC name: Methyl-1-[(butyl) carbamoyl]-H-benzimidazol-2-ylcarbamate)
Molecular formula: $C_{14}H_{18}N_4O_3$
Toxicity class: USEPA: IV; WHO: III

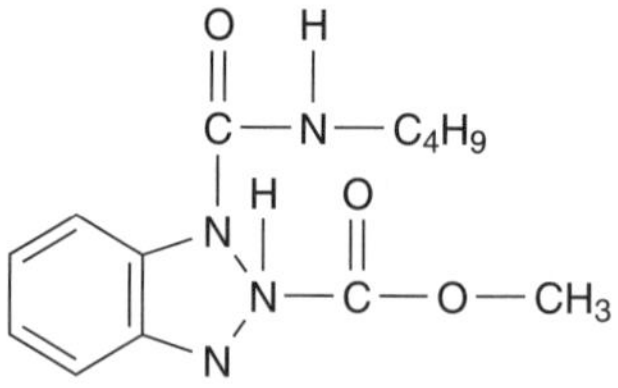

**Figure 6-16.** Structural formula of benomyl.

***Uses*** Benomyl is a crystalline solid compound. It has little or no odor, and the USEPA has grouped it under GUP. Benomyl is a systemic benzimidazole fungicide that is selectively toxic to microorganisms and invertebrates, especially earthworms. It is used against a wide range of fungal diseases of field crops, fruits, nuts, ornamentals, mushrooms, and turf. Formulations of benomyl include wettable powder, dry flowable powder, and dispersible granules.[12,13,17]

***Toxicity*** Benomyl has very low acute toxicity to mammals. The acute oral $LD_{50}$ is greater than 10,000 mg/kg in rats, and the acute dermal $LD_{50}$ is greater than 3,400 mg/kg (using a 50% wettable powder formulation). Skin irritation may occur in workers exposed to benomyl. Benomyl is readily absorbed into the body by inhaling the dust, but there are no reports of toxic effects to humans by this route of exposure. The acute inhalation $LC_{50}$ in rats is greater than 2 mg/L. When rats were fed diets containing approximately 150 mg/kg/day for 2 years, no toxic effects were observed. However, dogs fed benomyl in their diets for 3 months indicated evidence of altered liver function at the highest dose (150 mg/kg). The damage progressed to more severely impaired liver function and liver cirrhosis after 2 years.[53–56]

### **Chlorothalonil** (Bravo, Bombardier)

IUPAC name: Tetrachloroisophthalonitrile
Molecular formula: $C_8Cl_4N_2$
Toxicity class: USEPA II; WHO: III

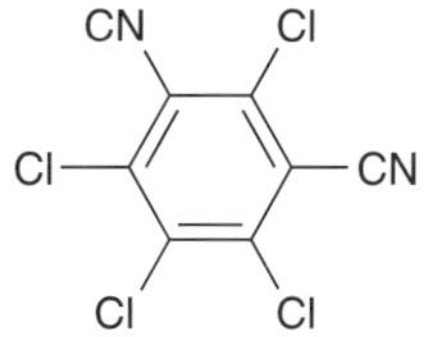

**Figure 6-17.** Structural formula of chlorothalonil.

***Uses*** Chlorothalonil is an aromatic halogen compound, a member of the chloronitrile chemical family, and the USEPA has grouped it under GUP. It is a grayish to colorless crystalline solid that is odorless to slightly pungent. Chlorothalonil is a broad-spectrum organochlorine fungicide that is used to control fungi that threaten vegetables, trees, small fruits, turf, ornamentals, and other agricultural crops. It also controls fruit rots in cranberry bogs. The compound can be found in formulations with many other pesticide compounds.[2,12,17]

***Toxicity*** Chlorothalonil is slightly toxic to mammals, but it can cause severe eye and skin irritation in certain formulations. Very high doses may cause a loss of muscle coordination, rapid breathing, nosebleeds, vomiting, hyperactivity, and death. Dermatitis, vaginal bleeding, bright yellow and/or bloody urine, and kidney tumors also may occur. The oral $LD_{50}$ is greater than 10,000 mg/kg in rats and

6,000 mg/kg in mice. The acute dermal $LD_{50}$ in both albino rabbits and albino rats is 10,000 mg/kg. In albino rabbits, 3 mg of chlorothalonil applied to the eyes caused mild irritation that subsided within 7 days of exposure.

In a number of tests of varying time, rats fed a range of chlorothalonil doses generally showed no effects on physical appearance, behavior, or survival. Skin contact with chlorothalonil may result in dermatitis or light sensitivity. Human eye and skin irritation is linked to chlorothalonil exposure; 14 of 20 workers exposed to 0.5% chlorothalonil in a wood preservative developed dermatitis. All workers showed swelling and inflammation of the upper eyelids. Allergic skin responses also have been noted in farm workers. Administration of higher doses of chlorothalonil daily over a lifetime to rats produced carcinogenic and benign kidney tumors, which showed enlargement, greenish brown color, and development of small grains.[52,54–56] In view of this, use of chlorothalonil requires care and caution.

## **Folpet** (Acryptan, Folpan)

IUPAC name: (N-[(Trichloromethyl)thio]phthalimide)
Molecular formula: $C_9 H_4Cl_3NO_2S$
Toxicity class: USEPA: IV; WHO: III

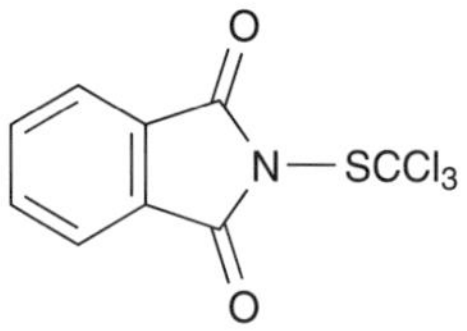

O
N—SCCl3
O

**Figure 6-18.** Structural formula of folpet.

***Uses*** Folpet is a foliage fungicide. It has several formulations, (e.g., Vinicoil, Captafol, Dinocap, and wettable powders and dusts). Folpet is no longer sold in the United States. Folpet is incompatible with strongly alkaline preparations such as lime sulfur.[2,12,17,50]

Folpet is a protective leaf fungicide. Its mode of action inhibits normal cell division of a broad spectrum of microorganisms. It is used to control cherry leaf spot, rose mildew, rose black spot, and apple scab. It is used on berries, flowers, ornamentals, fruits and vegetables, and for seed- and plant-bed treatment. Folpet also finds use in paints, plastics, and treatment of internal and external structural surfaces of buildings to control fungal attack.[50,52]

***Toxicity*** The acute oral $LD_{50}$ for male and female rats is more than 10 g/kg. The acute oral $LD_{50}$ for mice is 2,440 mg/kg. The acute dermal $LD_{50}$ for rabbits is greater than 5 g/kg. Primary eye irritation in rabbits causes reversible corneal opacity, which is prevented by immediately washing the exposed eye. It is not considered an eye irritant to rabbits and is a dermal sensitizer in the guinea pig. Folpet does cause irritation of the skin and mucous membranes in rabbits. Acute inhalation exposure to folpet may cause irritation of the mucous membranes. The acute $LC_{50}$ (2 hours) by inhalation for rats was greater than 5 mg/L. The acute $LC_{50}$ (30 minutes) for the mouse was more than 6 mg/L. Folpet is considered slightly toxic by ingestion. Inhalation of dust or spray mists and contact with the eyes can also result in local irritation.[57]

Chronic inhalation exposure to folpet showed an increase of fetal mortality in an inhalation study of pregnant mice exposed daily to 491 mg/L (4 hours) for 8 days. Folpet was found to be positive in producing developmental effects in both rabbits and rats. All these studies have indicated the effect of very high doses in animals and require prudent evaluation.[17,47,54] In a carcinogenicity study in mice, folpet was found to be a carcinogen with a dose-related increased incidence of adenocarcinomas in the duodenum in CD-1 mice. Similarly, folpet was a positive carcinogen in mice with a dose-related increased incidence of adenocarcinomas in the duodenum at different doses (142.9, 714.3, and 1,428.6 mg/kg/day).[57,58]

### **Mancozeb** (Dithane M-45)

IUPAC name: Manganese ethylenebis dithiocarbamate polymeric
Toxicity class: USEPA: IV; WHO: III

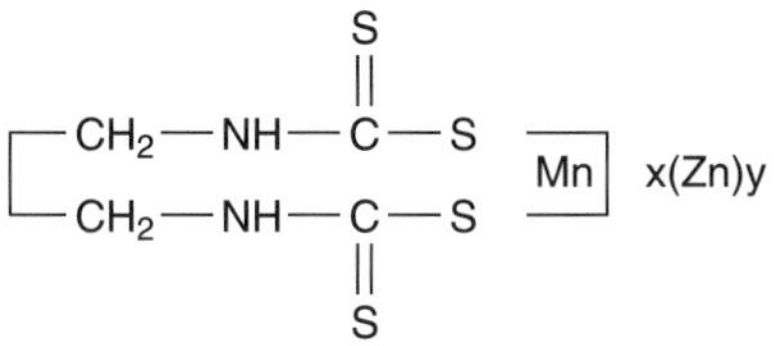

S
||
$CH_2$—NH—C—S
Mn x(Zn)y
$CH_2$—NH—C—S
||
S

**Figure 6-19.** Structural formula of mancozeb.

***Uses*** Mancozeb is a grayish yellow powder.

***Toxicity*** Mancozeb is practically nontoxic via the oral route, with an acute oral $LD_{50}$ of greater than 5,000 mg/kg to greater than 11,200 mg/kg in rats. The acute dermal $LD_{50}$ is greater than 10,000 mg/kg in rats and greater than 5,000 mg/kg in rabbits. It is a mild skin irritant and sensitizer and a mild to moderate eye irritant in rabbits. Workers with occupational exposure to mancozeb have developed sensitization rashes.[2,17,52]

No toxicological effects were apparent in rats fed dietary doses of 5 mg/kg/day in a long-term study.[55] No confirmatory reports are available regarding the mutagenic potential of mancozeb.[50,59,60]

## **Thiram** (Pomarsol, Rhodiason)

IUPAC name: Tetramethylthiuram disulfide
Molecular formula: $C_6H_{12}N_2S_4$
Toxicity class: USEPA: III; WHO: III

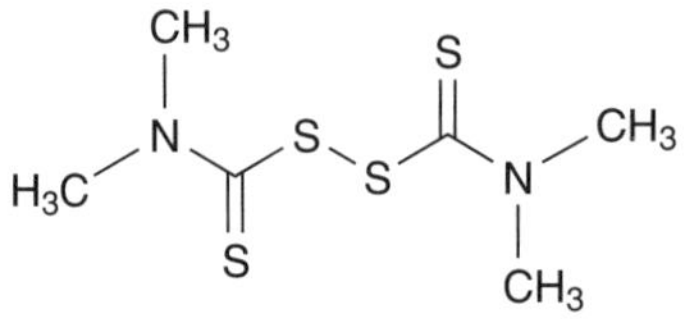

**Figure 6-20.** Structural formula of thiram.

***Uses*** Thiram is a dimethyl dithiocarbamate compound and appears as a white to yellow crystalline powder with a characteristic odor. Thiram is used to prevent crop damage in the field and to protect harvested crops from deterioration in storage or transport. Thiram also is used as a seed protectant and to protect fruits, vegetables, ornamentals, and turf crops from a variety of fungal diseases. In addition, it is used as an animal repellent to protect fruit trees and ornamentals from damage by rabbits, rodents, and deer. Thiram is available as dust, flowable, wettable powder, water-dispersible granules, water suspension formulations, and in mixtures with other fungicides. Thiram has been used in the treatment of human scabies, as a sunscreen, and as a bactericide applied directly to the skin or incorporated into soap.[17,50,52]

***Toxicity*** Thiram is slightly toxic by ingestion and inhalation, but it is moderately toxic by dermal absorption. The acute oral $LD_{50}$ for rats ranges from 620 to 1,900 mg/kg, for mice from 1,500 to 2,000 mg/kg, and for rabbits 210 mg/kg. In contrast, the acute dermal $LD_{50}$ is greater than 1,000 mg/kg in rabbits and rats.[2,52] The acute inhalation $LC_{50}$ (4 hours) for rats is more than 500 mg/L. Acute exposure in humans may cause headaches, dizziness, fatigue, nausea, diarrhea, and other gastrointestinal complaints. In rats and mice, large doses of thiram produced muscle incoordination, hyperactivity followed by inactivity, loss of muscular tone, labored breathing, and convulsions.[52]

Thiram is an irritant to the eyes, skin, and respiratory tract of humans. It is a skin sensitizer. Symptoms of acute inhalation exposure to thiram include itching, scratchy throat, hoarseness, sneezing, coughing, inflammation of the nose or throat, bronchitis, dizziness, headache, fatigue, nausea, diarrhea, and other gastrointestinal complaints. Persons with chronic respiratory or skin disease are at increased risk from exposure to thiram. Ingestion of thiram and alcohol together may cause stomach pains, nausea, vomiting, headache, slight fever, and possible dermatitis.

## **Zineb** (Dithane Z-78, Amitan)

IUPAC name: Zinc ethylenebisdithiocarbamate
Molecular formula: $C_4H_6N_2S_4Zn$
Toxicity class: USEPA: IV; WHO: III

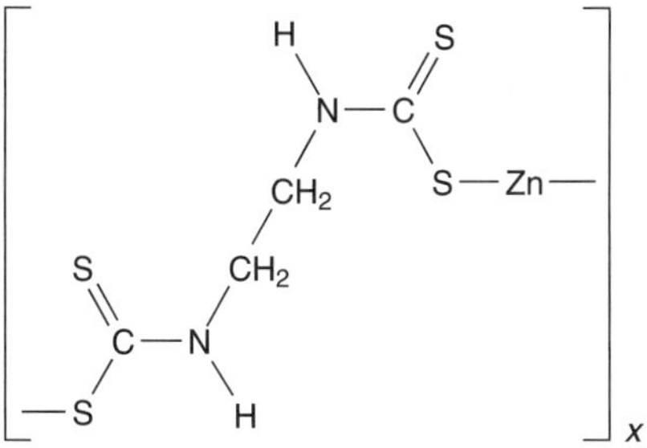

**Figure 6-21.** Structural formula of zineb.

***Uses*** Zineb is a light-colored powder or crystal. It is a polymer of ethylene (bis) thiocarbamate units linked with zinc. Zineb is used to prevent crop damage in the field and to protect harvested crops from deterioration during storage or transport. It was used to protect fruit and vegetable crops from a wide range of foliar and other diseases. It was available in the United States as wettable powder and dust formulations. Zineb can be formed by combining nabam and zinc sulfate in a spray tank.[7,12,13,19]

***Toxicity*** The acute oral $LD_{50}$ for rats ranges from 1,850 to 8,900 mg/kg, in mice from 7,600 to 8,900 mg/kg, and in rabbits 4,450 mg/kg. The acute dermal $LD_{50}$ in rats is more than 2,500 mg/kg—the highest dose that it is possible to administer. The acute inhalation $LC_{50}$ (4 hours) in rats is 0.8 mg/L. Zineb is slightly toxic when ingested. Following a single large dose of zineb, rats and mice exhibited incoordination, hyperactivity followed by inactivity, loss of muscle tone, and loss of hair. Experimental sheep died within 3 weeks of being given oral doses of 500 mg/kg of zineb. In spray or dust forms, zineb is moderately irritating to the skin, eyes, and respiratory mucous membranes. It also may be a dermal sensitizer, with possible cross-sensitization to maneb and mancozeb. This irritation may result in itching, scratchy throat, sneezing, coughing, inflammation of the nose or throat, and bronchitis. Early symptoms from exposure of humans of zineb include tiredness, dizziness, and weakness. More severe symptoms include headache, nausea, fatigue, slurred speech, convulsions, and unconsciousness.

These effects may be exacerbated with concurrent exposure to alcohol. Acute neurotoxic effects probably are due to carbon disulfide, a metabolite of zineb. Animal studies indicate that changes in the thyroid may occur following a single, large dose, but that these changes may be reversible.

Ethylene thiourea (ETU), a potentially toxic metabolite of zineb, may be involved in thyroid effects. Occupational inhalation of zineb can lead to changes in liver enzymes, moderate anemia and other blood changes, increased incidence of poisoning symptoms during pregnancy, and chromosomal changes in the lymphocytes. Liver function was affected in workers exposed to zineb. Repeated or prolonged dermal exposure may cause dermatitis or conjunctivitis. Farm workers who were repeatedly exposed to zineb in fields sprayed with 0.5% suspension of the fungicide reported severe and extensive contact dermatitis. ETU formation during the metabolism of zineb or other EBDC pesticides may potentially result in goiter, a condition in which the thyroid gland is enlarged.[61]

### **Ziram** (Fuklasin)

IUPAC name: Zinc bisdimethyldithiocarbamate
Molecular formula: $C_6H_{12}N_2S_4Zn$
Toxicity class: USEPA: III; WHO: III

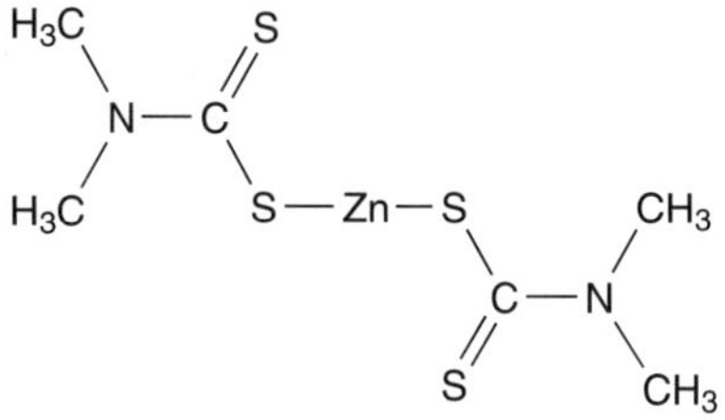

**Figure 6-22.** Structural formula of ziram.

***Uses*** Ziram is an odorless powder at room temperature. Ziram is grouped under class III and GUP by the USEPA. Ziram is slightly to moderately toxic. Ziram is used for the control of a variety of plant fungi and diseases that affect the foliage of plants, soil, and seeds. Ziram is used primarily on almonds and stone fruits. It also is used as an accelerator in rubber manufacturing, packaging materials, adhesives, and textiles. Ziram also is used as a bird and rodent repellent.[12,13,17,52.]

***Toxicity*** The acute oral $LD_{50}$ of ziram varies in different species of animals, for instance, in rats, mice, and rabbits as 1,400, 480, and 400 mg/kg, respectively. Ziram is quite toxic to guinea pigs, with the acute $LD_{50}$ 100 to 150 mg/kg. The acute dermal $LD_{50}$ in rats is greater than 6,000 mg/kg.[2,55] Acute exposure among industrial and farm workers in the former Soviet Union caused irritation of the skin, nose, eyes, and throat.

The prolonged exposure period of animals to ziram caused adverse effects. For instance, female rats administered with low doses (2.5 mg/kg/day) in diets for 9 months showed a decreased antibody formation. Exposure of rats and dogs for a prolonged period with lower doses (1 to 5 mg/kg/day) of ziram showed no adverse effects, whereas treatment of pregnant rats with ziram (12.5 to 100 mg/kg/day) during the organogenesis period demonstrated embryotoxicity.

A dose of 25 mg/kg/day caused embryotoxic effects but no terata formation. The higher doses were effective. Ziram caused growth-inhibiting effects on embryos and maternal toxicity. Different studies have shown ziram as mutagenic; it caused chromosome changes in bone marrow cells in mice treated with oral doses of 100 mg/kg/day.[2,55] Chronic exposure (3 to 5 years) of occupational workers to ziram has caused chromosomal changes.[49] Ziram, exposed to rats and mice for 103 weeks, was found to be carcinogenic to male rats, causing an increase in thyroid cancer, while no such effects were observed in female rats or male mice. More confirmatory studies are required in this direction. The primary target organ was determined to be the thyroid, since occupational workers exposed to ziram showed thyroid enlargement.[61–63]

## CONCLUSION

Improved crop varieties and protection methods have demanded the application of many herbicides and fungicides to attain increased crop yields, decreased food costs, and enhanced appearance of food. Without proper controls, however, the residues of some pesticides on foods can create potential health risks. Proper application of herbicides and fungicides (as discussed previously) has not only lead to increased crop yield but also has reduced the hazards of persistent pesticides, such as organochlorine insecticides. However, further studies are needed to arrive at meaningful conclusions before establishing a safety *in toto* to biological systems and the environment.

Before 1910, no legislation existed to ensure the safety of food and feed crops sprayed and dusted with pesticides. In 1910, the first pesticide legislation was designed to protect consumers from impure or improperly labeled products. During the 1950s and 1960s, pesticide regulation evolved to require maximum allowable residue levels of pesticides on foods and to deny registrations for unsafe or ineffective products. During the 1970s, acting under these strengthened laws, USEPA removed DDT and several other highly persistent pesticides from the marketplace. In 1996, the Food Quality Protection Act set a stricter safety standard and required the review of older allowable residue levels to determine whether they were safe. In 1999, federal and state laws required that pesticides meet specific safety standards. Thus, the USEPA reviews and registers each herbicide and fungicide before it can be used and sets levels and restrictions on each product intended for food or feed crops.

## REFERENCES

1. F.M. Ashton, and A.S. Crafts, *Mode of Action of Herbicides*, John Wiley & Sons, New York, 1973.
2. U.S. National Library of Medicine. *Hazardous Substances Databank*, Bethesda, MD, 1995.

3. D.A. Blum, J.A. Carr, R.K. Davis, and D.T. Pederson, Atrazine in a stream-aquifer system transport of atrazine and its environmental impact near Ashland, *Nebraska Ground Water Monitor Remed.* **13**(2):125, 1993.
4. R.C. Brian, The history and classification of herbicides, in *Herbicides: Physiology, Biochemistry, Ecology*, Vol. I, Audus, L.J., ed., Academic Press, London, 1976.
5. G. Jager, Herbicides, in *Chemistry of Pesticides*, Buchel, K.H., ed., John Wiley & Sons, New York, 1983.
6. P.C. Kearney, and D.D. Kaufman, eds., *Herbicides: Chemistry, Degradation, and Mode of Action*, 2nd ed., Vol. I, Marcel Dekker, New York, 1975.
6a. World Health Organization, Glyphosate. *Environ. Health Criteria* **159**: IPCS, WHO, Geneva, Switzerland, 1994.
6b. C. Cox, Herbicide Factsheet: Glyphosate, Part I: Toxicology. *J. Pesticide Reform.* **15**:3, September 1995.
6c. Pesticide Action Network, Glyphosate fact sheet. *Pesticides News* No. 33, 28–29, United Kingdom, September 1996.
7. J. Hassink, A. Klein, W. Kordal, and W. Klein, Behaviour of herbicides in noncultivated soils. *Chemosphere* **28**(2):285, 1994.
8. L. Hardel, and M. Ericksson, A case control study of non-Hodgkin lymphoma and exposure to pesticides. *Cancer* **85**(6): 1999.
9. K.P. Cantor, Farming and mortality from non-Hodgkin's lymphoma, A case control study. *Int. J. Cancer* **29**:239–247.
10. A.S. Felsot, *Hazard Assessment of Herbicides Recommended for Use by the King County Noxious Weed Control Program, Department of Crop and Soil Sciences/Food and Environmental Quality Lab*, Washington State University, Richland, WA, 99352, 1998.
11. Forest Service, *Pesticide Background Statements*, Vol. I, Herbicides, U.S. Department of Agriculture, Agriculture Handbook No. 633, 1984.
12. R.T. Meister, ed., *Farm Chemicals Handbook '92*, Meister Publishing Co., Willoughby, Ohio, 1992.
13. Weed Science Society of America, *Herbicide Handbook*, 7th ed. Champaign, IL, 10–59, 1994.
14. U.S. Environmental Protection Agency, *Pesticide Fact Sheet Number 97.1: Alachlor*, Office of Pesticides and Toxic Substances, Washington, DC, 10–63, 1987.
15. U.S. Environmental Protection Agency, *Health Advisory: Alachlor*, Office of Drinking Water, Washington, DC, 10–61, 1987.
16. Monsanto Company, *Material Safety Data Sheet, Alachlor Technical (94%)*, St. Louis, MO, 1998.
17. J.T. Stevens, and D.D. Sumner, Herbicides, in *Handbook of Pesticide Toxicology*, Hayes, W.J., Jr. and Laws, E.R., Jr., eds. Academic Press, New York, 1991.
18. U.S. Environmental Protection Agency, Amitrole: Preliminary Determination to Terminate Special Review. *Fed Reg.* **57**:46448, 1992.
19. U.S. National Library of Medicine, *Hazardous Substances Databank*, Bethesda, MD, 8–17, 1995.
20. U.S. Environmental Protection Agency, *Environmental Fate On-Line Summary: Triademifon. Environmental Fate and Effects Division*, Washington, DC, 1988.

21. U.S. Environmental Protection Agency, *Amitrole: Pesticide Registration Standard and Guidance Document*, Washington, DC, 8–19, 1984.
22. Loren Buhle, EPA Sets Regulatory Review for Triazine Herbicides, University of Pennysylvania, 1994.
23. U.S. Department of Agriculture, *Pesticide Background Statements, Vol. I, Herbicides*, Forest Service, U.S. Department of Agriculture, Agriculture Handbook No. 663, 1984.
24. Environmental Protection Agency, *Guidance for the Reregistration of Pesticide Products Containing Atrazine as the Active Ingredient*, Office of Pesticide Programs, Environmental Protection Agency, Washington, DC, 1983.
25. Environmental Protection Agency, *Guidance Package for Reregistration of Pesticide Products in Compliance with Bromacil and Its Lithium Salt Registration Standard*, Office of Pesticide Programs, Environmental Protection Agency, Washington, DC, EPA Report No. 540/RS-86-165, 1982.
26. Environmental Protection Agency, *Guidance for the Reregistration of Pesticide Products Containing Dicamba as the Active Ingredient*, Office of Pesticides and Toxic Substances, U.S. Environmental Protection Agency, Washington, DC, EPA Publication No. 540/RS-83-018, 1988.
27. Environmental Protection Agency, *Pesticide Fact Sheet: Dicamba*, Office of Pesticide Programs, U.S. Environmental Protection Agency, Washington, DC, 1988.
28. Environmental Protection Agency, *Pesticide Fact Sheet: Glyphosate*, Office of Pesticide Programs, U.S. Environmental Protection Agency, Washington, DC, EPA Publication No. 540/FS-88-124, 1986.
29. Environmental Protection Agency, *Registration Standard for Pesticide Products Containing Glyphosate as the Active Ingredient*, Office of Pesticides and Toxic Substances, U.S. Environmental Protection Agency, Washington, DC, EPA Publication No. 540/RS-86-156, 1986.
30. Grossbard, E. and D. Atkinson, eds., *The Herbicide Glyphosate*, Butterworths, Boston, MA, 1985.
31. Monsanto Company, *Toxicology of Glyphosate and Roundup Herbicide*, Department of Medicine and Environmental Health, St. Louis, MO, 1985.
32. U.S. Environmental Protection Agency, Health Advisory, Glyphosate, Office of Drinking Water, Washington, DC, 10, 1987.
33. U.S. Environmental Protection Agency, Pesticide Tolerance for Glyphosate. *Fed. Reg.* **57**(49): 8739-40-10-98, 1992.
34. J. Malik, G. Barry and G. Kishore, Minireview: The herbicide glyphosate. *BioFactors* **2**(1):17, 1989.
35. U.S. Environmental Protection Agency, *Pesticide Fact Sheet: Simazine*, Office of Pesticide Programs, Washington, DC, 1984.
36. U.S. Environmental Protection Agency, *Health Advisory Summary: Simazine*, Office of Drinking Water, Washington, DC, 8–11, 1988.
37. U.S. Environmental Protection Agency, *Guidance for the Reregistration of Pesticide Products Containing Picloram as the Active Ingredient*, Office of Pesticides and Toxic Substances, Washington, DC, EPA Publication No. 540/RS-88-132, 1988.
38. U.S. Environmental Protection Agency, *Pesticide Fact Sheet: Picloram*, Office of Pesticide Programs, Washington, DC, EPA Publication No. 540/FS-88-133, 1988.

39. D. Neubert, and I. Dillman, Embryotoxic effects in mice treated with 2,4,5-trichlorophenoxyacetic acid an 2,3,7,8-tetrachlorodibenzo-p-dioxin. *Arch. Pharmacol.* **272**:243, 1972.

40. R.N. Schlop, M.H. Hardy, and M.T. Goldberg, Comparison of the activity of topically applied pesticides and the herbicide 2,4-D in two short-term *in vivo* assays of genotoxicity in the mouse. *Fundam. Appl. Toxicol.* **15**:666–675, 1990.

41. S.K. Hoar, S. Zahm, D.D. Weisenburger, P.A. Babbitt, A case-control study of non-Hodgkin's lymphoma and the herbicide 2,4-dichlorophenoxyacetic acid (2,4-D), in Eastern Nebraska. *Epidemiology* **1**:349–356, 1990.

41a. R.E. Gosselin, R.P. Smith, and H.C. Hodge, *Clinical Toxicology of Commercial Products*, Williams and Wilkins, Baltimore, MD, 1984.

41b. U.S. Environmental Protection Agency, *Proposed Rules. Fed. Regist.* **55**:24116–17, 1990.

42. S.C. Buzik, *Toxicology of 2,4-Dichlorophenoxyacetic Acid (2,4-D)—A Review*, Toxicology Research Center of University of Saskatchewan, Saskatoon, Canada, 1992.

43. IARC, *Monograph on the Evaluation of Carcinogenic Risk of Chemicals in Man, Vol. 15, Some Fumigants, the Herbicides 2,4-D and 2,4,5-T, Chlorinated Dibenzodioxins and Miscellaneous Industrial Chemicals*, International Agency for Research on Cancer, Lyon, France, 1974.

44. ICI Americas Inc., *Material Safety Data Sheet: Fusilade 2000*, Wilmington, DE, 7–38, 1992.

45. U.S. Environmental Protection Agency, *Pesticide Environmental Fate On-Line Summaries: Devrinol*, Environmental Fate and Effects Division, Washington, DC, 10–104, 1991.

46. U.S. Environmental Protection Agency, *Environmental Effects Branch, Chemical Profile: Napropamide*, Environmental Fate and Effects Division, Washington, DC, 10–105, 1984.

47. U.S. Environmental Protection Agency, *Pesticide Environmental Fate On-Line Summaries: Napropamide*, Environmental Fate and Effects Division, Washington, DC, 10–106, 1991.

47a. U.S. Environmental Protection Agency. *Integrated risk information system database*, Washington, DC, 10–14, 1995.

48. R. Doyle, *Dalapon Information Sheet*, Food and Drug Administration, Washington, DC, 6–40, 1984.

49. U.S. Environmental Protection Agency, *Health Advisory Summary: Dalapon*, Washington, DC, January 1989.

50. U.S. Environmental Protection Agency, *Final Rule: Diquat, tolerances and exemptions from tolerances for pesticide chemicals in or on raw agricultural commodities. Fed. Reg.* **46**:30342, 10–91, Washington, DC, 1981.

51. U.S. Environmental Protection Agency, *Health Advisory Draft Report: Paraquat*, Office of Drinking Water, Washington, DC, 10–112, 1987.

52. U.S. Environmental Protection Agency, *Pesticide Fact Sheet Number 131: Paraquat*, Office of Pesticides and Toxic Substances, Washington, DC, 10–113, 1987.

53. H. Kidd, and D.R. James, eds., *The Agrochemicals Handbook, 3rd ed., Royal Society of Chemistry Information Services*, Cambridge, UK, 10-2, 991 (As Updated).

54. World Health Organization, *Environmental Health Criteria, No. 148, Benomyl*, Geneva, Switzerland, 1993.
55. I.R. Edwards, D.G. Ferry, and W.A. Temple, Fungicides and related compounds, in *Handbook of Pesticide Toxicology*, W.J. Hayes, and E.R. Laws, eds., Academic Press, New York, 1991.
56. A.M. Cummings, M.T. Ebron McCoy, J.M. Rogers, B.D. Barbee, and S.T. Harris, Developmental effects of methyl benzimidazole carbamate following exposure during early pregnancy. *Fundam. Appl. Toxicol.* **18**:288, 1992.
57. U.S. Environmental Protection Agency, *Chlorothalonil Health Advisory, Draft Report*, Office of Drinking Water, Washington, DC, 6–36, 1987.
58. U.S. Environmental Protection Agency, Pesticide tolerance for chlorothalonil. *Fed. Reg.* **50**:26592, Washington, DC, 6–37, 1985.
59. B.H. Chin, R.D. Heilman, R.T. Bachand, G. Chernenko, J. Barrowman, Absorption and biliary excretion of chlorothalonil and its metabolites in the rat. *Toxicol. Lett.* **5**(1):150, 1980.
60. U.S. Environmental Protection Agency, *Pesticide Fact Sheet Number 215: Folpet*, Office of Pesticides and Toxic Substances, Office of Pesticide Programs, Washington, DC, June 1987.
58. Occupational Health Services Database, *Occupational Health Services, Inc. MSDS for Captafol, OHS Inc.* Secaucus, NJ, December 1993.
59. U.S. Environmental Protection Agency, *Pesticide Fact Sheet Number 125: Mancozeb, Office of Pesticides and Toxic Substances*, (4–10) Washington, DC, 1987.
60. E.I. DuPont de Nemours, *Technical Data Sheet for Mancozeb*, Biochemicals Department, Wilmington, DE, 4–33, 1983.
61. U.S. Environmental Protection Agency, Ethylene bisdithiocarbamates (EBDCs); Notice of intent to cancel and conclusion of Special Review. *Fed. Reg.* **57**:7434, 1992.
62. National Toxicology Program, *Carcinogenesis Bioassay of Ziram (CAS No. 137-30-4) in F344/N Rats and B6CF1 Mice (Feed Study), (Technical Report No. 238)*, National Institutes of Health, Bethesda, MD, 1983.
63. Material Safety Data Sheet, Ziram, FMC Corporation, Philadelphia, PA, 4–44, 1991.

CHAPTER 7

# Carbamates

## INTRODUCTION

Carbamates are of relatively recent origin and constitute another important group of pesticides used as crop protectants. The toxicologic effects of the carbamate group of chemicals are comparable to those of organophosphorous chemicals. For instance, they have structural similarity with acetylchlorine (ACh) as well as the attraction of the C=O group to the site OH of the acetylcholinesterase (AChE) enzyme. The following discussion covers only a few selected members of the carbamate group, namely, aldicarb, carbonyl, carbofuran, methomil, propoxur, aminocarb, isoprocarb, and pirimicarb. Information on other carbamate pesticides is available in the literature.[1,2]

## CARBAMATES

### Aldicarb (Temik, Sanacarb)

IUPAC name: 2-methyl-2-(methylthio) propionaldehyde-O-methylcarbamoyl-oxime
Molecular formula: $C_7H_{14}N_2O_2S$
Toxicity class: USEPA: I; WHO: Ia

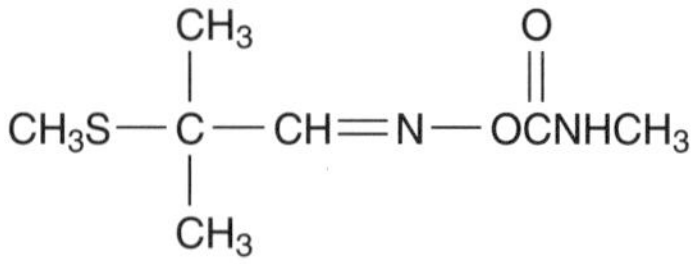

$$CH_3S-\overset{\displaystyle CH_3}{\underset{\displaystyle CH_3}{\overset{|}{\underset{|}{C}}}}-CH{=}N-O\overset{\displaystyle O}{\overset{\|}{C}}NHCH_3$$

**Figure 7-1.** Structural formula of aldicarb.

***Uses*** Aldicarb is a white crystalline solid. It is formulated as a granular mix (10% to 15% active ingredient) because it is extremely toxic. It is not compatible with alkaline materials and is noncorrosive to metals and plastics. The systemic

*Industrial Guide to Chemical and Drug Safety*, By T.S.S. Dikshith and Prakash V. Diwan
ISBN 0-471-23698-5 

toxicity of aldicarb is extremely severe. It is used to control mites, nematodes, and aphids, applied directly to the soil. It is used widely on cotton, peanut, and soybean crops.[1,2] In the mid-1980s, there were highly publicized incidents in which misapplication of aldicarb-contaminated cucumbers and watermelons led to adverse effects in people. In 1990, the manufacturer of Temik (aldicarb) announced a voluntary halt on its sale for use on potatoes because of concerns of groundwater contamination. Aldicarb is metabolized in the liver of mammals, first into aldicarb sulfoxide and later into aldicarb sulfone (Fig. 7-2).

```
         O   CH3                   O
         ||   |                    ||
CH3 —  S  —  C  — CH — N — O  —  C  — N — CH3
         ||   |                          |
         O   CH3                         H
```

**Figure 7-2.** Structural formula of aldicarb sulfone.

***Toxicity*** The primary route of human exposure to aldicarb is consumption of contaminated food and water from wells. Occupational exposure to high levels of aldicarb is due to product handling; most cases of aldicarb poisoning occur from loading and application of the pesticide. Aldicarb is extremely toxic through both the oral and dermal routes. Absorption from the gut is rapid and almost complete. When administered in oil or other organic solvents, aldicarb is absorbed rapidly through the skin. Its skin toxicity is roughly 1,000 times that of other carbamates. In humans, the onset of symptoms is rapid (15 minutes to 3 hours). Symptoms disappear in 4 to 12 hours.

The acute oral $LD_{50}$ of aldicarb in rats, mice, guinea pigs, and rabbits ranges from 0.5 mg/kg to 1.5 mg/kg when administered in a liquid or oil form. The toxicities of the dry granules are distinctly lower ($LD_{50}$ 7.0 mg/kg), although still highly toxic. Aldicarb is a cholinesterase inhibitor and can result in a variety of symptoms including weakness, blurred vision, headache, nausea, tearing, sweating, and tremors. Very high doses of aldicarb cause paralysis of the respiratory system, eventually leading to death.[3,4]

There is very little evidence of chronic effects from aldicarb exposure. Rats and dogs fed low doses of aldicarb for 2 years showed no significant adverse effects.[5] One epidemiological study suggested a possible link between low-level exposure and immunologic abnormalities.[5] The result of this study, however, has been widely disputed. Aldicarb administered to pregnant rats at very low levels (0.001 to 0.1 mg/kg/day) depressed AChE activity more in the fetus than in the mother. The aldicarb also was retained in the mother's body for longer periods than in nonpregnant rats.[6] A three-generation study (at doses of 0.05 and 0.10 mg/kg/day) produced no significant toxic effects, and in another study,[4] a dose of 0.70 mg/kg/day produced no adverse effects. Thus, adverse effects on reproductive systems in humans are unlikely at expected exposure levels. Since aldicarb is extremely toxic to animals and humans, it must be handled by trained and qualified workers.

### **Bendiocarb** (Ficam, Garvox, Seedox, Dycarb, Multamat)

IUPAC name: 2,3-isopropylidenedioxyphenyl methylcarbamate
Molecular formula: $C_{11}H_{13}NO_4$
Toxicity class: USEPA: II; WHO: II

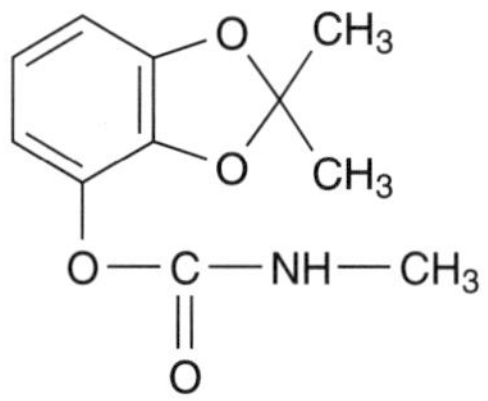

**Figure 7-3.** Structural formula of bendiocarb.

***Uses*** Bendiocarb is an odorless, white crystalline solid. It is stable under normal temperatures and pressures, but should not be mixed with alkaline preparations. Thermal decomposition products may include toxic oxides of nitrogen. It is noncorrosive. Formulations of bendiocarb are classified as general use pesticides (GUP), with the exception of Turcam and Turcam 2.5 G, which are classified as restricted use pesticides (RUP). In view of this, the chemical should be purchased and used only by certified and trained applicators.

Bendiocarb as a carbamate insecticide is effective against a range of insects that cause nuisances and act as disease vectors. It is used to control mosquitoes, flies, wasps, ants, fleas, cockroaches, silverfish, ticks, and other pests in homes, industrial plants, and food storage sites. In agriculture, it is used against a variety of insects, especially those in the soil. Bendiocarb also is used as a seed treatment on sugar beets and corn, and against snails and slugs. Pesticides containing bendiocarb are formulated as dusts, granules, ultra-low volume sprays, and wettable powders.[1,2]

***Toxicity*** Bendiocarb is moderately toxic if it is ingested or absorbed through the skin. Skin absorption is the most likely route of exposure. It is a mild irritant to the skin and eyes.[4] Like other carbamate insecticides, bendiocarb is a reversible inhibitor of cholinesterase, an essential nervous system enzyme. Symptoms of bendiocarb poisoning include weakness, blurred vision, headache, nausea, abdominal cramps, chest discomfort, constriction of pupils, sweating, muscle tremors, and decreased pulse.

Severe poisoning may include symptoms of twitching, giddiness, confusion, muscle incoordination, slurred speech, low blood pressure, heart irregularities, and loss of reflexes. Death can result from discontinued breathing, paralysis of respiratory system muscles, intense constriction of the lung openings, or all three.[4] In one case of exposure while applying bendiocarb, the victim experienced symptoms of severe headache, vomiting, and excessive salivation, and his cholinesterase level was depressed by 63%.[2,4] He recovered from these symptoms in less than 3 hours with no medical treatment, and his cholinesterase level returned to normal within 24 hours. In another case, poisoning occurred when an applicator who was not wearing protective equipment attempted to clean contaminated equipment. The victim experienced nausea, vomiting, incoordination, pain in arms, hands, and legs, muscle spasms, and breathing difficulty.[2] These symptoms abated within 2 hours after decontamination and treatment with atropine. The victim fully recovered by the following day. The oral $LD_{50}$ for bendiocarb is 34 to 156 mg/kg in rats, 35 to 40 mg/kg in rabbits, and 35 mg/kg in guinea pigs. The dermal $LD_{50}$ is 566 mg/kg in rats. The acute inhalation $LC_{50}$ (4 hours) is 0.55 mg/L air in rats.[2]

A 2-year study with rats exposed to high doses of bendiocarb (10 mg/kg/day) showed a range of adverse effects in organ weights, blood, and urine characteristics, as well as an increased incidence of stomach and eye lesions. In a three-generation study with rats, fertility and reproduction were not affected by bendiocarb at dietary doses of up to 12.5 mg/kg/day.[4] Very high doses (40 mg/kg/day) during prenatal and postnatal periods caused toxic effects to rats, as well as reduced pup weight and survival rates. No effects were seen at 20 mg/kg/day. Thus, no reproductive effects are likely to occur in humans at expected exposure levels. No teratogenic effects were seen in the offspring of rats given 4 mg/kg/day or in rabbits given 5 mg/kg/day during gestation. Numerous studies show that bendiocarb is not mutagenic. Bendiocarb was not carcinogenic in 2-year studies of rats and mice.[4]

Bendiocarb is absorbed through all the normal routes of exposure (oral, dermal, and inhalation), but dermal absorption is especially rapid. Carbamates generally are excreted rapidly and do not accumulate in mammalian tissue. If exposure does not continue, cholinesterase inhibition and its symptoms reverse rapidly. In nonfatal cases, the illness generally lasts less than 24 hours.[7] Bendiocarb is moderately toxic to birds. The $LD_{50}$ in mallard ducks is 3.1 mg/kg, and in quail is 19 mg/kg.[8] Bendiocarb is moderately to highly toxic to fish. The $LC_{50}$ (96 hours) for bendiocarb in rainbow trout is 1.55 mg/L.[2]

### **Carbaryl** (Sevin, Cabramec, Efaryl, Karl)

IUPAC name: 1-napthyl methylcarbamate
Molecular formula: $C_{12}H_{11}NO_2$
Toxicity class: USEPA: I; WHO: II

***Uses*** Carbaryl is a solid and varies from colorless to white or gray, depending on the purity of the compound. The crystals are odorless. Carbaryl is stable in

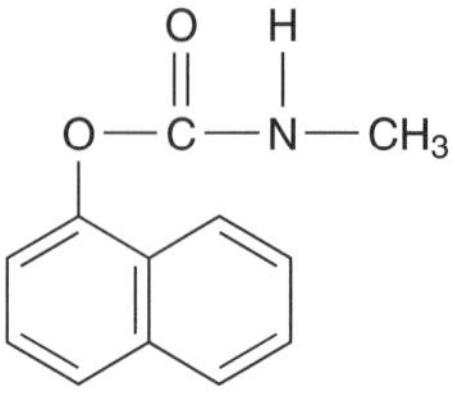

Figure 7-4. Structural formula of carbaryl.

heat, light, and acids. It is not stable under alkaline conditions. It is noncorrosive to metals, packaging materials, and application equipment. Carbaryl is a GUP. However, formulations vary widely in toxicity. Carbaryl is a wide-spectrum carbamate insecticide that controls more than 100 species of insects on citrus, fruit, cotton, forests, lawns, nuts, ornamentals, shade trees, and other crops, as well as on poultry, livestock, and pets. It also is used as a molluscicide and an acaricide. Carbaryl works whether it is ingested into the stomach of the pest or absorbed through direct contact. It is available as bait, dusts, wettable powders, granules, dispersions, and suspensions.[1,2]

***Toxicity*** Carbaryl is moderately to very toxic. It can produce adverse effects in humans by skin contact, inhalation, or ingestion. The symptoms of acute toxicity are typical of other carbamates. Direct contact of the skin or eyes with moderate levels of this pesticide can cause burns. Inhalation or ingestion of large amounts can be toxic to the nervous and respiratory systems, resulting in nausea, stomach cramps, diarrhea, and excessive salivation. Other symptoms at high doses include sweating, blurred vision, incoordination, and convulsions. The only documented fatality from carbaryl poisoning was through intentional ingestion. The acute oral $LD_{50}$ of carbaryl ranges from 250 mg/kg to 850 mg/kg in rats and from 100 mg/kg to 650 mg/kg in mice.[9,10] The inhalation $LC_{50}$ in rats is greater than 206 mg/L. Low doses of carbaryl cause minor skin and eye irritation in rabbits. The acute dermal $LD_{50}$ of carbaryl to rabbits is measured greater than 2,000 mg/kg. In a 90-day feeding study, carbaryl did not cause any significant adverse effects in rats.[11] There are no other reports on chronic toxicity of carbaryl in animals.

No reproductive or fetal effects were observed during a long-term study of rats fed high doses of carbaryl.[12] The evidence for teratogenic effects due to chronic exposure is minimal in test animals. Birth defects in rabbit and guinea pig offspring occurred only at dosage levels that were highly toxic to the mother. Carbaryl has been shown to affect cell division and chromosomes in rats. However, numerous studies indicate that carbaryl poses only a slight mutagenic risk.[9,13] There is a possibility that carbaryl may react in the human stomach to form a more mutagenic compound, but this has not been demonstrated. Evidence suggests that carbaryl is unlikely to be mutagenic to humans.[13,14] Technical-grade carbaryl has not caused tumors in long-term and lifetime studies of mice and rats. Rats were administered high daily doses of the pesticide for 2 years and mice received it for 18 months, with no signs of carcinogenicity.[15] Although N-nitrosocarbaryl, a possible by-product, has been shown to be carcinogenic in rats at high doses, this product has not been detected. Thus, the evidence indicates that carbaryl is unlikely to be carcinogenic to humans.[16]

Ingestion of carbaryl affected the lungs, kidneys, and liver of experimental animals. Inhalation of carbaryl caused adverse effects to the lungs.[4,17] High doses of carbaryl for a prolonged period caused nerve damage in rats and pigs.[18] Several studies indicate that carbaryl can affect the immune system in animals and insects.[17] Male human volunteers who consumed low doses of carbaryl for 6 weeks did not exhibit symptoms of toxicity, although they did show changes in body chemistry.[9,17] A 2-year study with rats revealed no effects at or below a dose of 10 mg/kg/day.[12] Carbaryl is practically nontoxic to wild bird species. The $LD_{50}$ values are greater than 2,000 mg/kg in mallards and pheasants, 2,230 mg/kg in quail, and 1,000 to 3,000 mg/kg in pigeons.[2]

### **Carbofuran** (Furadan, Agrofuran, Carbodan, Carbosip, Cekufuran, Chinufur, Furacarb, Terrafuran)

IUPAC name: 2,3-dihydro-2,2-dimethylbenzofuran-7-yl methylcarbamate
Molecular formula: $C_{12}H_{15}NO_3$
Toxicity class: USEPA: I (Formulation, Furadan 4F), II (Furadan G); WHO: Ib

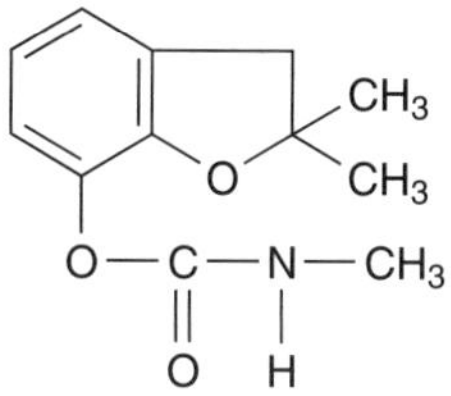

**Figure 7-5.** Structural formula of carbofuran.

***Uses*** Carbofuran is an odorless, white crystalline solid. Exposure to heat breaks down carbofuran, which then releases toxic fumes. It is used for the control of soil-dwelling and foliar-feeding insects. It also is used for the control of aphids, thrips, and nematodes that attack vegetables, ornamental plants, sunflowers, potatoes, peanuts, soybeans, sugarcane, cotton, rice, and variety of other crops.

Areas treated with carbofuran, coupled with the occurrence of fires and the runoff from fire control, release irritating or poisonous gases. Closed spaces (e.g., storage) should be aired before entering. In view of toxicity, it is advisable to enter storehouses or carbofuran-treated closed spaces with caution. Following a special review, the United States Environmental Protection Agency (USEPA) initiated a ban on all granular formulations of carbofuran, effective September 1, 1994. Before 1991, 80% of the total use of carbofuran was in granular formulation. The ban was established to protect birds and was not related to human health concerns. Bird kills occurred when birds ingested carbofuran granules, which resemble grain seeds, and when predatory or scavenging birds ingested small birds or mammals that had eaten carbofuran pellets. There is no ban on liquid formulations of carbofuran. Liquid formulations of carbofuran are classified as restricted use pesticides (RUP) because of their acute oral and inhalation toxicity to humans. Granular formulations also are classified under RUP.

***Toxicity*** The acute oral $LD_{50}$ of carbofuran for male and female rats is approximately 8 mg/kg, whereas the acute dermal $LD_{50}$ for rats is more than 3,000 mg/kg. Carbofuran is mildly irritating to the eyes and skin of rabbits. The acute inhalation toxicity ($LC_{50}$, 4 hours) is 0.075 mg/L to rats. As with other carbamate compounds, carbofuran's cholinesterase-inhibiting effect is short-term and reversible. Symptoms of carbofuran poisoning include nausea, vomiting, abdominal cramps, sweating, diarrhea, excessive salivation, weakness, imbalance, blurred vision, breathing difficulty, increased blood pressure, and incontinence. Death may result at high doses from respiratory system failure associated with carbofuran exposure. Complete recovery from an acute poisoning by carbofuran, with no long-term health effects, is possible if exposure ceases and the victim has time to regain his or her normal level of cholinesterase to recover from symptoms.[2,4]

Rats given very high doses (5 mg/kg/day) of carbofuran for 2 years showed decreases in weight. Similar tests with mice gave the same results. Prolonged or repeated exposure to carbofuran may cause the same effects as an acute exposure. Consuming high doses over long periods caused damage to testes in dogs, but carbofuran did not have any reproductive effects on rats or mice. Available studies indicate that carbofuran is unlikely to cause reproductive effects in humans at expected exposure levels.[4] Studies indicate carbofuran is not teratogenic. No significant teratogenic effects have been found in offspring of rats given carbofuran (3 mg/kg/day) on days 5 to 19 of gestation. No effects were found in offspring of mice given as much as 1 mg/kg/day throughout gestation. In rabbits, up to 1 mg/kg/day on days 6 to 18 of gestation was not teratogenic. Weak or no mutagenic effects have been reported in animals and bacteria. Carbofuran is most likely nonmutagenic. Data from animal studies indicate that carbofuran does not pose a risk of cancer to humans.[4]

### Methomyl (Lannate, Dunet, Methavin, Methomex, Methosan, Nudrin, Pilarmate, Sathomin)

IUPAC name: S-methyl N-(methylcarbamoyloxy) thioacetimidate
Molecular formula: $C_5H_{10}N_2O_2S$
Toxicity class: USEPA: IV; WHO: Ib

***Uses*** Methomyl is a white, crystalline solid with a slight sulfurous odor. It was introduced in 1966 as a broad-spectrum insecticide. It also is used as an acaricide to control ticks and spiders. It is used for foliar treatment of vegetable, fruit and field crops, cotton, commercial ornamentals, in and around poultry

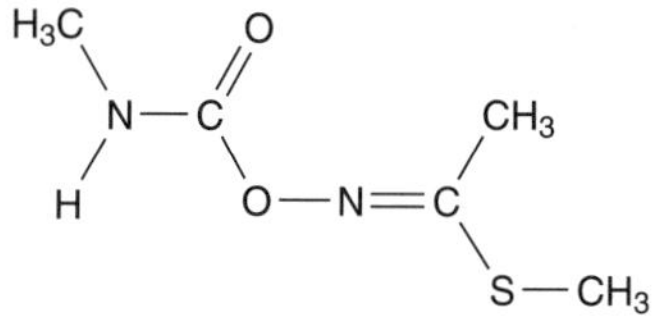

**Figure 7-6.** Structural formula of methomyl.

houses and dairies, and as fly bait. Methomyl is effective in two ways: (1) as a "contact insecticide" because it kills target insects upon direct contact; and (2) as a "systemic insecticide" because of its capability to cause overall "systemic" poisoning in target insects after it is absorbed and transported throughout the pests that feed on treated plants. It is capable of being absorbed by plants without being "phytotoxic" or harmful to the plant.[4]

***Toxicity*** Methomyl is highly toxic via the oral route and is classified as a RUP. The acute oral $LD_{50}$ for rats ranges from 17 to 24 mg/kg in rats, 10 mg/kg in mice, and 15 mg/kg in guinea pigs. The acute dermal $LD_{50}$ for rabbits is more than 5,000 mg/kg. The acute $LC_{50}$ (4 hours) for rats is 0.3 mg/kg. The symptoms of methomyl-induced toxicity are similar to those caused by other carbamates and cholinesterase inhibitors. These include weakness, blurred vision, headache, nausea, abdominal cramps, chest discomfort, constriction of pupils, sweating, muscle tremors, and decreased pulse. In cases of severe poisoning, symptoms of twitching, giddiness, confusion, muscle incoordination, slurred speech, low blood pressure, heart irregularities, and loss of reflexes also may be experienced. Death results from discontinued breathing, paralysis of respiratory system muscles, intense constriction of lung openings, or all three.

Inhalation of dust or aerosol causes irritation, lung and eye problems, and symptoms of chest tightness, blurred vision, tearing, wheezing, and headaches appearing upon exposure. Other systemic symptoms of cholinesterase inhibition may appear within a few minutes to several hours of exposure.[1,2,4] Methomyl is slowly absorbed through the skin. However, if sufficient amounts are absorbed through the skin, symptoms similar to those induced by ingestion or inhalation will develop. Within 15 minutes to 4 hours of exposure, the immediate area of contact may show localized sweating and uncoordinated muscular contractions.

In rabbits, application of methomyl resulted in mild eye irritation. Pain, short-sightedness, blurred distance vision, tearing, and other eye disturbances may occur within a few minutes of eye contact with methomyl. Prolonged or repeated exposure to methomyl may cause symptoms similar to the pesticide's acute effects. Repeated exposure to small amounts of methomyl may cause an unsuspected inhibition of cholinesterase, resulting in flu-like symptoms, such as weakness, lack of appetite, and muscle aches. Cholinesterase inhibition may persist for 2 to 6 weeks. This condition is reversible if exposure is discontinued. Since cholinesterase is increasingly inhibited with each exposure, severe cholinesterase inhibition symptoms may be produced in a person who has had previous methomyl exposure, whereas a person without previous exposure may not experience any symptoms at all. In a 24-month study with rats fed doses of 2.5, 5, or 20 mg/kg/day, effects were only observed at the highest dose tested: 20 mg/kg/day.[19] At this very high dose, red blood cell counts and hemoglobin levels were significantly reduced in female rats. In a 2-year feeding study of dogs, 5 mg/kg/day caused no observed adverse effects.[4,19,20]

It is not likely that chronic effects due to methomyl would be seen in humans unless exposures were unexpectedly high, as with chronic misuse. Methomyl fed to rats at dietary doses of 2.5 or 5 mg/kg for three generations caused no adverse effect on reproduction, nor was there any evidence of congenital abnormalities. No fetotoxicity was observed in offspring of pregnant rats given 33.9 mg/kg/day on days 6 to 21 of gestation. Based on these data, it appears unlikely that methomyl will have reproductive effects. No teratogenic effects were found in the fetuses of female rabbits that were fed approximately 15 to 30 mg/kg/day during days 8 to 16 of gestation. In rats, no embryonic or teratogenic effects were observed at the highest dietary dose administered, approximately 34 mg/kg/day. Thus, methomyl does not appear to be teratogenic. In a number of assays (including Ames test, a reverse mutation assay, a recessive lethal assay, three DNA damage studies, an unscheduled DNA synthesis assay, *in vivo* and *in vitro* cytogenetic assays), methomyl was not mutagenic. There is no evidence that methomyl is mutagenic or genotoxic. There was no evidence of carcinogenicity in either rats or dogs that ingested high doses of methomyl in 2-year feeding studies.[19] Methomyl was not carcinogenic in 22- and 24-month studies with rats fed doses of up to 20 mg/kg, or in a 2-year study with mice fed dietary doses of up to 93.4 mg/kg/day. Evidence suggests that methomyl is not carcinogenic. Lungs, skin, eyes, gastrointestinal tract, kidneys, spleen, and blood-forming organs have been affected in various experiments, depending on route of entry, duration of exposure, and dosage.[4,9,19]

### **Propoxur** (Baygon, Unden, Mitoxur, Proper)

IUPAC name: 2-isopropoxyphenyl methylcarbamate
Molecular formula: $C_{11}H_{15}NO_3$
Toxicity class: USEPA: II; WHO: II

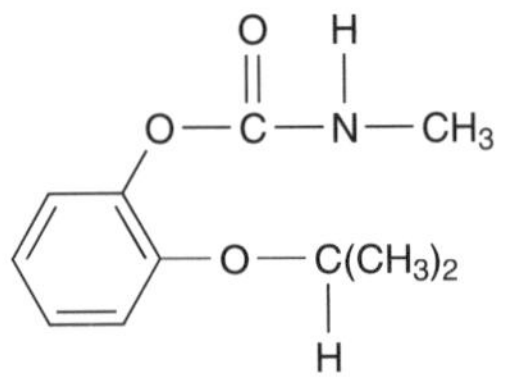

**Figure 7-7.** Structural formula of propoxur.

***Uses*** Technical propoxur is a white to cream-colored crystalline solid. Propoxur is a highly toxic compound. The USEPA has classified various formulations under different categories. It is a GUP, although some formulations may be for professional use only. Labels for pesticide products containing propoxur must bear the signal word "DANGER," "WARNING," or "CAUTION," depending on the type of formulation.[2]

Propoxur is a nonsystemic insecticide that was introduced in 1959. It is compatible with most insecticides and fungicides except alkalines and may be found in combination with azinphosmethyl, chlorpyrifos, cyfluthrin, dichlorvos, disulfoton, or methiocarb. It is used on a variety of insect pests, such as chewing and sucking insects, ants, cockroaches, crickets, flies, and mosquitoes, and may be used for control of these in agricultural or (as Baygon) nonagricultural (e.g., private or public facilities and grounds) applications. Agricultural applications include cane, cocoa, fruit, grapes, maize, rice, sugar, vegetables, cotton, lucerne, forestry, and ornamentals. It has contact and stomach action that is long-acting when it is in direct contact with the target pest. Propoxur is available in several types of formulations and products, including emulsifiable concentrates, wettable powders, baits, aerosols, fumigants, granules, and oil sprays.

***Toxicity*** Propoxur is highly toxic via the oral route. The acute oral $LD_{50}$ is 50 mg/kg in rats and mice and 40 mg/kg for guinea pigs. Propoxur is only slightly toxic via the dermal route, with acute dermal $LD_{50}$ of more than 5,000 mg/kg in rats. The acute $LC_{50}$ for rats (4 hours) is more than 0.5 mg/L. Studies have shown that propoxur does not cause skin or eye irritation in rabbits.

Like other carbamates, propoxur can inhibit the action of cholinesterase and disrupt nervous system function. Depending on the severity of exposure, this effect may be short-term and reversible.

Signs of propoxur intoxication include nausea, vomiting, abdominal cramps, sweating, diarrhea, excessive salivation, weakness, imbalance, blurring of vision, breathing difficulty, increased blood pressure, incontinence, or death. In rats, propoxur poisoning resulted in brain pattern and learning ability changes at lower concentrations than those that caused cholinesterase inhibition or organ and weight changes. During widescale spraying of propoxur in malarial control activities conducted by the World Health Organization (WHO), only mild cases of poisoning were noted. Applicators who used propoxur regularly showed a pronounced daily fall in whole blood cholinesterase activity and a distinct recovery after exposure stopped. No adverse cumulative effects on cholinesterase activity were demonstrated.[2,4,9]

Human adults have ingested single doses of 50 mg of propoxur without apparent symptoms. Prolonged or repeated exposure to propoxur may cause symptoms similar to acute effects. Propoxur is very efficiently detoxified (transformed into less toxic or practically nontoxic forms), making it possible for rats to tolerate long periods of daily doses approximately equal to the $LD_{50}$ of the insecticide, provided that the dose is spread out over the entire day, rather than ingested all at once. In female rats given high dietary doses of approximately 18 mg/kg/day of propoxur as a part of a three-generation reproduction study, reduced parenteral food consumption, growth, lactation, and litter size were observed.[4] At 25 mg/kg/day administered to pregnant rats, there was a decrease in the number of offspring.

Dietary doses of approximately 2.25 mg/kg/day did not affect fertility, litter size, or lactation, but parenteral food intake and growth were depressed in the exposed group. This evidence suggests that reproductive effects in humans are unlikely at expected exposure levels. Offspring of female rats fed 5 mg/kg/day of propoxur during gestation and weaning exhibited reduced birth weight, retarded development of some reflexes, and evidence of central nervous system (CNS) impairment. In another rat study, growth reduction was observed in the offspring of pregnant rats given doses of 3, 9, and 30 mg/kg/day, but no other physiological or anatomical abnormalities were observed.[4] Evidence suggests that teratogenic effects will occur only at high doses. Propoxur was not found to cause mutations in six different types of bacteria. Evidence also indicates that propoxur is not mutagenic. No carcinogenic effects have been reported for propoxur. As determined in animal tests and data from autopsies of poisoned humans, the CNS and liver are the organs principally affected by propoxur. During widescale spraying of propoxur in malarial control activities conducted by the WHO, only mild cases of poisoning were noted. Applicators who used propoxur regularly showed a pronounced daily fall in whole blood cholinesterase activity and a distinct recovery after exposure stopped. No adverse cumulative effects on cholinesterase activity were demonstrated. Human adults have ingested single doses of 50 mg of propoxur.[4,9]

### **Aminocarb** (Matacil)

IUPAC name: 4-dimethyamino-m-tolyl methylcarbamate
Molecular formula: $C_{11}H_{16}N_2O_2$
Toxicity class: USEPA: II; WHO: II

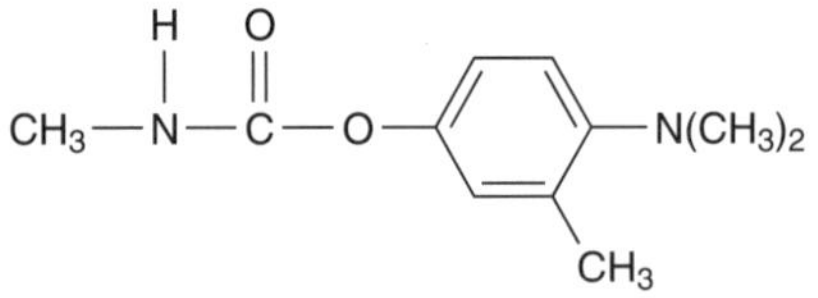

$CH_3$—N(H)—C(=O)—O—$C_6H_3(CH_3)$—$N(CH_3)_2$

**Figure 7-8.** Structural formula of aminocarb.

Aminocarb is a nonsystemic insecticide. It controls lepidopterous insects, aphids, and soil mollusks. The acute oral $LD_{50}$ for rats is 50 mg/kg, and the acute dermal $LD_{50}$ for rats is 275 mg/kg, suggesting that toxicity to skin is appreciably severe compared with carbamate pesticides.[2]

### **Isoprocarb** (Etrofolan, Isso, Mipcin)

IUPAC name: o-Cumenyl methylcarbamate
Molecular formula: $C_{11}H_{15}NO_2$
Toxicity class: USEPA: II; WHO: II

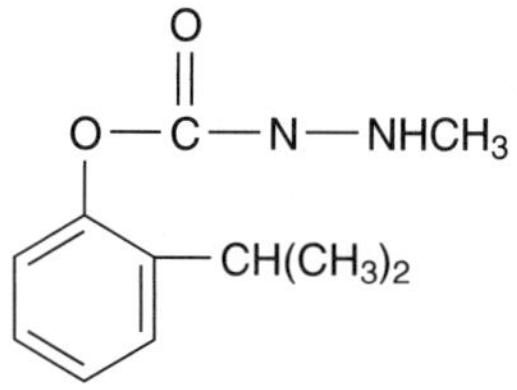

**Figure 7-9.** Structural formula of isoprocarb.

***Uses*** Isoprocarb is a colorless crystal. It is used as an insecticide for aphids, bugs, and leafhoppers that attack and infect crops such as rice, sugarcane, and cocoa. Formulations of isoprocarb include wettable powder, emulsion concentrate, and granules.[1,2]

***Toxicity*** The acute oral $LD_{50}$ of isoprocarb for rats is 450 mg/kg, and acute dermal $LD_{50}$ for rats is more than 500 mg/kg. The acute $LC_{50}$ (4 hours) for rats is 0.5 mg/L. It is slightly irritating to the skin of rabbits.

### **Pirimicarb** (Aphox, Pirimor, Pilly, Pirimisct)

IUPAC name: 2-dimethylamino-5,6-dimethylpyrimidin-4-yl dimethylcarbamate
Molecular formula: $C_{11}H_{18}N_4O_2$
Toxicity class: USEPA: II; WHO: II

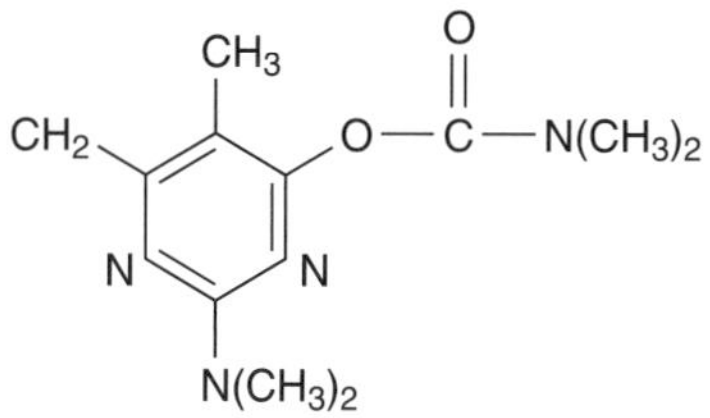

**Figure 7-10.** Structural formula of pirimicarb.

***Uses*** Pirimicarb is a colorless solid material. It is a selective systemic insecticide with contact effects on stomach (poison) and respiratory system (lungs). It is extensively used for the control of pests that infect a variety of crops (e.g., ornamentals, oilseeds, vegetables, cereals, greenhouse crops). Formulations include wettable powder, emulsion concentrate, and wettable granules. Pirimicarb is generally used for the control of aphids that are resistant to organophosphorous pesticides. It is a fast-acting fumigant agent as well.[1,2]

***Toxicity*** Pirimicarb has an acute oral $LD_{50}$ of 147 mg/kg in rats, 107 mg/kg in mice, and 100 to 200 mg/kg in dogs. The acute dermal $LD_{50}$ for rats and rabbits is more than 500 mg/kg. The acute $LC_{50}$ (6 hours) for rats is 0.3 mg/L. Pirimicarb is a mild irritant to eyes of rabbits. In view of its high toxicity and fumigant action, this insecticide should be handled with care by trained and qualified workers under proper supervision.

## REFERENCES


1. C.D.S. Tomlin, ed., *A World Compendium: The Pesticide Manual*, 11th ed., The British Crop Protection Council, Farnham, Surrey, UK, 1997.
2. H. Kidd, and D.R. James, eds., *The Agrochemicals Handbook*, 3rd ed., Royal Society of Chemistry Information Services, Cambridge, UK, 1991 (as updated).
3. P.H. Howard, *Handbook of Environmental Fate and Exposure Data for Organic Chemicals: Pesticides*, Lewis Publishers, Chelsea, MI, 1991.
4. R.L. Baron, Carbamate insecticides, *in Handbook of Pesticide Toxicology*, W.J. Hayes Jr. and E.R. Laws Jr., eds., Academic Press, New York, 1991.
5. U.S. Environmental Protection Agency, *Health Advisories for 50 Pesticides*. Office of Drinking Water, Washington, DC, 1987.
6. C. Chambon, C. Declune, and R. Derach, Effects of the insecticidal carbamate derivatives (carbofuran, pirimicarb, and aldicarb) on the activity of acetylcholinesterase in tissue from pregnant rats and fetuses. *Toxicol. Appl. Pharmacol.* **49**:203–208, 1979.
7. L. Townsend, and T. Potter, *Earthworms: Thatch-Busters*, College of Agriculture, University of Kentucky, Lexington, KY, 1995.
8. R. Tucker, *Handbook of Toxicity of Pesticides to Wildlife*, U.S. Department of Interior, Fish and Wildlife Service, Washington, DC, 1970.
9. U.S. National Library of Medicine, *Hazardous Substances Databank*, Bethesda, MD, 1995.
10. U.S. Environmental Protection Agency, *Health Advisory Draft Report: Carbaryl*, Office of Drinking Water, Washington, DC, 1987.
11. T.S.S. Dikshith, P.K. Gupta, J.S. Gaur, K.K. Datta, and A.K. Mathur, Ninety day toxicity of carbaryl in male rats. *Environ. Res.* **12**:161–170, 1976.
12. M.F. Cranmer, Carbaryl: A toxicological review and risk assessment. *Neurotoxicology* **7**(1):247–332, 1986.
13. D. Siebert, and G. Eisenbrand, Induction of mitotic gene conversion in *Saccharomyces cerevisiae* by N-nitrosated pesticides. *Mutat. Res.* **22**:121–132, 1974.
14. R. Elespuru, W. Lijinski, and J.K. Setlow, Nitrosocarbaryl as a potent mutagen of environmental significance. *Nature (London)* **247**:386–387, 1974.
15. National Technical Information Service, *Evaluation of Carcinogenic, Teratogenic, and Mutagenic Activities of Selected Pesticides and Industrial Chemicals, Carcinogenic Study*, Washington, DC, Vol. 1, 1968.

16. J.D. Regan, R.B. Setlow, A.A. Francis, and W. Lijinsky, Nitrosocarbaryl: Its effect on human DNA. *Mutat. Res.* **38**:293–301, 1976.
17. World Health Organization, *Carbaryl, Environ. Health Crite.* **153**, 1994.
18. W.W. Johnson, and M.T. Finley, *Handbook of Acute Toxicity of Chemicals to Fish and Aquatic Invertebrates*, U.S. Department of the Interior, Washington, DC, 3–19, 1980.
19. U.S. Environmental Protection Agency, *Health Advisory Summary: Methomyl*, Office of Drinking Water, Washington, DC, 3–40, 1987.
20. E.I. DuPont de Nemours, *Technical Data Sheet for Methomyl*, Agricultural Products Division, Wilmington, DE, 3–12, 1989.

CHAPTER 8

# Synthetic Pyrethroids, Fumigants, and Rodenticides

## INTRODUCTION

The application of synthetic pyrethroids in the pest management program has gained importance in recent years. In view of their moderate toxicity, these chemicals have been used for the control of field craft pest in abundance. Because of the low residual effects, these compounds have not caused any serious or undesirable health effects in agricultural farmers, occupational workers, or household workers. More information on synthetic pyrethroids, fumigants, and rodenticides is available in the literature.[1,2]

## SYNTHETIC PYRETHROIDS

Synthetic pyrethroids are produced to duplicate or improve the biological activity of the active principles of the pyrethrum plant, a flowering plant of class crasanthemum species. Pyrethrum is a natural botanical chemical, the active principal of which is extracted from flowers of the pyrethrum plant and are known collectively as "pyrethrins."

Natural pyrethrins (brand names include Blitz, Drione, etc.) are botanical pesticide poisons extracted from the daisy species, *Chrysanthemum cinerariaefolium* or *C. coccineum*, which can bring on allergic reactions, asthma attacks, or dermatitis, and interfere with nervous system function. They often are combined with organophosphates to kill. Inhalation of pyrethrins per the Extension Toxicology Network, a California public interest research group, can cause asthmatic breathing, sneezing, stuffiness, headaches, tremors, convulsion, burning, and itching. They are especially toxic to fish and other aquatic organisms. Synthetic pyrethroids are a diverse class of more than 1,000 powerful broad-spectrum insecticide poisons. Pyrethroid toxicity is highly dependent on stereochemistry, the three-dimensional configuration of the molecule. Each isomer (molecules consisting of the same atoms but a different stereochemistry) has its own toxicity.

*Industrial Guide to Chemical and Drug Safety*, By T.S.S. Dikshith and Prakash V. Diwan
ISBN 0-471-23698-5 

Some pyrethroids have as many as eight different isomers, and there are several different types. Acute toxicity of a mixture of two isomers depends on the ratio of the amounts of isomers in the formulation. For example, the female rat's acute oral $LD_{50}$ of permethrin increases from 224 mg/kg to 6,000 mg/kg as the proportion of transisomer increases from 20% to 80%. The route of exposure also is critical in assessing the acute toxicity of a synthetic pyrethroid. Like DDT and many other registered insecticides, naturally occurring pyrethrins and the synthetic pyrethroids are nerve poisons.

The synthetic pyrethroid principal mechanism of action is believed to be the disruption of the permeability of nerve membranes to sodium atoms.[3] Organophosphates and carbamates also are nerve poisons, but they do not attack the human peripheral (in addition to our central) nervous system, as do DDT and synthetic pyrethroids. The half-life of pyrethroids in soils ranges from 1 day to 16 weeks. (It is amazing that people apply them annually for termite control, even though permethrin is supposedly effective against termites in the very same soil for 1 to 5 years).

Mammalian toxicity levels to pyrethroid pesticides are extremely low, and chances for poisoning would be almost nonexistent unless animals ingested formulations that had not yet been mixed for application.

The United States Environmental Protection Agency (USEPA) has already suspected some synthetic pyrethroids for their carcinogenic properties. Long-term or chronic exposure to pyrethrum caused liver damage, especially when used with the synergists and Freon propellants. The chemical has caused allergic reactions and is found to be a neurotoxin. Synthetic pyrethroids have a very complex chemistry; most are primarily used as termite repellents. It has been observed that termites tunnel through pyrethroid-treated soil by lining their tunnels with clean soil particles. It also has been demonstrated that pyrethroids should not be used for longer than 3 to 4 months in a commercial kitchen area since it results in resistance in the pest population.

### **Bifenthrin** ((2-methyl[1,1-biphenyl]-3-yl)-methyl-3-(2-chloro-3,3,3-trifluoro-1-propenyl)-2,2-dimethyl cyclopropanecarboxylate)

This chemical is used for the control of cone worms, seed bugs, seed worms, and other insects and mites. The USEPA has classified bifenthrin as toxicity class II, meaning it is moderately toxic.[4,5] Bifenthrin is a member of the pyrethroid chemical class. It is an insecticide and acaricide that affects the central nervous system (CNS) and causes paralysis in insects.[6,7] It is highly toxic to fish and aquatic organisms.

***Toxicity*** Bifenthrin has an acute oral $LD_{50}$ in female rats of 53.8 mg/kg and 70 mg/kg in males. The acute dermal $LD_{50}$ of bifenthrin was greater than 2,000 mg/kg for rabbits. The chemical did not show any skin or eye irritation in tests with rabbits. The $LC_{50}$ in rats was 1.86 mg/L for Capture 2EC (the formulation). Large doses may cause incoordination, tremor, salivation, vomiting, diarrhea, and irritability to sound and touch. In large doses, bifenthrin may cause incoordination, tremor, salivation, vomiting, diarrhea, and irritability to sound and touch; hence, it requires caution and protection when handling. Bifenthrin did not demonstrate any teratogenic or mutagenic effects in animals. In a 2-year study in rats fed up to 200 ppm, bifenthrin showed no evidence of carcinogenicity. In a 87- to 92-week study in mice, there was an increase in the number of bladder tumors at the highest dose tested (600 ppm in the feed).[8] The USEPA classified bifenthrin as a class C carcinogen, which is a possible human carcinogen.[8]

### **Deltamethrin** (cyano(3-phenoxy-phenyl)methyl;2-(2,2dibromoethenyl)-2,2-dimethyly loropane carboxylate)

Deltamethrin is a crystalline powder, white or slightly beige in color. The formulations include emulsifiable concentrates, wettable powders, ultra-low volume and flowable formulations, and granules. There are no known incompatibilities with other common insecticides and fungicides. It is used as a contact poison to control apple and pear suckers, plum fruit moths, caterpillars on brassicas, pea moths, aphids (apples, plums, hops), winter moths (apples and plums), and codling and tortrix moths (apples). It also is used in control of aphids, mealybugs, scale insects, and whiteflies on glasshouse cucumbers, tomatoes, peppers, potted plants, and ornamentals.[9–11]

***Toxicity*** Deltamethrin produces typical type II motor symptoms in mammals. Type II symptoms include a writhing syndrome in rodents as well as copious salivation. The acute oral $LD_{50}$ in male rats ranged from 128 mg/kg to greater than 5,000 mg/kg, depending on the carrier and conditions of the study. The $LD_{50}$ for female rats was 52 mg/kg; other published values range from 31 to 139 mg/kg. Values ranging from 21 to 34 mg/kg were obtained for mice, whereas dogs had a reported $LD_{50}$ of 300 mg/kg. The intravenous $LD_{50}$ in rats and dogs was 2 to 2.6 mg/kg, and the dermal $LD_{50}$ was greater than 2,940 mg/kg. The acute percutaneous $LD_{50}$ was reported to be greater than 2,000 mg/kg for rats, greater than 10,000 mg/kg for quail, and greater than 4,640 mg/kg for ducks. The acute dermal $LD_{50}$ for rabbits was greater than 2,000 mg/kg. No skin irritation but slight eye irritation were reported. Another study indicated skin irritation in rats and guinea pigs.[4,5,8]

### **Permethrin** (cyclopropanecarboxylic acid, 3-(2,2-dichloroethenyl)2,2-dimethyl-(3-phenoxyphenyl)methyl ester)

Permethrin is classified as a restricted use pesticide (RUP) because of high toxicity to fish and aquatic organisms. Permethrin is used to control Nantucket pine tip moths, coneworms, seed bugs, and range caterpillars.[4,5]

***Toxicity*** Permethrin has an acute oral $LD_{50}$ of a wide range (430 to 4,000 mg/kg) in rats. The acute dermal $LD_{50}$ is greater than 2,000 mg/kg in rabbits. It has demonstrated mild skin irritation in rabbits and slight irritation to eyes in rabbits. The acute $LC_{50}$ for rats is 23.4 mg/L by inhalation. Permethrin showed no evidence of carcinogenicity in two studies in rats treated with up to 2,500 mg/kg in the diet for up to 104 weeks. In three studies, mice treated with up to 5,000 mg/kg diet for up to 104 weeks showed an increase in the number of lung tumors in females one strain at the highest dose tested.[12]

## FUMIGANTS

Several chemicals are used in the control of insect pests and rodents (field rats). Some are used in the form of fumigants and others as solid chemicals. Fumigant chemicals are extremely volatile and highly toxic to animals and humans. Fumigants are used to control microbial infestation and insect pests, and to check rodent population, specifically field rats that attack field crops and grain storage houses. These chemicals also are used for the purposes of fumigation of soil, warehouses, storage facilities, greenhouses, barns, freight cars, cloth, and abandoned areas. The chemicals include aluminum phosphide, zinc phosphide, sodium fluoroacetate, warfarin, and ANTU (alpha naphthyl-2-thiourea). The fumigants control rodents by selective action with the peculiar physiology of rodents.

Other chemicals used as fumigants include acrylonitrile, carbondisulfide, carbontetrachloride, p-dichlorobezene, dioxane, ethylene dibromide (EDB), ethylenedichloride, ethyleneoxide, hydrogen cyanide, methyl bromide, methylene chloride, naphthalene, sulfur dioxide, tetrachloroethane, and trichloroethylene. Because of the highly volatile nature of these chemicals, occupational workers in chemical industries require elementary knowledge, training, and management supervision during their handling and disposal.

### Phosphine

The most important fumigant used in test control operations and grain storage management is phosphine ($PH_3$), which is released from aluminum phosphide (AlP), magnesium phosphide ($Mg_3P_2$), and zinc phosphide ($P_2Zn_3$). Phosphine is a colorless, odorless, flammable gas. The technical-grade chemical has a garlic or rotting fish odor. Phosphine is released rapidly when aluminum phosphide, magnesium phosphide, and zinc phosphide react with atmosphere moisture. Based on the association with (1) aluminum, (2) magnesium, and (3) zinc, the phosphine-releasing commercial product has several common names: (1) Al-phos, Celphide, Celphos, Phostek, Photoxin, Quickphos, Shaphos; (2) Magtoxin and Magnaphos; and (3) Agzinphos, Ratol, Commando, Denkarin Grains, Rattekal—plus, Ridall zinc, Zinc-tox, and Zawa.

Use of aluminum phosphide is more recent in India as compared with other fumigants like methyl bromide (MB), EDB, and ethylene dibromide carbon tetrachloride mixture (EDCT). It is now being used on a large scale by the Food Corporation of India, the Central Warehousing Corporation, and State Warehousing Corporation for food grain fumigation stationed in different parts of India.

Aluminum phosphide tablets are very easy to handle; no cumbersome equipment is necessary for its use, and it is safe to handle. The decomposition of tablets is gradual. The general decomposition and release of phosphine gas facilitates gradual distribution of a predetermined dosage with maximum safety to operators. It does not affect germination of feeds or baking quality of flour.

***Toxicity*** The acute oral $LD_{50}$ for rats shows a wide difference in the three phosphine-releasing commercial products: AlP is 8.7 mg/kg, $Mg_3P_2$ is 11.2 mg/kg, and $P_2Zn_3$ is 45.7 mg/kg. Ingestion of phostoxin and other phosphine-releasing tablets or inhalation of phosphine reacts with the stomach acids of animals and humans, rapidly releasing the highly toxic phosphine gas into the system. The gas is then absorbed through the lung epithelium and bloodstream.[13,14]

Symptoms of mild to moderate acute aluminum phosphide toxicity include nausea, abdominal pain, tightness in chest, excitement, restlessness, agitation, and chills. Symptoms of more severe toxicity include diarrhea, cyanosis, difficulty breathing, pulmonary edema, respiratory failure, tachycardia (rapid pulse), hypotension (low blood pressure), dizziness, or death. Convulsions have been reported in lab animals exposed to high concentrations of phosphine. Acute toxicity, resulting from aluminum phosphide exposure, is apparent most immediately in the heart and lungs; it also may affect the CNS, liver, and kidneys. A postmortem examination of test animals revealed microscopic lesions in the outer cortex of rat kidneys exposed to 15 $mg/m^3$, but not at lower exposure levels. All of these effects were apparently reversible following a 4-week recovery period.[13–15]

No evidence was available regarding the ability of aluminum phosphide or phosphine to cause mutations or increase the mutation rate. Studies of human lymphocyte cultures exposed under laboratory conditions showed significant increases in phosphine-induced total chromosomal aberrations (e.g., gaps, deletions, breaks, exchanges) with increasing phosphine concentrations. No data are currently available; it is possible that some testing on oncogenicity may be initiated in the near future.

Special cautions while handling phosphine:

- Rubber gloves and protective clothing must be worn when tablets or pellets are being dispensed by hand.
- Respirators equipped with canisters designed for protection against phosphine should always be on hand in case of emergency (for respiratory protection).
- Avoid exposure for those individuals with diseases of blood or lung.
- Adequate ventilation in a downward direction because phosphine is heavier than air.
- Do not smoke or touch food at any time during the application of this insecticide.
- Odor of the fumigant can be relied on as an indication of whether the operator is breathing poisonous concentrations.

- Phosphine or materials containing reactive phosphides must be stored in a cool, dry, isolated area, away from acute fire hazards and powerful oxidizing materials.

### Ethylene Dibromide

Ethylene dibromide (EDB) is used extensively as a soil and postharvest fumigant for crops and as a quarantine fumigant for citrus, tropical fruits, and vegetables. It also may be used as a gas in termite and Japanese beetle control, beehive and vault fumigation, and spot fumigation of milling machinery.

The acute exposure of EDB caused toxic effects in animals and produced severe skin burning. The acute oral toxicity to the male rat is approximately 146 mg/kg, and the acute vapor toxicity is 200 ppm. Prolonged exposure to vapor of EDB (210 $mg/m^3$) of 7 hours/day, 5 days/week for 6 months was found to be tolerated by rats. EDB of single application on the skin has caused severe burning effects. The acute oral toxicity ($LD_{50}$) in males was found to be 146 mg/kg, and the acute inhalation of the vapor was 200 ppm.

The subacute and chronic doses (210 $mg/m^3$) of ethylene bromide were tolerated by experimental rats (7-hours exposures for 5 days/week, for 6 months). Repeated exposure to EDB may cause injury to the lungs, liver, or kidneys. Adverse effects, including abnormalities in offspring, mutations, and stomach cancer, have been found in animals following exposure to EDB.[16,17] The relevance of these findings to humans has not yet been established.

In a bioassay conducted by the National Cancer Institute (NCI), EDB was found to be carcinogenic to rats and mice when fed by gavage. The compound induced squamous cell carcinomas of the forestomach in rats (of both sexes), hepatocellular carcinomas in female rats, and hemangiosarcomas in male rats. In mice of both sexes, the compound induced squamous cell carcinomas of the forestomach as well as alveolar and broncheolar adenomas. The nature of adverse effects included abnormalities in offspring, mutations, and stomach cancer.[4,17] The available data suggest that additional information is needed.

### Methyl Bromide

Methyl bromide (MB) has severe eye and skin irritation properties and is classed under category I. MB is chiefly used as a gas soil fumigant against insects, termites, rodents, weeds, nematodes, and soilborne diseases.[4,10,11] It has been used to fumigate agricultural commodities, grain elevators, mills, ships, clothes, furniture, and greenhouses. Approximately 70% of the MB produced in the United States goes into pesticidal formulations.

Acute oral toxicity of MB in rats is 214 mg/kg, and the compound belongs to toxicity category II. In view of its acute inhalation toxicity in humans and its history of use, it is grouped under category I. Accordingly, the $LD_{50}$ in rats is 2,700 ppm for a 30-minute exposure, whereas in humans, a dose of 1,583 ppm (6.2 mg/L) for approximately 15 to 20 hours was found to be lethal. A much higher concentration (as high as 7,890 ppm; 30.9 mg/L) was lethal to humans within 90 minutes of exposure.[10,11] The principal route of exposure of MB is inhalation. Toxicity data regarding subchronic and chronic adverse effects are limited. MB seems to cause no teratogenic or mutagenic effects in rats. Further, preliminary studies (as required under the data call-in program for grain fumigants), findings of reproduction, and oncogenicity studies are negative. However, adequate data on mutagenicity, rabbit teratology, subchronic inhalation in the rat and rabbit, and chronic feeding studies in the rat and dog are required to complete the toxicology database for MB.[18]

### Other Chemicals

A number of other chemicals are used for the control of crop pests. Presently, only selected chemicals are discussed here. These include synthetic pyrethroids, plant regulators, and a rodenticide.

## RODENTICIDE

### Strychnine

Strychnine is an alkaloid grain that is absorbed when eaten by the animal. It attacks the CNS by antagonizing the action of glycine, an amino acid responsible for transmitting inhibitory nerve impulses that control muscle contraction. In addition, there is evidence of an increase in brain levels of glutamic acid, an amino acid that acts as a transmitter for excitatory nerve impulses that excite muscle contraction. As a result, the skeletal muscles become hyperexcitable and contract simultaneously without the normal restraints; this is called a convulsion or seizure. Convulsions prevent respiration, a process that depends on a controlled rhythm of contraction and relaxation, causing the animal to suffocate.

***Toxicity*** The acute oral $LD_{50}$ in two studies of rats treated with strychnine alkaloid is 2.2 mg/kg and 5.8 mg/kg, respectively, in females, and 6.4 mg/kg and 14 mg/kg, respectively, in males. The acute dermal $LD_{50}$ in one study was greater than 2,000 mg/kg in albino rabbits.[20,21] Dermal irritation was minimal, with absent or transient erythema (skin irritation). In white mice, dry powder was not absorbed through the skin, but the mice died after dermal application of 1:200 to 1:1,000 concentrations of strychnine in mixtures of alcohol and essential oils, eugenol, or anethole, which aid skin penetration. In rabbits exposed to strychnine, there were no corneal opacities in the eyes, but iritis occurred in one but cleared after 2 days. Slight conjunctival irritation persisted for 4 days. Four animals died, however, indicating that a lethal dose had been absorbed through the eye. Intravitreous (i.e., into the fluid within the eyeball) injection in rabbits did not affect optic nerve function. Strychnine is toxic by inhalation with a permissible exposure limit (PEL) of 0.15 $mg/m^3$ of air. There is no evidence on carcinogenicity.[19–23]

## REFERENCES


1. W.J. Hayes, Jr., ed., *Pesticides Studied in Man*, Willams & Wilkins, Baltimore, MD, 1982.
2. C.D.S. Tomlin, ed., *A World Compendium: The Pesticide Manual*, 11th ed., British Crop Protection Counci, Farnham, Surrey, UK, 1997.
3. W.N. Aldridge, An assessment of the toxicological properties of pyrethroids and their neurotoxicity. *Toxicology* **21**(2):89, 1990.
4. U.S. National Library of Medicine. *Hazardous Substances Databank*, Bethesda, MD, 1995.
5. I.R. Edwards, D.G. Ferry, and W.A. Temple, Fungicides and related compounds, in *Handbook of Pesticide Toxicology*, Hayes, W.J. and Laws, E.R., eds., Academic Press, New York, 1991.
6. W.N. Aldridge, An assessment of the toxicological properties of pyrethroids and their neurotoxicity. *Toxicology* **21**(2):89, 1990.
7. Forest Service, *Pesticide Background Statements, Vol. I, Herbicides. United States Dept. of Agriculture*, Agriculture Handbook No. 633, 1984.
8. U.S. Environmental Protection Agency, *Office of Pesticide Programs, Pesticide Fact Sheet: Bifenthrin, No. 177*, Washington, DC, 1988.
9. R.T. Meister, ed., *Farm Chemicals Handbook '92*, Meister Publishing Co., Willoughby, Ohio, 1992.
10. J.T. Stevens, and D.D. Sumner, Herbicides, in *Handbook of Pesticide Toxicology*, W.J. Hayes Jr., and E.R. Laws Jr., eds., Academic Press, New York, 1991.
11. H. Kidd, and D.R. James, eds., *The Agrochemicals Handbook*, 3rd ed., Royal Society of Chemistry Information Services, Cambridge, UK, 10–2, 1991 (As Updated).
12. World Health Organization, *International Programme on Chemical Safety, Permethrin., Environmental Health Criteria 94*, Geneva, Switzerland, 1990.
13. Department of Health and Human Services/NIOSH, *Packet Guide to Chemical Hazards*, 180, 90–117, 1990.
14. World Health Organization, Environmental Health Criteria 73, Phosphine and Selected Metal Phosphides, WHO, Geneva, Switzerland, 1988.
15. P.E. Newton, R.E. Shroeder, J.B. Sullivan, W.M. Busey, and D.A. Banas, Inhalation toxicity of phosphine in the rat: Acute, subchronic and developmental. *Inhal. Toxicol.* **5**(2):223, 1993.
16. U.S. Department of Health and Human Services, *File: Aluminum Phosphide Hazardous Substance Data Base (HSDB)*, HHS, Washington, DC, 1994.
17. L.C. Wong, J.M. Winston, C.B. Hong, and H. Plotnick, Carcinogenicity and toxicity of 1,2-dibromoethane in the rat. *Toxicol. Appl. Pharmacol.* **63**(2):155, 1982.
18. G.A. Letz, S.M. Pond, J.D. Osterloh, R.L. Wade, and C.E. Becker, Two fatalities after acute occupational exposure to ethylene dibromide. *JAMA*. **252**:2428, 1984.
19. L.H. Danse, F.L. Van Velsen, and C.A. Vander Heijden, Methylbromide: Carcinogenic effects in the rat forestomach. *Toxicol. Appl. Pharmacol.* **72**:262–271, 1984.
20. J. Evans, *Efficacy and Hazards of Strychnine Baiting for Forest Pocket Gophers*, in *Animal Damage Management in Pacific Northwest Forests*. Pullman, WA, p. 81, Washington State University, 1987.

21. H. Interdonati, Inc., *Strychnine Alkaloid NF Powder. Material Safety Data Sheet*, Cold Spring Harbor, NY, 1994.
22. H.R. Ludwig, S.G. Cairelli, J.J. Whalen, *Documentation for Immediately Dangerous to Life or Health Concentrations (IDLHs)*, U.S. Department of Health and Human Services, National Institute for Occupational Safety and Health, Cincinnati, Ohio. NTIS Publication No. PB-94–195047, Strychnine p. 435, 1994.
23. U.S. Environmental Protection Agency, *Office of Pesticides and Toxic Substances*, Pesticide Fact Sheet No. 178: Strychnine Update. Publication 540/FS-88-129, 1988.

CHAPTER 9

# Industrial Solvents

## INTRODUCTION

Process technology and industrial operations of a number of industries are closely associated with the use of organic solvents. A majority of industrial solvents are volatile and distribute rapidly in the work environment, storage, and disposal areas. The common term *volatile organic compounds* (VOCs) covers an extensively large group of compounds. All organic chemicals contain the element carbon (C) and are derived from living things, such as coal, petroleum, and refined petroleum products. Many of the organic chemicals we use do not occur in nature, but are synthesized by chemists in laboratories. Volatile chemicals produce vapors readily at room temperature and normal atmospheric pressure. Industrial solvents include gasoline, benzene, toluene, xylene, and tetrachloroethylene. Many volatile organic chemicals also are hazardous air pollutants.

Industrial solvents have applications in the manufacture of a variety of products (e.g., cosmetics, detergents and soaps, drugs, dyes, pigments, explosives, fertilizers, petrochemicals, inks, paints of different kinds, pesticides, plastics, synthetic fibers). Thus, the role played by organic solvents in the development of worldwide modern industry is large. Along with extensive use, the degree and pattern of exposure of occupational workers and the general public is large. Because many organic solvents have the potential to induce adverse effects in animals, humans, and the environment, great care is needed during use and disposal. Solvents are defined as substances that have the intrinsic property to dissolve. A solvent may be an organic agent or an aqueous substance. A host of industrial solvents have contributed in two-phase systems, either as carriers for substrates or as substrates themselves, such as alcohols, ketones, ethers, and alkanes. Many organic solvents are in use in household and industrial activities. Industrial solvents may be categorized (very broadly) as follows:

*Industrial Guide to Chemical and Drug Safety*, By T.S.S. Dikshith and Prakash V. Diwan
ISBN 0-471-23698-5 

- Aliphatic hydrocarbons
- Aromatic hydrocarbons
- Phenols
- Aromatic amines
- Aromatic halogenated hydrocarbons
- Nitro compounds
- Gases and vapors

There are a number of individual solvents in each category. The following pages, however, briefly discuss selected chemicals. Information regarding the toxicity, manner of use, and disposal may help guide occupational workers to achieve chemical safety in a working environment. The large group of industrial solvents is separated into small subcategories for purposes of easy understanding to technicians, students, and industrial workers with limited knowledge, to know each solvent fully well before use and the possible adverse effects it may cause. Since these solvents are highly toxic, proper care must be taken during use and disposal to certain chemical hazard and to improve human safety.

## ALIPHATIC HYDROCARBONS

Aliphatic hydrocarbons are an important group of compounds in the chemical industry. Saturated aliphatic hydrocarbons are present naturally in swamp gas, natural gas, paraffin, and crude oil fractions. It also is found available in coal, natural plant resins, and animal fats. These are released into the environment in the exhaust of gasoline and diesel engines, in the flue gas of municipal waste incinerators, and from vulcanization and extrusion processing operations. There are many industrial applications of paraffin wax (e.g., fuels, solvents, lubricants, degreasers, protective coatings, refrigerants, propellants, application processes of pesticides, intermediates in the synthesis of organic chemicals, food additives).[1] Industrial applications of aliphatic hydrocarbons include alkanes, alkenes (olefins), and alkynes, and are not within the scope of these discussions; details may be found in the literature.

The most common members of aliphatic hydrocarbons are methane, ethane, n-propane, n-butane, n-pentane, n-hexane, n-heptane, n-octane, n-nonane, and n-decane. In general, after repeated exposures, these compounds cause nausea, vomiting, abdominal discomfort, asphyxia, and chemical pneumonitis. In high concentrations as gas or vapor, these compounds trigger central nervous system (CNS) depression and axonopathy.[1–3] In keeping with the essential requirements of chemical safety, the American Conference of Governmental Industrial Hygienists (ACGIH) and the Occupational Health and Safety Administration (OSHA) have set the threshold limits for many aliphatic hydrocarbons.[4,5] In view of the industrial and occupational exposures, the following pages briefly discuss the safety and toxicity profile of selected aliphatic hydrocarbons.

### Methane

Molecular formula: $CH_4$

Methane is a natural gas present in coal mines, marsh gas, and sludge degradations. Although at low concentrations it causes no toxicity, high doses lead to asphyxiation in animals and humans.

### Ethane

Molecular formula: $CH_3CH_3$

Ethane is a flammable gas present in the exhausts of diesel and gasoline engines, municipal incinerators, and the combustion of gasoline. Inhalation and other exposures cause CNS depression in mammals. Ethane in liquid form results in frostbite,[3] and the United States Environmental Protection Agency (USEPA) has listed it under chemical inventory and the test substance database.

### Propane

Molecular formula: $CH_3CH_2CH_3$

Propane is released in the environment from automobile exhausts, burning furnaces, natural gas sources, and combustion of polyethylene and phenolic resins.

Propane is both highly inflammable and explosive and requires proper care and management of workplaces. Its use in industry includes a source for fuel and propellant for aerosols. Occupational workers exposed to liquified propane have demonstrated skin burns and frostbite. Propane also causes CNS depression.[3]

### Butane

Molecular formula: $CH_3(CH_2)_2CH_3$

Butane is found in exhausts of gasoline engines and in waste disposal sites. Butane as a gas is highly inflammable and explosive; pure butane has several applications in industries and processing associated with aerosol propellants, fuel source, solvents, rubber, plastics, food additive, and refrigeration. Occupational exposure to liquefied butane by direct contact results in severe adverse effects (e.g., burns or frostbite to skin, eyes, and mucous membrane, as well as CNS depression).

### n-Pentane

Molecular formula: $CH_3(CH_2)_3CH_3$

n-Pentane is a flammable liquid. It has applications in industry as an aerosol propellant and as an important component of engine fuel. n-Pentane is a CNS depressant. Studies with dogs have indicated that it induces cardiac sensitization. In high concentrations, it causes incoordination and inhibition of the righting reflexes. The National Institute for Occupational Safety and Health (NIOSH) has recommended limits of n-pentane for working areas.[6]

### n-Hexane

Molecular formula: $CH_3(CH_2)_4CH_3$

n-Hexane is a highly flammable liquid usually isolated from crude oil. It has extensive industrial applications as a solvent in adhesive bandage factories and other industries. It is highly toxic, triggering several adverse effects (e.g., nausea, skin irritation, dizziness, numbness of limbs, CNS depression, vertigo, respiratory tract irritation to animals and humans). Occupational exposure of industrial workers has demonstrated motor polyneuropathy. People who sniffed glue for prolonged periods showed adverse effects in the form of degeneration of axons and nerve terminals.[7–9]

### n-Heptane

Molecular formula: $CH_3(CH_2)_5CH_3$

n-Heptane is a flammable liquid present in crude oil and widely used in the automobile industry, as a solvent, gasoline knock testing standard, automotive starter fluid, and paraffinic naphtha. n-Heptane also causes adverse effects in workers (e.g., CNS depression, skin irritation, pain).[8] Other compounds such as n-Octane ($CH_3(CH_2)_6CH_3$), n-Nonane ($CH_3(CH_2)_7CH_3$), and n-Decane ($CH_3(CH_2)_8CH_3$) also have many industrial applications. Workers exposed to these compounds also experience adverse health effects. In principle, management of these aliphatic compounds requires proper handling and disposal to avoid health problems and to contain chemical safety to workers and the environment.

## AROMATIC HYDROCARBONS

Aromatic hydrocarbons are a group of chemicals with extensive industrial applications as raw materials, in oil and rosin extraction, in dry cleaning, printing, metal processing, in automotive gasolines, in pharmaceutical industry, and as solvents. The aromatic hydrocarbons may be categorized in three groups: (1) the alkyl group; (2) the aryl group; and (3) the polycyclic group, which have all been characterized by the benzene nucleus alone, substituted, fused, or joined. The most common chemical is benzene, a nonsubstituted ring system. When one methyl group is attached to the ring, it results in toluene; with two methyl groups, it forms xylene. These compounds are available in all the three physical states, namely, as liquid, solid, and vapor. In fact, lower molecular weight derivative compounds have larger absorbability, higher vapor pressure, greater volatility, and increased solubility, all of which result in severe toxicity to exposed occupational workers.

Because aromatic hydrocarbons are rapidly absorbed by the system, resulting in adverse effects, the worker must be aware of protective measures during handling. For instance, these compounds have caused defatting, skin dehydration, and dermatitis. Their affinity to blood-forming tissue and myelotoxic reaction prolonged exposure results in carcinogenicity. Further, these compounds have caused corneal irritation, severe burns, and cataracts. Entry of aromatics by repeated and continuous inhalation has resulted in depression, pulmonary edema, respiratory paralysis, pneumonitis, and sudden death.[10] These compounds also cause pathological lesions on different vital organs of the system, including liver, kidney, spleen, brain, thymus, and bladder.

Exposure of workers to petroleum refining differs under work different conditions, but petroleum products are transported from place to place in closed systems. In contrast to transport workers, refinery workers suffer from many health problems, including arterial diseases, digestive cancer, neurologic disorders, and death. This chapter briefly discusses the uses and toxicity of selected aromatic hydrocarbons.

### **Benzene** (Benzol, Cyclohexatriene)

Molecular formula: $C_6H_6$

***Uses*** Benzene is a clear, colorless, sweet-smelling liquid. Commercial benzene often consists of toluene, xylene, phenol, and traces of carbon disulfide. Benzene in large volumes is produced by fractionation distillation from crude oil, solvent extraction, and as a by-product of coke-oven processing. Benzene is found in coal tar distillates, petroleum naphtha, and gasoline.[1]

***Toxicity*** Exposure to benzene by ingestion or inhalation has caused toxicity in animals and humans. The acute oral $LD_{50}$ in rats ranged from 1,000 to 5,600 mg/kg, in dogs as high as 2,000 mg/kg, and in rabbits from 150 to 300 mg/kg. This indicates that rabbits are more susceptible to benzene-induced toxicity.[3] Signs and symptoms of toxicity include vomiting, loss of gait, staggering, rapid pulse, CNS depression, visual disturbances, fatigue, and loss of consciousness. In severe cases of poisoning, individuals exhibit major blood changes and reach a state of delirium, coma, and death. Humans exposed in work areas for a prolonged period have demonstrated excretion of peak levels of phenylsulfate within 4 to 8 hours.[11] Extensive studies have been carried out by researchers regarding the toxicologic profile of benzene.[9,12–14]

Earlier reports of Bowditch and Elkins[14a] on occupational workers engaged in operations such as coating, compounding, cementing, and point removing in industries of artificial leather manufacturing, shoe repairing, and telephone exchange for extended periods have suffered severe poisoning and fatalities.

### **Toluene** (Methyl Benzene)

Molecular formula: $C_6H_5CH_3$

***Uses*** Toluene is a clear, colorless, flammable, and sweet- pungent-smelling liquid. It is extensively used as a solvent in different industries, (e.g., rubber chemical manufacture, drugs and pharmaceuticals, thinner for inks, paints dyes, perfume manufacture).

***Toxicity*** Toluene is known to cause mild irritation, headache, nausea, and CNS effects in animals and humans. Prolonged exposure to high concentrations of toluene has caused disturbances in vision, dizziness, nausea, CNS depression, paresthesia, and sudden collapse. The acute oral $LD_{50}$ in laboratory rats ranges from 636 to 7,300 mg/kg. Toluene has caused rapid and severe corneal damage and conjunctiva inflammation. The acute dermal $LD_{50}$ in rabbits was found to be between 1,200 and 1,400 mg/kg.[3,15]

## Xylene

Molecular formula: $C_6H_4(CH_3)_2$

***Uses*** Xylene occurs in the manufacture of different petroleum products and as an impurity in benzene and toluene. It is a colorless and flammable liquid. Commercial xylene is a mixture of three isomers, namely, ortho-, meta-, and paraisomer. It is extensively used in industries associated with paints, rubber, inks, resins, adhesives, paper coating, solvents, and emulsifiers. It also is used in the manufacture of plasticizers, glass-reinforced polyesters, and alkyl resins. Xylene is used as an important raw material.

***Toxicity*** Acute and chronic exposure to xylene induces adverse effects on the skin and respiratory system of animals and humans. Prolonged exposure to xylene demonstrated a burning effect, drying, defatting of skin, eye irritation, lung congestion, CNS excitation, depression, mucosal hemorrhage, and mild liver damage.[16–18]

## Styrene (Cinnamene, Cinnamol, Phenyl Ethyl Benzene, Phenylethane, Styrol, Styrolene)

Molecular formula: $C_6H_5\text{-}CH{=}CH_2$

***Uses*** Styrene is a colorless or yellow, sweet-smelling liquid. It is produced during the alkylation of benzene and ethylene. It is highly reactive and polymerizes rapidly with a violent explosive reaction. This chemical demands proper handling, transportation, and storage by adding polymerization inhibitors in adequate quantities during these operations.[19,20] A styrene monomer has been used extensively in the manufacture of chemical intermediates, filling components, plastics, resins, and stabilizing agents.

***Toxicity*** Styrene induces adverse effects, which include irritation to eyes and mucous membranes, loss of appetite, vomiting, and nausea. Prolonged exposure results in skin damage in the form of dermatitis or rough and fissured skin.[1] Laboratory studies have demonstrated that styrene has carcinogenic potential to rodents; accordingly, the International Agency for Research on Cancer (IARC) has classified the chemical as a 2B carcinogen, a possible human carcinogen.[21,22]

## Petroleum Products

***Uses*** The results of exposure to petroleum and petroleum products during refining vary under different conditions. Petroleum products are transported from place to place in closed systems, but transport workers and refinery workers suffer from many health problems, including arterial diseases, digestive cancer, neurologic disorders, and death. Petroleum distillates and fractionation yield several compounds with multiple uses (Tables 9-1 and 9-2).

## Crude Oil (Earth Oil, Petroleum)

***Uses*** Crude oil is a complex mixture of organic and inorganic materials and contains paraffinic, aromatic, naphthenic, sulfur, and nitrogen-related compounds

**TABLE 9-1 Different Petroleum Distillates and Common Uses**

| Distillates | Uses |
|---|---|
| Asphalt (tar) | Construction; road oils; sealant |
| Naphthas (ligroin) | Thinner; polish |
| Gasolines (petroleum spirits) | Fuel thinner |
| Mineral spirits (white spirit) | Degreasing; dry cleaning; solvent |
| Kerosenes (coal oil) | Fuel; lighter fluid |
| Fuel oil (diesel oil) | Fuel |
| Lubricating oil (oil) | Lubrication |
| Petroleum (petroleum jelly) | Laxative; ointment |
| Paraffin wax (paraffin) | Sealant; polish |

**TABLE 9-2 Petroleum Fractionation and Common Uses**

| Fraction | Uses |
|---|---|
| Natural gas | Fuel chemical |
| Liquified or bottled gas | Fuel gas; petrochemical; synthesis of rubber products |
| Petroleum ether | Solvent |
| Gasolines | Aviation fuel; motor gasoline |
| Naphthas | Cleaning fluid; refining stock; solvent |
| Kerosenes | Jet and turbofuel; stove oil; tractor and gas turbine fuel |
| Gas oil | Furnace oil; diesel oil |
| Lubricating stocks | White oil; greases; lubricating oil |
| Waxes | Sealing wax; food component |
| Bottoms | Heavy fuel oil; road oil; asphalt |

in addition to different metals like boron, chromium, cobalt, manganese, nickel, sulfur, vanadium, and uranium. It is extensively used in the manufacture of gasoline and other lubricating oils.

***Toxicity*** Prolonged exposure to crude oil, particularly in work areas, triggers a variety of adverse effects among workers (e.g., depression of the CNS, chemical pneumonitis, skin irritation). Workers exposed repeatedly for prolonged periods have demonstrated vomiting, anorexia, weight loss, plasma glucose decrease, and mental depression. It is advised that during handling and distribution of crude oil workers should always protect themselves with proper protection to avoid spillage and inhalation.[7,23,24]

### **Gasoline** (Benzin, Petrol, Automotive or Aviation Fuel)

Molecular formula: $C_4H_{10}$ to $C_{13}H_{28}$

***Uses*** Gasoline is a flammable liquid obtained in petroleum fractionation. The majority of components include paraffins, olefins, naphthenes, aromatics, and approximately 10% to 40% ethyl alcohol.[1] Gasoline sniffing has caused morbidity and mortality because of acute, prolonged inhalation in workplaces.

***Toxicity*** Acute exposure to gasoline produces a spectrum of neurologic effects in workers. The pattern of adverse effects is associated with the dose and duration of exposure to gasoline in workplaces. The worker exposed experiences severe headaches, flushed face, giddiness, dizziness, blurred vision, vertigo, nausea, and numbness. Some studies revealed that workers suffer from cerebellar dysfunction, convulsions, and hallucination.[25,26] The neuropathologic lesions in patients include neuronal loss, gliosis in the cerebral cortex, cerebellum, and brainstem, including reticular formation. Prolonged gasoline-exposed workers develop leukemia and kidney cancer, as revealed by the epidemiologic study of petroleum refinery workers.

Gasoline rapidly sensitizes the myocardium and results in CNS depression, cyanosis, mild excitation, loss of consciousness, convulsions, and respiratory failure. Gasoline-poisoned workers also have demonstrated drowsiness, confusion, hyperemia of the conjunctiva, and fever. These symptoms can occur specifically among workers involved in cleaning storage tanks while using proper respiratory protections.[2] Complaints from gasoline pump workers in India include headache, fatigue, sleep disturbances, loss of memory, increased levels of urinary phenol excretion, and hepatic and CNS effects.[25] In view of the adverse effects, workers should not be allowed to work in areas with gasoline levels greater than 500 ppm and without proper protectives.[26]

### **Petroleum Ether** (Ligroin, Petroleum Benzene)

***Uses*** Petroleum ether is a flammable liquid used as a universal solvent and extractant during the processing of different chemicals like fats, waxes, paints, varnishes, furniture polish thinning, detergents, and fuel.[23,27,28] The majority of components include paraffins, olefins, naphthenes, aromatics, and approximately 10% to 40% ethyl alcohol.

***Toxicity*** Acute, prolonged exposure in workplaces has caused a variety of health disorders in workers (e.g., erythema, edema, skin peeling, loss of appetite, muscle weakness, paresthesia, CNS depression, peripheral nerve disorders, skin and respiratory irritation, chemical pneumonia in children).[7] In addition, rubber solvents, varnish, thinners, and petroleum spirits cause skin irritation, respiratory problems, and hematologic effects in workers; all demand proper handling and chemical safety.

### **Kerosene** (Stove Oil)

***Uses*** Kerosene is a white to pale yellow, flammable liquid in wide use in household and industrial activities (e.g., heating, cooking fuel, cleaning, degreasing, as a solvent for paints, enamels, polishes, and varnishes, and in asphalt coating).

***Toxicity*** Kerosene toxicity is variable and is based on the composition. It is rapidly absorbed by the skin, and accidental ingestion results in mucous membrane irritation, gastrointestinal irritation, vomiting, diarrhea, pneumonitis, CNS depression, drowsiness, coma, and may lead to death. Prolonged contact also is known to cause skin blisters and dermatitis.[3] Studies with nonhuman primates have demonstrated that aerosols and kerosene aspiration into the lungs causes cellular damage.[29,30] Because of the easy availability to household activities and possible long-term exposure, proper care is necessary to avoid skin contact and damage. Kerosene should never be sucked by mouth.

## ALIPHATIC AND ALICYCLIC AMINES

Aliphatic and alicyclic amines form one of the most extensively used categories of compounds in a variety of manufacturing industries. These compounds are first-phase materials for chemical synthesis, intermediates, pharmaceuticals, pesticides, soil sterilizers, and several kinds of solvents, rubber products, rocket propellants, and plastic monomers. This chapter discusses the safety and toxicity of selected compounds handled by workers in different industries. More information and extensive literature are available in several study reports on members of this important group.[31,32]

### **Methylamine** (Aminomethane, Methanamine, Monomethylamine)

Molecular formula: $CH_3NH_2$

***Uses*** Methylamine is a colorless, fishlike-smelling gas at room temperature. It is used in a variety of industries such as manufacture of dyestuffs, treatment of cellulose, acetate rayon, fuel additive, rocket propellant, and leather-tanning processes.

***Toxicity*** Methylamine is known to cause irritation to eyes, nose, and throat in workers. Studies have indicated that the compound causes injury to eyes through corneal opacities and edema hemorrhages in the conjunctiva, and injury to liver.[10] Studies of Guest and Varma[33] indicated no significant deleterious effects on internal organs or skeletal deformities in experimental mice.

**Dimethylamine** (DMA, N-Methylmethanamine)

Molecular formula: $(CH_3)_2NH$

***Uses*** Dimethylamine is a colorless, inflammable gas at room temperature. It is used in the manufacture of several products such as detergent soaps, in leather tanning, in pharmaceutical manufacture, and for cellulose acetate rayon treatment.

***Toxicity*** Dimethylamine induces adverse effects such as irritation to skin and lungs. Repeated exposure to the chemical has caused corneal injury in experimental guinea pigs and rabbits. Studies of Hollinger and Rowe,[34] Coon et al.,[35] and others[36] revealed inflammatory changes in the lungs, ulcerative rhinitis, nasal turbinates, and corneal ulceration in laboratory rabbits and nonhuman primates.

**Ethylamine** (Aminoethane, MEA, Monoethylamine)

Molecular formula: $CH_3CH_2NH_2$

***Uses*** Ethylamine is a colorless, inflammable gas. A variety of manufacturing industries have been associated with this compound (e.g., dyestuff industry, pharmaceuticals, rubber latex stabilizers, oil refining).

***Toxicity*** Ethylamine causes severe irritancy to exposed skin, eyes, and mucous membranes. Direct contact with the skin of experimental animals resulted in skin burns, scarring, and necrosis.[37–39] The chemical also caused adverse effects and degenerative changes in heart, liver, lung, kidney, and associated injury to the endocrine system of animals.[40]

**Diethylamine** (Aminoethane, DEA, Monoethylamine)

Molecular formula: $CH_3CH_2NH_2$

***Uses*** Diethylamine is a colorless, inflammable, strongly alkaline, fishy-smelling liquid. It has several applications in industries such as organic synthesis of resins, rubber accelerator, pharmaceuticals, pesticides, dyes, electroplating operations, and as a polymerization inhibitor.[41]

***Toxicity*** Diethylamine is a strong irritant to skin, eyes, and mucous membrane. The acute oral $LD_{50}$ and acute dermal $LD_{50}$ in rats and rabbits are 540 mg/kg and 580 mg/kg, respectively. The acute inhalation $LC_{50}$ (4-hour) for rats is 4,000 ppm.[38,42,43] The pathomorphologic changes observed in lungs, liver, and kidneys included cellular infiltration, bronchopneumonia, parenchymatous degeneration, and nephritis.

## BUTYLAMINES

Butylamines have played a versatile role in the manufacturing industries of different day-to-day products. These include industries associated with textiles, leather tanning, pesticides, antioxidants and pharmaceuticals, photography, plastics, dyestuffs, and many other emulsifying agents.

### **n-Butylamine** (1-Aminobutane, 1-Butanamine, MNBA)

Molecular formula: $CH_3CH_2CH_2NH_2$

***Uses*** n-Butylamine has multiple roles as a solvent and intermediate in a number of manufacturing industries associated with leather and synthetic tanning, pesticides, antioxidants and pharmaceuticals, photography, plastics, dyestuffs, and many other emulsifying agents.[41,44]

***Toxicity*** Studies by Cheever et al.[31,41,45] have indicated that n-Butylamine has caused toxicity to animals. The acute oral $LD_{50}$ of male and female Sprague Dawley rats is 371 mg/kg. The vapors of n-Butylamine cause irritation to eyes and mucous membranes, as well as irritation and blistering effects to the skin of animals and humans.[31,41,45] Workers exposed by direct skin contact to the liquid form demonstrated severe irritation of the skin in addition to deep second-degree burns and blistering.

## ALLYLAMINES

Allylamines include monoallylamines, diallylamines, and triallylamines. All have applications in industries associated with the manufacture of chemicals and pharmaceuticals. Although monoallylamine is highly toxic through all the three major routes of exposure (oral, dermal, and inhalation), diallylamines and triallylamines are relatively less toxic with larger acute oral $LD_{50}$ values (Table 9-3). The vapors, fumes, and dusts created during industrial processing (e.g., grinding,

**TABLE 9-3 Acute Toxicity of Allylamines in Rats and Rabbits**

| | Oral ($LD_{50}$ mg/kg) | Dermal ($LD_{50}$ mg/kg) | Inhalation ($LC_{50}$ ppm) |
|---|---|---|---|
| Monoallylamines | 106 | 35 | 286 |
| Diallylamines | 578 | 356 | 2,755 |
| Triallylamines | 1,310 | 2,250 | 828 |

polishing, sanding) with allylamines have induced irritation to eyes, mucous membranes, and skin, skin allergy, and sensitization effects in workers.

### Allylamine (Monoallylamine, 3-Aminopropylene, 3-Aminopentene)

Molecular formula: $CH_2{=}CHCH_2NH_2$

***Uses*** Allylamine is applied in different industries as a solvent. It also is used in the preparation of diuretics, sedatives, and antiseptics.[32]

***Toxicity*** Allylamine has caused adverse effects in animals and humans with irritation to eyes, skin, and mucous membranes. Reports on cardiovascular toxicity due to allylamine exposure in species of animals is not confirmatory.[31,46–48]

### Cyclohexylamine (Aminocyclohexane, CHA, Cyclohexanamine, Hexahydroaniline)

Molecular formula: $C_6H_{11}NH_2$

***Uses*** Cyclohexylamine is a strong inflammable liquid with a fish-like odor. It has many applications in both household and industrial processing (e.g., a corrosion inhibitor in water boilers, the synthesis of pesticides, dry-cleaning soaps, the manufacture of plasticizers, textile chemicals, cyclamates (artificial sweeteners), dyestuffs).[41,49]

***Toxicity*** Experimental animals exposed to cyclohexylamine have demonstrated severe toxicity and death. Administration of the compound through different routes (e.g., intraperitoneal, intravenous, intramuscular injections) has caused convulsion and death in laboratory mice, rats, and rabbits. Studies of workers have demonstrated that cyclohexylamine has mutagenic, embryogenic, and tumorigenic potentiality in experimental animals.[31,49–52] The IARC has reported no evidence of the teratogenic or carcinogenic effects of cyclohexylamine.[49] Extensive studies by Lorke et al.[31,53–55] using animal species and nonhuman primates suggested conflicting data on the abovementioned adverse effects, suggesting that more information is a must before arriving at a pragmatic conclusion.

**Ethylenediamine** (1,2-Diaminoethane, Dimethylenediamine, 1,2-Ethanediamine)

Molecular formula: $H_2NCH_2CH_2NH_2$

***Uses*** Ethylenediamine is a colorless, flammable, hygroscopic, corrosive, and fuming thick liquid. It has a musty, irritating odor. It is used as a solvent, in the manufacture of pharmaceuticals fungicides and synthetic waxes, and as a gasoline additive.[31]

***Toxicity*** Ethylenediamine has caused injury to animals and humans because of its irritating vapors and corrosion property. Spillage of liquid on skin or other body parts has resulted in skin corrosion, corneal injury, and irritation to mucous membranes and the respiratory tract.[31,41] Laboratory studies with guinea pigs showed that the compound induces hypersensitivity reaction in animals.[56] Exposure to ethylenediamine for prolonged periods has caused injury to liver, kidney, and lungs, as well as CNS depressant action.[57] Reports by Slesinski et al. and DePass et al.[58–60] indicated that ethylenediamine has no genotoxic or carcinogenic potential as demonstrated by both *in vivo* and *in vitro* studies.[58–60]

**Triethanolamine** (Triethyloamine, Trolamine, Trihydroxytriethylamine)

Molecular formula: $(HOCH_2CH_2)_2N$

***Uses*** Triethanolamine is a pale yellow, viscous liquid. It is hygroscopic with an irritant, ammoniac odor. The industrial and domestic applications of this compound are multiple and extensive. Use includes manufacture of toilet products, cosmetics formulations,[61] solvents for waxes, resins, dyes, paraffins, and polishes, herbicides, and lubricants for textile products.[41] In the pharmaceutical industry, triethanolamine is used as a nonsteroidal antiinflammatory agent,[62,63] emulsifier, and alkylating agent.

***Toxicity*** Triethanolamine, in contrast to other compounds, demonstrated low toxicity to animals with acute oral $LD_{50}$ in rats and guinea pigs as high as 8,000 to 9,000 mg/kg. The compound was found to be a moderate eye irritant, and a 5% to 10% solution did not induce skin irritation or sensitization.[31] Studies of Inoue et al.[64] and other researchers[65] have indicated the absence of the mutagenic potential of triethanolamine as evidenced by both *in vivo* and *in vitro* studies (*Salmonella typhimurium* tests, Chinese hamster ovary cells, and rat liver chromosome analysis). Further, extensive studies have demonstrated the absence of potential carcinogenicity of triethanolamine in rats and mice,[66] suggesting the low (or lack of) acute and chronic chemical toxicity to the mammals.

## AROMATIC NITRO AND AMINO COMPOUNDS

Present-day industrial operations and processes use different aromatic nitro compounds, (e.g., intermediates in the production of dyes and pigments, pharmaceutical products, polymers, products of rubber, paper, and textiles). The aromatic nitro amino compounds cause adverse effects in animals and humans by converting hemoglobin to methemoglobin (the oxidized ferric form). These compounds induce adverse effects like corneal opacity, skin irritation, disturbed metabolism, pathomorphologic lesions in the liver, kidney, spleen, and bladder, and bladder tumors. Tumor formation or carcinogenicity of the nitro and amino compounds is traced to the metabolic activation of the chemical to a more reactive electrophilic species, eventually targeting the genetic material. Three aromatic amines with sufficient evidence of carcinogenicity are 4-aminodiphenyl, benzidine, and beta-naphthylamine.[67]

Workers are potentially exposed to a variety of substances during daily industrial operations. The OSHA estimates that a large segment of workers are simultaneously exposed to more than one substance and suffer from adverse effects. Adverse effects include many acute and chronic diseases (e.g., allergic sensitization, cancer, cardiovascular diseases, dermatitis, erythema, edema, irritation of the mucous membrane, irritation of the skin, kidney disease, liver disease, metabolic interferences, narcosis, neuropathy, ocular effects, odor effects, respiratory diseases, systemic toxicity). These should be properly addressed by concerned management and the individual worker to improve chemical safety. Workers come in contact directly or indirectly during different work conditions; hence it is important to understand the possible adverse effects that these chemicals may cause vis-à-vis chemical safety.

### **Picric Acid** (Picronitric Acid, 2,4,6-Trinitrophenol, Trinitrophenol)

Molecular formula: $HOC_6H_2(NO_2)_3$

***Uses*** Picric acid is a white to yellowish, highly flammable crystalline substance. It is used in the manufacture of fireworks, matches, electric batteries, colored glass, explosives, and disinfectants. Pharmaceutical, textile, and leather industries also make use of picric acid.

***Toxicity*** Picric acid causes adverse effects on the skin of animals and humans like allergies, dermatitis, irritation, and sensitization. Absorption of picric acid by the system causes headache, fever, nausea, diarrhea, and coma. In high concentrations, picric acid is known to cause damage to erythrocytes, kidney, and liver.[68,69]

### **Aniline** (Aminobenzene, Aniline Oil, Phenylamine, 2,4,6-Trinitrophenol)

Molecular formula: $C_6H_5NH_2$

***Uses*** Aniline is an oily liquid used in the manufacture of dyestuffs, intermediates for dyestuffs, and manufacture of rubber accelerators and antioxidant substances. Aniline has been extensively used as an intermediate in the manufacture of plastics, pharmaceuticals, pesticides, isocyanates, and hydroquinones. Occupational exposure to aniline is extensive and as diverse as its industrial uses. Workers associated with the manufacturing of acetanalide bromide, coal tar, colors and dyes, leather, disinfectants, nitraniline, perfumes, rubber, and photographic materials become victims of adverse effects from aniline.

***Toxicity*** Aniline is highly toxic to animals and humans. The acute oral $LD_{50}$ in laboratory rats is 440 mg/kg. Humans exposed to aniline have demonstrated severe headache, narcosis, body tremors, cardiac arrhythmia, coma, and death. Bladder tumors associated with amines and dye industry workers is not well established, since the IARC has classified aniline as a group 3 carcinogen (not classifiable as to its carcinogenicity). NIOSH has labeled the chemical with sufficient evidence to recommend it as a potential occupational carcinogen.[70,71]

### Diphenylamine (Anilinobenzene, N-Phenylbenzeneamine, Phenylaniline, DPA)

Molecular formula: $(C_6H_5)_2NH$

***Uses*** Diphenylamine is a colorless, monoclinic leaflet substance. It is used in the manufacture of a variety of substances, (e.g., dyestuffs and their intermediates pesticides, antihelmintic drugs, as a reagents in analytical chemistry laboratories).

***Toxicity*** Diphenylamine is highly toxic and is rapidly absorbed by skin and through inhalation. It has caused anorexia, hypertension, eczema, and ladder symptoms. Experimental animals exposed to diphenylamine demonstrated cystic lesions but failed to demonstrate cancerous growth.[72–74]

### 4-Aminophenyl Ether (4,4-Oxydianiline, 4,4-Diaminodiphenyl Oxide, N-Phenylbenzeneamine)

Molecular formula: $(H_2N_6H_4)_2O$

***Uses*** 4-Aminophenyl ether is a resin used in the manufacture of a variety of industrial products, (e.g., in insulating varnishes, flame-retardant fibers as wire enamels, coatings, films). It also is used in the manufacture of other industrial fire-resistant products.

***Toxicity*** 4-Aminophenyl ether is highly toxic to animals. It shows sufficient evidence as a carcinogen and has caused adenomas and carcinomas in the thyroid and liver of experimental rats.[75,76]

**Nitrobenzene** (Nitrobenzol, Oil of Mirbane, Essence of Mirbane)

Molecular formula: $C_6H_5NO_2$

***Uses*** Nitrobenzene is a colorless to pale yellow liquid. It has been used extensively in a variety of industries, (e.g., the manufacture of aniline dyes and soaps, as solvent for paints, for refining lubricating oils, as shoe polish, floor polish, dressings for leather products, the manufacture of explosives).

***Toxicity*** Nitrobenzene is highly toxic to animals and humans. The acute oral $LD_{50}$ in rats has been reported as 640 mg/kg. Workers exposed to nitrobenzene are prone to adverse effects on skin and mucous membranes. Other adverse effects include headache, fatigue, giddiness, vomiting, weakness, tachycardia, depression, and coma. Studies have indicated that nitrobenzene could trigger pathomorphologic lesions in liver and spleen among workers. Severe disturbances during embryogenesis and organogenesis have been reported in nitrobenzene-exposed animals.[77–82]

**Benzidine** (4,4-Biphenyldiamine, 4,4-Diaminobiphenyl, p-Diaminodiphenyl)

Molecular formula: $NH_2C_6H_4NH_2$

***Uses*** Benzidine is a white or slightly reddish powder. It has several applications in industry. It is used for the synthesis of dyes like Congo Red and dye intermediates and as a hardener in rubber manufacturing industries.

***Toxicity*** Benzidine is a well-known carcinogen in animals and humans. It has been classified by the IARC as a group 1 carcinogen. Workers exposed to benzidine have demonstrated increased risk of bladder cancer.[67]

**O-Toluidine** (O-Methylamine, O-Aminotoluene, 2-Methylaniline)

Molecular formula: $CH_3C_6H_4NH_2$

***Uses*** O-Toluidine is a light yellow to reddish-brown liquid. The compound quickly turns to a dark color on exposure to light and atmospheric air. The compound has extensive use in a number of industries around the world (e.g., as an intermediate in the manufacture of azo and indigo dyes, pigments, sulfur dyes, pesticides, pharmaceutical products, rubber and vulcanizing chemicals).

***Toxicity*** O-Toluidine is highly toxic to animals and humans and is rapidly absorbed by oral, dermal, and inhalation routes. The acute oral $LD_{50}$ in rats ranges from 900 to 940 mg/kg.[78,83] The compound is known to cause adverse effects in workers that include headache, irritation of skin, eye, kidneys, and bladder, and hematuria. O-Toluidine has caused hepatocellular adenomas and carcinomas in experimental mice and rats.

Workers exposed to this compound also have demonstrated bladder cancer, although the role of aniline cannot be ruled out.[84,85] However, the IARC (because of insufficient data) classifies O-Toluidine as a group 2B agent, meaning possibly carcinogenic to humans.[75] NIOSH classifies this compound as an occupational carcinogen[86], and the ACGIH labels it as a suspected human carcinogen under the A2 class.[83]

### **O-Nitrotoluene** (Nitrotolul, Methylnitrobenzene, 2-Nitrotoluene, 2-NT)

Molecular formula: $CH_3C_6H_4NO_2$

***Uses*** O-Nitrotoluene is a yellow liquid. The compound is used for the synthesis of a variety of industrial products (e.g., azo dyes, agricultural chemicals, explosives, sulfur dyes, and rubber chemicals).

***Toxicity*** O-Nitrotoluene causes adverse effects to animals and humans. Acute and chronic exposure causes irritation to skin and mucous membranes, hypoxia, anemia, and depression in workers.[86] More reports are available regarding the genotoxicity, carcinogenicity, and reproductive effects of O-Nitrotoluene in experimental animals.[87,88]

### **Dinitrotoluene** (DNT)

Molecular formula: $C_6H_3CH_3(NO_2)$

***Uses*** The technical-grade dinitrotoluene is an oily liquid and is easily combustible. Essentially, dinitrotoluene is used in large volume during the manufacture of polyurethane foams and polymers, in the production of toluene diisocyanates, and in manufacturing explosives.

***Toxicity*** Dinitrotoluene is known to cause a variety of adverse effects in animals and humans. Workers come in contact with dinitrotoluene through all three major routes (oral, dermal, inhalation). The acute oral $LD_{50}$ in rats ranges from 568 to 650 mg/kg; female rats are more resistant to the chemical. The acute inhalation $LC_{50}$ (1 hour) in rats is less than 2 mg/L. Acute and repeated exposures to the compound are known to cause hypoxia, dyspnea, headache, dizziness, joint pain, optic neuritis, cyanosis, jaundice, and anemia among workers.[78,83]

Studies have indicated that repeated exposure of dinitrotoluene has resulted in testicular atrophy and disturbances in the spermatogenesis cycle in experimental mice, rats, and dogs. Female mice also revealed nonfunctioning ovaries.[89,90] Conflicting reports regarding the potential reproductive toxicity of dinitrotoluene among workers demand more confirmatory data.[91,92] NIOSH has classified technical dinitrotoluene as a human reproductive health hazard in industrial workplaces. Isomers of dinitrotoluene have caused complete liver cancers in animals, but technical-grade dinitrotoluene failed to induce any kind of hepatic cancer in humans.[37,93]

**Ethylene Glycol Dinitrate** (Ethylene Dinitrate, Glycol Dinitrate, Nitroglycol, EGDN)

Molecular formula: $O_2NOCH_2CH_2ONO_2$

***Uses*** Ethylene glycol dinitrate is a yellow, highly explosive liquid. It is extensively used in the manufacture of commercial dynamites and blasting gelatin.

***Toxicity*** Workers exposed to ethylene glycol dinitrate have demonstrated a variety of adverse effects such as headache, nausea, vomiting, respiratory arrest, angina, cyanosis, coma, and death. Absorption of the compound through skin, digestive tract, and lungs is fast and results in severe toxicity. Workers without proper protection have shown vasodilatation, causing a fall in blood pressure and an increase in heart rate.[94,95]

**Nitroglycerin** (NG, GTN, Trinitroglycerol, Trinitrin, Blasting Oil, Blasting Gelatin)

Molecular formula: $C_3H_5(ONO_2)_3$

***Uses*** Nitroglycerin (or glyceryltrinitrate) is a pale yellow, oily liquid also available in the form of rhombic crystals. It is highly explosive. It is used in combination with ethylene glycol dinitrite in the manufacture of dynamites, explosives, rocket propellants, smokeless powders, and guncotton.

***Toxicity*** Nitroglycerin is highly toxic to humans; it causes rapid vasodilatation and severe throbbing. It is rapidly absorbed through skin and respiratory routes. Repeated exposures and high concentrations cause vomiting, nausea, cyanosis, coma, and death. It causes increased heart rate, myocardial contraction, and angina pectoris. Occupational exposure to nitroglycerin also has been traced to work adaptation to the chemical and the consequences of withdrawal syndrome. Nitroglycerin on skin contact causes violent headache. Therefore, whenever the spirit glyceryl trinitrate is spilled, a solution of sodium hydroxide should be added immediately to check the chemical reaction of high explosion. The worker must remember that on evaporation of alcohol, glyceryl trinitrate results in a severe explosion. This requires care and caution in the proper management of workplaces and worker health.[2,9]

## OTHER ORGANIC SOLVENTS

A host of organic solvents are used for a variety of industrial applications. Workers associated with the manufacturing of industrial products (e.g., resins, fumigants, antiknock compounds, rubber and related products, metal-cleaning products, degreasers, varnishes and paints, dye makers, organic chemical synthesizers) are often exposed to adverse effects of organic solvents. The preceding list has already addressed specific organic solvents under the groups aliphatic hydrocarbons, aromatic hydrocarbons, aliphatic and alicyclic amines, and aromatic nitro and amino compounds. The following are other compounds that should be taken care of during use, handling, storage, and disposal in industrial environments to achieve chemical safety and protect human health.

## Alicyclic Hydrocarbons

***Uses*** The alicyclic hydrocarbons have numerous industrial applications. Cyclopropane ($C_3H_6$) is used as an anesthetic. Cyclohexane ($C_6H_{12}$) is used as a chemical intermediate; as an organic solvent for oils, fats, waxes, and resins; and for the extraction of essential oils in perfume manufacturing industries. Cyclohexene ($C_6H_{10}$) is used in the manufacture of maleic acid, cyclohexane carboxylic acid, and adipic acid. Methyl cyclohexane ($C_7H_{14}$) is used for the production of organic synthetics such as cellulose ethers. These compounds are used in different industries such as adipic acid makers, benzene makers, fat processors, fungicide makers, lacquerers, nylon makers, oil processors, paint removers, plastic molders, resin makers, rubber makers, varnish removers, and wax makers.

***Toxicity*** Potential symptoms of overexposure to alicyclic hydrocarbons are irritation of eyes and respiratory system, drowsiness, dermatitis, narcosis, and coma. It also causes adverse effects on the CNS of animals and humans. Acute and repeated exposure to these organic solvents induces symptoms of excitement, loss of equilibrium, stupor, coma, and respiratory failure, which may also lead to fatalities. In animal studies, alicyclic hydrocarbons have caused pathomorphologic effects in the heart, lung, and liver, as well as brain degeneration. It is important that workers be trained in proper handling of these solvents and use protective coverings.[2,96]

## 1,3-Butadiene (Biethylene, Bivinyl, Butadiene Monomer, Divinyl, Erythrene, Methylallene, Pyrrolylene, Vinylethylene)

Molecular formula: $H_2C{=}CH{-}CH{=}CH_2$

***Uses*** 1,3-Butadiene is a colorless, flammable gas with a pungent, aromatic odor. It is used essentially as a principal monomer in the manufacture of synthetic rubber, plastics, and resins. It also is used extensively as rocket fuel.

***Toxicity*** 1,3-Butadiene causes irritation to the eyes, nose, and respiratory system. Exposure to evaporating gas causes dermatitis and frostbite. The symptoms of toxicity include fatigue, drowsiness, headache, loss of consciousness, respiratory paralysis, and death. Workers should use protective clothing to avoid adverse effects.[96–98]

## Naphtha (Petroleum Naphtha, Ligroin, Benzine, Petroleum Ether, Petroleum Benzine)

Molecular formula: $C_4H_{10}$ to $C_8H_{18}$

***Uses*** Naphtha is a mixture of aromatic hydrocarbons essentially consisting of toluene, xylene, and cumene. Coal naphtha consists of benzene in appreciable amounts. Naphthas are used as organic solvents for dissolving or softening rubber, oils, greases, bituminous paints, varnishes, and plastics. Heavy naphthas are used for the manufacturing of pesticides, and the lower fractions are used in dry cleaning. Occupational exposure to naphthas occurs in chemical laboratory, dry cleaning, fat processing, metal degreasing, oil processing, manufacturing of paints, rubber coatings, stainers, varnishes, and waxes.

***Toxicity*** The naphthas cause severe irritation to skin, eyes, and the mucous membrane of the respiratory tract. Prolonged exposure to naphthas causes photosensitivity. It also causes skin burn. Workers exposed for prolonged periods suffer from headache, nausea, dizziness, convulsions, unconsciousness, and CNS depression. Naphthas containing higher amounts of benzene are known to produce hematologic and nephritic changes. Workers could avoid adverse health effects by using protective clothing and working under proper supervision.[99]

## Paraffin (Paraffin Wax, Hard Paraffin)

Molecular formula: $C_nH_{2n}{}^{+2}$

***Uses*** Paraffin wax is a colorless or white odorless mass. It consists of a mixture of solid aliphatic hydrocarbons. Paraffin is used in the manufacturing of paraffin papers, candles, food packaging materials, varnishes, floor polishes, extraction of perfumes from flowers, and in lubricants and cosmetics. It also is used in waterproofing wood and cork.

***Toxicity*** Prolonged exposure to paraffin causes chronic dermatitis, wax boils, folliculitis, comedones, melanoderma, papules, and hyperkeratoses. Studies by Hendricks et al.[100] have shown carcinoma of the scrotum in workers exposed to crude petroleum wax. The carcinoma of the scrotum in workers began with a normal hyperkeratotic nevus-like lesion, which subsequently resulted in squamous cell carcinoma.[101–103]

## Turpentine (Gum Turpentine, Spirit of Turpentine, Gum Spirit, Wood Turpentine)

Molecular formula: $C_nH_{2n}{}^{+2}$

***Uses*** Turpentine is oleorosin extracted from the plant Pinus. It is a yellowish, opaque, sticky mass with a characteristic odor. It is extensively used in different industries associated with the manufacturing of polishes, grinding fluids, paint thinners, resins, degreasing solutions, clearing materials, and ink.

***Toxicity*** Turpentine is absorbed through the skin, lungs, and intestine. The vapor of turpentine causes severe irritation to the nose, eyes, and respiratory system. Aspiration of liquid turpentine causes direct irritation to lungs and results in pulmonary edema and hemorrhage. It also causes dermatitis, eczema, and hypersensitivity among workers. Splashing liquid turpentine in the eyes causes corneal burns. Turpentine also is known to cause skin eruption, irritation to gastrointestinal tract, kidney and bladder damage, delirium, ataxia, and benign skin tumor. It is, therefore, important that workers should always use protective clothing, rubber gloves, and facemasks to avoid adverse effects.[104]

## ALCOHOLS

Alcohols are hydrocarbons with one or more hydrogen atoms substituted by hydroxyl ($^-OH$) groups. Compounds with one hydroxyl group are called alcohols, those with two are called glycols, and those with three hydroxyls are called glycerols. Alcohols are used extensively in industries as solvents for the manufacture of a variety of products. Generally, all alcohols cause irritation to the mucous membranes with mild narcotic effect. There are important classes of alcohols, namely, allyl alcohol, amyl alcohol, n-butyl alcohol, methyl alcohol, ethyl alcohol, and propyl alcohol.

### **Allyl Alcohol** (Vinyl Carbinol, Propenyl Alcohol, 2-Propeno-1, Propenol-3)

Molecular formula: $H_2C{=}CHCH_2OH$

***Uses*** Allyl alcohol is used in the manufacture of allyl esters, as monomers and prepolymers for the manufacture of resins and plastics. It has large use in the preparation of pharmaceutical products, in organic synthesis, and as a fungicide and herbicide. Workers engaged in industries such as pharmaceuticals, pesticides, allyl esters, organic chemicals, resins, war gas, and plasticizers are often exposed to this alcohol.

***Toxicity*** Exposure to vapors of allyl alcohol causes irritation to the eyes, skin, and upper respiratory tract. Laboratory studies with animals have shown symptoms of local muscle spasms, pulmonary edema, tissue damage to liver and kidney, convulsions, and death. In view of this, workers should be instructed to wear protective clothing.[2,105,106]

### **Amyl Alcohol** (Pentanols, Pentyl Alcohols, Fusel Oil, and Potato Spirit)

Molecular formula: $C_5H_{11}OH$

***Uses*** Amyl alcohol is produced during the fermentation of grains, potatoes, and beets. It is produced during the acid hydrolysis of petroleum fraction. Application of amyl alcohol in industries is very large including manufacturing of lacquers, paints, varnishes, perfumes, pharmaceuticals, plastics, rubber, explosives, hydraulic fluids, extraction of fats, and petroleum refinery industries.

***Toxicity*** Vapors of amyl alcohol cause mild irritation to mucous membranes of the eyes, nose, throat, upper respiratory tract, and to the skin. Acute and long-term exposure to amyl alcohol causes nausea, vomiting, headache, vertigo, and muscular weakness. Prolonged exposure also may cause narcotic effects.

**n-Butyl Alcohol** (n-Butanol, Butyl Hydroxide, n-Propylcarbinol, Butyric Hydroxybutane)

Molecular formula: $CH_3CH_2CH_2CH_2OH$

***Uses*** n-Butyl alcohol is used extensively in a number of industries. For instance, it is used as a solvent in industries associated with the manufacture of paints, varnishes, synthetic resins, gums, pharmaceuticals, vegetable oils, dyes, and alkaloids. n-Butyl alcohol also finds use in the manufacture of artificial leather, rubber, and plastic cements, shellac, raincoats, perfumes, and photographic films.

***Toxicity*** n-Butyl alcohol is a highly refractive liquid and burns with a strongly luminous flame. Exposure to n-Butyl alcohol causes irritation to the eyes, nose, throat, and respiratory system. Prolonged exposure results in symptoms of headache, vertigo, drowsiness, corneal inflammation, blurred vision, photophobia, and cracked skin. It is advised that workers coming in contact with n-Butyl alcohol should use protective clothing and barrier creams.[2,107,108]

**Ethyl Alcohol** (Ethanol, Grain Alcohol, Spirit of Wine, and Cologne Spirit)

Molecular formula: $CH_3CH_2OH$

***Uses*** Ethyl alcohol is a very common solvent used in the chemical synthesis of a variety of products (e.g., manufacturing of pharmaceuticals, plastics, lacquers, polishes, plasticizers, perfumes, adhesives, rubber accelerators, explosives, synthetic resins, nitrocellulose, inks, preservatives, and as a fuel).

***Toxicity*** Prolonged exposure to ethyl alcohol vapors causes irritation to the eyes and upper respiratory tract in addition to causing headache, drowsiness, fatigue, and mild to severe tremor.[1]

**Methyl Alcohol** (Methanol, Carbinol, Wool Alcohol, Wood Spirit)

Molecular formula: $CH_3OH$

***Uses*** Methyl alcohol is a clear, colorless liquid with a slight alcoholic odor. It is used in the synthesis of formaldehyde, methylamine, ethylene glycol, methacrylates, and as an industrial solvent for a number of products (e.g., inks, resins, adhesives, dyes for straw hats). Methyl alcohol is an important ingredient commonly used to prepare grease and dirt remover. It also is used in the manufacture of photographic films, plastics, celluloid, textile soaps, wood stains, coated fabrics, paper coatings, artificial leather, and other industrial products.

***Toxicity*** Methyl alcohol vapor may cause irritation to the nose, throat, and eyes. It causes headache, vomiting, nausea, dizziness, drunkenness, blurred vision, blindness, insomnia, abdominal pains, skin irritation, impaired vision, unconsciousness, and death. Methyl alcohol also can cause poisoning through skin absorption. Direct and repeated contact with methyl alcohol produces mild dermatitis and defatting. Workers exposed to methyl alcohol complain of optic nerve damage and blindness. The toxic effect has been due to the metabolic oxidation of the chemical. Oral ingestion of large amounts of methyl alcohol has caused nausea, giddiness, and loss of consciousness in humans.[1,109,110]

### Propyl Alcohol (n-Propyl Alcohol, 1-Proponal, Isopropyl Alcohol, Isoproponal, 2-Proponal)

Molecular formula: $CH_3CH_2CH_2OH$ — (n-propyl alcohol);
$CH_3CHOHCH_3$ — (isopropyl alcohol)

***Uses*** Propyl alcohol has two isomers, namely, n-Propyl alcohol and isopropyl alcohol. These alcohols have extensive use in a variety of industries (e.g., manufacturing, pharmaceuticals, perfumes, cosmetics, skin lotions, hair topics, mouthwashes, liquid soaps). It also is used in lacquers, dental lotions, polishers, and surgical antiseptics.

***Toxicity*** Propyl alcohol is not known to cause toxicity to animals and humans unless it is used improperly. Propyl alcohol vapor in high concentrations causes mild irritation to eyes, conjunctiva, and mucous membranes of the upper respiratory tract, and CNS depression.[111]

## PHENOL AND PHENOLIC COMPOUNDS

Phenols are the simplest group of compounds. Phenols have a wide application in pharmaceutical manufacturing industries. Phenols, in principle, are a group of chemicals causing severe irritation to the body system. They cause severe irritation to the eyes, skin, nose, respiratory tract, and mucous membranes. They are highly corrosive to skin and tissues. Creosate — a mixture of phenolic and aromatic substances — is well known as a carcinogenic agent to the skin. Phenols also cause adverse effects to CNS, cardiovascular, renal, and hepatic systems of animals and humans. In view of this, proper use, disposal, and management is a must for all workers.

### **Cresol** (Cresylic Acid, Cresylol, Hydroxytoluene, Methyl Phenol, Oxytoluene, Tricresol)

Molecular formula: $CH_3C_6H_4OH$

***Uses*** Cresol is a mixture of three isomeric forms, namely, ortho-, meta-, and paracresol. Cresol is a colorless, yellowish, brownish yellow, or pinkish liquid with a phenolic odor. It is used as an ore flotation agent and as an intermediate in the manufacture of chemicals, dyes, plastics, and antioxidants. It also is used in the manufacture of dyes, paint removers, plastics, stains, resins, chemical disinfectants, flotations, foundries, wool scours, and insulation enamels.

***Toxicity*** Cresol is very corrosive to all body tissues and may cause burns; with extensive exposure, death may result. Gangrene may develop on contact with the skin, and contact with eyes may lead to blindness. Toxic effects of cresol are weakness of the muscles, headache, dizziness, rapid breathing, mental confusion, and loss of consciousness.[112]

### **Creosote** (Creosotum, Cresote Oil, Brick Oil)

***Uses*** Creosote is a flammable, oily liquid with a characteristic smoky smell, caustic burning taste, but colorless in pure form. It is primarily used as a wood preservative and as a waterproofing agent, animal dip, an ingredient in fuel oil, and in the manufacture of chemicals and lampblack. It is extensively used in the pharmaceutical industry as an antiseptic, disinfectant, antipyretic, astringent, styptic, germicide, and expectorant.

***Toxicity*** Prolonged exposure to creosote liquid and vapors causes irritation to skin, burning, itching, pigmentation, vesiculation, ulceration, and gangrene. Occupational exposures in industrial environments may cause contact dermatitis, eye injuries like keratitis, conjunctivitis, and permanent corneal scars, and photosensitization to workers. Laboratory animal studies have shown symptoms of salivation, vomiting, vertigo, headache, loss of papillary reflexes, hypothermia, cyanosis, convulsions, respiratory difficulties, and death. In view of this, workers should be instructed to wear protective clothing and full-face masks under proper supervision.[113]

### **Hydroquinone** (Quinol, hydroquinol, p-diphenol, dihydroxybenzene, 1,4-benzenediol)

Molecular formula: $C_6H_4(OH)_2$

***Uses*** Hydroquinone, a colorless, hexagonal prism, has been reported to be a good antimitotic and tumor-inhibiting agent. It is a reducing agent used in a photographic developer, which polymerizes in the presence of oxidizing agents. In the manufacturing industry, it may be used in production of bacteriostatic agent, drugs, fur processing, motor fuels, paints, organic chemicals, plastics, stone coatings, and styrene monomers.

***Toxicity*** Ingesting a large quantity of hydroquinone may produce blurred speech, tinnitus, tremors, sense of suffocation, vomiting, muscular twitching, headache, convulsions, dyspnea and cyanosis from methemoglobinemia, coma, and collapse from respiratory failure. Workers should be instructed to wear protective clothing and full-face dust masks or goggles under proper management.[114–116]

**Phenol** (Carbolic Acid, Phenic Acid, Hydroxybenzene, Phenyl Hydrate)

Molecular formula: $C_6H_5OH$

***Uses*** Phenol is a white crystalline substance with a distinct aromatic, acrid odor. Phenol is used in the production of explosives, fertilizer, coke, illuminating gas, lampblack, paints, paint removers, rubber, perfumes, asbestos goods, wood preservatives, synthetic resins, textiles, drugs, and pharmaceutical preparations. It also is extensively used as a disinfectant in the petroleum, leather, paper, soap, toy, tanning, dye, and agricultural industries.

***Toxicity*** Phenol is absorbed rapidly through skin, lungs, and stomach. Prolonged exposure to phenol may cause chronic poisoning and include vomiting, difficulty in swallowing, diarrhea, lack of appetite, headache, fainting, dizziness, mental disturbances, and skin rash. Direct contact with phenol causes burning in the mouth, and irritation to the eyes, nose, and dermatitis. Liver and kidney damage and discoloration of the skin also may occur. In view of this, workers should be instructed to wear protective clothing, rubber boots, and goggles.[69,117,118] Exposure to phenol in different concentrations is known to cause mental disturbances, CNS depression, and coma.[116]

**Quinone** (Benzoquinone, Chinone, p-Benzoquinone, 1,4-Benzoquinone)

Molecular formula: $C_6H_4O_2$

***Uses*** Quinone exists as a large yellow, monoclinic prism with an irritating odor. It is extensively used in the dye, textile, chemical, tanning, and cosmetic industries. In the chemical synthesis of hydroquinone and other chemicals, quinone is used as an intermediate. It also is used in manufacturing industries and chemical laboratories associated with protein fiber, photographic film, hydrogen peroxide, and gelatin making.

***Toxicity*** Quinone vapor is highly irritating to the eyes and may be followed by corneal opacities, structural changes in the cornea, and loss of visual acuity. Solid quinone may produce discoloration, severe irritation, swelling, and formation of papules and vesicles.[114]

### **Benzyl Chloride** (Alpha-Chlorotoluene)

Molecular formula: $C_6H_5CH_2Cl$

***Uses*** Benzyl chloride is a colorless liquid with an irritating odor. It is used in the manufacture of plastics, dyes, synthetic tannins, perfumes, resins, gasoline additives, germicides, perfumes, photographic developers, rubbers, wetting agents, drugs, and pharmaceuticals. It also is used in production of benzal chloride, benzyl alcohol, and benzaldehyde.[116]

***Toxicity*** Benzyl chloride is a severe irritant to the eyes and respiratory tract. Contact with skin may cause dermatitis, and liquid contact with the eyes produces severe irritation and may cause corneal injury.[116,119] Benzyl chloride is a potential agent known to cause pulmonary edema in animals and humans. Experimental studies have shown development of sarcoma in animals.[119a] It is, therefore, important that workers be instructed to use and handle this chemical under proper management.

### **Chlorodiphenyls** (Chlorobiphenyls, Polychlorinated Diphenyl, PCB)

Molecular formula: $C_{12}H_{10-x}Cl_x$

***Uses*** Chlorodiphenyls are used in combination with chlorinated naphthalenes, which are stable, thermoplastic, nonflammable, and are used in electric cables and wires in the production of electric condensers, additives for extreme pressure lubricants, and as a coating in foundry. Chlorodiphenyls are widely used in the manufacture of herbicides, lacquers, paper, plasticizers, resins, rubbers, textiles, wood preservatives, and electric equipments.

***Toxicity*** Acute and chronic exposure can cause liver damage. Signs and symptoms include edema, jaundice, vomiting, anorexia, nausea, abdominal pains, and fatigue. Accidental oral ingestion indicates that chlorodiphenyls are embryotoxic, causing stillbirth, characteristic gray-brown skin, and increased eye discharge in infants born to women exposed during pregnancy.[120,121] Workers, particularly women of childbearing age, must show care and caution when handling.

### **Chlorobenzenes**

***Uses*** Chlorobenzene is used as a solvent and an intermediate in dyestuffs. o-Dichlorobenzene is used as a solvent, fumigant, insecticide, and chemical intermediate. Paradichlorobenzene is used as an insecticide, chemical intermediate, disinfectant, and moth-preventive agent.

***Toxicity*** On exposure, chlorobenzenes are known to cause irritation to the skin, conjunctiva, and mucous membranes of upper respiratory tract. Prolonged and repeated exposure with liquid chlorobenzene causes skin burns. Acute exposure to chlorobenzenes causes drowsiness, incoordination, gait disturbances, and unconsciousness. In view of this, workers are advised to handle the chemical with proper protective clothing and under qualified supervision.[122,123]

### Chlorinated Naphthalenes (Chloronaphthalenes)

Molecular formula: $C_{10}H_8{-}xCl$

***Uses*** Chlorinated naphthalenes appear as mobile liquids or waxy solids. Chlorinated naphthalenes have a variety of industrial uses such as in electric equipment manufacture, petroleum industries, plasticizer industries, and rubber industries.

***Toxicity*** Chronic exposure to chlorinated naphthalenes causes chloracne, erythematous eruptions with pustules, and papules. Symptoms of poisoning by chlorinated naphthalenes include headache, fatigue, vertigo, and anorexia. The chemicals are also known to cause liver damage and jaundice. Workers using proper protective clothing should avoid skin contact. Use of face masks and respirators in work areas is advised.[124]

### 1,1,1-Trichloroethane (Methylchloroform, Chlorothene)

Molecular formula: $C_2H_3Cl_3$, $CH_3CCl_3$

***Uses*** 1,1,1-Trichloroethane is a colorless, nonflammable liquid with a sweet smell similar to that of chloroform. It is absorbed by the system mainly by inhalation but also can penetrate the skin. 1,1,1-Trichloroethane does not occur naturally in the environment. It is found in many common products such as glue, paint, industrial degreasers, and aerosol sprays. In view of its impact on the ozone layer, all attempts have been made to stop its production in different countries of the world.[116,116a,125]

***Toxicity*** Any acute exposure to vapors of 1,1,1-trichloroethane leads to irritation of the nose, throat, and eyes, and results in headaches. Exposure to high concentrations or vapors of 1,1,1-trichloroethane is known to cause damage to the CNS, leading to behavioral disorders, dizzy spells, sleepiness, and, in some cases, coma; reversible injuries to the liver and kidneys also have been observed. Overexposure of 1,1,1-trichloroethane in occupational/work environments causes headache, CNS depression, irritation to eyes, dermatitis, and cardiac arrhythmias. Effects of repeated or long-term exposure to the solvent causes visual problems, loss of coordination, reduction of the tactile sensitivity of the skin, trembling, giddiness, anxiety, and slowing of the pulse rate.

### Carbon Tetrachloride

Molecular formula: $CCl_4$

***Uses*** Carbon tetrachloride is a clear, colorless, volatile liquid with a characteristic sweet odor. It is miscible with most aliphatic solvents and is itself a solvent. Its solubility in water is low. Carbon tetrachloride is nonflammable and is stable in the presence of air and light. In the presence of flame or hot metal, carbon tetrachloride partially converts into phosgene — a highly poisonous war gas. Carbon tetrachloride is a solvent for fats and oils, and is extensively used in a variety of products such as soaps and detergents, textiles, and rubber cements. Decomposition may produce phosgene, carbon dioxide, and hydrochloric acid.

***Toxicity*** Carbon tetrachloride is well absorbed from the gastrointestinal and respiratory tracts in animals and humans. Dermal absorption of liquid carbon tetrachloride is possible, but dermal absorption of the vapor is slow. Carbon tetrachloride is distributed throughout the whole body, with highest concentrations in liver, brain, kidney, muscle, fat, and blood. The liver and kidney are target organs for carbon tetrachloride toxicity. Short-term and prolonged exposure to carbon tetrachloride causes nausea, vomiting, headache, irritation to eyes and respiratory tract, pallor, and weak pulse. Chronic exposure to the chemical causes abdominal pain and jaundice. Carbon tetrachloride can induce embryotoxic and embryolethal effects, but only at doses that are maternally toxic, as observed in inhalation studies in rats and mice.[126,127] Carbon tetrachloride is not teratogenic.[1,126,127]

## CONCLUSION

Aromatic hydrocarbons have been supporting important industries around the globe as raw materials and solvents, and in the manufacture of commercial products. Because these compounds are extremely toxic to humans, particularly industrial workers who are constantly exposed through all major routes (oral, dermal, inhalation), it is necessary to educate workers about the adverse effects, proper training, and management. Direct spillage, contact, and accidental ingestion result in respiratory and blood-forming system damage in humans. Members of aliphatic and alicyclic amine groups, and their applications in industrial environment, are large. Their role in causing acute and chronic toxicity and adverse effects (particularly skin irritation, dermatitis, skin burns, and respiratory effects in workers) demands proper handling and storage in all work areas.

## REFERENCES

1. R.F. Gould, ed., *Refining Petroleum for Chemicals*, American Chemical Society, Washington, DC, 1970.

2. E.W. Flick, *Industrial Solvents Handbook*, 4th ed., Noyes Publications, Park Ridge, NJ, 1998.
3. National Library of Medicine, *Hazardous Substance Data Bank, TOXNET System*, Bethesda, MD, 1992.
4. Occupational Health and Environmental Control, Subpart G., 1910.93, Air Contaminants. *Fed. Reg.* **39**(125):23540, 1974.
5. American Conference of Governmental Industrial Hygienists, *Threshold Limit Values for Chemical Substances and Physical Agents and Biological Exposure Indices for 1991–1992*, ACGIH, Cincinnati, OH, 1991.
6. National Institute for Occupational Safety and Health, *Recommendations for Occupational Safety and Health Standards*, NIOSH/Communicable Disease Center, Atlanta, GA, 1988.
7. V.B. Guthrie, Ed., *Petroleum Products Handbook*, McGraw-Hill, New York, 1960.
8. F.A. Patty and W.P. Yant, *Odor Intensity and Symptoms Produced by Commercial Propane, Butane, Pentane, Hexane, and Heptane Vapor*, Reprinted from the United States Bureau of Mines Report of Investigation No. 2979, 1929.
9. L.S. Andrews and R. Snyder, *Toxic effects of solvents and vapors, in Casarett and Doull's Toxicology*, 4th ed., M.O. Amdur, J. Doull, and C.D. Klassen, eds., Pergamon Press, New York, pp. 681–722, 1991.
10. H.W. Gerade, *Toxicology and Biochemistry of Aromatic Hydrocarbons*, Elsevier, London, UK, 1960.
11. National Research Council, *Review of the Health Effects of Benzene*, National Research Council, Committee on Toxicology, National Academy of Sciences, Washington, DC, 1975.
12. C.E. Searle, *Chemical Carcinogens*, American Chemical Society, Washington, DC, 1976.
13. Agency for Toxic Substances and Disease Registry, *Toxicological Profile for Benzene (update)*, Division of Toxicology, Toxicology Information Branch, U.S. Department of Health and Human Services, Atlanta, GA, 1998, PB/98/101157/AS.
14. U.S. Department of Health and Human Services, *Criteria for Recommended Standard—Occupational Exposure to Benzene*, U.S. Department of Health, Education, and Welfare, Cincinnati, OH, 1974.

14a. M. Bowditch, and H.B. Elkins, Chronic exposure to benzene 1. The industrial aspects. *J. Ind. Hyg. Toxicol.* **21**:32, 1939.

15. Agency for Toxic Substances and Disease Registry, *Toxicological Profile for Toluene (update)*, U.S. Department of Health and Human Services, Atlanta, GA, 2000, PB/2000/108028.
16. W.J. Hayes, *Clinical Handbook on Economic Poisons*, U.S. Government Printing Office, Washington, DC, 1971.
17. U.S. Department of Health, Education and Welfare, *Criteria for a Recommended Standard—Occupational Exposure to Xylene*, U.S. Department of Health, Education, and Welfare, National Institute for Occupational Safety and Health, Washington, DC, 1975.
18. Agency for Toxic Substances and Disease Registry, *Toxicological Profile for Total Xylenes (update)*, U.S. Department of Health and Human Services, Atlanta, GA, 1995, PB/95/264404.

19. American Petroleum *Institute*, *Toxicological Review of Styrene*, API, New York, 1962.
20. Chemical Safety Data Document, *Chemical Safety Data SD-37, Properties and Essential Information for Safe Handling and Use of Styrene Polymer*, Manufacturing Chemists' Association, Inc., Washington, DC, 1971.
21. Agency for Toxic Substances and Disease Registry, *Toxicological Profile for Styrene*, U.S. Department of Health and Human Services, Atlanta, GA, 1993, PB/93/110849/AS.
22. International Agency for Research on Cancer, *IARC Monographs on the Evaluation of the Carcinogenic Risk of Chemicals to Humans.* Overall Evaluations of Carcinogenicity—An Updating of the IARC Monographs, Vols. 1–42 (suppl. 7), Lyon, France, pp. 136–137, 1987.
23. National Cancer Institute, *Report on NCI Bioassay of a Solution of Beta-Nitro-Styrene and Styrene for Possible Carcinogenicity—TRI70*, Office of Cancer Communications, National Cancer Institute, Bethesda, MD, U.S. Printing Office, 1979.
24. National Institute of Occupation and Safety Health, *Criteria for a Recommended Standard—Occupational Exposure to Refined Petroleum Solvents*, U.S. Department of Health Education, and Welfare, Public Health Service, NIOSH, Washington, DC, 1977.
25. L.S. Goodman and Gilman, *The Pharmacological Basis of Therapeutics*, 4th ed., Macmillan, New York, 1971.
26. K.P. Pandya, Bio-hazards of benzene. *J. Sci. Indust. Res.* **44**:613–621, 1985.
27. F.A. Patty, ed., *Industrial Hygiene and Toxicology*, Vol. II, Wiley-Interscience, New York, 1963.
28. Hazardous Materials Regulations, Department of Transportation Bureau. *Fed. Reg.* **41**:5708, 1976.
29. S. Budavari, M.J. O'Neil, A. Smith, P.E. Heckelman, and J.F. Kinneary, eds., *Beryllium*, in *The Merck Index, An Encyclopedia of Chemicals and Biologicals*, 12th ed., Merck Research Laboratories, Merck & Co., Inc., Whitehouse Station, NJ., pp. 195–197, 1997.
30. American Petroleum Institute, *Toxicological Review of Kerosene*, API, New York, 1967.
31. R.R. Beard and J.T. Noe, Aliphatic and alicyclic amines, Chapter 44, in *Patty's Industrial Hygiene and Toxicology*, 3rd ed., Vol. 2B, G.D. Clayton and F.E. Clayton, eds., John Wiley & Sons, New York, pp. 3135–3173, 1981.
32. T.J. Benya, and R.D. Harbison, Aliphatic and alicyclic amines, Chapter 17, in *Patty's Industrial Hygiene and Toxicology*, 4th ed., Vol. 2B, G.D. Clayton and F.E. Clayton, eds., John Wiley & Sons, New York, pp. 1087–1175, 1994.
33. I. Guest, and D.R. Varma, Developmental toxicity of methylamines in mice. *J. Toxicol. Environ. Health* **32**:319–330, 1991.
34. R.L. Hollinger, and Rowe, Report of Dow Chemical Company (Unpublished data), 1964.
35. R.A. Coon, et al., Animal inhalation studies on ammonia, ethylene glycol, formaldehyde, dimethylamine, and ethnol. *Toxicol. Appl. Pharmacol.* **16**:646–655, 1970.

36. Steinhagen, et al., Acute inhalation toxicity and sensory irritation of dimethylamine. *Am. Ind. Hyg. Assoc. J.* **43**:411–417, 1982.
37. W.K. Anger, and B.L. Johnson, Chemicals affecting behaviour, Chapter 3, in *Neurotoxicity of Industrial and Commercial Chemicals*, Vol. I, J.O'Donoghue, ed., CRC Press, Boca Raton, FL, pp. 51–148, 1985.
38. N.H. Proctor, J.P. Hughes, M.L. Fischman, eds., *Chemical Hazards of the Work Place*, 2nd ed., Lippincott, Philadelphia, PA, 1988.
39. H. Brieger and W.A. Hode, Toxic effects of exposure to vapors of aliphatic amines. *Arch. Ind. Hyg. Occup. Med.* **3**:287–291, 1951.
40. W.E. Ribelin, The effects of drugs and chemicals upon the structure of the adrenal gland. *Fundam. Appl., Toxicol.* **4**:105–119, 1984.
41. M. Windholz, S. Budavari, R. Blumetti, and E. Otterbein, eds., *The Merck Index*, 10th ed., Merck & Co., 1983.
42. H.F. Smyth, Jr., C.P. Carpenter,, and C.S. Weil, Range finding toxicity data, List IV. *Arch. Ind. Hyg. Occup. Med.* **4**:119–122, 1951.
43. D.W. Lynch, et al., Subchronic inhalation of diethylamine vapor in Fischer-344 rats: Organ system toxicity. *Fundam. Appl. Toxicol.* **6**:559–565, 1986.
44. R.T. Foley, and B.F. Brown, Corrosion and corrosion inhibitors, in *Kirk-Othmer Encyclopedia of Chemical Technology*, 3rd ed., Vol. 7, H.F. Mark, D.F. Othmer, C.G. Overberger, and G.T. Seaborg, eds., John Wiley & Sons, New York, pp. 113–142, 1979.
45. K.L. Cheever, et al., The acute oral toxicity of isomeric monobutylamines in the adult male and female rat. *Toxicol. Appl. Pharmacol.* **63**:150–152, 1982.
46. J.G. Wagner, Introduction to biopharmaceutics, Chapter 1, *in Biopharmaceutics and Relevant Pharmacokinetics*, J.G. Wagner, ed., Drug Intelligence Publications, pp. 1–5, 1971.
47. D. Kumar, et al., Allylamine and beta-aminopropionitrile induced vascular injury: An in vivo and in vitro study. *Toxicol. Appl. Pharmacol.* **103**:288–302, 1990.
48. P.J. Boor, and V.J. Ferrans, Ultrastructural alterations in allylamine induced cardiomyopathy—early lesions. *Lab. Invest.* **47**:76, 1982.
49. International Agency for Research on Cancer, *IARC Monographs on the Evaluation of the Carcinogenic Risk of Chemicals to Humans, Some Non-nutritive Sweetening Agents*, Vol. 22, pp. 64–109, Lyon, France, 1980.
50. S. Kojima, and H. Ichibagase, Studies on synthetic sweetening agents, VIII. Cyclohexylamine, a metabolite of sodium cyclamate. *Chem. Pharmacol. Bull.* **14**:971–974, 1966.
51. M.S. Legator, et al., Cytogenetic studies in rats of cyclohexylamine, a metabolite of cyclamate. *Science* **165**:1139–1140, 1969.
52. J.M. Price, et al., Bladder tumors in rats fed cycloxylamine or high doses of a mixture of cyclamate and saccharin. *Science* **167**:1131–1132, 1970.
53. K. Takano and M. Suzuki, Cycloxylamine, a chromosome-aberration inducing substance: No teratogenicity in mice (Jpn). *Congenit. Anomal.* **11**:51–57, 1971.
54. J.G. Wilson, Use of primates in teratological investigations, in *Medical Primatology*, E.I. Goldsmith and J. Moor-Jankowski, eds., Karger, Basel, Switzerland, 1972.
55. International Agency for Research on Cancer, *IARC Monographs on the Evaluation of the Carcinogenic Risk of Chemicals to Humans. Overall Evaluations of*

*Carcinogenicity: An Updating of the IARC Monographs*, Vols. 1–42, (suppl. 7), pp. 178–182, Lyon, France, 1987.

56. C. Babiuk, et al., Induction of ethylenediamine hypersensitivity in the guinea pig and the development of ELISA and lymphocyte blastogenesis techniques for its characterization. *Fundam. Appl. Toxicol.* **9**:623–634, 1987.
57. M.N. Perkins, and T.W. Stone, Comparison of the effects of ethylenediamine analogues and gamma amino butyric acid on cortical and pallidal neurons. *Brit. J. Pharmacol.* **75**:93–99, 1982.
58. R.S. Slesinski, et al., Assessment of genotoxic potential of ethylenediamine: In vitro and in vivo studies. *Mut. Res.* **124**:299–314, 1983.
59. L.R. DePass, et al., Dermal oncogenicity studies on ethylenediamine in male C3H mice. *Fundam. Appl. Toxicol.* **4**:641–645, 1984.
60. R.S. H. Yang, et al., Chronic toxicity/carcinogenicity study of ethylenediamine in Fischer 344 rats. *Toxicologist* **4**:53, 1984.
61. H. Isacoff, *Cosmetics, in Kirk-Othmer Encyclopedia of Chemical Technology*, 3rd ed., Vol. 7, H.F. Mark, D.F. Othmer, C.G. Overberger, and G.T. Seaborg, eds. John Wiley & Sons, New York, pp. 143–176, 1979.
62. American Hospital Formulary Service, *Drug Information*, American Society of Hospital Pharmacists, 1992.
63. United States Adapted Names (USAN) and the USP Dictionary of Drug Names, United States Pharmacopeial Convention, Inc., 1992.
64. K. Inoue, et al., Mutagenicity tests and *in vitro* transformation assays on triethanolamine. *Mut. Res.* **101**:305–313, 1982.
65. A. Maekawa, H. Onodera, H. Tanigawa, K. Furata, J. Kanno, T. Matsuoka, Y. Ogui, and Y. Hayashi, Lack of carcinogenicity of triethanolamine in F344 rats. *J. Toxicol. Environ. Health* **19**:345–357, 1986.
66. Y. Konishi, et al., Chronic toxicity carcinogenicity studies of triethanolamine in B6C3F1 mice. *Fundam. Appl. Toxicol.* **18**:25–29, 1992.
67. U.S. Department of Health and Human Services, *Sixth Annual Report on Carcinogens Summary*, U.S. Public Health Service, Atlanta, GA, pp. 79–80, 1991.
68. R.D. Kimbrough, K.R. Mahaffey, P.Grandjean, S.H. Sandoe, and D.D. Rutstein eds., *Clinical Effects of Environmental Chemicals*, Hemisphere, New York, 1989.
69. NIOSH, *Occupational Disease, A Guide to Their Recognition*, Communicable Disease Center, Department of Health, Education and Welfare, Public Health Service, Publ. No. 7–181, 1977.
70. F.A. Lowenheim, and M.K. Moran, eds., *Faith, Keyes and Clark's Industrial Chemicals*, 4th ed., Wiley-Interscience, New York, 1975.
71. D.G. Goodman, et al., Splenic fibrosis and sarcomas in F344 rats fed diets containing aniline HCl, *p*-chloroaniline, azobenzene, *o*-toluidine HCl, 4,4-sulfonyl dianiline, or D + C Red No. 9. *J. Natl. Cancer Inst.* **73**(1):265–273, 1984.
72. J.O. Thomas, et al., Chronic toxicity of diphenylamine to albino rats. *Toxicol. Appl. Pharmacol.* **10**:362–374, 1967.
73. J.O. Thomas, et al., Chronic toxicity of diphenylamine for dogs. *Toxicol. Appl. Pharmacol.* **11**:184–194, 1967.
74. S.D. Lenz, and W.W. Carlton, Diphenylamine-induced renal papillary necrosis and necrosis of the pars recta in laboratory rodents. *Vet. Pathol.* **27**:71–178, 1990.

75. E.K. Weisburger, et al., Neoplastic response of F344 rats and B6C3F1 mice to the polymer and dyestuff intermediates 4,4-Methylenebis (N,N-dimethyl)-benzenamine, 4,4-Oxydianiline and 4,4-Methylenedianiline. *J. Natl. Cancer Inst.* **72**:1457–1463, 1984.
76. ACGIH, *Documentation of the Threshold Limit Values and Biological Exposure Indices*, 6th ed., American Conference of Governmental Industrial Hygienist, Inc., 1991.
77. N.H. Proctor, J.P. Hughes, M.L. Fischman, eds., *Chemical Hazards of the Work Place*, 2nd ed., Lippincott, Philadelphia, 1988.
78. J.A. Bond, et al., Induction of hepatic and testicular lesions in Fischer 344 rats by single oral doses of nitrobenzene. *Fundam. Appl. Toxicol.* **1**:389–394, 1981.
79. M.A. Medisky, et al., in *Toxicity of Nitroaromatic Compounds*, D.E. Rickett, ed., Hemisphere, New York, 1985.
80. R.W. Tyl, et al., Development toxicity evaluation of inhaled nitrobenzene in CD rats. *Fundam. Appl. Toxicol.* **8**:482–492, 1987.
81. D.E. Dodd, et al., Reproduction and fertility evaluation in CD rats following nitrobenzene inhalation. *Fundam. Appl. Toxicol.* **8**:493–505, 1987.
82. ACGIH, *Documentation of the Threshold Limit Values and Biological Exposure Indices*, 5th ed., American Conference of Governmental Industrial Hygienist, Inc., 1986.
83. E. Ward, et al., Excess number of bladder cancers in workers exposed to o-toluidine and aniline. *J. Natl. Cancer Inst.* **83**:501–506, 1991.
84. D.G. DeBord et al., Binding characteristics of ortho-toluidine to rat hemoglobin and albumin, **66**:231–236, 1992.
85. National Institute for Occupational Safety and Health, *Pocket Guide to Chemical Hazards*, U.S. Department of Health and Human Services, Public Health Service, Centers for Disease Control NIOSH, Publication Number 90–117, June, 1990.
86. National Toxicology Program, *Annual Plan, Fiscal Year 1991*, U.S. Department of Health and Human Services, June, 1991.
87. H.V. Ellis III et al., Chronic toxicity of 2,4-dinitrotoluene in the rat. *Toxicol. Appl. Pharmacol.* **45**:245–246, 1978.
88. National Institute for Occupational Safety and Health, *Current Intelligence Bulletin*, 44, Publ. No. 85–109, NTIS Publ. No. PB-86-105-913, 1985.
89. S.C.J. Reader, and R.M.D. Foster, The *in vitro* effects of four isomers of dinitrotoluene on rat sertoli and sertoli-germ cell co-cultures: Germ cell detachment and lactate and pyruvate production. *Toxicol. Appl. Pharmacol.* **106**:287–294, 1990.
90. P.V.V. Hamill, et al., The epidemiologic assessment of male reproductive hazard from occupational exposure to toluenediamine and dinitrotoluene. *J. Occup. Med.* **24**:985–991, 1982.
91. D.E. Rickert, ed., *Toxicity of Nitroaromatic Compounds*, Hemisphere, New York, 1985.
92. J.A. Popp and T.B. Leonard, *The hepatocarcinogenicity of dinitrotoluenes, Chapter 4, in Toxicity of Nitroaromatic Compounds*, D.E. Ricke, ed., Hemisphere, New York, pp. 53–60, 1985.
93. R.J. Levine et al., Heart disease in workers exposed to dinitrotoluene. *J. Occup. Med.* **28**:811–816, 986.

94. K.D. Bergert et al., Cardiovascular diseases in exposure to the saltpeter acid ester ethylene glycol dinitrite. *Z. Ges. Int. Med.* **42**:581–583, 1987.
95. A. Ben-David, Cardiac arrest in an explosives factory worker due to withdrawal from nitroglycerine exposure. *Am. J. Ind. Med.* **15**:719–722, 1989.
96. I.P. Batkina, Maximum permissible concentration of divinyl vapor in factory air (o presdel 'no dopustimoi kontsentratsii parov divinika v vozdukhe radiochikh pomeshchenii). *Hyg. Sanit.* **31**:334, 1966.
97. C.P. Carpenter, C.B. Shaffer, and H.F. Smyth, Jr., Studies on the inhalation of 1,3-dutadiene; with a comparison of its narcotic effect with benzol, toluol, and styrene, and a note on the elimination of styrene by the human. *J. Ind. Hyg. Toxicol.* **26**:69, 1944.
98. *Patty's Industrial Hygiene and Toxicology*, Vol. 2B, G.D. Clayton, and F.E. Clayton, eds., 4th ed., John Wiley & Sons, New York, p. 1242, 1250–1252, 1994.
99. L.D. Pagnotto, H.B. Elkins, H.G. Brugsch, and J.E. Walkley, Industrial benzene exposure from petroleum naphtha, 1, Rubber coating industry. *Am. Ind. Hyg. Assoc. J.* **22**:417, 1961.
100. N.V. Hendricks, C.M. Berry, J.G. Lione, and J.J. Thorpe, Cancer of the scrotum in wax pressmen, I. Epidemiology. *AMA Arch. Ind. Health* **19**:524, 1959.
101. W.C. Hueper, and W.D. Conway, *Chemical Carcinogens and Cancers*, Charles C. Thomas, Springfield, IL, p. 27, 1964.
102. J.G. Lione, and J.S. Denholm, Cancer of the scrotum in wax pressmen, II. Clinical observations. *AMA Arch. Ind. Health* **19**:530, 1959.
103. F. Urbach, S.S. Wine, W.C. Johnson, and R.E. Davies, Generalized paraffinoma (sclerosing lipogranuloma). *Arch. Dermatol.* **103**:227, 1971.
104. R.G. Gosselin, et al., *Clinical Toxicology of Commercial Products*, R.G. Gosselin et al., eds., 4th ed., Williams & Wilkins, Baltimore, MD, Section III, pp. 315–317, 1976.
105. M.K. Dunlap, D.K. Kodama, J.S. Wellington, H.H. Anderson, and C.J. Hine, The toxicity of allyl alcohol. *AMA Arch. Ind. Health* **18**:303, 1958.
106. T.R. Torkelson, M.A. Wolf, F. Oyen, and V.K. Rowe, Vapor toxicity of allyl alcohol as determined on laboratory animals. *Am. Ind. Hyg. Assoc. J.* **20**:224, 1959.
107. J.H. Sterner, H.C. Crouch, H.G. Brockmyre, and M. Cusack, A ten year study of butyl alcohol exposure. *Am. Ind. Hyg. Assoc. J.* **10**:53, 1949.
108. I.R. Tabershaw, J.P. Fahy, and J.B. Skinner, Industrial exposure to butanol. *J. Ind. Hyg. Toxicol.* **26**:328, 1944.
109. J.E. Crook, and J.S. McLaughlin, Methyl alcohol poisoning. *J. Occup. Med.* **8**:467, 1966.
110. R.L. Kane, W. Talbert, J. Harlan, G. Sizemore, and S. Cataland, A methanol poisoning outbreak in Kentucky. *Arch. Environ. Health* **17**:119, 1968.
111. E.V. Henson, The toxicology of some aliphatic alcohols, Parts I and II. *J. Occup. Med.* **2**:442, 497, 1960.
112. American Industrial Hygiene Association, Community air quality guides, phenol and cresol. *Am. Ind. Hyg. Assoc. J.* **30**:425, 1969.
113. A.J. Arieff, Acute, toxic, polioencephalitis (creosote). *JAMA.* **193**:745, 1965.

114. B. Anderson, and F. Oglesby, Corneal changes from quinone-hydroquinone exposure. *Arch. Ophthalmol.* **59**:495, 1958.
115. E. Seutter, and A.H.M. Sutorius, Quantitative analysis of hydroquinone in urine. *Clin. Chim. Acta.* **38**:231, 1972.
116. National Institute for Occupational Safety and Health, *Pocket Guide to Chemical Hazards*, U.S. Department of Health and Human Services, Public Health Service, Centers for Disease Control, NIOSH, Publication Number 90–117, p. 128, 1990.
116a. E.M. Adams, et al., Vapor toxicity of 1,1,1-trichloroethane (methylchloroform) determined by experiments on laboratory animals. *Arch. Ind. Hyg. Occup. Med.* **1**:225, 1950.
117. S.J. Evans, Acute phenol poisoning. *Br. J. Ind. Med.* **9**:227, 1952.
118. J.K. Piotrowski, Evaluation of exposure of phenol: Absorption of phenol vapor in the lungs and through the skin and excretion of phenol in urine. *Br. J. Ind. Med.* **28**:172, 1971.
119. T.V. Mikhailova, Comparative toxicity of chloride derivatives of toluene: benzyl chloride, benzal chloride, and benzotrichloride. *Gig. Tr. Prof. Zabol.* **8**:14, 1965.
119a. U.S. Environmental Protection Agency, Integrated risk information system (IRIS) on benzyl chloride. Environmental Criteria and Assessment Office, Office of Health and Environmental Assessment, Office of Research and Development, Cincinnati, OH, 1993.
120. J.W., Meigs, J.J. Albom, and B.L. Kartin, Chloracne from an unusual exposure to arochlor. *JAMA.* **154**:1417, 1954.
121. D.B. Peakall, Polychlorinated diphenyls: Occurrence and biological effects. *Residue Rev.* **44**:1, 1972.
122. R.L. Hollingsworth, V.K. Rowe, F. Oyen, T.R. Tokelson, and E.M. Adams, Toxicity of o-dichlorobenzene, Studies on animals and industrial experience. *AMA Arch. Ind. Health* **17**:180, 1956.
123. S.P. Varshavskaya, Comparative toxicological characteristics of chlorobenzene and dichlorobenzene (ortho- and para- isomers) in relation to the sanitary protection of water bodies. *Hyg. Sanit.* **33**:17, 1968.
124. M. Kleinfeld, J. Messite, R. Swencicki, Clinical effects of chlorinated naphthalene exposure. *J. Occup. Med.* **14**:377, 1972.
125. Agency for Toxic Substances and Disease Registry, *Toxicological Profile for 1,1,1-Trichloroethane*, U.S. Department of Health and Human Services, Public Health Service, Atlanta, GA, 1995.
126. Agency for Toxic Substances and Disease Registry, *Toxicological Profile for Carbon Tetrachloride*, U.S. Department of Health and Human Services, Public Health Service, Atlanta, GA, 1994.
127. World Health Organization, Carbon tetrachloride. *Environ. Health Crit.* **208**: WHO, Geneva, Switzerland, 1999.

CHAPTER 10

# Food Additives and Food Contaminants

## INTRODUCTION

Food safety is an important worldwide issue, involving a large group of the population over a long period of time. Different aspects of food safety are discussed often in both an irrational and an emotional manner by different sections of the global society. This is an unscientific, improper, unpragmatic way to solve the problem. Food safety is always modulated by human physiology, detoxification processes, and metabolic and excretory patterns. Food safety is a matter based on risk and benefit more than any other measurement. The food and beverage industry is a large component of activities. Effective food production and protection is carried out by a group that includes agriculturists, storage management, and food commodities distribution. During this complex management of food preparation and distribution, it can become contaminated in a variety of ways (Fig. 10-1) (e.g., by microorganisms, pesticide residues, heavy metals). Eighty to 90% of food articles are normally contaminated with heavy metals (e.g., mercury, lead, arsenic, cadmium, selenium), residues of pesticides, and radionuclides at different levels.

Safety is a judgment made by regulatory authorities and is based on toxicity and risk assessment data. Safety is not measurable. It is impossible to evaluate a chemical for absolute safety, since biological risk is most often a statistical event in a mixed human population. Further, because of sophisticated analytical tools, even minute traces of chemicals can now be quantified. However, this does not mean these chemicals should invoke toxicological suspicion because people have been exposed to and ingested these for a long time without obvious adverse effects.

The Food and Agricultural Organization (FAO) and the World Health Organization (WHO) define a food additive as, "any substance not normally consumed as a food by itself and not normally used as a typical ingredient of the food, whether it has nutritive values or not, the intentional addition of which to food

*Industrial Guide to Chemical and Drug Safety*, By T.S.S. Dikshith and Prakash V. Diwan
ISBN 0-471-23698-5 

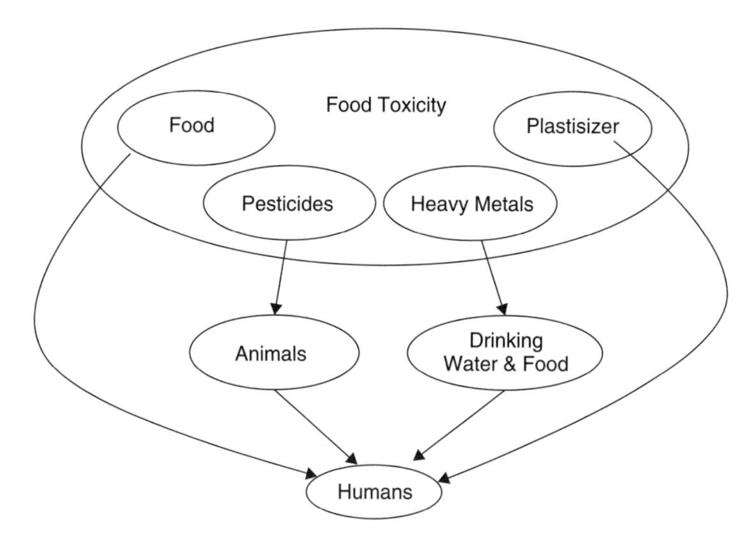


**Figure 10-1.** Schematic diagram of food toxicity.

for a technological (including organoleptic) purpose in the manufacture, processing, preparation, treatment, packing, packaging, transport or holding of such food, results or may be reasonably expected to result directly or indirectly in it or in its by-products becoming a component of or otherwise affecting the characteristics of such foods." The term *food additive* does not describe contaminants or substances added to food for maintaining or improving nutritional qualities.

Essentially, a food additive brings some benefit to the consumer and is not used to disguise faulty food processing or handling techniques. A food additive does not substantially reduce the nutritive value of the food and does not pose any health hazard to the consumer. Thus, food additives should be used for the preservation of food, to maintain its nutritive quality and storage quality, and for organoleptic acceptance by the consumer.

The present-day food world is complex; food additives are added to many items. Examples include, anticaking agents, antioxidants, bleaching and maturing agents, buffering and sequestering agents, emulsifying agents, flavoring agents, flavor boosters, food colorants, nonnutritive agents, nutrient supplements, preservatives, and special dietary sweeteners. Information on food additives and possible adverse effects, particularly to growing children, has received global attention.[1–3]

## FOOD ADDITIVES

### Anticaking Agents

Anticaking agents include carbonates, phosphates, silicates, and fatty acid salts of calcium and magnesium. Levels of anticaking agents in food products are determined by manufacturing, based on good practices.

### Antioxidants

Antioxidants include ascorbic acid, lecithin, and tocopherol and are added to all food products. Other antioxidants, such as butylated hydroxy anisole (BHA), butylated hydroxy toluene (BHT), and gallates, are permitted for use only in specific concentrations of food products. BHA is permitted for use in many countries, but BHT is not permitted because of possible toxicity, such as tumorigenicity.

### Bleaching and Maturing Agents

Bleaching and maturing agents include ascorbic acid, salts of calcium, potassium bromate, and benzoyl peroxide and are used in the manufacture of food products.

### Buffering and Sequestering Agents

Buffering and sequestering agents used in different food products include organic acids, such as acetic acid, citric acid, carbonates, fumaric acid, malic acid, and phosphoric acids. These agents have not imparted any adverse effects in food products.

### Flavoring Agents and Flavor Boosters

Addition of flavoring agents to different foods has become common, although restrictions are imposed on use. In fact, individual flavors are not allowed to be used in foods exceeding 300 ppm. Ethyl maltol and maltol are flavoring agents permitted globally within specific levels, particularly in biscuits and baked foods. Addition of monosodium glutamate is allowed in meat products up to a maximum of 500 ppm but is not permitted in other meat products meant for infants younger than 1 year of age.

Solvents, such as diethyl ether, diethylene glycol, hexylene glycol, monoethyl ether, 1,2 dichloroethane, and 1,1,2-trichloroethylene, are not permitted for flavors. However, solvents including ethyl alcohol, propylene glycol, glycerol, isopropyl alcohol, propane diol, and liquid paraffin are permitted.

Among flavoring agents, vanillin occupies an important position in food industry. Vanillin is used in the manufacture of chocolate, beverages, confectionaries, custards, ice creams, puddings, and cakes. In the modern food industry, vanillin is produced from aniline, sulfite liquor, and isoeugenol. Safrol, guiacol, and other lignin-containing substances are used as base materials in the manufacture of vanillin. Because of the extensive use of vanillin, toxicologic studies were conducted in animal species. Accordingly, the acute oral $LD_{50}$ in mice, rats, and rabbits showed a variation ranging from 1,580 mg/kg to 2,800 mg/kg.[4–6] In contrast, the acute oral $LD_{50}$ for the guinea pig is reported as 1,400 mg/kg.[6] Vanillin-treated laboratory animals demonstrated toxic symptoms including respiratory distress, muscular weakness, lacrimation, depression, coma, and death.[7] Studies by Makaruk[5,8] indicated that acute doses of vanillin induced adverse effects in blood index, central nervous system (CNS), cardiovascular system, hepatotoxicity, and nephrotoxicity in rats, but the inhalation of saturated solutions or repeated oral doses failed to demonstrate any adverse effects in animals. Several studies with vanillin did not indicate any potential carcinogenicity, mutagenicity, or teratogenicity effects in animals.[9–11] It is significant that workers associated with the manufacture of vanillin did not demonstrate adverse effects, suggesting the safety of the additive whenever handled properly.[12]

## Food Colorants

Food colorants may be of natural origin or synthetic (e.g., coal tar dyes). Natural food colorants include curcumin or turmeric, saffron, carotene, caramel, chlorophyll, riboflavin, and ratanjot. Natural colors are unstable to heat and are expensive. Synthetic food colors include Amaranth, Brilliant blue FCF, Carmoisine Fast red E, Indigo Carmine, Green and Fast green, Ponceau 4 R, Sunset yellow FCF, and Tartrazine. Although there is no permissible limit for natural food colorings, the maximum limit of synthetic colors in different food products is 0.2 g/kg (Table 10-1).[13]

**TABLE 10-1 Food Colors**

| Name | Food Colors | | |
|---|---|---|---|
| Brilliant blue FCF | FD & C | Blue | No. 1 |
| Indigotine | FD & C | Blue | No. 2 |
| Guinea green B | FD & C | Green | No. 1 |
| Bright green FS | FD & C | Green | No. 2 |
| Fast green FCF | FD & C | Green | No. 3 |
| Yellow AB | FD & C | Yellow | No. 3 |
| Yellow OB | FD & C | Yellow | No. 4 |
| Tartrazine | FD & C | Yellow | No. 5 |
| Sunset yellow FCF | FD & C | Yellow | No. 6 |
| Orange 1 | FD & C | Orange | No. 1 |
| Orange F | FD & C | Orange | No. 2 |
| Ponceau 3 R | FD & C | Red | No. 1 |
| Amaranth | FD & C | Red | No. 2 |
| Erythrosine | FD & C | Red | No. 3 |
| Ponceau SX | FD & C | Red | No. 4 |
| Oil Red XO | FD & C | Red | No. 32 |
| | FD & C | Violet | No. 1 |
| | Citrus | Red | No. 2 |
| | Ponceau | MX | |
| | Ponceau | 4R | |
| | D & C | Red | No. 9 |
| | D & C | Red | No. 10 |

*Source*: Radomski, 1974.[13]

### Nonnutritive Dietary Sweeteners

Food items have contained a variety of artificial sweeteners (e.g., saccharin, sorbitol, cyclamate). These sweeteners are permitted for use by the United States Food and Drug Administration (USFDA) based on their grouping under "generally regarded as safe" (GRAS). However, reports and regulatory decisions on the use of artificial sweeteners are inconclusive and often contradictory, and therefore require reevaluation.

### Nutrient Supplements

A variety of nutrient supplements are added to foods to improve their nutritive value and enhance the shelf-life of the food product. Lysine is a common example of such a supplement.

### Preservatives

There are two kinds of preservatives, class I and class II. Class I preservatives include common salt, sugar, glucose/sucrose syrup, acetic acid or vinegar, spices, and wood smoke. The addition of these to foods is not restricted. Class II preservatives include benzoic acid and its salts, nitrates and nitrites, sorbic acid and its salts, and sulfurous acid and its salts. Their addition to food is, however, restricted.

## FOOD CONTAMINANTS

### Mycotoxins

The majority of toxicants in foods are contaminants. (e.g., microbial toxins, pesticide residues, leachable chemicals from packaging materials, food coatings, traces of heavy metals). However, the major issue in food safety is the contamination of food by mycotoxins in items such as milk and milk products, meat and meat products, and peanuts (groundnuts). Aflatoxin is highly toxic and lethal, and its carcinogenic potential is well established, even at doses as low as 0.05 μg. Mycotoxins also infect food products like rice, pulses, tapioca, and betelnuts. (Table 10-2).

Food also is contaminated with bacteria, such as *Salmonella* species, and other organisms. These organisms and microbial infections cause gastroenteritis in humans. Essentially, the majority of the food poisoning observed in adults and children occurs due to bacterial infection. Food poisoning from bacterial infection is more frequent and sudden than poisoning from metal contamination, but both can occur. Food preparations using enamelware utensils often contain traces of antimony, a cadmium metal leading to poisoning. Cyanides are well known poisons that can contaminate food material whenever the metal is used as a silverware cleaner or polisher.

**TABLE 10-2 Mycotoxin and Food Contamination**

| Toxin | Organism | Food |
|---|---|---|
| Aflatoxins | *Aspergillus flavus* | Peanuts, peanut meal, rice. |
| | *A. parasiticus* | rye, sorghum, red clover |
| Ergot alkaloids | *Claviceps purpurea* | Many cereal foods |
| | *Rhizoctonia legminicola* | |
| Slaframine | | |
| Zearalenone | *Fusarium roseum* | Kernels of corn, barley |
| Cyclopiazonic acid | *Penicillium cyclopium* | Wheat and other grains |
| Rubratoxins | *P. rubrum* | Animal feeds |
| Citreoviridin | *P. citreoviride* | Rice, yellow rice |
| | *P. islandicum* | |
| Ipomeamarone | *F. solani* | Sweetpotato |
| Ipomeamoronol | | |
| Ochratoxins | *A. ochraceus* | Grains |

## Food Allergy

Food allergy is an adverse reaction to food normally observed in susceptible individuals. This adverse effect or allergy is mediated by a classical immune mechanism. The allergy is very specific to the food in question. Thus, the "true" allergic reaction to any food is the result of an oversensitive reaction of the body's immune system. Food intolerances are non–immune-mediated adverse reactions to a specific food (Fig. 10-2).[14]

## Food Poisoning

Food poisoning has become a common occurrence worldwide. Reports are available about food poisoning among schoolchildren, industrial workers, and unprivileged

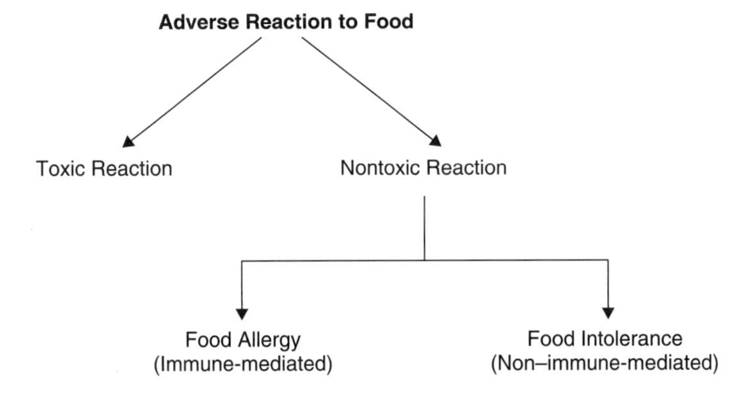


**Figure 10-2.** Allergy reaction to food in humans.

groups who live in remote parts of towns and villages in different countries.[2] Food poisoning occurs when chemicals are accidentally introduced into the food stocks or leached into foods from stored containers. This has resulted in varieties of ill health and human fatalities. Well-known examples of food poisoning may be seen in the reports of pesticide poisoning (fallidol) in the Kerala state of India during the early 1960s. The metal antimony is normally leached from enamel through the storage of acidic foods. An example is storage of lemon punch or pickles in chipped gray enamel vessels, which has resulted in adverse effects to consumers. A variety of insecticides are used to control household pests, such as cockroaches, flies, and rodents. In such situations, there are instances of accidental food and water contamination that can lead to adverse reactions. Metals, such as mercury, lead, and cyanide, have caused fatalities because of their contamination to household articles and daily foods. Lead is often leached from improperly glazed pottery vessels and utensils when combined with acidic beverages. Similarly, methyl mercury has caused blindness, paralysis, and death among people who consumed contaminated fish and shellfish in Japan.

## Food Storage

Food storage as a raw material or prepared wholesome food is an important aspect of food safety management. Food is stored under either dry conditions or refrigeration. Food refrigeration has been in practice for centuries. In this method, food is stored normally below 1° to 2°C, which stops degeneration and contamination. Different food items require refrigerated storage in a range of (Table 10-3).[15] Frozen foods such as fish, shellfish, and leftovers are often prone to bacterial contamination. Consumption of these contaminated foods results in an outbreak of food poisoning. Therefore, it is necessary to use proper defrosting and adequate cooking methods before the item is consumed.

## Dry Storage of Food

Food stored as raw material in the form of grains, pulses, and seeds is often attacked by stored-grain pests. To contain proper quality of dry food and prevent

**TABLE 10-3 Storage and Shelf-Life of Foods**

| Food Item | Period of Storage | Temperature (°F) |
|---|---|---|
| Butter | 6 months | 35 |
| Cheese | 15 months | 32–34 |
| Cream | 4 months | 5–10 |
| Eggs (fresh) | 2 months | 38–45 |
| Eggs (frozen) | 18 months | 0–5 |
| Milk | 5 days | 35–40 |

*Source*: Modified from Koren H, and Bisesi M, 1999.[15]

insect attack, pesticide use has become an easy tool. Proper care should be taken in the use of agrochemicals and their management to avoid food poisoning.

### Food Preparation and Management

Food is consumed by everyone, so its preparation and management requires careful attention. Some important aspects should be remembered during its preparation and distribution.

- Use only good quality food.
- Always keep foods clean and free from insects, rodents, and other household pests.
- Food preparation and serving areas should be maintained very clean.
- Use well-constructed and well-ventilated areas for food preparation.
- Equipment, such as utensils and refrigerators, should be kept clean and hygienic.
- Workers associated in food preparation must be healthy and clean.
- Always cook food items long enough to kill organisms.

In essence, food safety may be achieved by following a sanitary program controlled or managed by food service supervisors and properly trained personnel. Scheduled cleaning operations, cleaning and sanitizing techniques, and proper use of equipment go a long way in food safety management. Supervisors of different operations should instruct all personnel on proper handling techniques and regularly inspect activities associated with food preparation and storage. An educational program for homemakers and individuals who prepare food helps to control food poisoning outbreaks and related health effects.

### Food Handling

Contamination of food, drink, and smoking materials is a potential route for exposure to toxic substances. Food should never be stored, handled, or consumed in any laboratory area.

1. Well-defined areas should be established for storage and consumption of food and beverages. No food should be stored or consumed outside this designated area.
2. Areas where food is permitted should be prominently marked and a warning sign displayed. All chemicals, chemical equipment, and laboratory coats must be excluded from such areas.
3. Consumption of food or beverages and smoking must be excluded from areas where laboratory operations are performed.
4. Glassware or utensils that have been used for laboratory operations should never be used to prepare food or beverages. Laboratory refrigerators, ice

chests, and cold rooms should not be used for food storage; separate equipment should be dedicated and prominently labeled for that use.

## CONCLUSION

The problems of food contamination, due to multiple processing and handling varieties of food, are of concern because they involve several different workers. Because food is an essential item of all people, it becomes important that food processing industries be cautious in manufacturing and distribution practices. However, it is also important that false health scares about food poisonings be curbed. Accordingly, machinery of food processing industries should be manned by well-trained, experienced workers. Basic knowledge about food colorings, adulterants, and mycotoxins would go a long way in maintaining the food quality required by the fast-growing population of the world.

## REFERENCES

1. C.L. Puertollano, C.B. Malcom, J. Banzon, and J.C. Melgae, Effect of changes on the formulation of soymilk on its acceptability to Filipino children. *Phillipp. Agric.* **54**:227–240, 1971.
2. A.V. Folia, Food additives as toxic factors in childhood. *Clin. Int. Barc.* **26**(9):396–403, 1976.
3. T. Larkin, Food additives and hyperactive children. *Cereals Foods World* **22**(11):582–584, 1977.
4. H.C. Hodge, and W.L. Downs, The approximate oral toxicity in rats of selected household products. *Toxicol. Appl. Pharmacol.* **3**(6):689–695, 1961.
5. C.L. Hake, and V.K. Rowe, *Ethers, in Industrial Hygiene and Toxicology*, F.A. Patty, ed., Vol. II, 2nd ed., Interscience Publishers, New York, p. 1695, 1963.
6. P.M. Jenner, E.C. Hagan, J.M. Taylor, E.L. Cook, and O.G. Fitzhugh, Food flavorings and compounds of related structure, I. Acute oral toxicity. *Food Cosmet. Toxicol.* **2**(3):327–343, 1964.
7. J.M. Taylor, M.P. Jenner, and W.I. Jones, A comparison of the toxicity of some alkyl, propenyl, and propyl compounds in the rat. *Toxicol. Appl. Pharmacol.* **6**(4):378–387, 1964.
8. M.I. Makaruk, On the toxicity of vanillin. *Gig. Sanit.* **6**:78–80, 1980.
9. G.D. Stoner, M.B. Shimkin, A.J. Kniazaff, J.H. Weisburger, E.K. Weisburger, and G.B. Gori, Test for carcinogenicity of food additives and chemotherapeutic agents by the pulmonary tumor response in strain A mice. *Cancer Res.* **33**:3069–3085, 1973.
10. V.S. Zhurkov, Investigation of the mutagenic activity of drug preparations and food additives in a culture of human lymphocytes. *Sov Genet.* **11**:528–530, 1975.
11. M. Verrett, S. Jacqeline, F. William, E.F. Reynaldo, E.K. Alterman, and C. Thomas, Toxicity and teratogenicity of food additive chemicals in the developing chicken embryo. *Toxicol App Pharmacol.* **56**(2):265–273, 1980.

12. Wealth of India, *A Dictionary of Indian Raw Materials and Industrial Products, Raw Materials.* Vol. 10, Council of Scientific and Industrial Research, Government of India, New Delhi, India, XXV, p. 591, 1976.
13. J.L. Radomski, Toxicology of food colours. *Ann Rev Toxicol Pharmacol.* **14**:127–137, 1974.
14. C. Bruijnzeel-Koomen, C. Ortolani, K. Aas, C. Bindslev-Jensen, B. Bjorksten, D. Moneret-Vautrin, and B. Wuthrich, Position paper of the European Academy of Allergy and Clinical Immunology on adverse reactions to food: Adverse Reaction to Food. *Allergy* **50**:623–635, 1995.
15. H. Koren, and M. Bisesi, *Handbook of Environmental Health and Safety: Principles and Practices*, Vol. I, National Environmental Health Association, CRC Press, Lewis Publishers, Boca Raton, FL, 1999.

CHAPTER 11

# Industrial Gases and Fumes

## INTRODUCTION

A number of toxic gases have caused adverse effects to animals, humans, and the environment. The most common instance is the release of gases from automobiles, which contain many hazardous substances (e.g., carbon monoxide, nitrogen dioxide, sulfur dioxide, suspended particles less than 10 microns in size, benzene, polycyclic hydrocarbons). Air pollutants are many, and the adverse effects they cause in animals, humans, and vegetation are numerous (Tables 11-1 to 11-3). All these gases are linked with the air quality issues to which all biological systems are exposed today. The reaction and interaction of toxic gases and adverse effects on humans and animals has gained importance. Details of each of the following could be studied separately: (1) acid rain and acid deposition; (2) smog; (3) greenhouse gases ($CO_2$, $CH_4$, $N_2O$); (4) chlorofluorocarbons (CFCs) and ozone depletion; and (5) particulate matter (cancer, respiratory problems, visibility, building damage). Greenhouse gases are carbon dioxide, methane, nitrous oxide, and CFCs. These gases act like the glass covering a greenhouse, letting sunlight in but blocking the infrared radiation from the Earth's surface that carries heat back into space. The gases act like a blanket wherever their concentration increases. Local concentrations increase heat, and increased differences between hotter and colder regions drive weather events to more extreme ranges.[1,2]

## KINDS OF GASES AND ADVERSE EFFECTS

### Ammonia

Ammonia ($NH_3$) is a colorless gas with a pungent odor that has many uses in industry. It is used in the manufacture of fertilizers, such as ammonium sulfate and ammonium nitrate. It also is used in the manufacture of nitric acid, synthetic urea, soda, synthetic fibers, plastics, and dyes. Ammonia also has extensive applications in the petroleum refining and pharmaceutical industries. Some occupations

*Industrial Guide to Chemical and Drug Safety*, By T.S.S. Dikshith and Prakash V. Diwan
ISBN 0-471-23698-5 

**TABLE 11-1 Air Pollutants and Their Adverse Effects**

| Chemicals | Formula | Effects |
|---|---|---|
| Sulfur dioxide | $SO_2$ ($SO_x$) | Vegetation, buildings, acid rain, respiratory |
| Nitric oxide | NO ($NO_x$) | Converted to $NO_2$ |
| Nitrogen dioxide | $NO_2$ ($NO_x$) | Toxic, acid rain, respiratory |
| Carbon monoxide | CO | Toxic |
| Carbon dioxide | $CO_2$ | Greenhouse gas |
| Ozone | $O_3$ | Reactive, vegetation, respiratory |
| Chlorofluorocarbons | $CFCl_3$,$CF_2Cl_2$ | Stratospheric ozone depletion |
| Hydrocarbons | $C_xH_x$ | Photochemical smog |

**TABLE 11-2 Toxic Effects of Carbon Monoxide***

| CO (ppm) | CO in Air (%) | Symptoms |
|---|---|---|
| 100 | 0.01 | No symptoms |
| 200 | 0.02 | Mild headache, few other symptoms |
| 400 | 0.04 | Headache after 1 to 2 hours |
| 800 | 0.08 | Headache after 45 minutes, nausea, collapse, and unconsciousness after 2 hours |
| 1,000 | 0.10 | Dangerous, unconsciousness after 1 hour |
| 1,600 | 0.16 | Headache, dizziness, nausea after 20 minutes. |
| 3,200 | 0.32 | Headache, dizziness, nausea after 5 minutes, unconsciousness after 30 minutes. |
| 6,400 | 0.64 | Headache, dizziness after 1 to 2 minutes, unconsciousness after 10 to 15 minutes, |
| 12,800 | 1.28 | Immediate unconsciousness, danger of death in 1 to 3 minutes, |

*Information taken from an old fire safety publication; original source unknown.

**TABLE 11-3 Status of Sulfur Dioxide in Different Countries**

| Country | Time Average | Concentration (ppm) |
|---|---|---|
| Australia | 1 year | 2 |
| Japan | 24 hours | 4 |
| | 1 hour | 10 |
| West Germany | 24 hours | 15 |
| | 30 minutes | 30 |
| United States | 24 hours | 14 |

*Source*: Pollution control in New South Wales: A progress report. Supplement to 1982–1983. Annual report of the State Pollution Control Commission, Sydney, Australia. 1983.

associated with human exposure to ammonia are aluminium metal extractions, chemical workers, dye makers, electroplate workers, fertilizer manufacturing, galvanizers, mirror silverers, refrigeration workers, tannery workers, and rayon industry workers.

Long-term exposure to ammonia causes irritation to the eye, nose, throat, and lungs. It causes dyspnea, chest pain, pulmonary edema, and frothy sputum. High concentrations of ammonia cause skin damage and skin burn.[3]

### Carbon Dioxide

Carbon dioxide ($CO_2$) at room temperature is a colorless, odorless gas. In industrial environments, it is found in liquid or solid ("dry ice") form. Inhalation is the main route of exposure. Eye or skin contact with the liquid or solid is also a health hazard; skin contact may cause frostbite. Inhalation of high concentrations may cause an increase in breathing rate, excitation, headaches, a feeling of suffocation (dyspnea), an increase in muscular tone, and, at very high concentrations, loss of consciousness and eventually death by asphyxiation. Prolonged exposure to carbon dioxide at low concentrations may cause signs of stress and emotional changes, such as increased irritability. At high concentrations, prolonged or repeated exposure produces the same effects as short-term exposure.[4] Some industries associated with carbon dioxide are aerosol packaging, blast furnace operations, brewery industry, carbonic acid manufacture, charcoal burning, explosive manufacture, fire extinguisher production, foundry operations, mining, refrigeration plants, tanneries, and the textile industry.[5,6]

### Carbon Monoxide

Carbon monoxide (CO) is a nonirritating, colorless, odorless, tasteless gas. Inhalation is the major route of entry into the body. In short-term exposure, its toxicity depends on the amount bound to the blood. Early symptoms of poisoning include headache and feelings of faintness. Serious exposures produce irregular heartbeat, unconsciousness, and death (Table 11-2). There is evidence that CO causes chronic heart disease in instances of repeated or long-term exposure. There is limited and contradictory evidence about effects on the arteries.[2] Some occupations associated with human exposure to carbon monoxide are in the manufacture of acetylene, operation and handling of blast furnaces, carbon black manufacture, coke oven operation, diesel engine operation, garage mechanics, mining operations, organic chemicals, and during the manufacture of pulp, paper, and steel.[6]

### Hydrogen Cyanide

Hydrogen cyanide (HCN) is a colorless gas that is extremely poisonous. It has the odor of bitter almond, and its presence normally goes unnoticed. Hydrogen cyanide is used for rodent control, fumigation of ships for pest control, and control of crop pests. It also is used in electroplating in the synthesis of acylates and nitriles, particularly acrylonitrile. Its use in steel hardening, electroplating, and the extraction of gold and silver has many advantages.

Hydrogen cyanide causes irritation to the eye, skin, nose, and throat. Repeated exposure to the gas form results to corneal ulceration, erythemia, and formation of vesicles on the skin. Hydrogen cyanide combines with certain enzyme systems, most importantly cytochrome oxidases, and interferes with the respiratory function of the organism. Inhalation, ingestion, or skin absorption of HCN is rapidly fatal. In large doses, HCN causes loss of consciousness, cessation of respiration, and death. Hydrogen cyanide must always be handled by qualified and trained personal under proper supervision to avoid human fatalities.[7] Some industries associated with human exposure to hydrogen cyanide are acid dipping operations, the manufacture of acrylate and ammonium salt, operation of a blast furnace, cellulose product treatment, operation of a coke oven, electroplating, steel production, fumigation, gold and silver extraction, and the synthesis of organic chemicals.[8]

### Hydrogen Sulfide

Hydrogen sulfide ($H_2S$) is a flammable, poisonous gas with a characteristic odor of rotten eggs. It burns in air with a pale blue flame. It is used extensively in the manufacture of inorganic sulfides, sulfuric acid, organic sulfur compounds, analytical reagents, and in metallurgy. Hydrogen sulfide is present in different concentrations in tanneries, breweries, fat rendering, and oil- and gas-drilling wells.

Hydrogen sulfide is highly toxic and leads to fatalities if ingested. It causes irritation to the eyes and respiratory tract, resulting in keratoconjunctivitis, photophobia, lacrimation, corneal opacity, rhinitis, laryngitis, and bronchopneumonia. Deep penetration of hydrogen sulfide into the lungs results in edema, respiratory failure, and death.[9,10] Repeated exposures to high concentrations of $H_2S$ in workplaces may result in systemic intoxication in workers, causing paralysis of the respiratory center of the brain, apnea, sudden collapse, and death.[11]

### Nitrogen Oxides

Nitrogen oxides ($NO_x$) include several chemicals, such as nitrogen oxide ($N_2O$), nitric oxide (NO), nitrogen dioxide ($NO_2$), nitrogen trioxide ($N_2O_3$), nitrogen tetroxide ($N_2O_4$), nitrogen pentoxide ($N_2O_5$), nitric acid ($HNO_3$), and nitrous acid ($HNO_2$). Nitric oxide is extremely toxic and is known to cause adverse effects. For example, overexposure results in irritation of voice, nose, and the respiratory system. It also may cause drowsiness and unconsciousness in exposed workers. Areas associated with the use and release of nitrous oxides should be properly managed by qualified and trained workers with adequate ventilation to avoid respiratory diseases.[12–14] Some areas associated with human exposure to nitrogen oxide are the manufacture of chemical dyes and fertilizers, the processing and operation of bleaching of food and textile products, working in the garage, gas and electric arc welding, metal cleaning, silo filling, the manufacture of sulfuric acid, and the synthesis of organic chemicals.

## Ozone

Ozone ($O_3$) is used extensively as an oxidizing agent in a variety of organic chemical industries. Ozone is an important air contaminant found in the industrial atmosphere encountered in arc welding, industrial waste treatment, oil bleaching, and cold storage. Ozone causes inflammation and congestion of the respiratory tract, pulmonary edema, hemorrhage, and death. More information on the adverse effects of ozone may be found in the literature.[15–17] Ozone is a deep lung irritant capable of causing death as a result of pulmonary edema. Studies with mice and guinea pigs have indicated an $LC_{50}$ (3 hours) of 20 ppm and 500 ppm, respectively. As discussed in earlier chapters, ozone-induced toxicity is modulated by several factors. Young mice are more susceptible than adults, and higher temperatures trigger increased toxicity versus ambient temperature. Studies have indicated that exposure to a lower concentration of ozone (1 ppm) resulted in the development of engorged blood vessels and excess leukocytes in lung capillaries of animals.[15] Along with the pulmonary edema, animals demonstrated desquamation of the ciliated epithelium throughout the lung airways. The alveolar damage included swelling and denudation of the cytoplasm of type I cells, swelling or rupture of the capillary endothelium, and lysis of erythrocytes. It has been reported that long-term exposure to ozone affects the morphology and function of mammalian lung and the acceleration of lung tumor formation.[16] Several laboratory animals exposed to ozone for prolonged periods have demonstrated bronchitis, bronchiolitis, fibrosis, and emphysematous.[1]

## Phosgene

Phosgene ($COCl_2$) is a colorless gas with a sweet and pleasant odor (in low concentrations). Its uses in industry are many. Phosgene is used in the organic synthesis of isocyanates and their derivates, carbonic acid esters (polycarbonates), acid chlorides, and for the preparation of organic chemicals such as a war gas. It also is used in metallurgy, in the manufacture of dyestuffs, in insecticides, and in the pharmaceutical industry. Phosgene produces conjunctivitis, lacrimation, and irritation of the upper respiratory tract.[17a] Acute exposure to phosgene results in pulmonary edema, followed by discomfort, cough, and viscous sputum. In chronic cases of phosgene exposure, irreversible pulmonary changes of emphysema and fibrosis have been reported.[17a] Since phosgene does not cause an immediate irritation when inhaled, it becomes risky and hazardous to identify the presence of this insidious poisonous gas. Exposure to phosgene results in irritation to the eye, causes severe pulmonary edema, and results in sudden fatality. Workers exposed to phosgene suffer from symptoms of choking, a constricted feeling in the chest, coughing, dyspnea or painful breathing, and release of bloody sputum.[18]

### Sulfur Dioxide

Sulfur dioxide ($SO_2$) has a number of industrial uses. It is used in the manufacture of sodium sulfite, sulfuric acid, sulfuryl chloride, thionyl chloride, organic sulfonate, disinfectants, fumigants, industrial and edible proteins, and other materials. Sulfur dioxide also is used extensively as a bleaching agent, particularly in bleaching of sugar beet, flour, straw, textiles, and wood pulp.[19] Sulfur dioxide has industrial use in leather tanning, brewing, and preserving. Sulfur dioxide is a colorless gas with a characteristic strong, suffocating odor. Gaseous sulfur dioxide is particularly irritating to mucous membranes of the upper respiratory tract. Chronic exposure to sulfur dioxide produces dryness of the throat, cough, rhinitis, conjunctivitis, corneal burns, and corneal opacity. Acute exposure to high concentrations of sulfur dioxide also may result in death due to asphyxia. In contrast, chronic exposure to sulfur dioxide leads to nasopharyngitis, fatigue, and disturbances of pulmonary function.[20] Animals exposed to chronic doses of sulfur dioxide have shown thickening of mucous layers in the trachea and hypertrophy of goblet cells and mucous glands, resembling the pathology of chronic bronchitis.[1,21] It has been found that penetration of sulfur dioxide into the lungs is greater during mouth breathing than during nose breathing. In fact, an increase in flow rate of the gas would markedly increase the penetration.[22] Human subjects exposed for very brief periods to sulfur dioxide showed alterations in pulmonary mechanics. More information on the adverse effects of sulfur dioxide and the manner of its potentiation in association with other chemicals may be found in the literature (Table 11-1).[1,23–27]

## REFERENCES


1. M.O. Amdur, Air pollutants, in *Casarett & Doull's Toxicology, The Basic Science of Poisons*, 2nd ed., John Doull, Curtis D. Klassen, and Mary O. Amdur, eds., Macmillan, New York, 1980.
2. R.F. Coborn, ed., Biological effects of carbon monoxide. *Ann. NY. Acad. Sci.* **174**:1–430, 1970.
3. NIOSH, *Pocket Guide to Chemical Hazards*, Department of Human Health Services, NIOSH, Bethesda, MD, p. 38, 90–117, 1990.
4. D.J. Cullen, E.I. Eger, Cardiovascular effects of carbon dioxide in man. *Anesthesiology*. **41**:345, 1974.
5. H.I. Williams, Carbon dioxide poisoning report of eight cases and two deaths. *Brit. Med. J.* **2**:1012, 1958.
6. B.C. Levin, M. Paabo, J.L. Gurman, S.E. Harris, E. Braun, Toxicological interactions between carbon monoxide and carbon dioxide. *Toxicology* **47**:135, 1987.
7. Patty's *Industrial Hygiene and Toxicology*, Vol. 2c, 3rd ed., G.D. Clayton and F.E. Clayton, eds., Wiley-Interscience, New York, pp. 4114–4128, 1982.
8. J.H. Wolfsie, C.B. Shaffer, Hydrogen cyanide—hazards, toxicity prevention and management of poisoning. *J. Occup. Med.* **1**:28, 1959.
9. C.L. Winek, W.D. Collom, C.H. Wecht, Death from hydrogen sulphide fumes. *Lancet*. **1**:1096, 1968.

10. R.E. Simson, G.R. Simson, Fatal hydrogen sulphide, poisoning associated with industrial waste exposure. *Med. J. Aust.* **1**:331, 1971.
11. R.J. Reiffenstein, et al., Toxicology of hydrogen sulfide. *Ann. Rev. Pharmacol. Toxicol.* **32**:109–134, 1992.
12. W.C. Cooper, I.R. Tabershaw, Biologic effects of nitrogen dioxide in relation to air quality standards. *Arch. Environ. Health.* **10**:455, 1965.
13. R. Morley, S.J. Silk, The industrial hazard from nitrous fumes. *Ann. Occup. Hyg.* **13**:101, 1970.
14. NIOSH, *Pocket Guide to Chemical Hazards*, Department of Human Health Services, NIOSH, Bethesda, MD, p. 164, 90–117, 1990.
15. H.E. Stokinger, Ozone toxicology, a review of research and industrial experience 1954–1964. *Arch. Environ. Health*, **10**:719–731, 1965.
16. L.S. Jaffe, The biological effects of ozone on man and animals. *Am. Ind. Hyg. Assoc. J.* **28**:267, 1967.
17. D.L. Dungworth, Short-term effects of ozone on lungs of rats, mice and monkeys. *Environ. Health Perspect.* **16**:179, 1976.
17a. World Health Organization, Phosgene, *Environ. Health Criteria* 193, IPCS, WHO, Geneva, Switzerland, 1997.
18. *Patty's Industrial Hygiene and Toxicology*, Vol. 2C, 3rd ed., G.D. Clayton, F.E. Clayton, eds., Wiley-Interscience, New York, pp. 4126–4128, 1982.
19. I.O. Skalpe, Long-term effects of sulphur dioxide exposure in pulp mill. *Brit. J. Ind. Med.* **21**:69, 1964.
20. B.G. Ferris, Jr., W.A. Burgess, J. Worcester, Prevalence of chronic respiratory disease in a pulp mill and a paper mill in the United States. *Brit. J. Ind. Med.* **24**:26, 1976.
21. U.S. Department of Health, Education and Welfare, Air quality criteria for sulfur dioxides. *Nat. Air Pollution Control Adm.* Publication No. AP-50, Washington, DC, January, 1977.
22. N.R. Frank, et al., $SO_2$ ($^{32}S$ labeled) absorption by the nose and mouth under conditions of varying concentrations and flow. *Arch. Environ. Health* **18**:315–322, 1969.
23. N.R. Frank, et al., Effects of acute controlled exposure to $SO_2$ on respiratory mechanics in healthy male adults. *J. Appl. Physiol.* **17**:252–258, 1962.
24. T. Toyama, and K. Nakamura, Synergistic response of hydrogen peroxide aerosols and sulphur dioxide to pulmonary airway resistance. *Indust. Health* **2**:34–45, 1964.
25. D.B. Menzel, and R.O. McClellan, Toxic responses of the respiratory system, in *Cassatett & Doull's Toxicology, The Basic Science of Poisons*, 2nd ed., John Doull, Curtis D. Klassen, and Mary O. Amdur, eds., Macmillan New York, Chapter 12, 1980.
26. H.N. MacFarland, Respiratory toxicology, in *Essays in Toxicology*, Vol. 7, Hayes, W.J. Jr., ed., Academic Press, New York, pp. 121–154, 1976.
27. U.S. Department of Health, Education and Welfare, Air quality criteria for carbon monoxide, *Nat. Air Pollution Control Adm.* Publication No. AP-62, Washington, DC, March, 1970.

CHAPTER 12

# Drugs: Discovery and Development

## INTRODUCTION

Essentially, all drugs are chemical molecules. Entry of drugs into the body initiates the process of action and interaction with other molecules. Drugs vary in their chemical nature. A *drug* has been defined by the World Health Organization (WHO) as "a substance or product that is used or intended to be used to modify or explore physiologic systems or pathologic states for the benefit of the recipient." The drug discovery process covers a range of therapeutic areas and treatment regimens. Further, drug discovery is a time-consuming, multifaceted, and expensive undertaking. The very purpose is to develop a new drug product with therapeutic benefits (efficacy) and few side effects (toxicity). The drug development process starts at the chemist's bench with isolation of a new chemical entity (NCE) and moves through efficacy pharmacology testing, using various *in vitro* and *in vivo* models. Next, the drug development process proceeds through an abbreviated toxicology profile, including pharmacological profiling (the determination of pharmacologic effects other than the desired therapeutic effect), based on the proposed clinical plan for first human dose (FHD).

The purpose of preclinical safety testing is to understand adverse effects of the candidate drug during clinical trials. Clinical safety, pharmacokinetic, and pharmacodynamic studies are initiated based on the acceptance of efficacy of the pharmacology and initial toxicology profiles. As human clinical trials progress, the candidate drug moves through definitive stages of evaluation, such as the subchronic, chronic, reproduction, and carcinogenicity tests. New drugs are evaluated based on certain parameters (Table 12-1).[1] For details, refer to Chapter 2. The risk of new drug development has been discussed by several authors.[2] The following pages briefly discuss some of these aspects.

The discovery, development, and registration of a pharmaceutical product is an expensive operation and represents a unique challenge for the chemical and pharmaceutical industry. In fact, of 10,000 active molecules specifically synthesized as potential therapeutics, only 100 will reach the human trial scale market. Further, only 10% will reach the market, and less than 20% of these would

*Industrial Guide to Chemical and Drug Safety*, By T.S.S. Dikshith and Prakash V. Diwan
ISBN 0-471-23698-5 

**TABLE 12-1 Safety Evaluation of New Drugs and Parameters**

| |
|---|
| Acute toxicity |
| Cumulative toxicity |
| Absorption from various routes |
| Elimination $t_{1/2}$ and accumulation in deep compartments |
| Penetration of barriers |
| Milk excretion |
| Teratogenicity |
| Mutagenicity |
| Carcinogenicity |
| Sensitization |
| Local irritation |

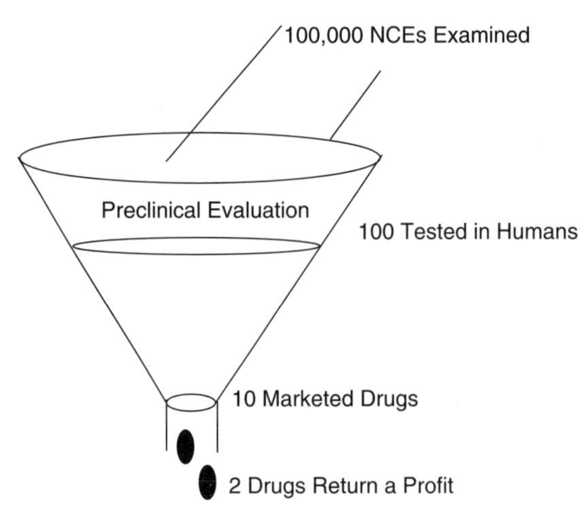


**Figure 12-1.** Attrition rate of new drug candidates drawn from information presented in Chien (1979).

reach a profitable level for the drug industry (Fig. 12-1).[3] Steps involved in drug discovery are shown in Table 12-2.

The framework for each drug toxicity test is sufficiently well defined. In fact, conduction of tests and data gathering in a similar manner is fully acceptable to various regulatory bodies in different countries. Growing demands for testing and evaluating the chemical substance toxicity will place increasing pressure on personnel and laboratory resources. A harmonized approach, promoting the scientific aspects of toxicity testing and ensuring a wide acceptability of test data for regulatory purposes, will avoid wasteful duplication of tests. Further, this practice will contribute to the efficient use of laboratory facilities and skilled personnel.[4–6]

**TABLE 12-2 Steps Involved in Drug Discovery**

| Preclinical Testing | Time | Remarks |
|---|---|---|
| Identification of new chemical entity | 1–5 years | Data submission of PCT for approval Phase I |
| Preclinical toxicity studies | | |
| Test in human beings | 2–10 years | Regulatory authority clearance for Phase II |
| Phase I—Normal volunteers | | |
| Tolerance, safety, metabolism, kinetics & drug interactions | | |
| Phase II—Selected diseased patients | | Regulatory authority clearance Phase III |
| • Therapeutic efficacy | | |
| • Dose determination | | |
| • Pharmacokinetics | | |
| Phase III—Multicentered trials (large number of patients) | | Marketing permission Phase IV |
| • Safety & efficacy | | |
| Phase IV—Postmarketing surveillance | | Data retrieval from market after the introduction |
| • Adverse reaction | | |
| • Toxic effects (if any) | | |
| • Drug utilization | | |
| • Additional use | | |

## Safety Evaluation

Essentially, all toxicologic tests are conducted using animal species *in vivo* and *in vitro*, and the results are subsequently extrapolated to humans.[7] There are limitations in conducting different toxicological tests and the type of findings. Extrapolation of data from animals to humans is not always accurate. Furthermore, since it is not possible to devise "standard" test methods appropriate to all chemicals and drugs, judgment must be exercised on a case-by-case basis to assess the suitability of a particular method. In many areas of toxicology, controversy exists concerning the appropriate experimental design to be used. For example, opinions differ on the duration of study, the number of animals to be used, and types of species/strains of test animals to be treated with the drug. Another important limitation in the assessment of drug hazard to humans comes from the manner or exposure. As a rule, single substances are tested by defined routes of exposure (e.g., oral, dermal, intraperitoneal, intravenous); under normal conditions, exposure to chemicals is a combination of substances, hence the exposure is by more than one route. Also, it should be obvious that due to size limitations of animal experiments, low incidence effects may not be easily recognized. Consequently, while toxicity testing results from laboratory animals provide a good indication of health hazards of the test substance, they do not eliminate the need for human studies.

Guidelines for conducting toxicological tests are adequate for the evaluation of chemicals and drugs. Any further elaboration or extension of a test should be done only with good reason. Scientific judgment is essential to determine the conduct of a particular test, so that a reasoned flexibility of approach is always necessary. The present guidelines have been developed considering a proper balance between resources and scientific requirements.

Toxicology is a developing experimental multidisciplinary science. Excessive rigidity or overdetailed specification of methods could lead to lack of scientific initiative and may be counterproductive. There must be provision for the exercise of toxicologic skill and judgment during the course of the study, even where this forms part of a prescribed set of test requirements. Therefore, all guidelines or similarly defined procedures should be based on the rationale for changes in procedure(s) and should have a scientific approach. The emphasis on a flexible approach should not be construed as a recommendation for a lack of order. Rather, it should be seen as creating a situation in which the examination of drug toxicity is conducted as a scientific exercise rather than a set of stereotyped routine tests (Fig. 12-2). Certain signs and symptoms of drug-induced toxicity in animals

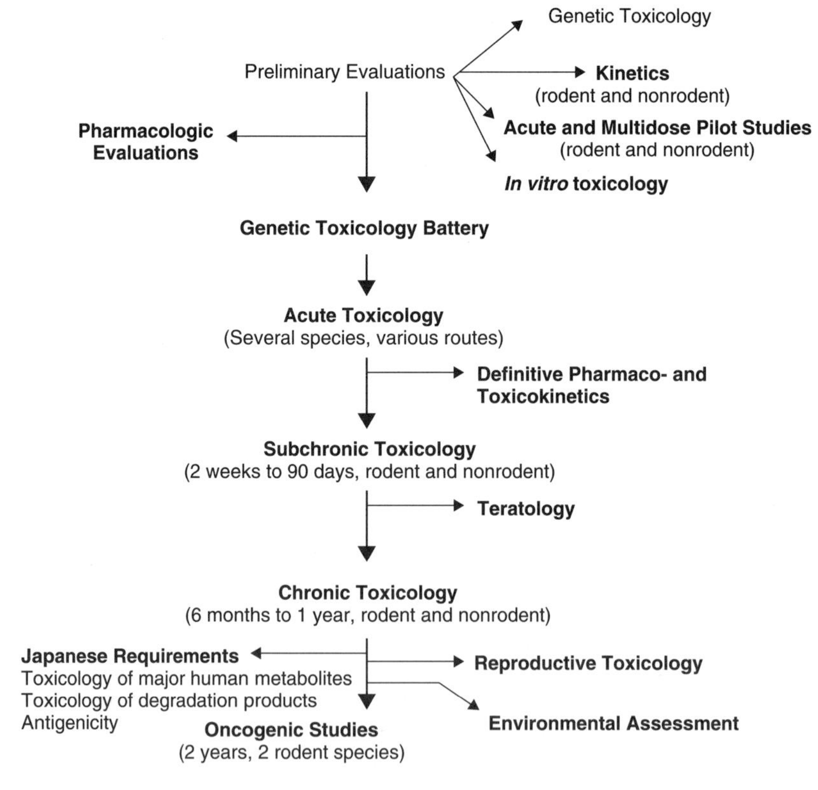


**Figure 12-2.** General approach to developing a toxicity profile for pharmaceutical agents.

**TABLE 12-3 Signs and Symptoms of Drug Toxicity**

| Clinical Side Effect | Predictable from Animal Studies (√/×)* |
|---|---|
| Drowsiness | √ |
| Hypertension | √ |
| Anorexia | √ |
| Nausea | × |
| Insomnia | √ |
| Depression | √ |
| Dizziness | × |
| Fatigue | × |
| Increased appetite | √ |
| Sedation | √ |
| Constipation | √ |
| Tremor | √ |
| Dry mouth | √ |
| Tinnitus | × |
| Perspiration | √ |
| Nervousness | √ |
| Weight gain | √ |
| Dermatitis | √ |
| Epigastric distress | × |
| Hypotension | √ |
| Headache | × |
| Vertigo | × |
| Vomiting | √ |
| Heartburn | × |
| Palpitation | √ |
| Weakness | √ |
| Diarrhea | √ |
| Blurred vision | √ |
| Skin rash | √ |
| Lethargy | √ |

*√—YES; ×—NO.

are not easily identifiable with humans (e.g., heartburn, headache) (Table 12-3). Toxicity tests are broadly divided into acute, subchronic, and chronic studies. For more details, refer to Chapter 2.

## CONSIDERATION IN TOXICITY TESTING

It is important that the identity of the test substance be defined in the first instance. The physical and chemical properties of the drug under test provide important information for the selection of the administration route, the study design, handling and storage of the test substance. It is important to characterize the test mixtures and to identify impurities known or likely to be present. Separate studies of impurities may provide useful evidence in the evaluation of the carcinogenicity of the mixtures.

### Choice of Test Animals

There is no experimental animal species that is identical to humans in structure or metabolism. However, there are obvious similarities in function between humans and some animal species. Even with fellow primates, it is not always possible to obtain straightforward extrapolations to humans. Interpretation of animal test results in the assessment of possible human health hazards remains a matter of skilled judgment.[7]

Accepting that no ideal animal analog of humans is available for testing, the choice of test species can be influenced by other logistic considerations, such as ease of breeding or purchasing, animal husbandry, speed of growth/development, and handling under experimental conditions. Rodents fulfill many logistic requirements and are used extensively in toxicological studies. The rat is the preferred species for acute oral, dermal, and inhalation studies with the option of the rabbit in the dermal study. In the latter context, the rabbit has the advantage of a larger size combined with a reasonable background of information on its behavior in dermal studies. From the point of view of comparing toxic effects by different routes and evaluating hazards, there is much to be said in favor of the rat.

### Animal Care

Stringent control of environmental conditions and proper animal care techniques are mandatory for meaningful results. Diet should meet all nutritional requirements of the species used in the tests. It is important to know the effect of the dietary regimen on metabolism and animal longevity as well as toxicity development. Variations in worldwide use patterns of industrial and agricultural chemicals preclude harmonization on one list of dietary contaminants.

### Number and Size of Groups

With the objective of an efficient approach to testing chemicals, there is no point in having more groups (or more animals per group) than are strictly necessary to attain the endpoint of reliable detection of toxic effects.

Toxicity studies are undoubtedly expensive in both financial and resource terms. Part of the cost is related to the number of animals and the extent of clinical, necropsy, and histopathologic investigations required. Considering the inherent variability of biological systems, there must always be a balance between the number of animals theoretically required to detect all effects (from the weakest upward) and the number required to detect significant toxic effects. In a well-conducted study that goes according to plan, it is possible to use fewer animals.

However, in tests of chemicals with unknown toxic characteristics, problems often arise because the actual responses of animals differ widely from those anticipated when the study was designed. To address this problem, it is prudent to increase the number of animals to ensure that animals are available at key points of a study to provide adequate information. In acute studies, the requirement for groups and number of animals in groups is related to the reliable determination of acute toxic effects and the estimation of a median lethal dose. In subchronic and chronic testing, the numbers are related to the detection of effects, providing sufficient animals for an acceptable investigation of toxic mechanisms and giving an indication of a "no-effect level."

A sufficient number of animals should be used so that at the end of the study enough animals are available for thorough biological evaluation. After considerable discussion, it was agreed that for rodents each dose group and concurrent control group should contain at least 10 animals of each gender. For nonrodent studies, a minimum of four animals of each gender is recommended.

### Blood Collection

To evaluate chemical and drug safety, different animal species are used. Blood is collected to estimate different parameters of drug toxicity under test. The collection of blood from different sites in test animals, along with specific needle size, has been recommended by regulatory bodies and international agencies (Table 12-4). Based on the size of the laboratory animal(s) and the requirement of the blood needed to generate quality data, blood is withdrawn from each test animal. This is to avoid withdrawal of a large volume of blood and to avoid injury to the test animal (Table 12-5).

**TABLE 12-4 Common Bleeding Sites of Laboratory Animals**

| Species | Heart | Saphenous Vein | Femoral Vein | Jugular Vein* | Ear Vein | Tail Vein | Needle Size (gauge) |
|---|---|---|---|---|---|---|---|
| Mouse | × | | | | | × | 23–25 |
| Rat | × | × | | | | × | 23–25 |
| Hamster | × | | | | | | 22–24 |
| Guinea pig | × | × | | | × | | 20–22 |
| Rabbit | × | | | | × | | 20–24 |
| Cat | × | | | × | | | 22–24 |
| Dog | | × | × | × | | | 20–22 |
| Monkey | | × | × | × | | | 20–22 |

*Under anesthetic conditions.

*Source*: Modified from Laboratory Animal Information Service and Centre, National Institute of Nutrition, Indian Council of Medical Research, Government of India, Hyderabad, India.

**TABLE 12-5 Practicable Volume of Blood (mL/kg) Obtainable from Various Animals**

| Animal Species | Total Blood Volume | Available Volume When Blood Out (rounded mean) | Maximum Safe Volume at One Bleeding | Practicable Volume (mL) for Diagnostic Use from a Normal Adult |
|---|---|---|---|---|
| Mouse | 8 | 3.5 | 1 | 0.10 |
| Rat | 50 | 20 | 5 | 0.25 |
| Hamster | 72 | 29 | 5.50 | 1 |
| Guinea pig | 75 | 35 | 7 | 0.50 |
| Rabbit | 70 | 35 | 7 | 1 |
| Cat | 75 | 35 | 7 | 1 |
| Dog | 90 | 45 | 9 | 2 |
| Monkey (Rhesus) | 75 | 36 | 6.60 | 2 |

*Source*: Modified from Laboratory Animal Information Service and Centre, National Institute of Nutrition, Indian Council of Medical Research, Government of India, Hyderabad, India.

## Termination of Animals

Recent research has indicated that any study using laboratory animals must be conducted based on principles of good laboratory practice. This is particularly important while handling and during termination of laboratory animals. All animals should be treated humanely (Tables 12-6 and 12-7). For more details, refer to Chapter 18.

For the safe evaluation of chemicals and drugs, several animal species are used (e.g., mice, rats, cats, dogs, nonhuman primates). Test chemical(s) are administered through different routes such as oral, dermal, intravenous, intraperitoneal,

**TABLE 12-6 Termination of Mouse and Albino Rat by Euthanasia (by gases)**

| Species | Carbon Monoxide | Carbon Dioxide | Carbon Dioxide and Chloroform | Nitrogen Flushing | Argon Flushing |
|---|---|---|---|---|---|
| Mouse | A | A | A | NR | NR |
| Rat | A | A | A | MR | MR |
| Hamster | A | A | A | NR | NR |
| Guinea pig | A | A | A | NR | NR |
| Rabbit | A | A | A | NR | NR |
| Cat | A | A | A | NR | NR |
| Dog | A | NR | NR | NR | NR |
| Monkey | A | NR | NR | NR | NR |

A — Acceptable; NR — Not Recommended.
*Source*: Modified from Laboratory Animal Information Service and Centre, National Institute of Nutrition, Indian Council of Medical Research, Government of India, Hyderabad, India.

**TABLE 12-7 Termination of Mouse and Albino Rat by Euthanasia (by parenteral administration of pharmacological agents)**

| Species | Barbiturate Overdose | Route | Chloralhydrate Overdose | Route | Ketamine Overdose* | Route |
|---|---|---|---|---|---|---|
| Mouse | A | IP | A | — | A | IM |
| Rat | A | IP | A | — | A | IM |
| Hamster | A | IP | A | — | A | IM |
| Guinea pig | A | IP | A | — | A | IM |
| Rabbit | A | IV, IP | NR | IV | A | IM |
| Cat | A | IV, IP | NR | IV | A | IM |
| Dog | A | IV, IP | NR | IV | A | IM |
| Monkey | A | IV, IP | NR | IV | A | IM |

A — Acceptable; IP — Intraperitoneal; IM — Intramuscular; IV — Intravenous; NR — Not recommended.
*Ketamine, produces rapid loss of consciousness; nevertheless, fairly high doses are required to kill.
*Source*: Modified from Laboratory Animal Information Service and Centre, National Institute of Nutrition, Indian Council of Medical Research, Government of India, Hyderabad, India.

**TABLE 12-8 Administration of Volume of Test Chemicals to Laboratory Mouse and Albino Rat**

| Route | Mouse Volume (in mL) | Albino Rat Volume (in mL) |
|---|---|---|
| IV | 0.4 | 2 |
| IP | 1 | 2 |
| SC | 0.4 | 1 |
| IM | 0.4 | 0.4 |
| Oral | 1 | 2.5 |

IM — Intramuscular; IP — Intraperitoneal; IV — Intravenous; SC — Subcutaneous.
*Source*: Modified from Laboratory Animal Information Service and Centre, National Institute of Nutrition, Indian Council of Medical Research, Government of India, Hyderabad, India.

subcutaneous, and respiratory (inhalation). In all cases, the volume of the drug under test should be based on the species and administration route (Table 12-8).

## ISSUES INVOLVED IN TOXICITY TESTING

Major issues stem from the fact that toxicology is in a stage of rapid development, and the harmonization of testing approaches rests on skill and judgment, not necessarily on purely scientific criteria. Accordingly, the guidelines represent an agreed basic approach, which must serve as a foundation for future development and refinement. Development can take place only as a result of experience, and it is important to first follow guideline methods and then ensure that their performance is evaluated, so that any refinements found can be introduced in an agreed and harmonized manner.

The preclinical assessment for the safety of potentially new pharmaceuticals represents a special case in the general practice of toxicology. This phase has its own peculiarities and considerations and differs in several ways from the practice of toxicology. Pharmaceuticals, unlike industrial chemicals, agriculture chemicals, or environmental agents, are intended for human use and systemic exposure. Therefore, pharmaceuticals are specific in action and have biological effects on those treated for different diseases.

Pharmacology and medicinal chemistry have been transformed from an intellectual exercise in diagnosis to a powerful force for the relief of human disease. Drug discovery is a challenging area of research, and there is a great need for new chemical entities as we do not have cures for many diseases or syndromes.

Drug discovery requires a host of inputs, new ideas, design and synthesis of substances, evaluation of preclinical toxicity tests in animals, clinical studies in human volunteers, permission to market the drug, postmarketing studies of safety, and comparison with other medicines. Drug development is highly technical and enormously expensive, with a success rate of 1 in 10,000 compounds.

## NEW AVENUES IN DRUG DEVELOPMENT AND SAFETY

### Techniques of Discovery

Molecular modeling, aided by three-dimensional computer graphics (including virtual reality), allows structure design based on new and known molecules to enhance their desired properties and to eliminate undesired properties, thus offering ways to create highly selective targeted compounds. Essentially, all molecular structures can be modeled and are capable of binding to a single high-affinity site.

Combinational chemistry involves the random mixing of many chemical building blocks (e.g., amino acids, nucleotides, simple chemicals) to produce "libraries" of possible combinations. This technology generates billions of new compounds that are initially evaluated using automated high-throughput devices capable of screening thousands of compounds a day. These screens use radiolabeled ligand displacement on single human receptor subtypes or enzymes on nucleated (eukaryotic) cells. If the screen records a positive response, the compound is further investigated using traditional laboratory methods, and the molecule is manipulated to enhance selectivity or potency.

### Proteins as Medicines

The targets for a majority of drugs are proteins (cell receptors and enzymes). It is only lack of technology that has prevented the exploitation of proteins (and peptides) as medicines. Although technology is now available, there are still great practical problems of sending proteins to the target site in the body. (They are digested when swallowed and cross cell membranes with difficulty.)

Biotechnology involves the use of recombinant DNA technology and or genetic engineering to clone and express human genes, such as in microbial *Escherichia coli*, or yeast cells. They also produce hormones and autacoids in commercial amounts (e.g., insulin and growth hormone, erythropoietins, cell growth factors and plasminogen activators, interferons, vaccines, and immune antibodies). Transgenic animals (that breed true for the gene) also are being developed as models for human diseases and for the production of medicines.

The polymerase chain reaction (PCR) is an alternative to bacterial cloning. This is a method of gene amplification that does not require living cells; it takes place *in vitro* and can produce (in a cost-effective way) commercial quantities of pure potential medicines.

### Genetic Medicines

Synthetic oligonucleotides are being developed to target defined sites on DNA sequences or genes (double-strand DNA: triplex approach) or messenger RNA (antisense approach) so that the production of disease-related proteins is blocked. These oligonucleotides offer prospects of treatment for cancers and viruses without harming healthy tissues.[8,9]

Gene therapy of human genetic disorders is a strategy in which nucleic acid, usually in the form of DNA, is administered to modify the genetic repertoire for therapeutic purposes, as in cystic fibrosis.[10]

### Immunopharmacology

Understanding the molecular basis of immune responses has allowed the definition of mechanisms by which cellular function is altered by local hormones or autacoids in infections, cancer, autoimmune diseases, and organ transplant rejection. These processes present targets for therapeutic intervention.

### Drug Delivery Systems

A number of drugs reach a delicate state of patent by 2005. Therefore, an alternative is required for new drug development: the formulation of drugs into separate delivery mechanisms, often known as drug delivery systems. Clearance for these innovative formulations as abbreviated new drug applications (ANDA) is easier, compared with a new drug entity that requires approval of the United States Food and Drug Administration (USFDA).

## PRECLINICAL STUDIES IN ANIMALS

For the generation of drug safety data, different animal species are used (e.g., mouse, rat, hamster, guinea pig, rabbit, cat, dog, nonhuman primate). In general, the following tests are undertaken:

- *Pharmacodynamics:* explore actions relevant to the proposed therapeutic use and other effects at that dose.
- *Pharmacokinetics:* discover how the drug is distributed and disposed of by the body.
- *Toxicology:* see if and how the drug causes injury using *in vitro* test systems and intact animal models.
  — single dose studies (acute toxicity)
  — repeat-dose studies (subacute, intermediate, and chronic or long-term toxicology)
  — the duration of repeat-dose studies with different ranges (Table 12-9).

Special toxicology involves areas in which a particularly toxic drug accident might occur on a substantial scale; all involve interaction with genetic materials or expression in cell division.

- Mutagenicity (genotoxicity). A bacterial mutagenicity test that demonstrates the induction of point (gene) mutations is always required. Some mutations result in the development of cancer.
- Definitive carcinogenicity (oncogenicity) tests are often not required prior to early human studies unless there is serious reason to be suspicious of the drug. For example, whenever a mutagenicity test is unsatisfactory, the molecular structure, including likely metabolites in humans, gives rise to suspicion or the histopathology in repeat-dose animal studies raises suspicions.

Full-scale (i.e., most of an animal's life) carcinogenicity tests will generally be required only 1) if the drug is to be given to humans for more than 1 year; 2) if it resembles a known human carcinogen; 3) if it is mutagenic (in circumstances relevant to human use); or 4) if it has major organ-specific hormonal agonist action.

It may now be asked why any novel compound should be given to humans before full-scale formal carcinogenicity studies are completed. The answers are

**TABLE 12-9 Comparative Duration Regimen of Repeated-Dose Studies in Humans and Animals**

| Intended Duration of Use in Humans | Duration of Animal Studies |
|---|---|
| Single dose (or several doses on 1 day) | 14 days |
| Up to 10 days | 28 days |
| Up to 30 days | 90 days |
| Beyond 30 days | 180 days |

that animal tests are uncertain predictors, and that such a requirement would make desirable drug development that is seriously expensive, or might even cause potentially valuable novel ventures to cease. For example, tests would be done on numerous compounds that are eventually abandoned for other reasons.

## PHARMACEUTICAL EVALUATION OF NEW DRUGS

The aim of preclinical testing is to understand the therapeutic usefulness and toxicity profile. It has been estimated that the cost of developing an NCE is $350,000,000. Presently in India, we have nearly 75,000 formulations concerned with various drugs. The basic question is, *Do we need drugs*?

The answer is **yes** because there are reported nearly 30,000 diseases or disorders, and only 10,000 of these can be presently treated. This more than justifies the need to develop new drugs. NCEs are required, as drug therapy is unsatisfactory for viral diseases, cancer, and various skin diseases.

Does this mean that we are "drugging ourselves to death," as there are thousands of formulations available in the market, some rational and most irrational. There is no single drug that is available as safe, and according to Paracelsus, "safe drug never was and never will be." Drug discovery has tried various approaches to find an NCE, for example:

| Various Approaches to Drug Discovery | |
|---|---|
| Chance observations | (Penicillin) |
| Chemical modification | (Thiazide) |
| Random | (Cyclosporine) |
| Rational drug design | (Cimetidine) |
| Me too drugs | (Prantosil, Barbitone) |
| Folk medicine | (Digitalis) |
| Molecular modeling | (Awaited) |
| Planned research | (Sulfonamides) |
| Natural product | (Neem, Antibiotics) |

Out of these approaches, imaginative concepts will yield a better NCE.

Identification of a new chemical entity with specific biological activity(ies) is an important goal of drug discovery. After identifying the NCE, the same is subjected to the critical and rigorous evaluations. This is conducted in two phases: (a) the evaluation with animals, and (b) the human study. Both studies are required before the drug qualifies for entry into the market of any country. Basically, the goal of drug therapy is to achieve a beneficial effect with minimum side effects and maximum efficacy. It has been characterized with a rational approach considering the pharmacokinetic and pharmacodynamic aspects. In the discovery and development of a new drug, the idea is important (e.g., $H_2$ receptor antagonists). Histamine is a potent stimulant of gastric acid, and $H_1$ blockers did not block this effect. Because of this property, it was thought that there may be very distinct blockers. Hence, $H_2$ blockers, such as cimetidine, ranitidine, and famotidine, came into existence as excellent antiulcer drugs.

Toxicity studies consist of acute, subacute, and chronic phases, which come under preclinical evaluation in species to asses the extent of toxicity by using various parameters (e.g., hematology, biochemical, histopathologic, body weight, food intake, water intake, general behavior). General requirements of a toxicity study are given in Appendixes I and II.

The pharmacologic screening is an important part of any NCE. The NCE can be rejected at various stages, as listed in Appendix III. The pharmacologic screening is essentially a scanning procedure to distinguish useful or nonuseful drugs as rapidly, comprehensively, and inexpensively as possible. These will definitely help to make a decision for further screening. Screening can be of pure chemical or of plant extract. Screening with plants is a mammoth task. A tiny portion of the plant kingdom has been investigated thoroughly, but 98% is still awaiting exploration and commercial exploitation. Same plants have emerged as successful drugs:

- Morphine (poppy seeds)
- Digitalis (*Digitalis purpurea L.*)
- Quinine (Cinchona bark)
- Penicillin (Penicillin mold)
- Reserpine (*Rawolfia serpentina*)
- Progesterene (Dioscorea — in Mexico)
- Cortisone (Dioscorea — in Mexico)
- Artemisin (Chinese herbal medicine)
- Ergometrine (*Claviceps purpura*)
- Vincristine and vinblastine (*Vinca rosea*)
- Taxol (*Taxes buccata*)

The pharmacologic screening must be performed by a battery of tests, which will prove its efficacy by various methods (e.g., antiinflammatory). These tests are necessary to ascertain the activity of rat paw edema, adjuvant arthritis, granuloma pouch, ascitis, and others.

Lack of suitable animal models often hinders assessment of drug action. Previously, there were no animal models for HIV infection; now, severe combined immunodeficient (SCID) mice are available. In these mice, human lymphoid or hemolymphoid tissue is injected to generate the disease. Apart from this example, efforts have been made to develop a transgenic mouse that can mimic various diseases (e.g., Alzheimer's disease, Parkinson's disease, hepatitis C virus). Overall, disease models are important as vital tools for molecular medicine. These can be used to study the safety and efficacy of new drugs during preclinical screening. The fact is that many novel drugs have failed in human clinical trials despite encouraging results in animals. Incorporation of animal models can be a significant and unavoidable contributor in the development of an NCE. Technologies, such as transgenic and knockout, have created a mouse model that will serve as an advantage to finding an NCE for deadly diseases.

The present status of drug screening has made a radical improvement in disease testing. The screening process has been accelerated by new technologies. The testing pathway throughout the globe remains the same. There have been many discussions on *in vitro* and *in vivo* screening methods, but overall the efficacy of drugs can be proved only in *in vivo* experimentation.

Diverse and focused approaches are likely to continue to provide complementary benefits. The portfolio of techniques will provide a more effective and comprehensive approach to the identification of an NCE for the number of drug targets being revealed through the Human Genome Mapping Project.

Natural products are the most consistently successful source of drug leads. Despite this, their use in drug discovery has fallen out of favor. Natural products continue to provide greater structural diversity than standard combinatorial chemistry, and they offer major opportunities for finding low molecular weight lead structures that are active against a range of assay targets. Because less than 10% of the world's biodiversity has been tested for biological activity, many more natural lead compounds are awaiting discovery. The challenge is how to access this natural chemical diversity.

Approximately 60% of the world's population relies almost entirely on plants for medication. Natural products have long been recognized as an important source of therapeutically effective medicines. Of the 520 new drugs approved between 1983 and 1994, 39% were natural products or derived from natural products. Sixty percent to 80% of antibacterials and anticancer drugs were derived from natural products.

In addition to this historical success in drug discovery, natural products are likely to continue as sources of new commercially viable drug leads. The chemical novelty associated with natural products is higher than that of any other source: 40% of the chemical scaffolds in a published database of natural products (*Dictionary of Natural Chemistry*). This is particularly important when searching for lead molecules against newly discovered targets. Despite commonly held assumptions, natural products can be a more economical source of chemical diversity than the synthesis of equivalent diverse chemicals.

### New Thinking in Research

We have read much about positive results, but do come across negative findings that we may hesitate to report. Negative findings are important to help the researcher to determine what *not* to do. This is a common problem for researchers who do not receive adequate funding in the initial stages of research.

If there is no money available for fundamental research, begin working on industrial research. If you proceed logically, you will soon be doing fundamental research, and you are sure to succeed. Down syndrome has emerged as one of the most exciting and fruitful areas of research.

The needs of drugs for the millennium are

- Virology
- Tumor diseases
- Autoimmune diseases
- Organ transplant
- Vaccines
- Genetic diseases
- Arteriosclerosis
- Rheumatic diseases
- Skin diseases

The future of drug research is at stake, as stated in *Molecular Medicine Today*, June 1999, "Disease models will not only be of educational value but will also stimulate the use and development of models that are truly relevant to human disease, which will eventually catalyze the development of safe and efficacious therapeutics for human use."

The process by which NCEs are discovered and developed is undergoing a dramatic change. The traditional process of discovery depended only on an empirical approach to screening compounds for a novel pharmacologic activity.

Because of the ever-increasing demand for a new drug and high-quality treatment with fewer side effects, pharmaceutical companies are facing increased pressures to develop drugs at faster rates, in greater numbers, and more cost effectively.

## REFERENCES

1. G. Zbinden, Acute toxicity testing, public responsibility and scientific challenges. *Cell Biol. Toxicol.* **2**(3):325–335, 1986.
2. R.E. Chien, *Issues in Pharmaceutical Economics*, Lexington Books, New York, 1979.
3. J.A. DiMasi, Success rates for new drugs entering clinical testing in the United States. *Clin. Pharmacol. Ther.* **58**:1, 1995.
4. J.S. Cohen, and M.E. Hogan, The new genetic medicines. *Sci. Am. Annual Index*, 50–55, Dec. 1994.
5. Bertram G. Katzung, and Barry A. Berkowitz, Basic and clinical evaluation of new drugs, in *Basic and Clinical Pharmacology*, 8th ed., Bertram G. Katzung, ed., McGraw-Hill, New York, 67–69, 2001.
6. R.L. Dixon, Extrapolation of laboratory toxicity data to man: Factors influencing the dose-toxic response relationship, Symposium, Edited by R.L. Dixon, *Fed. Proc.* **39**: 53, 1980.
7. W.R. Chappell, and J. Mordenti, Extrapolation of toxicological and pharmacological data from animals to humans. *Adv. Drug Res.* **20**:1, 1991.
8. F.R. Jelovsek, D.R. Mattison, and J.J. Chen, Prediction of risk for human developmental toxicity: How important are animal studies? *Obstet. Gynecol.* **74**:624, 1989.
9. A. Dorato, and M.J. Vodicnik, *Principles and Methods in Toxicology*, 3rd ed., A. Wallace, ed., Raven Press, New York, 1994.

CHAPTER 13

# Drugs and Human Diseases

## INTRODUCTION

Drugs are essential to maintaining the health and longevity of animal and human life. Drugs are required to combat many diseases and infections beginning from birth (even before birth). All drugs, depending on the dosage, route of exposure, age of the individual, and gender, behave differently. It is therefore important to know the toxicologic profile of different drug categories, which are often required for the improvement and maintenance of health status. In fact, drugs structurally related to different compounds are known to cause a sort of dependence and liability in humans. Further, drugs used to medicate children and infants require greater attention, and adequate information is a must to avoid adverse effects and maintain safety. The following pages briefly discuss different categories of essential and nonessential drugs often used by humans.

## ANALGESICS AND ANTIINFLAMMATORY AGENTS

Rheumatoid arthritis is a disease that involves inflammation in joints and causes severe pain and immobility. The worldwide scenario indicates that the elderly population suffers the most from this disease. The drugs of choice for treatment of this condition are nonsteroidal antiinflammatory drugs (NSAIDs), which include acetylsalicylic acid, diclofenac sodium, piroxicam, nimesulide, celecoxib, and refecoxib. Therapeutic doses, adverse effects, and precautions are given in the following pages.

NSAIDs are groups of unrelated organic acids such as mefenamic acid, which have analgesic, antiinflammatory, and antipyretic activities. NSAIDs are used to relieve mild to moderate pain, raised temperature, and acute and chronic inflammatory disorders such as osteoarthritis, rheumatoid arthritis, juvenile chronic arthritis, and ankylosing spondylitis. NSAIDs are applied locally for the relief of muscular pain and are useful in ophthalmic preparations for inflammatory eye disorders.[1–4] NSAIDs have been used to relieve pain, particularly in cancer patients, forming a cornerstone in pain management. NSAIDs also are used

*Industrial Guide to Chemical and Drug Safety*, By T.S.S. Dikshith and Prakash V. Diwan
ISBN 0-471-23698-5 

for the prevention of Alzheimer's disease, dysmenorrhea, hemorrhage, ovulation, preeclampsia, osteoarthritis, and acute rheumatic fever.[5,6]

The major problem of NSAIDs is gastrointestinal (GI) irritation, leading to ulceration. Recent drugs such as celecoxib (a cyclooxygenase-II inhibitor) have a mild effect on the GI tract and have proven to be the drugs of choice for the treatment of rheumatoid arthritis.

## Aspirin

Aspirin is chemically named acetylsalicylic acid (ASA) and is the most commonly used drug for pain relief and for rheumatoid arthritis. Aspirin hypersensitivity is usually caused in middle age, in females, or with diagnoses of asthma.[7,8] Aspirin is contraindicated in children suffering from chicken pox, should be avoided in diabetes, and should be withdrawn 1 week before surgery. Administration of aspirin during pregnancy leads to low birth weight of a child; at higher doses, it may cause tinnitus.

Mild antiinflammatory doses will produce salicylate intoxication, also known as salicylism. The symptoms of this syndrome include dizziness, tinnitus, vertigo, reversible impairment of hearing and vision, excitement and mental confusion, hyperventilation, and electrolyte imbalance. Aspirin poisoning is an acute emergency and may result in death. In case of poisoning, the drug must be eliminated from the body as early as possible.

Aspirin may cause gastric irritation leading to gastric ulcer; hence, it is advisable to take aspirin with food. High doses may precipitate hemolytic anemia in patients with glucose-6-phosphate dehydrogenase deficiency. The risk of Reye's syndrome is seen in children younger than 12 years of age after aspirin administration. The drug is contraindicated in mothers who are breast-feeding. Prolonged use in the elderly may lead to GI bleeding and gastric ulcers.

Aspirin and other salicylates are used in the treatment of mild to moderate pain. Aspirin is commonly used to treat headache (including migraine and tension type) and to decrease elevated body temperature. Salicylic acid, as a 10% to 20% solution in alcohol or propylene glycol, is used for the treatment of corns.

## Paracetamol

Paracetamol (acetaminophen) has an analgesic, antipyretic, and weak antiinflammatory action. It is safely used in patients with asthma and gastric ulcers, and for relieving pain in children where salicylates are contraindicated because of Reye's syndrome. Low back pain (lumbago) is a common complaint in the industrialized world; only few patients suffer from a known organic disease. To avoid chronic conditions, immediate treatment is necessary and begins with paracetamol and later NSAIDs.[9] Paracetamol is used to create experimental liver damage in the study of hepatoprotective agents.[11,12]

Paracetamol causes skin rashes, blood dyscrasia, and liver damage and nephrotoxicity in overdose. Large doses of paracetamol cause nephritis and pancreatitis.[10,11] Paracetamol causes hypersensitivity reactions such as urticaria, dyspnea, and hypotension.[12,13]

Overdoses of paracetamol can be very dangerous, as the drug has a narrow therapeutic index and may cause hepatic and renal necrosis. Nausea, vomiting, lethargy, and sweating are the early overdose symptoms. Paracetamol must be given with caution in alcoholics and patients with liver and kidney damage.

### Diclofenac

Diclofenac is a phenylacetic acid derivative used mainly as sodium salt for the treatment of various pain and inflammation. Intramuscular injection occasionally causes tissue damage at the injection site. Suppositories may cause local irritation; transient burning and stinging are reported when used for the eye and large doses can cause aplastic anemia.[14]

In the patient with contact lenses, it is advised not to use diclofenac preparations for ophthalmic treatment. Diclofenac also is contraindicated for intravenous administration in patients with renal impairment, hypovolemia, dehydration, asthma, or cerebrovascular bleeding.

## SEDATIVES, HYPNOTICS, AND ANTIPSYCHOTICS

Many drugs have a capacity to depress central nervous system (CNS) function. A sedative is a drug that calms the patient and reduces activity and excitement. A hypnotic drug reduces drowsiness and produces sleep. For some drugs, these actions are dose-dependent, particularly drugs such as benzodiazepines, and are used to produce sedation during diagnostic and operative procedures. Certain barbiturates are used in high doses to induce surgical anesthesia.

Diazepam is a long-acting benzodiazepine with anticonvulsant, anxiolytic, sedative, and muscle-relaxant properties. Drugs that belong to the diazepam group are alprazolam, chlordiazepoxide, diazepam, flunitrazepam, flurazepam, lorazepam, nitrazepam, oxazepam, temazepam, and triazolam. This group is used in the treatment of severe anxiety as a hypnotic. Diazepam is administered orally, parenterally, and rectally, and is known to cause drug dependence. Hence, it is required to be withdrawn gradually. It also is used in preanesthetic medication during surgery. Diazepam is used with a variety of seizures. Alcohol withdrawal syndrome can be controlled by diazepam. Common adverse effects include drowsiness, sedation, and ataxia. Some less frequent adverse effects are headache, confusion, mental depression, slurred speech, tremors, and visual disturbances. Thrombophlebitis may occur with some intravenous formulations of diazepam. Overdose can produce CNS depression and coma, but death is very rare. Congenital malformation in infants may be observed in the first trimester of pregnancy of mothers taking diazepam. In case of CNS depression and coma, diazepam should be avoided. Caution should be exercised while administering the drug to elderly patients or patients with kidney and liver damage or muscle weakness.

Patients who are receiving diazepam therapy should not operate machinery or drive motor vehicles.

Diazepam may precipitate or initiate suicidal tendencies and should not be used alone when treating patients with depression. Many manufacturers of diazepam and other benzodiazepines advise against use of the drug in patients with glaucoma, but this fact is not established. Treatment of overdose may be performed by gastric lavage. The specific benzodiazepine antagonist, flumazenil, may be used in the diagnosis of unclear cases of overdosage, but expert advice is essential, since severe side effects are associated with its use. Continuous intravenous infusions through plastics may cause problems, as 50% of solution may be absorbed into the wall of the polyvinyl chloride (PVC) infusion bag, causing significant alteration in the drug concentration and resulting in a subtherapeutic dose. This may be prevented by avoiding cellulose propionate and minimizing PVC use in infusion bags or biomedical devices. Preferably, polypropylene, glass, or polyethylene should be used, and care should exercised by drug industry operators in this regard. Prolonged use of high-dose intravenous infusions of diazepam containing benzyl alcohol can result in benzyl alcohol poisoning.[15–26]

## DRUGS USED IN GOUT

Gout is a metabolic disease in which there is a overproduction of purines. It is characterized by intermittent attacks of acute arthritis produced by the deposition of sodium urate crystals in the synovial tissue of joints. Drugs used for treating gout are allopurinol, probenecid, colchicine, and NSAIDs.

### Colchicine

Colchicine is an alkaloid obtained from various *Colchicum* species. It is used for the relief of acute gout and as a prophylaxis of acute attacks. Colchicine is used in several other conditions including amyloidosis, Behçet's syndrome, pyoderma, and gangrenosis.

Most common adverse effects include nausea, vomiting, diarrhea, abdominal pain, bone marrow depression with agranulocytosis, thrombocytopenia, and aplastic anemia. Cumulative toxicity is possible in elderly patients, hence it should be used cautiously. Care also should be exercised in patients with cardiac, hepatic, and renal dysfunctions. Colchicine causes teratogenicity in animals, and there are evidences of the risk of fetal chromosomal damage in humans. Colchicine should not be administered by the parenteral route as it causes severe local irritation.

### Allopurinol

Allopurinol is used in the treatment of hyperuricemia, which is associated with chronic gout and in cancer chemotherapy. Allopurinol has been used in renal calculi caused by the deposition of calcium oxalate and of 2,8-dihydroxy-adenine. Allopurinol treatment does cause hypersensitivity reaction, which may be fatal.

The special risk is observed in patients with hepatic or renal impairment. It is not advised to use allopurinol in acute attacks of gout, but it is useful in chronic gout. Excretion of allopurinol and its active metabolite oxypurinol is primarily via the kidneys and therefore the dosage should be reduced if renal function is impaired. The adverse effects have been reported in patients receiving allopurinol with thiazide diuretics, particularly in patients with impaired renal function. The metabolism of azathioprine and mercaptopurine is inhibited by allopurinol and their doses should be reduced to one-quarter to one-third of the usual dose when either of them is given with allopurinol to avoid potentially life-threatening toxicity.[27–29]

## VITAMINS

These are organic compounds belonging to nonenergy-producing agents that are essential for normal human metabolism and required in small quantities. Vitamins are used as drugs in the case of deficiency and diseases that occur because of inadequate intake, malabsorption, increased excretion, genetic abnormalities, and drug-vitamin interactions. Chemical forms and daily recommended doses of vitamins are given in Table 13-1. Vitamins are often used as tools in self-medication by a large population. Vitamins are subdivided into two categories, fat-soluble (e.g., vitamins A, D, E, K) and water-soluble (B complex, vitamin C).

### Vitamin A

Vitamin A is found in nature and is available in several forms. Retinol (vitamin A) is an unsaturated alcohol containing an ionone ring and can be obtained from fish liver oil (Table 13-2), egg yolk, milk, and butter. Vitamin $A_2$ (dehydroretinol) is present in freshwater fishes. $\beta$-carotene, a carotenoid, is the most important precursor of this vitamin.

Vitamin A is stored in the liver; symptoms of deficiency are observed only after a long period of deprivation. Manifestations are night blindness, xerosis, softening of cornea, corneal opacities leading to total blindness, sterility due to faulty spermatogenesis, abortions, and fetal malformation. Vitamin A is useful in skin diseases like psoriasis, ichthyosis, and acne, and as a prophylaxis in pregnancy, infancy, and during lactation. Regular use of liquid paraffin for smooth defecation causes vitamin A deficiency. Oral neomycin interferes with vitamin A absorption.

Vitamin A (in excessive doses) produces toxicity such as nausea, vomiting, erythema, dermatitis, hair loss, bone and joint pains, loss of appetite, bleeding, and chronic liver diseases. Excess vitamin A is teratogenic in animals and humans; daily intake should not exceed 20,000 IU.

### Vitamin D

Vitamin D is the collective name for vitamins $D_1$, $D_2$, and $D_3$. It is used as prophylaxis for nutritional vitamin D deficiency. It also is used to treat rickets,

**TABLE 13-1 Chemical Forms and Daily Allowance of Vitamins**

| Vitamin | Chemical Forms | Daily Allowance (Adult Males) |
|---|---|---|
| | Retinol (A) | *Fat-Soluble Vitamins* |
| A | Dehydroretinol ($A_2$) | 1000 μg |
| | $\beta$-Carotene (pro-vit) | (4000 IU) |
| | Calciferol ($D_2$) | 5 μg |
| D | Cholecalciferol ($D_3$) | (200 IU) |
| | Calcitriol | 1 μg |
| E | $\alpha$-Tocopherol | 10 mg |
| | Phytonadione ($K_1$) | |
| | (Phylloquinone) | |
| K | Menaquinones ($K_2$) | 50–100 μg |
| | Menadione ($K_3$) | |
| | Acetomenaphthone | |
| | | *Water Soluble Vitamins* |
| $B_1$ | Thiamine | 1.5 mg |
| $B_2$ | Riboflavin | 1.7 mg |
| | Nicotinic acid | 20 mg |
| $B_3$ | Nicotinamide | |
| | Tryptophan (pro-vit) | |
| | Pyridoxine | 2 mg |
| $B_6$ | Pyridoxal | |
| | Pyridoxamine | |
| | Pantothenic acid | 4–7 mg |
| | Biotin | 0.1–0.2 mg |
| | Folic acid | 0.2 mg |
| | Folinic acid | |
| $B_{12}$ | Cyanocobalamin | 2 μg |
| C | Ascorbic acid | 60 mg |

hypoparathyroidism, and Fanconi syndrome. Hypervitaminosis may occur due to chronic administration of large doses. High doses can cause hypercalcemia, weakness, fatigue, sluggishness, polyurea, albuminaria, renal stones, hypertension, growth retardation in children, and may produce coma. Chronic use of liquid paraffin can reduce vitamin D absorption.

## Vitamin E

Vitamin E ($\alpha$-tocopherol) is most abundantly available in wheat germ oil and in small quantities in cereals, nuts, spinach, and egg yolk. Deficiency of vitamin E normally does not occur. Large doses of vitamin E are used to reduce the toxicity of vitamin A. Prolonged use causes delay in wound healing but significant

**TABLE 13-2 Vitamin Content in Fish Liver Oil**

| Fish | Vitamin A IU/g | Vitamin D IU/g | Daily Dose |
|---|---|---|---|
| Shark liver oil | 6,000 | — | 0.2–1 g |
| Cod liver oil | 600 | 85 | 4–12 g |
| Halibut liver oil | 30,000 | 300 | 0.2–0.5 g |

toxicity due to vitamin E has not been observed or reported among humans even with large doses for longer periods.

### Vitamin K

Vitamin K is an essential cofactor for the synthesis of prothrombin and other blood-clotting factors. Vitamin K deficiency occurs due to liver disease, long-term antimicrobial therapy, and malabsorption. Vitamin K deficiency can lead to hemorrhages in newborns and development of hypoprothrombobinemia. Rapid intravenous injection of emulsified vitamin K produces flushing, breathlessness, hypotension, and may lead to death.

### Vitamin $B_1$ (Thiamine)

Vitamin $B_1$ is present in pulses, nuts, green vegetables, yeast, eggs, and meat. Vitamin $B_1$ is used for prophylactic purposes in infants, during pregnancy, for chronic diarrhea, and in alcoholics. Thiamine is nontoxic, but parenteral injection may cause some reactions. Vitamin $B_1$ deficiency causes beriberi in dry and wet forms. Alcoholism is a common cause of thiamine deficiency and is called *alcoholic neuritis*; the dose requirement is 40 mg/day. Cardiovascular disease of nutritional origin is observed in chronic alcoholics and pregnant women. During pregnancy, the thiamine requirement is slightly more.

### Vitamin $B_2$ (Riboflavin)

Vitamin $B_2$ is found in milk, eggs, green leafy vegetables, and grains. Vitamin $B_2$ deficiency occurs along with other vitamin deficiencies. Characteristic lesions are seen in sore and raw tongue, lips, throat, and ulcers in the mouth.

### Vitamin $B_3$ (Niacin)

Vitamin $B_3$ is a pyridine compound found in fish, meat, liver, cereal husks, nuts, and pulses. Niacin deficiency produces "pellagra," which manifests as dermatitis, diarrhea with enteritis, nausea and vomiting, dementia with headache, insomnia, loss of memory, and disturbances of motor and sensory neurons. Anemia and hypoproteinemia are common in pellagra; chronic alcoholics have more tendency to suffer from pellagra. Corn eaters suffer from pellagra because corn flour is low in tryptophan, and it is postulated that it contains a niacin antagonist. Vitamin $B_3$ is used as a prophylactic and in the treatment of pellagra is given 200 to 500 mg/day. Niacin has many toxic effects such as dyspepsia, vomiting, diarrhea, hyperpigmentation, liver dysfunction and jaundice, hyperglycemia, and hyperuricemia; and it may worsen peptic ulcer with prolonged use.

### **Vitamin $B_6$** (Pyridoxine)

Vitamin $B_6$ is present in dietary sources such as soybeans, vegetables, whole grains, livers, meats, and eggs. Vitamin $B_6$ deficiency causes growth retardation, mental confusion, convulsions, neuritis, and anemia. Vitamin $B_6$ deficiency occurs along with other B-complex vitamin deficiencies. Neurologic disturbances caused by isoniazid, hydralazine, and cycloserine can be treated with vitamin $B_6$ (10 to 50 mg/day). Women taking contraceptives who suffer from loss of memory can be treated with a dose of 50 mg/day. Anemia due to defective heme synthesis and homocystinuria are rare genetic disorders and are treated with large doses of vitamin $B_6$ (50 to 200 mg/day). It also is useful in infants and children who suffer from convulsions. Vitamin $B_6$ is prophylactically used in a dose of 2 to 5 mg/day.

### **Vitamin C** (Ascorbic Acid)

The richest sources of vitamin C are citrus fruits (e.g., lemon, oranges), tomatoes, potatoes, green chilies, and human milk. Severe deficiency causes scurvy and is prevalent in malnourished infants, children, adults, alcoholics, and drug addicts. Symptoms such as bleeding gums, deformed teeth, brittle bones, impaired wound healing, anemia, and growth retardation are observed.

It is used in the treatment of scurvy, postoperative cases, and healing bedsores and chronic leg ulcers. Vitamin C increases the absorption of iron during anemia and is frequently combined with ferrous salts. It is used in urinary tract infections to acidify urine. Large doses of vitamin C have been tried to cure everything from the common cold to cancer, with not much success. The usefulness of vitamin C in asthma, cancer, atherosclerosis, psychologic symptoms, and fertility is doubtful. Ascorbic acid is well tolerated in large doses and may cause rebound scurvy on withdrawal. There is a possibility of forming urinary stones.

## DIABETES

Diabetes mellitus is a disease related to carbohydrate metabolism in which insulin is absent, low in quantity, or a combination of both. It is characterized by hyperglycemia. Progress of the disease causes tissue or vascular damage, leading to diabetic complications such as retinopathy, neuropathy, cardiovascular disease, and foot ulcerations.

Diabetes mellitus is classified in two major types: as type-1 insulin-dependent diabetes mellitus (IDDM) and type-2 non–insulin-dependent diabetes mellitus

(NIDDM). The disease can be diagnosed by measuring blood glucose concentration during fasting and 1.5 hours after eating. Confirmation of diabetes can be done by an oral glucose tolerance test (GTT). Various drugs such as oral hypoglycemics (such as sulfonylureas and biguanides), insulin, acarbose, troglitazone, and parenteral exogenous insulin are used for the management of diabetes mellitus.

### Insulin

Insulin is an endogenous hormone produced by $\beta$-cells of islets of Langerhans of the pancreas, which consist of two chains of amino acids. It is required to be administered by a parenteral routes as it is destroyed when given orally. Insulin is used for the control of IDDM and in the emergency management of diabetic ketoacidosis.[30] Insulin promotes the intracellular uptake of potassium and is used in hyperkalemia. Baker et al.[31] have used insulin and glucagon in the treatment of liver disorders. Recent evidence indicates that the effects of insulin with glucose and potassium in ischemic heart disease have proved beneficial.[32] It also is used in acute myocardial infarction.[32]

Increased doses of insulin are required during infection, emotional stress, surgical or accidental trauma, and puberty. It should be used cautiously during pregnancy. Insulin doses must be adjusted properly or high doses cause a sudden fall in blood sugar and lead to hypoglycemic complications. Factors such as a hot water bath (or sauna) have been reported to accelerate absorption of subcutaneous insulin; hence there is a risk of hypoglycemia.[33] Smoking has been reported to decrease insulin absorption.[34]

### Glibenclamide

Glibenclamide belongs to the class of sulfonylureas. Borderline diabetic patients are advised to control blood sugar levels by diet restrictions. If this fails, drugs such as glibenclamide are administered. Some patients do not respond to sulfonylureas; these patients are given another class of drug (e.g., metformin or acarbose) before changing to insulin therapy.

Sulfonylureas should not be used in IDDM. Use in NIDDM is contraindicated in patients with ketoacidosis, stress, severe infection, and trauma. This class of drugs should be avoided during breast-feeding, as these drugs are excreted in milk. A drug such as chlorpropamide or glibenclamide can produce hypoglycemia, and they should be avoided in renal and hepatic dysfunction. In the United Kingdom, patients with diabetes mellitus are required to declare their condition to the vehicle-licensing center, who then assess their fitness to drive.[35]

Adverse effects of glibenclamide include GI disturbances such as nausea, vomiting, heartburn, anorexia, diarrhea, and metallic taste. Hypersensitive reactions include cholestatic jaundice, leukopenia, thrombocytopenia, aplastic anemia, and photosensitivity.

### Acarbose

Acarbose is a glucopyranose derivative that acts by inhibiting intestinal $\alpha$-glucosidase. This delays carbohydrate absorption and reduces the postprandial (1.5 hours after food) blood glucose levels and is used in combination with other sulfonylureas. Acarbose may cause GI disturbances, flatulence, abdominal distortion, diarrhea, and pain. Acarbose should be avoided during pregnancy, as it affects the fetus. Acarbose is contraindicated in inflammatory bowel disease and hepatic dysfunction.

### Troglitazone

Troglitazone is useful in insulin resistance. This drug was withdrawn because of hepatotoxicity, but many derivatives have been prepared and are at various stages of clinical trials. Perhaps this will be a future drug for the treatment of diabetes. Patients who take antidiabetic drugs should be frequently monitored for blood glucose levels. Shorter intervals of food intake in small quantities are beneficial for known diabetic patients to avoid sudden hypoglycemia.

## DRUGS FOR MALARIA

More than 90 million cases of malaria occur each year, and it is one of the most important transmittable parasitic diseases. Malaria is caused by species of *Plasmodium* through mosquitoes as vectors. *P. falciparum* is the most serious form of malaria, which can be rapidly fatal in nonimmune individuals if not treated properly. The other three species, *P. vivax*, *P. malaria*, and *P. ovale*, cause benign malaria. Clinical symptoms of malaria are variable and nonspecific, but can be generalized as fever, fatigue, malaise, headache, myalgia, and sweating. Anemia is a common complication due to hemolysis. In *P. falciparum* malaria, serious complications such as renal failure, pulmonary edema, and cerebral dysfunction can occur. Antimalarial drug treatment should not be withheld in the absence of a positive blood smear. Drugs used as antimalarials are quinine, quinoline, and artemisin derivatives.[36]

### Chloroquine

Chloroquine is a 4-aminoquinoline used in the treatment and prophylaxis of malaria and hepatic amebiasis, as well as rheumatoid arthritis. Adverse effects are generally less common and less severe. Frequent effects include headache, GI disturbances, and diarrhea. Large doses may cause blurred vision and difficulty focusing. A common adverse effect on the eye is retinopathy. Parenteral therapy with chloroquine can be hazardous, and rapid intravenous injections may result in cardiovascular toxicity. Acute overdose is extremely dangerous; death may occur within a few hours. Chloroquine should be used cautiously in patients with liver and kidney impairment. Chloroquine may aggravate the condition of myasthenia gravis, and care is needed. In the case of an existing eye problem, the drug should not be administered.

### Quinine

Quinine is a cinchona alkaloid that acts rapidly against all four species of *Plasmodium*. It is used to treat protozoal infections and leg cramps, and as a bitter and flavoring agent. However, the drug is not used prophylactically for malaria. Quinines are contraindicated in patients with a history of hypersensitivity to quinine or quinidine. They should not be used in the presence of hemolysis and should be used with caution in patients with atrial fibrillation, cardiac conduction defects, or heart block. Quinine administration in myasthenia gravis may aggravate the disease, hence it should be avoided. Quinine can be used in pregnancy.[37] Intravenous infusion of quinine should be slow, and the patient should be monitored for cardiotoxicity.[38] Cinchonism, which is characterized by tinnitus, GI disturbances, and impaired vision may occur with therapeutic doses of quinine.[39]

### Artemisinin

Artemisinin is a sesquiterpene lactone isolated from *Artemisia annua*, a traditional herb used in China. It is active against *P. vivax* and both chloroquine-sensitive and chloroquine-resistant strains of *P. falciparum*.[40] Adverse effects of artemisinin are nausea, vomiting, itching, drug fever, abnormal bleeding, and leukopenia. At high dosage, it causes severe neurotoxicity in animals. It should be used with caution in pregnant women.

## BRONCHODILATORS AND ANTIASTHMATIC DRUGS

Asthma is a chronic inflammatory disease in which the patient suffers episodes of reversible airway obstruction due to bronchial hyperresponsiveness. The adult population (5%) and children (10%) suffer from asthma. Fatalities can occur, and mortality has significantly increased.[41,42] The etiology of asthma is poorly understood. Asthma is characterized by dyspnea (breathlessness), wheezing, cough, chest tightness, and nocturnal awakening.[43] Life-threatening features are exhaustion, cyanosis, bradycardia, hypotension, confusion, and coma.[44] Textiles and dry nitrodes are more prone to dust diseases, which can aggravate asthma.

### Aminophylline

Aminophylline is used in the treatment of asthma (as a bronchodilator) and in chronic obstructive pulmonary disease. In acute bronchospasm, aminophylline should not be given by the intravenous route; the oral route is preferred. Parenteral administration can cause pain and is not recommended. Aminophylline is used as a cosmetic to remove fat from the thigh.[45]

### Salbutamol

Salbutamol is a $\beta_2$-receptor antagonist and a bronchodilator. Salbutamol and other $\beta_2$-receptor antagonists should not be administered in the presence of hyperthyroidism, myocardial insufficiency, arrythmias, hypertension, and diabetes mellitus. Salbutamol is considered a drug of abuse.[46] Salbutamol may cause skeletal muscle tremors, palpitation, muscle cramps, and hypersensitivity reactions. An overdose of salbutamol can cause tachycardia, CNS stimulation, tremor, hypokalemia, and hyperglycemia.[47]

## ADRENOCORTICAL STEROIDS

The adrenal cortex synthesizes corticosteroids (glucocorticoids and mineralocorticoids), which differ in activities. In humans, cortisol is the main glucocorticoid, and aldosterone is a main mineralocorticoid. Steroid therapy causes severe potential side effects, hence a careful consideration is always exercised before starting therapy. These are used in variety of disorders such as rheumatic disorder, renal disease, allergic manifestation, bronchial asthma, skin diseases, infectious diseases, malignancy, and hepatic diseases.

Corticosteroids have a range of activity. They have potent antiinflammatory and immunosuppressive activity. Many synthetic drugs are available as corticosteroids. In appropriate doses, these are used as replacement therapy in adrenal insufficiency. The topical application of corticosteroids is safer when compared with systemic use. Corticosteroids should be used in smaller doses for the shortest duration of time. A high dose may be used for life-threatening syndromes or diseases. A tapering pattern of withdrawal should be followed to avoid complications of sudden withdrawal. Systemic therapy is indicated in a variety of conditions. These are administered by intraarticular injections with aseptic conditions for rheumatoid arthritis and osteoarthritis. In skin diseases, such as eczema, contact dermatitis, and psoriasis, corticosteroids are used topically. In some cases, steroids are combined with antimicrobial substances such as neomycin.

Corticosteroids should be used cautiously in the presence of congestive heart failure, myocardial infarction, hypertension, diabetes mellitus, epilepsy, glaucoma, hepatic disorders, osteoporosis, peptic ulceration, and renal impairment. Children are more susceptible to these adverse effects. To avoid cardiovascular collapse, steroids must be given slowly by intravenous injection. Large doses produce Cushing's syndrome (with moon face and sometimes hirsutism).

Glucocorticoids are more potent and widely used. Systemic glucocorticoids are advocated in the treatment of severe chronic asthma.[48] Aerosol formulations of glucocorticoids are very safe.[49] Glucocorticoids are known to suppress growth in children and cause peptic ulceration and Cushing's syndrome. Potential adverse reactions are associated with inhaled glucocorticoids. Adverse effects reported are bone resorption, skin thinning, and purpura.[50]

## ANTICANCER DRUGS

Cancer is a disease in which cells multiply and spread within the body in abnormal forms. It is one of the major causes of death in developed nations. At least 20% of the population in Europe and North America die of cancer. The terms *cancer*, *malignant neoplasm*, and *malignant tumor* are synonymous and are distinguished from benign tumors. There are three main approaches to treating cancer: 1) surgical excision; 2) irradiation; and 3) chemotherapy. The chemotherapeutic agents are used as adjuvants, along with surgical and irradiation procedures of the tumor. Other approaches for cancer treatment are immunotherapy, gene therapy, and the use of biological response modifiers (e.g., interferons).

The chemotherapeutic agent does not differentiate between cancerous and normal cells, and hence causes severe toxicity. Toxic effects are leukopenia, thrombocytopenia, ulceration, diarrhea, azoospermia, infertility, premature menopause, alopecia, and vomiting. On prolonged use, these agents may cause gonadal damage and teratogenicity.

Anticancer drugs cause bone marrow depression, which is a major side effect and warrants regular monitoring of hematological parameters. Care must be taken when administering anticancer drugs by intravenous injection, and contact should be avoided with skin and eyes. Inhalation of these compounds is dangerous.[51]

Drugs used in cancer chemotherapy are cytotoxic drugs, hormones, plant derivatives, radioactive isotopes, and miscellaneous agents (e.g., procarbazine, hydroxyurea, mitotane). The plant-based drugs vincristine, vinblastine, vinorelbine, etoposide, and campothecins. Radioactive isotopes, such as 131 iodine (131 I), are used in the treatment of thyroid tumors. Cytotoxic drugs (e.g., cisplatin, cyclophosphamide, 6-mercaptopurine, 5-fluorouracil, and methotrexate are used for the treatment of cancer.

## CARDIOVASCULAR DRUGS

Hypertension is a common cardiovascular disease in which blood pressure is above 140/90. The constant increased blood pressure damages the blood vessels of the kidney, heart, and brain, and leads to diseases relating to the heart (e.g., coronary disease, renal failure, stroke). Antihypertensive therapy has shown a lower incidence of these diseases by lowering blood pressure and keeping it under control. There are effective drugs available for treatment, such as methyldopa, clonidine, $\beta$-blockers, calcium channel blockers (e.g., verapamil, diltiazem), and angiotensin-converting enzyme (ACE) inhibitors (e.g., enalapril). Ischemic heart disease is common in Western countries; atheromatous obstruction in coronary arteries is the major cause. Vasodilators are useful in the treatment of angina pectoris. Commonly used treatment drugs are nitrates-nitrites and calcium channel blockers.

### Cardiac Glycosides

Cardiac glycosides are obtained from plants of the foxglove family. Their effectiveness for cardiac failure was first described by William Withering. Cardiac glycosides are still widely used for the treatment of congestive heart failure. The digitalis group of drugs is used in congestive heart failure and the control of fibrillation or flutters. It is the drug of choice for various cardiac arrhythmias.

### Glyceryltrinitrates

Glyceryltrinitrates are used for the treatment of angina pectoris, heart failure, and myocardial infarction. Other indications include hypotension and controlling hypertension during surgery. The drug is administered in emergency either intravenously or sublingually to prevent an acute anginal attack. For this purpose, aerosol spraying or buccal tablets are given to produce a rapid and effective treatment and provide relief of anginal pain.

Care should be exercised when treating patients with hypothyroidism, malnutrition, hypothermia, and impaired renal or hepatic dysfunction. Glyceryltrinitrates should not be given to patients with severe hypotension, hypovolemia, marked anemia, or heart failure. Adverse effects of glyceryltrinitrates include flushing of the face, dizziness, tachycardia, and throbbing headache. Overdose may induce vomiting, restlessness, blurred vision, severe hypertension, syncope, impairment of respiration, and bradycardia.

Topical application of glyceryltrinitrates has been reported to cause contact dermatitis and local irritation. Chronic poisoning may occur in workers in manufacturing units. Development of tolerance and nitrate dependence also can lead to severe withdrawal symptoms after regular handling if these workers are abruptly removed from chronic exposure.

### Angiotensin-Converting Enzyme Inhibitors

ACE inhibitors are used in the treatment of hypertension, heart failure, and diabetic neuropathy. These drugs should not be used in patients with renovascular disease. Anaphylactic reactions may occur in patients. It has been reported that ACE inhibitors affect the fetus when administered to pregnant animals.[52] ACE inhibitors can cause injury and even death to the developing fetus in the second and third trimesters of pregnancy.[53–55]

Adverse effects of ACE inhibitors include hypotension, dizziness, headache, fatigue, GI disturbances, bad taste in the mouth, persistent dry cough, skin rashes, renal impairment, hypokalemia, and blood disorders. ACE inhibitors also can cause chest pain, palpitation, tachycardia, abdominal pain, cholestatic jaundice, alopecia, mood and sleep disturbances, and impotence.

### Beta-Blockers

Beta ($\beta$)-blockers (e.g., propranolol, atenolol, oxyprenolol, pindolol) are used for treating hypertension, cardiac arrhythmias, angina pectoris, and myocardial infarction. These drugs have proven important in the management of alcohol withdrawal and hypothyroidism.[56] $\beta$-blockers also are used as prophylactics in migraine headaches and are the drugs of choice for ocular hypertension and glaucoma.[57] Adverse effects of $\beta$-blockers are heart failure, heart block, bronchospasm, fatigue, and coldness of the extremities. $\beta$-blockers should not be used in asthmatic patients. $\beta$-blockers may increase the symptoms of hyperthyroidism and hypoglycemia. The dose in renal and hepatic dysfunction should be reduced. Abrupt withdrawal sometimes results in angina, myocardial infarction, arrhythmias, and death. Administration of $\beta$-blockers during pregnancy and before delivery results in bradycardia and other adverse effects, such as hypoglycemia and hypertension, in neonates.

## DISEASES OF THE CENTRAL NERVOUS SYSTEM (CNS)

The understanding of drugs acting on the CNS is a complex phenomenon. CNS drugs are of special significance to clinical and therapeutic practice. Life-threatening neurodegenerative diseases (e.g., Alzheimer's, Parkinson's, Huntington's, amytrophic lateral sclerosis) have no effective therapy. Despite fast-track research to find the remedy for these diseases, treatment remains unsatisfactory. The pharmacologic agent available for these diseases is symptomatic and does not alter the disease status.

### Alzheimer's Disease

Alzheimer's disease, which is characterized by a loss of memory, occurs in elderly patients and was initially thought of as presenile dementia. The incidence of the disease drastically increases with age (from approximately 5% at 65 to 90% or more at 95 years of age). Studies have revealed a specific molecular and genetic mechanism in causing Alzheimer's disease.[58,59]

Despite recent advances in understanding the mechanism at molecular levels, suitable drugs are not available for treatment. Most promising drugs are in experimental stage, yet to be tested in humans. Recently, tacrine was introduced in the United States, and trials have shown modest improvement in 40% of patients. Tacrine produces side effects such as nausea, abdominal pain, and hepatotoxicity. A more effective drug, donepezil, has an advantage over tacrine and has been introduced in the United Kingdom. Use of NSAIDs has proven useful in Alzheimer's disease.[60]

### Parkinson's Disease

Parkinson's disease occurs mainly in the elderly. Parkinsonism is characterized by rigidity, bradykinesia, tremor, and postural defects. Parkinson patients experience movements with a characteristic gait; standing posture is difficult, and patients cannot stop quickly or change direction. This disease is progressive and leads to increased disability if the treatment is not appropriate. The drug of choice for the disease is dopamine, but it does not cross the blood-brain barrier, hence its precursor (L-dopa) is used. Best results are obtained within a few years of treatment.

Later, the disease does not respond to the drug and doses are required to be given in combination with carbidopa. Levodopa is effective in relieving bradykinesia and other disorderly voluntary movements. Parkinson's disease is not a hereditary disease. Drugs such as levodopa, carbidopa, benserazide, bromocriptine, pergolide, selegiline, and amantadine are used as therapeutic agents.[61]

Untoward effects of levodopa are dyskinesia and rapid fluctuations related to rigidity, which may suddenly worsen. Psychologic effects such as confusion, disorientation, insomnia, and nightmares are common in 20% of patients. Various transplantation approaches have been tried based on the injection of dissociated fetal cells directly into the substantia nigra. An alternative approach includes the use of genetically modified nonneuronal cells (e.g., fibroblasts), so that they will secrete missing mediators such as dopamine and growth factors.[62]

### Huntington's Disease

Huntington's disease is an inherited disorder characterized by gradual onset of motor incoordination. The symptoms relate to movement disorders with jerk of the extremities, trunk, face, and neck. The progressive disease causes loss of memory (particularly recognition of friends and family) and is fatal.[63] This disease is treated only symptomatically. Drugs such as fluoxetine are used to treat depression and irritability. Carbamazepine also is useful in treating associated depression.

## ANTIMICROBIAL DRUGS

Antimicrobial drugs have revolutionized the therapy of infectious diseases. These belong to the curative category of drugs. Chemotherapy means treatment of systemic infections with specific drugs that selectively suppress or kill the infecting microorganism without significantly affecting the host. Antibiotics are the substances produced by microorganisms, which suppress the growth of or kill other microorganisms at very low concentrations. These drugs cause gastric irritation and pain when given orally. Intramascular and intravenous injections may cause abscess and thrombophlebitis, respectively. Some antibiotics, namely, tetracycline, erythromycin, and chloramphenicol, are irritants.

Organ toxicity is produced by most antimicrobial agents if administered parenterally. Antibiotics such as penicillins, some cephalosporins, and erythromycin have a high therapeutic index, whereas aminoglycosides (kidney toxicity), tetracyclines (liver and kidney damage), chloramphenicols (bone marrow depression), polymyxin B (neuronal and renal toxicity), vancomycins (kidney damage and hearing loss), and amphoterecin B (neurologic toxicity, bone marrow depression, kidney damage) have low therapeutic indexes. Drugs such as penicillins, cephalosporins, and sulfonamides cause hypersensitivity reactions, ranging from skin rash to anaphylactic shock. All antimicrobial agents are capable of causing hypersensitive reactions and are unpredictable but not dose-related.

Drug resistance is a common feature of antibiotics, and there are two types (i.e., natural resistance and acquired resistance). Acquired resistance to antibiotics is a major clinical problem, as some microorganisms are notorious in rapid acquisition of resistance. Mutation and gene transfer are two main factors behind drug resistance. Mutation is stable, and inherited genetic changes occur spontaneously and randomly among microorganisms. Gene transfer can occur by conjugation, transduction, and transformation.

It is well documented that antimicrobial agents cause cross-resistance. Use of antibiotics causes alterations in normal microbial flora of the body. The suprainfection is a common problem associated with antibiotic therapy and is very difficult to treat. Prolonged uses of antibiotics alter the microflora of the intestine and cause vitamin deficiency. Neomycin causes morphologic abnormalities in the intestinal mucosa.

### Sulfonamides

Sulfonamides are antimicrobial agents that are effective against pyrogenic bacterial infections. These drugs are used extensively in developing countries because they are economical. Presently, their use is limited because of the development of resistance and the availability of more powerful antimicrobial agents. Sulfonamides are commonly used in urinary tract infections, meningitis, gum infections, conjunctivitis, and in malaria along with antimalarial drugs. Sulfonamides in high doses cause nausea, vomiting, and epigastric pain. Hypersensitivity reactions can occur in the form of drug fever, rashes, and urticaria.

### Penicillins

Penicillin was first used in 1941 as an antibiotic. Because of its low toxicity, it was called a miracle drug. It was used in infections such as staphylococcus, streptococcus, pneumococcus, meningococcus, gonorrhea, syphilis, diphtheria, and tetanus. Furthermore, penicillin also has prophylactic uses for some diseases.

Hypersensitivity reactions are major problems with penicillins. Frequent manifestations are skin rashes, itching, and urticaria, which may lead to anaphylaxis and eventual death. Parenteral administration causes severe pain. Toxicity to the brain may be manifested as confusion, muscular twitching, convulsions, and coma. Superinfection is rare with penicillin.

### Cephalosporins

Cephalosporins are semisynthetic antibiotics obtained from the fungus *Cephalosporium*. These bactericidal agents act in a similar way to that of penicillins. There are different types of cephalosporins available: first-generation (cefazolin, cefadroxil); second-generation (cefuroxime, cefaclor); third-generation (cefotaxime, cefoperazone); and fourth-generation (cefpirome).

Cephalosporins are extensively used as the antibiotics of choice and alternatives to penicillins. They are used for respiratory, urinary, and soft tissue infections caused by gram-negative organisms, in surgical prophylaxis, for meningitis caused by *Hemophilus influenzae*, and for gonorrhea and typhoid.

Although cephalosporins are more toxic than penicillin, they are well tolerated. Parenteral injection may cause pain when given intramuscularly and may cause thrombophlebitis when given intravenously. The oral cephalosporin administration causes diarrhea by altering the gut ecology. Hypersensitivity reactions are caused and are similar to those of penicillins. Cephaloridine causes nephrotoxicity, but presently available cephalosporins have less renal toxicity.

### Tetracyclines

Tetracyclines are classified as broad-spectrum antibiotics obtained from soil actinomycetes. All tetracyclines are slightly bitter solids and are weakly soluble in water. Although tetracyclines develop organism resistance, they are the drugs of first choice in venereal disease, atypical pneumonia, cholera, brucellosis, and plague. Tetracyclines also are used in urinary tract infections, amebiasis, acne, and as prophylaxis against meningitis.

Side effects of tetracyclines include nausea, vomiting, diarrhea, and epigastric pain. Tetracyclines are known to cause dose-related toxicity, such as liver and kidney damage. It is well known that these drugs cause phototoxicity when the patient is exposed to sunlight. Administration of tetracyclines during pregnancy or childhood causes temporary suppression of bone growth. These are the drugs that cause superinfection to various antibiotics.

Tetracyclines should not be used during pregnancy, lactation, or in children. They should be used cautiously in patients with renal and hepatic insufficiency. The drug should never be used after the expiration date, as it becomes toxic. Tetracyclines should not be administered along with milk, as it chelates calcium.

### Aminoglycosides

Aminoglycosides belong to a group of natural and semisynthetic antibiotics. Streptomycin was the first member. It was discovered by Waksman and colleagues and used for the treatment of tuberculosis.

Aminoglycosides are most active against gram-positive organisms, and their activity is enhanced when given along with penicillins. Streptomycin develops rapid resistance; hence, it has limited use. Neomycin antibiotics cannot be given systemically because of nephrotoxicity and ototoxicity. Paromomycin is used for the treatment of intestinal amebiasis and leishmaniasis. Kanamycin is less toxic and can be used for systemic infection and action against penicillin-resistant gonorrhea. Gentamicin and tobramycin are the drugs of choice in the treatment of life-threatening infections.

Toxic effects, which depends on dose and duration of treatment, mainly manifest as ototoxicity. Aminoglycosides also may cause nephrotoxicity and are reversible if they are withdrawn. They are known to cause neuromuscular blockade; hence, care is necessary when used along with neuromuscular-blocking agents. Other reactions include allergy and cross-reactivity infections, as well as resistance to other antibiotics. Most of these drugs develop resistance. Amikacin and netilmicin should be reserved for severe infections that are resistant to other aminoglycosides. Chronic administration of aminoglycosides causes mild renal impairment and is reversible.

## ANTIVIRAL DRUGS

Viruses are small infective agents consisting essentially of nucleic acid (either RNA or DNA enclosed in a protein coat). Some viruses contain additional lipoproteins, which may contain antigenic viral glycoproteins. Viruses are intracellular parasites with no metabolic machinery of their own. To replicate, they must attach to and enter the living host cell animal, plant, or bacteria and use its metabolic process.

### Human Immunodeficiency Virus

In 1997, it was estimated that 30 million adults were infected with the human immunodeficiency virus (HIV) worldwide, with increments of five people infected every minute. It is estimated that approximately 7% of the population of sub-Saharan Africa has been infected. The incubation of the disease is 7 to 8 years. Currently available drugs for acquired immunodeficiency syndrome (AIDS) and HIV are zidovudine, didanosine, lamivudine, and stavudine. The causative agent for AIDS is generally an HIV virus, which is transmitted by sexual contact, blood and blood products, the use of contaminated drug needles, and from mother to fetus.

### Acyclovir

Acyclovir is used in herpes simplex infection and should be administered as early as possible after the appearance of symptoms by the intravenous, oral, or topical route. Prolonged treatment reduces the recurrence but sudden withdrawal is dangerous, as the disease may recur. Acyclovir improves the healing of *herpes zoster* lesions and reduces pain.

Acyclovir should be administered with caution in the presence of renal dysfunction. Rapid or bolus injection should be avoided, and adequate hydration should be maintained. Acyclovir should be administered intravenously with caution in cases of neurologic abnormalities or serious hepatic and electrolytic abnormalities. When administered to nursing mothers, care should be exercised.

Acyclovir is a well-tolerated drug. Intravenous administration may cause local reaction at the injection site with inflammation and phlebitis. Renal impairment does occur, but it is reversible. Intravenous administration may occasionally increase serum bilirubin and liver enzymes and cause hematologic changes, skin rashes, fever, headache, dizziness, nausea, vomiting, and diarrhea. Neurologic effects such as confusion, hallucinations, agitation, tremors, convulsions, and coma have been reported.[64]

### Indinavir Sulfate

Indinavir sulfate is a protease inhibitor and is used in combinations for the treatment of viral infection. During the high risk of HIV infection, indinavir is combined with zidovudine and lamivudine.[65] Indinavir sulfate should be used with caution in patients with hepatic impairment and avoided in patients with severe liver damage. Caution is needed in diabetic patients and in patients with hemophilia. Adverse effects of indinavir sulfate include nausea, vomiting, diarrhea, fatigue, dizziness, headache, skin rashes, and allergic reactions (hematuria).

### Interferon

Interferons are proteins or glycoproteins that are produced either by animal cells or plant cells in response to stimuli or DNA recombinant technology. These drugs are active against malignant neoplasms and have immunomodulating effects. These are useful in chronic hepatitis B, hepatitis C, hairy cell leukemia, myeloid leukemia, follicular lymphoma, carcinoid tumor, multiple myeloma, renal cell carcinoma, multiple sclerosis, chronic granulomatous diseases, blood disorders, common cold, herpes simplex, inflammatory bowel disease, and leishmaniasis.

Interferons should be used with caution in patients with severe renal and hepatic impairment, cardiac disorders, diabetes mellitus, coagulation disorders, epilepsy, and psychiatric disorders. Interferons produce influenza-like symptoms (e.g., fever, chills, headache, myalgia, arthralgia). Other side effects include anorexia, weight loss, bone marrow depression, alopecia, and taste alteration.

## ANTIFUNGAL AGENTS

Antifungal agents are used for superficial and systemic fungal infections. Ketoconazole opened a new era for antifungal therapy. Later progress in antifungal drug development includes amphotericin B and griseofulvin formulations, which are less toxic. Many topical antifungal agents have become available since the antiseptic era. Recently, compounds like terbinafine have been added to antifungal drugs. Ideally, antifungal treatment should be started after identifying the fungi, but in most cases, this care is not taken.

### Amphotericin B

Amphotericin B is obtained from *Streptomyces nodosus*. It is fungistatic and administered intravenously as an infusion in the treatment of severe systemic fungal infections. It also is used for the local treatment of superficial candidiasis. Test-dose administration is advised to confirm adverse reactions. The amphotericin infusion should be slow to prevent the risk of irritation and infusion-related adverse effects. The drug is used in pregnancy without any adverse side effects.[66]

Adverse effects reported are headache, nausea, vomiting, chills, fever, muscle and joint pain, anorexia, hypertension, hypotension, cardiac arrhythmias, cardiac arrest, skin rashes, anaphylactic reactions, GI bleeding, and convulsions.[66a–66d]

All patients receiving amphotericin intravenously suffer from nephrotoxicity. In nonconventional dosage forms, such as liposomal formulations, the adverse effects are similar but less toxic when compared with conventional dosage.

### Fluconazole

Fluconazole is used for superficial mucosal candidiasis and fungal skin infections. It should be used with caution in patients with renal and hepatic dysfunction. It is reported that teratogenicity is observed in animals; hence it is not recommended for use in pregnant women.[66e–66g] Adverse effects include abdominal pain and diarrhea. Alopecia has occasionally been reported in patients during prolonged use.[67]

### Ketoconazole

Ketoconazole is administered topically and orally, and is extensively used for candidiasis of the GI tract. Liver function tests should be performed before starting treatment. It has been shown to be teratogenic in animals and hence is not recommended during pregnancy. GI disturbances are frequent following oral administration of ketoconazole. Other adverse effects include allergic reaction such as urticaria and angioedema, pruritus, alopecia, headache, and dizziness. On topical application, irritation and dermatitis or burning sensation occur.

## ANTHELMINTICS

Worms that commonly cause infection in humans include cestodes, tapeworms, flukes, nematodes, or roundworms. Anthelmintics are drugs that either kill or expel infesting helminths. Helmintic infections are found in a large population of the world, mainly the tropical regions. These mainly contribute to the prevalence of malnutrition, anemia, eosinophilia, pneumonia, and they pose a threat to public health in developing countries. Helmentic infections cause lymphatic filariasis, river blindness, and schistosomiasis. The WHO is making efforts to control infections, both in individuals and the population by using chemotherapeutic agents and preventing transmission by food preparation, hygiene, adequate sanitation, and sewage treatment.

### Albendazole

Albendazole is a benzimidazole carbamate anthelmintic. It not only has broad-spectrum activity and excellent tolerability like mebendazole, but also has the advantage of single-dose administration. Albendazole is used in the treatment of cestode infections in relatively high doses. Albendazole also is used in ascariasis, enterobiasis, hookworm, and strongyloidiasis. It has caused abnormalities in liver function tests during prolonged therapy.[68] Albendazole is teratogenic in animals. The use of albendazole is contraindicated in pregnant women.

### Levamisole Hydrochloride

Levamisole hydrochloride is used in the treatment of helminth infections and as an adjuvant in malignant disease. It is active against intestinal nematode worms and appears to act by paralyzing and eliminating susceptible worms. It has proven valuable in the treatment of ascariasis and hookworm infections. The use of levamisole should be avoided in patients with preexisting blood disorders.

It is generally well tolerated; side effects are vomiting, diarrhea, abdominal pains, dizziness, and headache. Prolonged use may cause hypersensitive reactions such as fever, arthralgia, muscle pains, and CNS effects like headache, insomnia, dizziness, and convulsions. Hematologic abnormalities such as granulocytosis, leukopenia, and thrombocytopenia may occur.

## LIPID-LOWERING AGENTS

It is well recognized that hypercholesterolemia is associated with the increased risk of atherosclerosis in humans, leading to ischemic heart disease. At low plasma cholesterol concentrations, the risk of ischemic heart disease is very low. Higher concentrations of low-density lipoprotein (LDL) cholesterol, and lower levels of high-density lipoprotein (HDL) increase the risk of ischemic heart diseases. These are not only risk factors for heart disease, but also others related to the cardiovascular system. Regular estimations of these lipids is not required in normal individuals, but therapy recommended for patients who have symptoms of ischemic heart disease.[69] The main aim therefore is to treat hypercholesterolemia and reduce the risk of the heart disease. For this purpose, lipid-lowering agents, such as clofibrate, gemfibrozil, and lovastatin, are used.

The total fat intake must be reduced to achieve weight reduction by diet therapy.[70] Dietary recommendations also include a reduction in saturated fatty acids. Drug therapy must be started only after confirmation of lipid levels such as cholesterol, triglycerides, LDL, and HDL. Adequate trial must be initiated with modification of dietary therapy, only failing which drug therapy must be started.

### Clofibrate

Clofibrate is a lipid regulatory drug that reduces plasma lipids. It also may be useful in hypertriglyceridemia. Clofibrate is administered as an oral dose (20 to 30 mg/kg body weight) three times a day. Clofibrate previously was used as a prophylatic for ischemic heart disease, but is no longer recommended, as it causes adverse effects after prolonged use.[71] Long-term adverse effects include gallstones, pulmonary embolism, and cardiovascular disorders other than ischemic disease.[72,73] Common side effects are anorexia, nausea, and gastric discomfort. Less frequently, it causes headache, dizziness, vertigo, fatigue, skin rashes, alopecia, impotence, and anemia. Clofibrate (when administered to diabetic patients) reduces the fasting blood glucose levels in prolonged therapy. Clofibrate is contraindicated in patients with liver and kidney damage, as well as gallbladder disorders.

### Simvastatin

Simvastatin acts by inhibiting 3-hydroxy-3-methyl glutaryl coenzyme (HMG COA) reductase enzyme. Simvastatin reduces triglycerides and increases HDL cholesterol. Statins are less effective in patients with homozygous familial hypercholesterolemia, in case of an absence of LDL receptors. A dose of 5 to 10 mg is administered orally in the evening and should not exceed 40 mg.[74,75] In general, statins cause side effects related to the GI tract, headache, skin rashes, blurred vision, insomnia, and dizziness. It should be avoided in pregnant patients and those with liver damage. It should be cautiously used in patients with severe renal impairment when given along with coumarin anticoagulants, as it increases bleeding and prothrombin time.[76]

### Probucol

Probucol is a lipid-lowering agent, but the results are not consistent with respect to LDL cholesterol. It lowers HDL cholesterol; hence it is not the first drug of choice in therapy. The ability of probucol to correct atherosclerosis has been attributed to its antioxidant properties.[77] The usual oral dose is 500 mg twice daily and is administered after food. Many experts use it as adjuvant therapy in familial hypercholesterolemia. The drug is well tolerated but causes GI side effects such as nausea and flatulence, headache, and dizziness. Patients taking probucol must be on a low-fat diet. Probucol should not be used in patients with recent myocardial infarction, and it should not be given to children or pregnant women.

## SEX HORMONES

Steroidal hormones (e.g., testosterone, progesterone, estradiol) are secreted by sex organs, the adrenal cortex, and the placenta and are useful for the development and maintenance of secondary sexual characters and reproduction. In males, testosterone is the main androgenic hormone responsible for the development and maintenance of sex organs and male characteristics. In females, estrogens (mainly comprising estradiol) are secreted in ovarian follicles for the development and maintenance of female sex organs, physical changes (development of mammary glands), and physiologic changes, such as menstruation and reproductive functions. Progesterone, secreted by the corpus luteum, plays a vital role in regulating female sexual functions and estrogens. During pregnancy, large quantities of progesterone are produced by the placenta to aid in reducing uterine motility and help the development of mammary glands. These hormones are used as contraceptives to prevent pregnancy. Oral hormonal contraceptives contain either an estrogen and progesterone combination or only progesterone. Female contraceptives, such as intrauterine devices (IUD), contain progesterone.

Side effects associated with contraceptive therapy include nausea, vomiting, hair loss, headache, weight gain, breast tenderness, breakthrough bleeding, and menstrual irregularity. It has been reported that these patients suffer from depression, mental changes, and cardiovascular diseases.[78] The oral combination therapy of contraceptives causes side effects such as hypertension and reduced glucose tolerance, and may cause cervical and breast cancer. The single contraceptive progesterone may induce nausea, vomiting, headache, breast discomfort, depression, skin disorders, increase in body weight, and menstrual irregularities.[78] Contraceptive drug therapy should be advocated only after thorough medical examination and regular periodical examination. Combined oral contraceptives are contraindicated in patients with liver damage and cardiovascular disease. Appropriate care should be exercised when treating a patient with diabetes mellitus, asthma, epilepsy, or mental depression. Therapy must be avoided in patients over 50 years of age. Contraceptives containing only progesterone are used when combined contraceptives are contraindicated. Therapy must be started cautiously in patients with heart disease, liver dysfunction, malabsorption syndrome, and a history of jaundice during pregnancy.[79]

### Hormone Replacement Therapy

Menopause age varies anywhere between 40 and 50 years. In many cases, menopause occurs at the age of 40 and is called premature menopause. Premature menopause causes palpitation, headache, backache, and psychological and symptoms (e.g., lack of concentration, tiredness, insomnia, depression, and irritability). These symptoms can be acute or long-term in nature. Most are treated with hormone replacement therapy (HRT) with either estrogen only or with combinations containing estrogen and progesterone. HRT produces side effects such as nausea, vomiting, depression, headache, weight changes, tenderness and enlargement of breasts, and liver dysfunction. Prolonged use of HRT causes more serious side effects, which include thromboembolic complications and breast cancer.[80–82] Careful evaluation (medical report and examination) must be carried out before and during HRT. Therapy must be used judiciously in women with thromboembolic complications, migraine, uterine fibroids, diabetes, asthma, epilepsy, hypertension, melanoma, estrogen-dependent neoplasma, or multiple sclerosis. HRT is contraindicated in pregnancy and breast-feeding women.

### Clomiphene Citrate

Clomiphene stimulates ovulation and acts as a profertility agent in humans and (surprisingly) as an antifertility agent in animals. In some cases, it is combined with gonadotropins for *in vitro* fertilization. This drug enhances spermatogenesis in males by stimulating gonadotropin release. The main side effects include reversible ovarian enlargement and cyst formation. It also may induce abnormal uterine bleeding, weight gain, breast tenderness, and headache. Tension, fatigue, insomnia, depression, dizziness, and convulsions are common effects on the CNS. HRT carries a risk of multiple births (rarely more than twins) and the possibility of congenital disorders. Its use is contraindicated in patients with liver damage and should be avoided during pregnancy. The patient should be informed about the possibility of multiple births during long-term therapy.

## ANTIULCER AGENTS

Peptic ulcer disease is a common condition in which damage of the GI mucosa — the stomach, duodenum, esophagus, jejunum, and ileum — occurs. Antiulcer drugs are used for the treatment and prevention of peptic and duodenal ulcers. The drugs used include (1) histamine-receptor antagonists such as cimetidine, ranitidine, and famotidine; (2) proton pump inhibitors such as omeprazole, lansoprazole, and pantoprazole; and (3) antacids containing magnesium and aluminium compounds. The role of *Helicobacter pylori* in the development of duodenal ulcer has gained more importance. Ninety percent of these bacteria are found in the region of the duodenum and 70% are found in cases of gastric ulcers. *Helicobacter pylori* are sensitive to amoxicillin, clarithromycin, tetracycline, and bismuth salts. For eradication of *H. pylori*, triple therapy with bismuth compounds, two antibacterials, and metronidazole has been successful.[83,84] Omeprazole with amoxicillin or clarithromycin also has been tried with success[85,86] and is more effective than triple therapy. The four-drug combination is used for therapy, in which omeprazole is included along with the triple therapy. Attempts are being made to develop a vaccine for *H. pylori*.[87–91]

### Cimetidine

Cimetidine is an $H_2$-receptor antagonist that inhibits gastric secretion and reduces peptone output condition. Cimetidine can be administered by the oral, nasal, or parenteral route. Cimetidine should be taken along with food. Adverse effects with cimetidine are rare and are reversed either by reducing the dose or by stopping cimetidine therapy. These effects include dizziness, tiredness, skin rashes, and very rarely, effects such as hypersensitive reactions, arthralgia, and myalgia. The drug is safe; no toxic manifestations are observed at a dose of 12 g/day. It should be used with caution and given in small doses in patients with kidney damage. Intravenous administration should be slow, and infusions are advised for patients with cardiovascular diseases.

### Omeprazole

Omeprazole inhibits the secretion of gastric acid and acts as a proton pump inhibitor. It is highly effective for ulcerative reflux esophagatis and peptic ulcer

(Zollinger-Ellison syndrome). Common side effects include diarrhea, skin rashes, headache, and CNS effects such as depression, hallucination (in severely ill patients), arthralgia, myalgia, blurred vision, peripheral edema, nephritis, and hepatotoxicity. Omeprazole may mask the symptoms of malignancy and delay diagnosis. It is extensively metabolized in the liver; hence doses should be reduced in patients with hepatic dysfunction. The use of drugs should be seen against the background that many ulcers heal spontaneously, especially by cessation of smoking.

## ANTIEPILEPTIC DRUGS

Epilepsy is a common disease with devastating effects, afflicting millions of people every year. Forty or more forms of epilepsy have been identified. Epileptic seizures often cause transient impairment of consciousness, and the individual is at risk of bodily harm. Therapies are symptomatic (i.e., they inhibit seizures but cannot be used as prophylactic or for the cure). Usually the therapy is long term with unwanted drug effects. Much effort is being devoted to newer approaches for epilepsy treatment. Initiation of epilepsy treatment does not depend on seizure occurrence; prognosis of epilepsy is frequently considered in implementing antiepileptic treatment, and the drug combination should be chosen carefully. Treatment with a single drug is preferable to the multiple-drug regimen. If this fails, an alternative single drug could be tried.[92] The first dose should be decreased simultaneously with a slow increase of the second drug. Various antiepileptic drugs available for the treatment of tonic and clonic seizures include carbamazepine, phenytoin, valproate, and phenobarbital. Primidone also may be used, but it may cause sedation.

### Carbamazepine

Carbamazepine is an antiepileptic drug used to control secondary generalized tonic and clonic seizures and partial seizures. It also is used as a prophylactic for manic depression that is unresponsive to lithium. Carbamazepine should be taken at a scheduled time to avoid variation in absorption. It also can be administered rectally. Carbamazepine is used in treating trigeminal neuralgia. The main side effects of carbamazepine are dizziness, drowsiness, ataxia, and rarely, GI symptoms such as dryness of mouth, abdominal pain, nausea, vomiting, anorexia, diarrhea, or constipation. Occasional reports on blood include aplastic anemia, agranulocytosis, eosinophilia, and thrombocytopenia.[92] Overdose may cause additional side effects such as convulsions, respiratory depression, coma, and death.[93] Congenital malformations have been reported when it is administered during pregnancy.[94,95] It should be avoided in patients with bone marrow depression. Carbamazepine should be administered with caution in patients with a history of cardiac, hepatic, or renal diseases. It also is suggested that the patient be monitored for ocular changes. Carbamazepine can be a drug of abuse if overdosed.[96]

### Phenobarbitone

Phenobarbitone belongs to the barbiturate class of drugs and is used as an antiepileptic. The plasma concentration of 15 to 40 μg/mL is necessary to control seizures. It can be given parenterally during emergency management of seizures and for status epilepticus. The most common and frequent side effect of phenobarbitone is its sedative effect. Typical of antiepileptics drugs, it also causes mood changes, impairment of memory, and depression. Long-term therapy may result in folate deficiency and (rarely) megaloblastic anemia. It may interfere with vitamin D metabolism. Overdose can induce many toxic effects, including severe respiratory depression and cardiovascular depression with hypotensive shock, leading to renal failure. Subcutaneous injection causes necrosis at the injection site, and it has high alkalinity. Intravenous injection can be hazardous and cause hypotension and shock apnea. Congenital malformations have been reported in mothers given phenobarbitone during pregnancy.[97] Barbiturate drugs are used cautiously in the elderly and in patients having severe pain. Care must be exercised in patients with impaired hepatic, renal, and respiratory functions and when withdrawing phenobarbitone therapy. Drug dependence develops rapidly with barbiturate use; on abrupt discontinuation, withdrawal symptoms appear.

### Acute Barbiturate Poisoning

Acute barbiturate poisoning is mostly suicidal but sometimes accidental. There is no specific antidote for barbiturate poisoning. Earlier analeptics such as pentylenetetrazol (Metrazol) and bemegride have been used for the treatment of poisoning.

## VACCINES

Vaccines and sera are biological products that impart active immunity. A foreign substance acts as an antigen, which specifically produces antibodies when administered to an individual. They are two types of immunization 1) active immunity; and 2) passive immunity. Active immunization is very effective and long-lasting when compared with passive immunization. Antisera and immunoglobulins impart passive immunity and are ready-made antibodies. These are biological products that are potentially dangerous and used in public health programs. These products are standardized by using various bioassays and stored at refrigerated conditions to retain their potency. There are three types of vaccines: 1) inactivated microorganisms killed either thermally or chemically; 2) attenuated vaccines, which contain live bacteria or viruses; and 3) toxoids, which are modified bacterial endotoxins with retained antigenicity. The important vaccines are for bacterial typhoid and paratyphoid, cholera, whooping cough, meningococcus, and plague. Bacillus Calmette-Guerin and viral vaccines are used to treat varieties of rabies, influenza, hepatitis B, poliomyelitis, mumps, rubella, and measles. Toxoids are used to treat diphtheria and tetanus and are available as combined vaccines, triple antigen (DPT), measles, mumps, rubella (MMR), typhoid-paratyphoid-cholera (TABC), and double antigen (DT-DA).

Cholera vaccines are produced by inactive bacteria and are administered subcutaneously, intramuscularly, or intradermally. Cholera vaccines should not be administered intradermally in children less than 5 years of age. The vaccination is particularly indicated for people living in highly endemic areas, as well as laboratory and medical personnel exposed to *Vibrio cholerae*. Diphtheria tetanus pertussis (DTP) vaccine can be made either as toxoids or inactivated whole bacteria. *Hemophillus influenzae* vaccine is a bacterial polysaccharide conjugated to proteins and is given as one intramuscular dose. A booster dose is not recommended. This vaccine is given to all children in cases such as plenia and other at-risk conditions.

Hepatitis vaccines such as hepatitis A (inactivated virus) and hepatitis B (an inactive viral antigen, usually recombinant), Hib-hepatitis B (bacterial glycoprotein, inactive antigen) and Hib-hep B (inactivated polio vaccine, recombinant) are all given intramuscularly. Hepatitis A vaccines are given to people exposed to hepatitis A–endemic areas, illicit drug users, those with chronic liver disease, homo- and bisexual individuals, people with occupational risk of infection, and individuals having sexual contact with hepatitis A–positive people. Usually booster doses are given within a period of 6 to 12 months. Hepatitis B vaccine is indicated for all infants and young adult hemophiliacs, hemodialysis patients, individuals with occupational and environmental risk, and for postexposure prophylaxis. Booster doses of the vaccines may not be required. Influenza vaccine is either an inactivated virus or its components and is administered intramuscularly to adults over 65 years of age who live in nursing homes and related areas. Individuals having or prone to high-risk conditions (e.g., asthma), health care professionals, and workers associated with different biological and chemical conditions suffering or prone to high-risk condition are also given influenza vaccines. Booster doses are given yearly with currently available vaccine. Vaccination for lymph disease is for individuals who live in high-risk areas or who engage in activity that exposes them to tick-infested habitats.

Various measles and mumps vaccines are live viruses given subcutaneously to all children or for postexposure prophylaxis in unimmunized persons. Meningococcal vaccine is a bacterial glycoprotein given subcutaneously that has one dose in primary immunization and booster doses every 3 to 5 years. If individuals are prone to continuing high-risk exposure, this vaccine is particularly preferred, especially for military persons, those exposed to epidemic meningococcal disease areas, and people with asplenia or complement deficiency or properdin deficiency. Pneumococcal vaccine is indicated for adults over 65 years of age or individuals who are at high risk for pneumococcal infections. Polio vaccine is given either by oral or subcutaneous routes for all children, unimmunized individuals at high risk for occupational or travel exposure, or children of parents who do not accept the recommended number of injections. Rabies inactivated virus is given prophylactically to persons at high risk for rabies virus infection. Booster doses are given for up to 2 years in high-risk individuals. Rubella live virus vaccination must be performed for adults born after 1956 without a history of rubella or of such vaccination.

Diphtheria vaccine is given to all individuals who were not immunized as children. Booster doses are given every 10 years. Typhoid vaccines are given either orally or intramuscularly to those who have risk of exposure to typhoid fever. Usually booster doses are also given. Varicella vaccine is a live virus given subcutaneously to all children, individuals with a risk of exposure, and those who are neither immunized nor infected before 13 years of age. Yellow fever vaccine is a live virus given subcutaneously to laboratory personnel who may be exposed to the virus or people who live in areas where the disease is endemic. The risk of immunization is higher, and it is the responsibility of the physician to inform patients and use it whenever it is needed. It is necessary to perform a skin test to assess hypersensitivity reactions. It is estimated that the incidence of paralytic poliomyelitis caused by oral vaccines is approximately 10% in the United States and 109,000 paralytic cases if vaccination is not done. The manufacturers advise adherence to the standards of biological production to avoid failure and legal problems. Every adult must be immunized against oxide, poliomyelitis, measles, and diphtheria. Smallpox vaccine is not required for traveling to many countries.[97,98]

## OVER-THE-COUNTER DRUGS

Individuals take drugs prescribed by physicians for various diagnoses. Some drugs can be taken without a prescription and are listed as safe to use. These drugs are called *nonprescription* or *over-the-counter* (OTC) *drugs*. There are many different drugs that contain aspirin, acetaminophen, and various NSAIDs (Table 13-3).

The misuse of OTC drugs may complicate medical treatment. Phenylpropanolamine is a part of a medicine for cold, allergy, and weight control, and is a drug of abuse that is often sold as a cocaine or amphetamine substitute. This drug may

**TABLE 13-3 Categories with Largest Numbers of Deaths**

| Category | Number | Exposure in Category (%) |
|---|---|---|
| Antidepressants | 194 | 0.497 |
| Analgesics | 186 | 0.104 |
| Stimulants and street drugs | 81 | 0.364 |
| Cardiovascular drugs | 80 | 0.301 |
| Alcohols/glycols | 59 | 0.117 |
| Gases and fumes | 42 | 0.144 |
| Asthma therapies | 35 | 0.198 |
| Chemicals | 24 | 0.046 |
| Pesticides | 20 | 0.028 |
| Cleaning substances | 19 | 0.010 |
| Anticonvulsants | 18 | 0.150 |

cause severe hypertension, seizures, and intracranial hemorrhage. Ephedrine HCl is an OTC drug for asthma patients and has a similar testing profile.

Certain ingredients in OTC products should be used with caution in selected patients. There are many potent OTC ingredients hidden in products that may cause problems for therapy. Many OTC products (e.g., cough syrups, decongestants) contain, appetizer certain sympathomimatic amines, which should be avoided in IDDM or patients with hypertension, angina, or hyperthyroidism. Cimetidine, an effective antiulcer agent, inhibits drug metabolism and increases the toxicity of drugs such as phenytoin, theophylline, and warfarin. OTC products containing aspirin, acetaminophen, neuroprotein, ibuprofen, and ketoprotein increase the risk of hepatosensitivity and GI hemorrhage in alcoholics.

For example, aspirin should be avoided in patients with ulceration and patients taking anticoagulation medications. It is possible that misuse or overdose of OTC products may develop into many medical problems. For example, improper use of antacids (e.g., aluminum hydroxide) causes constipation.

Laxative abuse in elderly patients results in abdominal pain, as well as fluid and electrolyte imbalance. The chronic misuse of analogs containing large amounts of caffeine may produce rebound headache. Antihistamines may cause sedation and drowsiness, and it is advised that patients do not drive after taking antihistamines. People who are involved in performing skilled jobs should restrict their use of antihistamines and sedatives.[99,100]

## DRUGS AND THE ELDERLY

The person 65 years of age and older is considered a geriatric patient. Important changes in response to drugs occur with increasing age. Use of drugs is important because a patient's increase in age leads to multiple diseases and nutritional problems. Hence, drug therapy is different in the elderly when compared with the adult population. It is common practice that all of us use drugs for ailments. Requirements are modulated by the health status of the child, adult, or elderly. High incidences of infection are observed in elderly patients, mainly due to reduced host defense. These incidences can lead to more serious infection and cancer. Major age-dependent changes relate to the decrease in renal function as most of the $\beta$-lactam, aminoglycoside, excretes fluoroquinolone antibiotics and changes the half-life of the drugs.

The intelligent use of drugs with elderly patients can increase their life span. Drug noncompliance by the elderly may result from forgetfulness, confusion, or even deliberate intent if prescribed with large number of drugs to be taken at different time intervals. Some errors in drug intake are caused by physical disabilities such as arthritis, tremors, blurred vision, or deafness.

The toxicity of aminoglycosides in the kidney and other organs is concentration-dependent. Antibiotics such as kanamycin and gentamycin have their half-lives doubled in elderly patients. The elderly commonly suffer from osteoarthritis and (less commonly) rheumatoid arthritis. NSAIDs must be carefully used in geriatric patients, as they cause GI toxicity. For example, aspirin causes GI irritation and bleeding, and leads to ulceration. Newer NSAIDs cause renal damage, which may be irreversible. An efficacious NSAID, benoxiprofen, was withdrawn from the market when elderly patients died of renal failure after taking the drug. Corticosteroids are useful in elderly patients who cannot tolerate NSAIDs. However, they cause dose- and duration-dependent osteoporosis, which is hazardous.

Clinical studies showed that elderly patients are more sensitive to drugs such as sedatives, hypnotics, and analgesics. The most extensive studies show a decrease in responsiveness to $\beta$-adrenoreceptors, stimulants, and blockers. Hypothermia is poorly tolerated by the elderly, as their body temperature regulation is impaired.

### Sedatives, Hypnotics, and Analgesics

Sedative and hypnotic drugs, such as benzodiazepines and barbiturates, increase the half-life from 50% to 150% in patients between 30 and 70 years of age. The most alarming toxicity is ataxia, which must be considered to avoid accidents. Ataxia is an alarming effect; hence the patient is advised not to drive if taking these drugs. Elderly patients are found to be more sensitive to opioids and should use them with caution.

### Antipsychotics, Antidepressants, and Drugs for Alzheimer's Disease

Antipsychotic agents, such as phenothiazines and haloperidol, have been extensively used in the elderly for the management of psychiatric disorders. These are useful in the treatment of schizophrenia and dementia, agitation, and paranoid syndrome.

Alzheimer's disease is characterized by progressive impairment of memory and cognitive functions, and may lead to a vegetative state. Such patients are exquisitely sensitive to CNS drugs.

### Cardiovascular Drugs

Systolic blood pressure is high in the elderly; thiazide diuretics are used to control blood pressure, which induces hypokalemia, hyperglycemia, and hyperuricemia. Congestive heart failure is a common and particularly lethal disease in the elderly, because of overdoses of cardiac glycosides. A high dose of digitalis will induce arrythmias in the elderly. Hence, drugs for cardiovascular diseases should be used with caution. Drug therapy has considerable potential for helpful and harmful effects in elderly patients.[101,102]

## DRUGS AND OBESITY

Obesity is defined as accumulation of fat in the body. Body weight that is 20% over the ideal body weight is considered obesity. Obese people are at risk for many complicated diseases, such as cardiovascular disease, diabetes mellitus, gallstones, respiratory disease, osteoarthritis, and some forms of cancer. This occurs when there is an imbalance between energy intake and expenditure. A commonly used measure for body fat is body mass index (BMI).

Obesity is a growing health problem in developed nations of the world. Approximately 33% of the population in the United States and 15% to 20% in Europe suffer from obesity.[103] Human obesity has been linked to genes and other factors relevant to energy balance.[104] The control of appetite and the mechanism of obesity are under investigation. The Ob-gene and its protein product, leptin, have been identified and appear to regulate food intake.

### Adipogenesis and Obesity

It is now believed that physical activity has a more positive role in reducing fat in an obese person. Drugs used for obesity, fenfluramine and dextrofenfluramine, were licensed and later withdrawn from the market because of the incidence of pulmonary hypotension and heart valve defects in patients. The drugs, that are under development are leptin, cholecystokinin promoters, and troglitazone. At present, there are no safe drugs that are effective in the treatment of obesity. All one can usefully say to the obese patient is, stick to the diet, keep jogging, and seek medical advice.

## CONTRAST MEDIA

Contrast media are the agents that enhance the images obtained from radiography (X-ray imaging), magnetic resonance imaging (MRI), or ultrasound imaging. The radiographic contrast media contains elements with high atomic numbers that absorb X-rays. These agents are iodinated organic compounds, barium sulfate, and other compounds (e.g., thorium and titanium dioxides). These latter compounds are toxic and generally not preferred over barium sulfate.

Radiographic techniques are used to visualize particular body structures such as in urography (kidney and ureters), angiography (circulatory systems), radiography (GI), myelography (structure of spinal cord), arthrography (joint capsule), bronchrography (bronchial tree), and hysterosalpingography (uterus and fallopian tubes).

### Barium Sulfate

Barium sulfate is used to visualize the GI tract.[105] Barium sulfate should be avoided in a patient with GI obstruction. It is insoluble and nontoxic but is a soluble compound, which is toxic and may lead to death. Constipation may be detected on oral and rectal administration, and is remedied if a large amount of water is given to the patient. A barium sulfate enema causes electrocardiogram abnormalities. Pneumonitis or granuloma formation are reported during accidental aspiration into lungs. Hypersensitivity reactions also have been reported.[106]

### Coloring Agents

It is common practice to use coloring agents in food, cosmetics, and medicines to make the product elegant and appealing. Coloring agents can be synthetic or natural (e.g., turmeric, caramel, caramine, chlorophyll, saffron). Other compounds, which may be used in cosmetics and food, are anthocyanins, carotenoids, aluminium, gold, indigo, patent blue, riboflavin, silver, titanium dioxide, and tartrazine. Pollock[107] has reported the influence of coloring agents on the behavior of children. The hypersensitivity reaction is well documented with the use of tartrazine.[108]

### Radiopharmaceuticals

Radioactive compounds have been used for radiotherapy and diagnosis in various diseases. Radioactive material is available as a sealed radioactive source. Unsealed sources are usually liquid, particulate, or gaseous. Utmost care is taken in the preparation, handling, and disposal of these radioactive materials, which are very hazardous.

Radiopharmaceuticals are widely used in medical practice either for diagnosis or for treatment of disease. These techniques are useful in case of failures with contrast media, ultrasound, computed tomography (CT), or external irradiation. Interestingly, monoclonal antibodies have been tagged with radionucleotides. To obtain the desired precision and accuracy of image measurement, only small quantities are used.

Commonly used radiopharmaceuticals are carbon 14 ($t_{1/2}$ 30 years), cobalt 57 ($t_{1/2}$ 271 days), cobalt-58 ($t_{1/2}$ 70.8 days), gold-198 ($t_{1/2}$ 2.7 days), iodine-123 ($t_{1/2}$ 12.3 hours), iodine-125 ($t_{1/2}$ 60 hours), iodine-131 ($t_{1/2}$ 8.04 days), and tritium ($t_{1/2}$ 12.3 years). The iodine radioisotope is used to study thyroid function and is used in the treatment of hyperthyroidism and thyroid carcinoma. Various monoclonal antibodies labeled with iodine-171 are used for the detection of malignant neoplasms. Genetic damage is a dangerous side effect of radioactive isotopes prior to and during the reproductive period. Exposure to large doses leads to leukopenia, anemia, skin inflammation, radiation sickness, and neoplasm.

## THERAPEUTIC GASES

A number of compressed and liquified gases are used as refrigerants and aerosol propellants. These include nitrous oxide, nitrogen, carbon dioxide, propane, and butane. The use of chlorofluorocarbons (CFCs) is restricted because of environmental pollution leading to health hazards. These have been replaced by hydrogenated fluorocarbons (HFCs), which are less likely to cause environmental pollution.

Refrigerants and aerosol propellants have been gases of abuse. Inhalation of high-concentration halogenated hydrocarbons causes euphoric effects and later

CNS depression, cardiac arrythmias, respiratory depression, and death. Halogenated hydrocarbons decompose when heated and become a toxic gas (e.g., hydrogen chloride, phosgene). Evaporation of halogenated hydrocarbons as propellants causes intense cold, which leads to numbness. These gases have been used as topical analgesics.

## Oxygen

A colorless, odorless, and tasteless gas, oxygen ($O_2$) is usually supplied in a metal cylinder as compressed air. Animal organisms require oxygen, water, and food, which are essential for the maintenance of life. Pure oxygen, administered at an ambient pressure greater than that of the atmosphere, has unique therapeutic implications but multiple toxic effects. Oxygen is of great value in the treatment of poisoning such as from carbon monoxide, cyanide, and dichloromethane. For intelligent therapeutic use of oxygen, a knowledge of oxygen deficiency is necessary. The term *hypoxia* is used to denote insufficient oxygenation of the tissues. Deprivation of oxygen leads to death.

Oxygen is used for the correction of hypoxia and as a diluent or carrier gas for vapors and gases, primarily anesthetic agents. The use of oxygen at increased pressure is termed *hyperbaric oxygen therapy*. This therapy is used in diverse conditions such as multiple sclerosis, traumatic spinal cord injury, cerebrovascular accidents, bone graft, fractures, and leprosy. However, there are no available authentic data, as there are no well-controlled clinical trials.

Any fire or spark is highly dangerous in the presence of oxygen, especially when it is under pressure. Combustible materials soaked in liquid oxygen are potentially explosive. The respiratory system is a prime target of toxicity. Humans exposed to 100% oxygen for 24 hours show symptoms such as nausea, vomiting, and anorexia. Death may result because of pulmonary edema or hypoxia. CNS toxicity does not occur if the pressure is less than 2 atmospheres, if more convulsions and muscle twitching are observed. Exposed newborns may develop retrolental fibroplasia. Oxygen-induced retinopathy is rare in adults.

## Carbon Dioxide

Carbon dioxide ($CO_2$) is a colorless, odorless gas that is soluble in water and 1.5 times heavier than air. Carbon dioxide is important in the regulation of many vital functions. Small changes in $CO_2$ in the body have marked physiologic effects. More than a 6% change causes headache, dizziness, palpitation, mental confusion, hypertension, and CNS depression; a 30% or higher change may cause convulsions. The inhalation of higher concentrations may produce respiratory acidosis. Abrupt withdrawal may cause hypotension, dizziness, severe headache, nausea, and vomiting.

Carbon dioxide is known to stimulate respiration but is seldom used for this purpose. Carbonated mixtures are used to mask the unpleasant taste of medicinal preparations. Solid carbon dioxide or dry ice ($-80\,^{\circ}C$) is used for treating warts.[109] Inhalation of carbon dioxide can increase the speed of anesthetic induction because of respiratory depression. Inhalation of carbon dioxide is suggested to control hiccoughs, and sudden deafness is treated with a mixture of carbon dioxide and oxygen.

### Nitric Oxide

Nitric oxide (NO) is produced by many cell types, including endothelial cells, and has functions ranging from neurotransmission to vasodilatation. NO may be useful to treat bronchodilatation in clinical conditions and is used successfully in patients with persistent fetal circulation and pulmonary hypertension. NO is also a toxic oxidant.

### Helium

Helium is a colorless, odorless gas that is not combustible and does not support combustion. It is always supplied as compressed air in cylinders. Due to the low solubility of helium, divers and people working under high pressures use a mixture of helium and oxygen to prevent the development of decompression sickness. Breathing helium causes increased vocal pitch and voice distortion. Helium is an inert gas with its low density, solubility, and high thermal conductivity, and is the basis for medical and diagnostic use. It is used in pulmonary function testing and laser surgery because of its high thermal conductivity.

## DISINFECTANTS AND PRESERVATIVES

Disinfectants are chemicals that destroy or inhibit the growth of pathogenic microorganisms but do not necessarily kill all microorganisms. An antiseptic is a disinfectant that is used for skin and other living tissues to prevent infection.

### Disinfectants

Disinfectants are used in industrial establishments, hospitals, and homes to prevent infection. Common disinfectants can kill some fungi, lipid-containing viruses, gram-negative bacteria, and mycobacteria, but bacterial spores are resistant to disinfectants. Factors affecting the effectiveness of disinfectants are contact line, concentration, pH, and the presence of interfering substances such as lipids, rubber, or plastics.

### Preservatives

These are chemical agents that are included in preparations to prevent deterioration of manufactured products. Antimicrobial preservatives are used in sterile preparations such as eye drops, parenteral injections, cosmetics, foods, and nonsterile pharmaceutical products (e.g., oral liquids and creams) to prevent microbial spoilage. Preparations that are injectables in cerebrospinal fluid (CSF), the eye, or the heart should not include preservatives.

Use of contact lenses increases the risk of corneal infection. The most common pathogens that cause eye infection are *Pseudomonas aeruginosa*, *Serratia marcescens*, *Staphylococcus aureus*, *S. epidermidis*, and *S. pneumoniae*. Fungi rarely cause eye infections. Commonly used disinfectants are benzalkonium chloride and chlorhexidine.

Equipment used in medical practice for disease diagnosis, which come into contact with the body, must be sterilized. Nonsterilizations can cause transmission of infection from one person to other. Commonly, 2% glutaraldehyde solution is used. Instruments must be kept in the solution for at least 3 hours, but high-level disinfection is achieved in 20 to 30 minutes for most. Glutaraldehyde is an irritant and may cause sensitization. Alternatively, paracetic acid, chlorine dioxide, and superoxidized water are used for this purpose.

### Disinfection of Water

Chlorine-releasing disinfectants (e.g., chloramine, halazone, sodium hypochlorite) are commonly used. Iodine-releasing disinfectants, such as tetraglycine hydroperiodide or iodine itself, also are used. Emergency treatment of drinking water with lemon juice has been suggested during epidemics of water-borne gastroenteritis. Effective disinfection can be achieved by raising water temperature, introducing ultraviolet light, and copper silver ionization.

The need of skin disinfectants before injection is controversial.[110] It is generally thought that the use of antiseptics may be ineffective and unnecessary.[111,112] Aseptic conditions are required for the use of catheters to minimize the infection. Povidone iodine and chlorhexidine are commonly used for the catheters.

### Chlorhexidine

Chlorhexidine is available in the form of acetate, gluconate, or hydrochloride. Chlorhexidine salts are incompatible with soaps and other anionic materials. Fabrics in contact with chlorhexidine solution may develop brown stains if bleached with hypochloride.

Chlorhexidine occasionally causes skin sensitivity. Hemolysis has been reported following accidental intravenous administration.[112a] It may damage safe tissues such as the eye. Chlorhexidine is an irritant and should not be used for sensitive tissues. Contact with the eyes should be avoided, as it will cause corneal damage. Syringes sterilized with chlorhexidine must be cleaned perfectly before using for other purposes. Aqueous solutions of chlorhexidine used for the storage of instruments should contain sodium nitrite (0.1%) to inhibit metal corrosion and should be required to be changed every week.

Chlorhexidine is available as lotions, creams, washes for disinfection and cleansing of skin and wounds, and as oral gels and mouthwashes. It also has been used in combination with neomycin. Chlorhexidine gluconate is used as a 1% dental gel to prevent plaque and gingivitis. A mild 0.02% solution may be used as bladder irrigation in the treatment of urinary tract infection. Chlorhexidine acetate also is used as an emergency disinfectant to clean instruments.

### Formaldehyde Solutions

Formaldehyde solutions contain 34.5% to 38% weight/weight, with methyl alcohol as a stabilizing agent to delay the polymerization of formaldehyde to solid formaldehyde. According to the United States Pharmacopeia, the formaldehyde solution content should not be less than 36.5% to 37.0% w/w with methyl alcohol. Formaldehyde is a clear, practically colorless liquid with a characteristic pungent and irritating odor. It should be stored at 15° to 25° in air-tight containers and should be protected from exposure to light.

Accidental ingestion of formaldehyde solution causes severe pain, ulceration, and necrosis of the mucous membranes. Formaldehyde (30 mL) ingestion will lead to death. Asthma has been reported after repeated exposures;[112b,112c] persons working in histopathologic laboratories must be careful, as they are extensively exposed to the solution.

Occupational exposure of medical personnel and industrial workers to formaldehyde and the risk of carcinogenicity cannot be ruled out. Ten percent buffered formalin is used to preserve organs for histopathology study. Formaldehyde solution is used to disinfect blankets and beddings. Formaldehyde solutions 3% w/v have been applied for the treatment of warts.

## DRUG POISONING

OTC drugs can be harmful when taken in large quantities without knowing the precautions of their toxicity, as in large doses of paracetamol, aspirin, and fat-soluble vitamins. Accidental poisoning can occur in children younger than 5 years of age if medicines or domestic chemicals (e.g., kerosene, detergents, bleach) are left within their reach (Table 13-3). Acute poisoning can be treated successfully with common sense and identification of poison. Some require only an observation in the hospital, whereas others require specific antidotes or intensive care (Table 13-4). It is important to identify the substance, dose, and time taken in accidental poisoning (Table 13-5). In this instance, rapid biological screens are available and can be performed if the patient is seriously ill or unconscious. Common poisoning can occur with paracetamol, aspirin, or sedatives such as benzodiazepines and phenobarbital. Response to specific antidotes may provide a diagnosis (e.g., dilated pupils and increased respiratory rate after intravenous opioid poisoning or arousal from unconsciousness in response to intravenous benzodiazepine poisoning.

To prevent further absorption of poison from the gut, oral absorbants (activated charcoal), gastric lavage, emesis, and cathartics are used. Chelating agents are

**TABLE 13-4 Specific Antidotes and Indications**

| Antidote | Indication |
|---|---|
| Acetylcysteine | Paracetamol, chloroform, carbon tetrachloride |
| Atropine | Cholinesterase inhibitors (e.g., organophosphorus insecticides, $\beta$-blocker poisoning) |
| Dicobalt edetate | Cyanide and derivates (e.g., acrylonitrile) |
| Digoxin-specific antibody | Digitalis glycosides |
| Ethanol | Ethylene glycol, methanol |
| Flumazenil | Benzodiazepines |
| Naloxone | Opioids |
| Penicillamine | Copper, gold, lead, elemental mercury (vapor), zinc |
| Propranolol | $\beta$-adrenoceptor agonists, ephedrine, theophylline, thyroxine |
| Protamine | Heparin |

**TABLE 13-5 Substances Most Frequently Involved in Human Poison Exposures**

| Substance | Number | %* |
|---|---|---|
| Cleaning substances | 196,022 | 10.5 |
| Analgesics | 178,284 | 9.6 |
| Cosmetics | 153,721 | 8.2 |
| Cough and cold preparations | 107,980 | 5.8 |
| Plants | 106,939 | 5.7 |
| Bites | 74,906 | 4.0 |
| Pesticides | 70,687 | 3.8 |
| Topicals | 70,458 | 3.8 |
| Hydrocarbons | 64,041 | 3.4 |
| Foreign bodies | 63,297 | 3.4 |
| Antimicrobials | 63,025 | 3.4 |
| Sedatives/hypnotics/antipsychotics | 58,582 | 3.1 |
| Chemicals | 52,499 | 2.8 |
| Food poisoning | 50,511 | 2.7 |
| Alcohols | 50,276 | 2.7 |
| Vitamins | 43,187 | 2.3 |

* Percentages are based on total number of known ingested substances rather than the total number of human exposure cases.

used to treat poisoning with heavy metals such as dimercaprol (British Antilevisite, BAL). For arsenic and other metals, unithiol (dimercaptopropanesulfonate, DMPS) effectively chelates lead and mercury, penicillamine chelates copper, and sodium calciumedetate chelates lead. Most substances used cause CNS and ANS dysfunction. Sedatives, opioids, and alcohol may cause respiratory depression, miosis, hypotension, and hypothermia. Amphetamines, cocaine, ephedrines, and theophyllines may cause symptoms such as tachycardia, hypertension, hyperthermia, sweating, mydriasis, agitation, and delusion.

### Poisoning by Biologic Substances

It is common to consume a variety of plants by eating or chewing. Sometimes we accidentally eat nonedible plants, which may contain dangerous chemicals or drugs. Datura causes dilated pupils, blurred vision, dry mouth, flushed skin, confusion, and delirium. Nicotinic acid, for example, from hemlock (Conium) and Laburnum, causes salivation, dilated pupils, vomiting, convulsions, and respiratory paralysis. Psilocybin-containing mushrooms have hallucinogenic properties. Water dropwort and cowbane contain related and very dangerous convulsant substances, such as oenanthotoxin and cicutoxin. Primula causes cutaneous irritation. GI symptoms, nausea, vomiting, diarrhea, and abdominal pain occur with the ingestion of numerous plants.

### Treatment of Plant Poisoning:

- Activated charcoal for toxins in GI tract.
- Ipecac for vomiting.
- Diazepam to control convulsions.
- Penicillin for mushroom poisoning (“death cap”).

## REFERENCES

1. Anonymous, Modifying disease in rheumatoid arthritis. *Drug Ther. Bull.* **36**:3–6, 1998.
2. J.F. Fries, et al., Reduction in long-term disability in patients with rheumatoid arthritis by disease modifying antirheumatic drug based treatment strategies. *Arthritis Rheum.* **39**:616–622, 1996.
3. J.J. Gomez-Reino, Long-term therapy for rheumatoid arthritis. *Lancet* 343–344, 347, 1996.
4. M.J.M. Wijnands, and P.L.C.M. Van Riel, Management of adverse effects of disease modifying anti-rheumatic drugs. *Drugs Saftey* **13**:219–227, 1995.
5. J.R. Mackenzie, and D.G. Monoz, *Neurology* **53**(1):197–203, 1999.
6. J.J. Starmek, and N.R. Cutler, *Drugs Ageing* **14**(5):359–373, 1999.
7. C.K. Kwoh, and A.R. Feinstein, Rates of sensitivity reactions to aspirin: Problems in interpreting the data. *Clin. Pharmcol. Ther.* **40**:494–505, 1986.
8. I. Power, Aspirin induced asthma. *Br. J. Anaesth.* **71**:619–621, 1993.
9. K. Bush, Lower back pain and sciatica: How best to manage. *Them. Br. J. Hosp. Med.* **51**:216–222, 1994.

10. T.W. Underwood, and C.B. Frye, Drug induced pancreatitis. *Clin. Pharmacol.* **12**:440–448, 1993.
11. L.F. Prescott, Paracetamol overdose, in *paracetamol (acetaminophen): A critical bibliographic review*. London: Taylor & Francis, 401–473, 1996.
12. B.H.C. Stricker, et al., Acute hypersensitivity reaction to paracetamol. *Br. Med. J.* **291**:938–939, 1985.
13. L. Van Diem, and J.P. Grilliant, Anaphylactic shock induced by paracetamol. *Eur. J. Clin. Pharmacol.* **38**:389–390, 1990.
14. The International Agranulocytosis and Aplastic Anemia Study, Risks to agranulocytosis and aplastic anemia, a first report of their relation to drug use with special reference to analgesics. *JAMA* **256**:1749–1757, 1986.
15. S.M. Juergens, Problems with benzodiazepines in elderly patients. *Mayo Clin Proc.* **68**:818–820, 1993.
16. J.C. Cloyd, et al., Availability of diazepam from plastic containers. *Am. J. Hosp. Pharmacol.* **37**:492–496, 1980.
17. W.A. Parker, and M.E. MacCara, Compatibility of diazepam with intravenous fluid containers and administration sets. *Am. J. Hosp. Pharmacol.* **37**:496–500, 1980.
18. E.A. Kowaluk, et al., Factor affecting the bio-availability of diazepam stored in plastic bags and administered though intravenous sets. *Am. J. Hosp. Pharmacol.* **40**:417–423, 1983.
19. H.J. Martens, et al., Sorption of verify drugs in polyvinyl chloride, glass, and polyethylene–lined infusion containers. *Am. J. Hosp. Pharmacol.* **47**:369–373, 1990.
20. H. Aston, The treatment of benzodiazepine dependence. *Addiction* **89**:1535–1541, 1994.
21. F.J. Tedesco, and L.R. Mils, Diazepam (Valium) hepatitis. *Dig. Dis. Sci.* **27**: 470–472, 1982.
22. C.H. Ashton, et al., Drug induce stupor and coma: Some physical signs and their pharmacological basis. *Adv. Drug React. Acute Poison. Rev.* **9**:1–59, 1989.
23. N.A. Buckeley, et al., Relative toxicity of benzodiazepines in over dose. *Br. Med. J.* **310**:219–221, 1995.
24. J. Lopez–Herce, et al., Alcohol poisoning following diazepam intravenous infusion. *Ann Pharmacol. Ther.* **29**:632, 1995.
25. B. Hemmelgarn, et al., Benzodiazepine use and the risk of motor vehicle cross in the elderly. *JAMA* **278**:27–37, 1997.
26. F. Barbone, et al., Association of road-traffic accidents with benzodiazepines use. *Lancet* **352**:1331–1336, 1998.
27. P. Chocair, et al., Low dose allopurinol plus a zathioprine/cyclosporin/prednisolone, a novel immunosuppressive regimen. *Lancet* **342**:83–84, 1993.
28. C.N. Chunge, et al., Visceral leishmaniasis unresponsive to antimonial drugs: Successful treatment using a combination of sodium stibogluconate plus allopurinol. *Trans R. Soc. Trop. Med. Hyg.* **79**:715–718, 1985.
29. B. Ettinger, et al., Randomized trial of allopurinol in prevention of calcium oxalate calculi. *N Engl. J. Med.* **315**:1386–1389, 1986.
30. R.N. Brogden, and R.C. Heel, Human insulins: A review of its biological activity, pharmacokinetic and therapeutic use. *Drugs* **34**:350–371, 1987.

31. A.L. Baker, et al., A randomized clinical trail of insulin and glucagons infusion for treatment of alcoholic hepatitis: Progress report in 50 patients. *Gastroenterology* **80**:1410–1414, 1981.
32. L.H. Opie, Glucose and the metabolism of ischaemic mycocardium. *Lancet* **345**:1552–1555, 1995.
33. D.J. Husband, and G.V. Gill, "Seizures": A hypoglycaemic hazard for insulin dependent diabetics. *Lancet* **ii**:1477, 1984.
34. L.G. Miller, Recent developments in the study of the effects of cigarette smoking on clinical pharmacokinetics and clinical pharmacodynamics. *Clin. Pharmacokinet.* **17**:90–108, 1989.
35. P.A.B. Raffle, Drugs and driving. *Prescribers' J.* **21**:197–204, 1981.
36. WHO, Practical Chemotherapy of malaria: Report of a WHO Scientific Group. *WHO Tech. Rep. Ser.* **805**: 1990.
37. S. Looareesuwan, et al., Quinine and severe falciparum malaria in late pregnancy. *Lancet* **ii**:4–8, 1985.
38. R.E. Phillips, et al., Hypoglycaemia and anti-malarial drugs: Quinidine and release of insulin. *Br. Med. J.* **292**:1319–1321, 1986.
39. A. Jaeger, et al., Clinical features and management of poisoning due to anti-malarial drugs. *Med. Toxicol.* **2**:242–273, 1987.
40. P.J. de Vries, and T.K. Dien, Clinical pharmacology and therapeutic potential of artemisinin and its derivatives in the treatment of malaria. *Drugs* **52**:818–836, 1996.
41. P. Javis, and J.A. Golish, Clinical management of asthma is the 1990's: Current therapy and new directions. *Drugs* **52**(suppl. 6):1–11, 1996.
42. E.R. McFadden, and I.A. Gilbert, Asthma. *N Engl. J. Med.* **327**:1928–1937, 1992. *Correction. ibid.* **328**:1640–1641, 1993.
43. J.R. Catterall, and C.M. Shapiro, Nocturnal asthma. *Br. Med. J.* **306**:1189–1192, 1993.
44. British Thoracic Society, et al., Guidelines on the management of asthma. *Thorax* **48**(suppl.):S1–S24, 1993. *Correction* **49**:386, 1994, (Summary Published in *Br. Med. J.* **306**:776–782, *Correction* **307**:1054, 1993).
45. B.I. Dickinson, and M.L. Gora Harper, Aminophylline for cellulite removal. *Ann. Pharmacother.* **30**:292–293, 1996.
46. P.O. Brennan, Inhaled salbutamol: A new form of drug abuse? *Lancet* **ii**:1030–1031, 1983.
47. L.D. Lewis, et al., A study of self-poisoning with oral salbutamol—Laboratory and clinical features. *Hum. Exp. Toxicol.* **12**:397–401, 1993; *Correction* **13**:371, 1994.
48. E.R. McFadden, Jr., Dosages of corticosteroids in asthma. *Am. Rev. Resp. Dis.* **147**:1306–1310, 1993.
49. B.J. Lipworth, Clinical pharmacology of corticosteriods in bronchial asthma. *Pharmacol. Ther.* **58**:173–209, 1993.
50. I. Pavord, and A. Knox, Pharmacokinetic optimizations of inhaled steroid therapy in asthma. *Clin. Pharmacokinet.* **25**:126–135, 1993.
51. U. Tirelli, et al., Cancer treatment and old people. *Lancet* **338**:114, 1991.

52. F. Broughton Pipkins, et al., Possible risk with captopril in pregnancy: Some animal data. *Lancet* **I**:1256, 1980.
53. S.L. Nightingale, Warnings on use of ACE inhibitors in second and third trimester of pregnancy. *JAMA* **267**:2445, 1992.
54. R.C. Parish, and L.J. Miller, Adverse effects of angiotensin converting enzyme (ACE) inhibitors: An update. *Drug Safety* **7**:14–31, 1992.
55. D.M. Coulter, and I.R. Edwards, Cough associated with captopril and enalapril. *Br. Med. J.* **294**:1521–1523, 1987.
56. R. Manchanda, Propranolol the wonder drug for psychiatric disorders? *Br. J. Hosp. Med.* **39**:267–271, 1988.
57. W.H. Frishman, et al., Topical ophthalmic $\beta$-adrenergic blockade for the treatment of glaucoma and ocular hypertension. *J. Clin. Pharmacol.* **34**:795–803, 1994.
58. D.J. Selkoe, Physiological production of the $\beta$-amyloid protein and the mechanism of Alzheimer's disease. *Trends Neurosci.* **16**:403–409, 1993.
59. D.J. Selkoe, Alzheimer's disease: Genotypes, phenotype and treatments. *Science* **275**:630–631, 1997.
60. G.W. Small, Treatment of Alzheimer's disease: Current approaches and promising developments. *Am. J. Med.* **104**(suppl. 4A):325–385, 1998.
61. M.B. Stern, Contemporary approaches to the pharmacotherapeutic management of Parkinson's disease. *Neurology* **49**(suppl. 1), 529, 1997.
62. C.W. Olanow, J.H. Kordower, and T.B. Freeman, Fetal nigral transplantation as a therapy for Parkinson's disease. *Trends Neurosci.* **19**:102–109, 1996.
63. P.S. Harper, The epidemiology of Huntington's disease. *Hum. Genet.* **89**:365–376, 1992
64. K.A. Arndt, Review to acyclovir, topical, oral, and intravenous. *J. Am. Acad. Dermatol.* **18**:188–190, 1988.
65. G.J. Moyle, and S.E. Barton, HIV-proteinase inhibitors in the management of the HIV infection. *J. Antimicrob. Chemother.* **33**:921–925, 1996.
66. M.A. Ismail, and S.A. Lerner, Disseminated blastornycosis in pregnant woman. *Am. Rev. Respir. Dis.* **126**:350–353, 1982.

66a. S.D. Goodwin, et al., Pretreatment regimens for adverse events related to infusion of amphotericin B. *Clin. Infect. Dis.* **20**:755–761, 1995.

66b. R.J. Hay, Liposomal amphotericin B, AmBisome. *J. Infect*, **28**(Suppl. 1):35–43, 1994.

66c. C.P. Thakur et al., Comparison of three treatment regimens with liposomal amphotericin B (AmBisome) for visceral leishmaniasis in India: A randomized dose-finding study. *Trans. R. Soc. Trop. Med. Hyg.* **90**:319–322, 1996.

66d. U. Eriksson, B. Seifert, A. Schaffner, What is the best way to give amphotericin B? *British Medical J.* **322**:579–582, 2001.

66e. T.J. Pursley et al., Fluconazole-induced congenital anomalies in three infants. *Clin. Infect. Dis.* **22**:336–340, 1996.

66f. G.G. Briggs, R.K. Freeman, and S.J. Yaffe, *Fluconazole, Drugs in Pregnancy and Lactation*, 5th ed., Williams & Wilkins, Baltimore, MD, 436–439, 1998.

66g. W. Inman, et al., Safety of fluconazole in the treatment of vaginal candidiasis: A prescription-event monitoring study, with special reference to the outcome of pregnancy. *Eur. J. Clin. Pharmacol.* **46**:115–118, 1994.

67. S.E. Weinroth, and C.S. Tuazon, Alopecia associated with fluconazole treatment. *Ann. Intern. Med.* **119**:637, 1993.
68. D.L. Morris, and P.G. Smith, Aldendazole in hydatid disease hepatocellular toxicity. *Trans. R. Soc. Trop. Med. Hyg.* **81**:3434, 1987.
69. The Long-Term Inverventions with Pravastatin in Ischaemic Disease (LIPID) Study Group, Prevention of cardiovascular events and death with pravastatin in patients with coronary heart disease and a broad range of initial cholesterol levels. *N. Engl. J. Med.* **339**:1349–1357, 1998.
70. Department of Health. Nutritional aspects of cardiovascular disease. *Report on Health and Social Subjects*, **46**, Her Majesty's Service Office, London, 1994.
71. M.F. Oliver, et al., Report, WHO cooperative trial on primary prevention of ischaemic heart disease using clofibrate to lower serum cholesterol: Final mortality follow-up. *Lancet* **ii**:600–604, 1984.
72. M.F. Oliver, et al., A cooperative trial in the primary prevention of ischaemic heart disease using clofibrate. *Br. Heart J.* **40**:1069–1118, 1978.
73. Report on the coronary drug project research group, Clofibrate and niacin in coronary heart disease. *JAMA* **231**:360–380, 1975.
74. G.L. Plosker, and D. McTavish, Simvastatin: A reappraisal of its pharmacology and therapeutic efficacy in hypercholesterolaemia. *Drugs* **50**:334–363, 1995.
75. G. Schectman, and J. Hiatt, Dose-response characteristics of cholesterol lowering drug therapies: Implications for treatment. *Ann. Intern. Med.* **125**:990–1000, 1996.
76. W.R. Garnett, Interactions with hydroxymethylglutaryl-coenzyme A reductase inhibitors. *Am. J. Health Syst. Pharmacol.* **52**:1639–1645, 1995.
77. J.L. Witztum, The oxidation hypothesis of atherosclerosis. *Lancet* **344**:793–795, 1994.
78. C.R. Kay, and P.C. Hannaford, Breast cancer and the pill—a further report from the Royal College of General Practitioners oral contraceptive study. *Br. J. Cancer* **58**:675–680, 1988.
79. L. Rama-Wilms, et al., Fetal genital effects of first trimester sex hormone exposure: A meta-analysis. *Obstet. Gynecol.* **85**:141–149, 1995.
80. M.P. Evans, et al., Hormone replacement therapy: Management of common problems. *Mayo Clin. Proc.* **70**:800–805, 1995.
81. T.A. Sellers, et al., The role of hormone replacement therapy in the risk for breast cancer and total mortality in women with a family history of breast cancer. *Ann. Intern. Med.* **127**:973–980, 1997.
82. E. Barrett–Connor, and T.L. Bush, Estrogen and coronary heart disease in women. *JAMA* **265**:1861–1867, 1991.
83. J.J.Y. Sung, et al., Antibacterial treatment of gastric ulcers associated with *Helicobacter pylori*. *N. Engl. J. Med.* **332**:139–142.
84. Anonymous, *Helicobacter pylori* infection—when and how to treat. *Drug Ther. Bull.* **31**:13–15, 1993.
85. D. Markham, and D. McTavish, Clarithromycin and omeprazole: As *Helicobacter pylori* associated gastric disorders. *Drugs* **51**:161–178, 1996.
86. J. Labenz, et al., Omeprazole plus amoxicillin: Efficacy of variants treatment regimes to eradicate *Helicobacter pylori*. *Am. J. Gastroenterol.* **88**:491–495, 1993.

87. P.S. Phule, et al., Are weak treatment for *Helicobacter pylori* infects: A randomized study of quadruple therapy versus triple therapy. *J. Antimicrob. Chemother.* **36**:1085–1088, 1995.

88. J.L. Telford, and P. Ghiora, Prospects for development of a vaccine against *Helicobacter pylori. Drugs* **52**:799–804, 1996.

89. C.K. Ching, and S.K. Lam, Drug therapy of peptic ulcer disease. *Br. J. Hosp. Med.* **54**:101–106, 1995.

90. G.M. Forbes, et al., Duodenal ulcer treated with *Helicobacter pylori* eradication: Seven year follow-up. *Lancet* **343**:288–360, 1994.

91. de Boerw, et al., Effect of anti-sopromim on efficacy of treatment for *Helicobacter pylori* infections. *Lancet* **20**:345–817, 1995.

92. M.A. Mikati, and T.R. Browne, Comparative efficacy of antiepileptic drugs. *Clin. Neuropharmacol.* **11**:130–140, 1988.

93. J.L. Sobotka, et al., A review of carbamazepine's hematologic reactions and monitoring recommendations. *DICP Ann. Pharmacother.* **24**:1214–1219, 1990.

94. F.W. Rosa, Spina bifida in infants of women treated with carbamazepine during pregnancy. *N. Engl. J. Med.* **324**:674–677, 1991.

95. A. Ornoy, E. Cohen, Outcome of children born to epileptic mothers treated with carbamazepine during pregnancy. *Arch. Dis. Child* **75**:517–520, 1996.

96. P.J. Crawford, and D.M. Fisher, Recreational overdosage of carbamazepine in paisley drug abusers. *Scot. Med. J.* **42**:44–45, 1997.

97. Bertram G. Katzung, and BarryA. Berkowitz, Basic and Clinical Evaluation of New Drugs, in *Basic and Clinical Pharmacology*, 8th ed., Bertram G. Katzung, ed., McGraw-Hill, New York 1114–1121, 2001.

98. Advice for travelers. *Med. Lett. Drugs Ther.* **41**:39, 1999.

99. P.K. Honig, and B.K. Gillepsie, Clinical significance of pharmacokinetic drug interactions with over-the-counter (OTC) drugs. *Clin. Pharmacokinet.* **35**:167, 1998.

100. P.H. Rheinstein, Prescription to over the counter analgesics. *Am. Fam. Phys.* **56**:1211, 1997.

101. W.B. Abrams, Cardiovascular drugs in the elderly. *Chest* **98**:980, 1990.

102. N. Bellamy, Treatment considerations in the elderly rheumatic patient. *Gerontology* **34**(supp. I):16, 1988.

103. P. Bjorntorp, Obesity. *Lancet* **350**:423–426, 1997.

104. B.M. Spiegelman, and J.S. Flier, Adipogenesis and obesity: Rounding out the big picture. *Cell* **87**:377–389, 1996.

105. W.L. Janower, Hypersensitivity reactions after barium studies of the upper and lower gastrointestinal tract. *Radiology* **161**:139–140, 1986.

106. A. Bentley, and K. Piper, Barium sulfate preparations. *Pharmacol. J.* **238**:138–139, 1987.

107. I. Pollock, and J.O. Warner, Effect of artificial food colors on childhood behaviour. *Arch. Dis. Child.* **65**:74–77, 1990.

108. Anonymous, Tatrazine: A yellow hazard. *Drug Ther. Bull.* **18**:53–55, 1980.

109. K.R. Beutner, and A. Ferenczy, Therapeutic approaches to genital warts. *Am. J. Med.* **102**(suppl. 5A):28–37, 1997.

110. G.A.J. Ayliffe, et al., *Chemical Disinfection in Hospitals*, 2nd ed., London, PHLS, 1993.

111. T.C. Dann, Routine skin preparation before injection: An unnecessary procedure. *Lancet* **ii**:96–98, 1969.

112. J. Liauw, and G.J. Archer, Swaboholics. *Lancet* **345**:1648, 1995.

112a. O. Mimoz, et al., Chlorhexidine compared with povidone-iodine as skin preparation before blood culture: A randomized, controlled trial. *Ann. Intern. Med.* **131**:834–837, 1999.

112b. Agency for Toxic Substances and Disease Registry (ATSDR), *Toxicological Profile for Formaldehyde* (Draft). Public Health Service, U.S. Department of Health and Human Services, Atlanta, GA, 1997.

112c. U.S. Environmental Protection Agency, *Integrated Risk Information System (IRIS) on Formaldehyde*. National Center for Environmental Assessment, Office of Research and Development, Washington, DC, 1999.

CHAPTER 14

# Nonmedical Use of Drugs

## INTRODUCTION

Drugs are chemicals used for the prevention, diagnosis, and treatment of human disease. In the process, drugs interact at the molecular level in the body and produce both beneficial and harmful effects. Certain drugs also affect human behavior. Some drugs are always taken in excess to experience euphoria, which leads to nonmedical use of drugs. Proper and judicious prescription of drugs makes complex surgery possible and relieves pain for millions of people. However, nonmedical use of drugs, such as opioids, central nervous system (CNS) depressants, and stimulants, is a serious health problem. Use of these drugs can lead to drug abuse and addiction, characterized by compulsive drug-seeking. A therapeutic dose does not cause addiction, but inappropriate use or misuse can lead to addiction. Illegal drugs, such as heroin, cocaine, and nicotine, cause health hazards as individuals misuse them.

## DRUG USE AND MISUSE

### Drug Abuse

Nonmedical use of drugs has become a common trend among teenagers worldwide, particularly in certain underdeveloped countries. The misuse of prescribed drugs may be the most common drug abuse in the elderly. Prescription medication is about three times more frequent in the elderly than in general population, and sometimes they do not comply with the direction of use. In general, men and women have similar rates of nonmedical drug use.

### Drug Dependence

Drugs are capable of influencing mood and are liable to be used repeatedly to attain euphoria (state of well-being), withdrawal from reality, and social adjustment. The term *drug dependence* is not specific and often is used vaguely.

*Industrial Guide to Chemical and Drug Safety*, By T.S.S. Dikshith and Prakash V. Diwan
ISBN 0-471-23698-5 

There are two types of dependence (i.e., psychological dependence and physical dependence).

Psychological drug dependence is said to develop when the individual seeks a state of well-being that is only achieved through certain CNS drugs. Initially, the drug may evoke desire for its intake, later progressing to compulsion. The intensity of psychological dependence may vary from desire to craving for the drug, ultimately leading to self-medication.

### Physical Dependence

Physical dependence for drugs is achieved by the repeated administration of a drug, which creates a necessity of its presence in the body for the individual to maintain physiological equilibrium. Discontinuation of the specific drug results in withdrawal symptoms or withdrawal syndrome. The process of CNS adaptation to the drug for normal functioning is known as "neuroadaptation." Drugs that commonly produce physical dependence are opioids and barbiturates. Other CNS depressants including alcohol, benzodiazepines, and stimulants (e.g., amphetamines, cocaine, LSD, cannabis) produce better or no physical dependence. Alcoholism, the chronic use of alcohol, may lead to suicidal tendency and memory impairment associated with specific brain damage. People with different backgrounds and nutritional deficiencies are commonly seen in alcoholism. Alcohol is toxic to many organs and causes adverse effects to the liver and cardiovascular system. It also causes endocrine and gastrointestinal effects, malnutrition, and CNS dysfunction. Ethanol readily crosses the placental barrier and produces fetal alcohol syndrome (FAS), a major cause of mental retardation.

### Drug Habituation

Drug habituation is a complex phenomenon often associated with different social orders. The drug user often exhibits mild discomfort to withdrawal of the drug. Consumption of tea, coffee, tobacco, or social drinking of alcohol is not a physical dependence. Principally, habituations and addiction imply different yardsticks of psychological dependence; it is difficult to draw a clear-cut distinction between addiction and habituation. Moreover, there is no substantial scientific logic, and terms are often interpreted differently. Therefore, it is advisable to avoid this term in describing drug dependence and related conclusions.

### Drug Addiction

Drug addiction is a phenomenon of compulsory drug use characterized by overwhelming involvement with the drug. This aspect (i.e., pursuit of the drug) becomes the primary activity of the individual, who spends time in procuring and using the drug. Even after withdrawal, most addicts tend to relapse. Physical dependence is not essential for addiction. *Drug abuse* refers to drug use as a self-medication, deviating from therapeutic doses. The term conveys social disapproval of the manner and purpose of drug use. There are many dangers in this experience; addicts believe they can experience and enjoy panacea with the first dose, which eventually hooks them.[1] Drugs of abuse and addiction reach higher plasma concentrations when taken intravenously or inhaled, compared with oral administration. This accounts for the "kick" most addicts experience. In the United Kingdom, an official list of drugs of addiction is available and permitted under law to be supplied with legal limitations.

## Nicotine

Nicotine has complex effects that result in self-administration. Cigarette smoking is the most popular method of inhaling nicotine since the beginning of the twentieth century. In 1998, 60 million Americans were cigarette smokers, which amounts to 28% of the total population. Cigarette smoking is a major cause of stroke and the third leading cause of death in the United States. It is not established that cigarettes and other forms of tobacco (e.g., cigars, pipe tobacco, chewing tobacco) are addictive.[1a,1b]

Nicotine is a CNS stimulant and depressant. It releases epinephrine from the adrenal cortex, and its effects are immediate. It enters the body in the form of cigarette, pipe, cigar, and chewing tobacco. Nicotine addiction results in withdrawal symptoms.[1c,1d] Chronic smokers, when deprived of cigarettes for even 24 hours, become aggressive and hostile. Nicotine use by women has many adverse effects, such as the onset of early menopause in women who smoke. The interaction between nicotine and contraceptives increases the risk of cardiovascular diseases. Pregnant woman who smoke risk having premature infants or infants with low birth weights. In addition to nicotine, cigarette smoke is primarily composed of dozen of gases, mainly carbon monoxide and tar, whose high concentration may lead to lung cancer, emphysema, and bronchial disorders. Carbon monoxide increases the risk of cardiovascular disease.

Withdrawal symptoms are less severe in people who quit smoking gradually. Rates of relapse are highest in the first few weeks and diminish considerably after 3 months. The Food and Drug Administration (FDA) has approved nicotine chewing gum for nicotine dependence. The results of its use are encouraging; it acts as a nicotine replacement to help quit smoking. Also available is a transdermal patch of nicotine, which has better results in combating nicotine addiction. The effective treatment for tobacco in the future may be the development of a nicotine vaccine, which may restrict nicotine's entry in the brain to reduce its effects. A recently developed vaccine contains the nicotine derivative attached to a large protein, which has been found to reduce 64% of nicotine reaching the brain. Clinical trials for the developed vaccines are scheduled in 2002.

Nicotine is highly addictive and is a CNS stimulant and depressant. Ingestion of nicotine results in an almost immediate "kick," because it causes a discharge of epinephrine from the adrenal cortex. This stimulates the CNS and other endocrine glands, which causes a sudden release of glucose. Stimulation is then followed by depression and fatigue, leading the user to seek more nicotine. Nicotine is absorbed readily from tobacco smoke in the lungs; it does not matter if the tobacco smoke comes from cigarettes. Cigarette smoking leads to nicotine addiction that is comparable to cocaine or amphetamine addiction, although the effects of nicotine are of lower magnitude. Cigarette addiction is influenced by multiple variables. Nicotine is absorbed readily when tobacco is chewed. With regular use of tobacco, nicotine levels accumulate in the body during the day and persist overnight. Thus, daily smokers or chewers are exposed to the effects of nicotine 24 hours each day. Nicotine taken in by cigarette or cigar smoking takes only seconds to reach the brain, but does not have a direct effect on the body for approximately 30 minutes.

Stress and anxiety affect nicotine tolerance and dependence. The stress hormone corticosteroid reduces the effects of nicotine; therefore, more nicotine is required to achieve the same effect. This increases tolerance to nicotine and leads to increased dependence. Animal studies have shown that stress can directly cause relapse of nicotine self-administration after a period of abstinence. Withdrawal symptoms of nicotine include irritability, impatience, hostility, anxiety, depressed mood, restlessness, increased appetite, and weight gain.[2–4]

## Opioids

The most common use of opioids is to relieve acute pain. Heroin is the most common opioid drug that is abused. It is available as an illegal drug. Injection of a heroin solution (made from opioids) produces a variety of sensations such as warmth, and high or intense pleasure. Heroin addicts are docile but they become aggressive on withdrawal. Heroin is processed from morphine (a naturally occurring substance extracted from the poppy seed) and appears as a white or brown powder. Drug users call it as smack, junk, “H,” or skag. It is also known as “Mexican black tar,” specific to its geographic area.

Opioids (heroin) are frequently used in combination with cocaine (speedball) by persons generally involved in crime. Early death may occur as a result of their use. Heroin addicts acquire bacterial infections producing skin abscesses, pulmonary infections, endocarditis, viral hepatitis, and acquired immunodeficiency syndrome (AIDS). There is a range of treatment options for heroin addiction, including medication and behavioral therapies. Methadone, a synthetic opiate medication, blocks the effects of heroin; its results are encouraging.

Euphoric effects are seen immediately after a single injection of heroin, which is accompanied by flushing of skin and dryness of mouth. The CNS will be depressed because of mental clouding. Regular use of heroin may affect blood vessels and cause cellulitis, liver diseases, and pneumonia. Regular use develops tolerance. Heroin abuse causes serious health problems including overdose, collapsed veins, infectious diseases, spontaneous abortions, AIDS, and hepatitis. Naloxone is used to treat heroin overdose by blocking the effects of opiates. There are some behavioral therapies available for heroin addiction. Contingency management therapy uses a voucher-based system, wherein patients earn “points” based on negative drug tests. Cognitive behavioral interventions are designed to help modify the thinking, behavior, and coping skills of addicts.

Heroin withdrawal is much less dangerous than alcohol and barbiturate withdrawal. Withdrawal symptoms are craving for opioid, restlessness, irritability, increased sensitivity to pain, nausea, cramps, muscle and bone aches, insomnia, anxiety, cold flashes with goose bumps (cold turkey), and movements (kicking the habit).[5]

## Cocaine

Cocaine is a very powerful drug that is commonly misused by teenagers and adults. It is taken by sniffing, snorting, injecting, and smoking. Snorting is the process of inhaling cocaine powder through the nose, where it is then absorbed into the bloodstream through the nasal tissues. It produces a sense of well-being and elevates confidence levels. Higher doses produce euphoria of a short duration, involuntary motor activities, stereotyped behavior; paranoia may occur after repeated doses. Cocaine is frequently used in combination with alcohol or heroin. Use of cocaine can cause addiction, cardiac arrhythmias, myocardial ischemia, myocardia, and seizures.

The street name for cocaine is "crack." It is processed from cocaine hydrochloride into a freebase for smoking. The term "crack" refers to the cracking sound heard when the mixture is smoked. Pure cocaine is obtained by purifying with sodium bicarbonate, ether, and ammonia. Smoking cocaine is dangerous, as high doses reach the brain quickly and immediate effects are seen. Sharing needles may lead to human immunodeficiency virus (HIV) infection.

Cocaine is a CNS stimulant that affects blood vessels and pupils, and increases body temperature, heart rate, and blood pressure. The euphoric effects of cocaine are quick and include reduced fatigue and mental clarity, as well as hyperstimulation. Research reports that the faster the absorption, the shorter the duration of action. The effects of cocaine in humans are variable (e.g., feeling of restlessness, irritability, and anxiety). Cocaine has powerful neuropsychological-reinforcing properties that are responsible for its repeated compulsive use. In some cases, the first dose may prove fatal. Cocaine-related death may be due to cardiac arrest or convulsion followed by respiratory arrest. In drug abuse, people mix cocaine with alcohol, leading to a chemical complex called cocaethylene, which intensifies the euphoria but can culminate in death.

Adverse effects of cocaine include constricted peripheral blood vessels, dilated pupils, and increased body temperature, heart rate, and blood pressure. Cocaine induces several immediate euphoric effects, such as hyperstimulation, reduced fatigue, and mental clarity, all of which depend on the administration route. The faster the absorption of cocaine, the more severe the effects. In contrast, faster absorption limits the duration of action. For example, the effect from snorting cocaine may last 15 to 30 minutes, whereas effects from smoking may last 5 to 10 minutes. Increased use can reduce the period of stimulation, as addicted humans may develop tolerance. In rare instances, sudden death may occur on the first use of cocaine or unexpectedly thereafter.

High doses or prolonged use of cocaine can trigger paranoia. Smoking crack cocaine can produce a particularly aggressive paranoid behavior. When addicted individuals stop using cocaine, they often become depressed. This also may lead to further cocaine use to alleviate depression. Prolonged cocaine snorting can result in ulceration of the mucous membranes of the nose and can damage the nasal septum enough to cause pathologic lesions. Cocaine-related deaths are often a result of cardiac arrest or seizures followed by respiratory arrest. Withdrawal symptoms of cocaine include depression, sleepiness, fatigue, cocaine craving, and bradycardia.

### Marijuana

Marijuana is a mixture of dried flowers and leaves from the plant *Cannabis sativa*. It is known by other names such as weed, pot, herb, Mary Jane, boom, gangster, and chronic. People use marijuana in the form of cigarette or in pipes. It also is often mixed with foods or brewed tea. Marijuana use affects the CNS as observed with memory and learning, difficulty in thinking, loss of coordination, increased heart rate, and anxiety.

The smoke from burning marijuana contains many chemicals such as tetrahydrocannabinol (THC), which produces characteristic pharmacologic effects and complex behavioral changes.

Long-term marijuana use also produces changes in the brain. It affects the lungs, which may lead to chronic bronchitis and more frequent chest cold. Regardless of THC content, the amount of tar inhaled and level of carbon monoxide absorbed by marijuana smokers are 3 to 5 times greater than those for tobacco smokers. Pregnant women addicted to marijuana impart effects to the fetus, leading to fetotoxicity.

Marijuana is the most commonly used illicit drug within a large U.S. population. More than 72 million Americans (33%) 12 years of age and older have tried marijuana at least once in their lifetime. Marijuana withdrawal symptoms are restlessness, irritability, mild agitation, insomnia, sleep disturbances, nausea, and cramping.[6–9]

### Lysergic Acid Diethylamide

Lysergic acid diethylamide (LSD) belongs to the class of hallucinogens. LSD was discovered in 1938 and is a potent mood elevator. This is found in ergot, a fungus that grows on rye and other grains. LSD is a colorless, odorless, bitter solid. It is available in various forms such as tablets, capsules, and liquid.

Interestingly, LSD is available in well-decorated absorbent paper and dispersed as small squares, facilitating its abuse by children and adults. The average dose is approximately 20 to 80 μg; higher dose limits are 100 to 200 μg. The effect of LSD depends on the amount of intake; effects appear within 30 to 90 minutes. Adverse effects include mydriasis, increased body temperature, blood pressure, and heart rate, loss of appetite, sleeplessness, dry mouth, sweating, and body tremors.

The drug produces delusions and visual hallucinations. The sensations vary; it may give "a feeling of seeing sounds and hearing color" or "a small water canal looking like a river" leading to a state of panic often termed a "trip" by users. Misuse of LSD results in insanity, fear, schizophrenia, depression, and may lead to death. LSD is not considered an addictive drug since it does not produce the same compulsive drug-seeking behavior as cocaine, amphetamines, heroin, nicotine, and alcohol. However, LSD does produce tolerance like other addicting drugs, which means that any increase in LSD dose is fatal. LSD is not linked with any withdrawal syndrome.

### Phencyclidine

Phencyclidine (PCP) was developed in late 1950 as an intravenous anesthetic agent. PCP produces anesthesia and analgesia with respiratory or cardiovascular depression. However, postoperatively, the drug produced psychotomimetic effects (e.g., delirium and hallucinations) and was subsequently withdrawn from the market.

PCP at low dose (50 μg/kg) produces emotional withdrawal, concrete thinking, and bizarre responses to projective testing. With high doses of PCP, abusers exhibit hostile and aggressive behavior.

PCP has been a "street drug" in the United States since the 1960s, called angel dust, ozone, wack, killer, and joints. Initially, it was not popular because of its variant absorption and devastating dysphoric effects. The drug is easily and cheaply synthesized with (piperdines and cyclohexane). Many PCP derivatives have been used as street drugs.

PCP typically causes tachycardia and hypertension. PCP is sold in the form of capsules, tablets, powder, as a solution in water or alcohol, or as rock salt. It may be sniffed, smoked, or injected through intravenous or subcutaneous injection.

PCP intoxication typically produces miosis, nystagmus, hypertension, tachycardia, salivation, flushing, sweating, ataxia, and CNS stimulation or depression. Overdose of PCP is dangerous, as the user becomes violent and emergency treatment is required. It is necessary to keep the user calm and not leave him alone. Withdrawal symptoms of PCP are tremor, seizures, diarrhea, piloerection, and vocalizations.

### Methamphetamine

Methamphetamine is an addiction-inducing drug that is closely related to amphetamine. However, methamphetamine shows greater CNS effects when compared with amphetamine. The therapeutic use of these drugs is limited to the treatment of obesity.

Methamphetamine is available on the streets and has been identified by names such as chalk and speed. Methamphetamine hydrochloride is a clear chunky crystal resembling ice that can be inhaled by smoking. It is referred to as ice, crystals, and glass.

Methamphetamine at low doses has prominent CNS stimulant effects without significant peripheral actions. Large doses produce a sustained rise in systolic and diastolic blood pressure. Methamphetamine enhances the mood and body movements by increasing the levels of dopamine, which is a neurotransmitter. Methamphetamine can decrease the levels of dopamine and cause Parkinson's disease.

Methamphetamine is taken by the oral, nasal, or intravenous route. The euphoric effects are observed immediately after smoking or administering intravenously, and they last for a few minutes. High doses of methamphetamine in animals have shown damage to the neuron cell endings. Even small amounts of the drug show effects, such as increase in wakefulness, increased physical activity, decreased appetite, hyperthermia, euphoria, and increased respiration. Other CNS effects include irritability, confusion, insomnia, tremors, convulsions, anxiety, aggressiveness, and even death. Methamphetamine can cause irreversible damage to blood vessels of the brain, produce strokes, and lead to death because of cardiovascular collapse.

The use of this drug is widespread in homosexual and bisexual populations. It also is reported that sexual and needle-use behaviors may lead to AIDS. There is no clear-cut physical withdrawal syndrome such as that with opiates. Large doses may cause high risk of acute toxicity and cardiovascular effects within a few days.

## Alcohol

Alcohol abuse and addiction (alcoholism) in America affects 5% to 10% of the population. The consumption of alcohol has become a status symbol among the business class, students, and high society. People always look for an occasion to celebrate with drinks and ultimately may become the victims of alcohol addiction.[10–15]

***Alcohol and Car Driving*** The effects of alcohol and driving are a subject of great importance, as many countries have made laws designed to prevent car accidents. Accidents involving alcohol number as many as 50% of total accidents (Fig. 14-1). Hence, it is necessary to perform breath or blood tests to determine a driver's alcohol content. In the United Kingdom, a blood-alcohol level exceeding 80 mg/100 mL is an offense; other countries such as the United States and Australia have lower limits.

Alcohol consumption causes peripheral vasodilatation; the person feels warm, body heat is slowly lost, and an overdose can cause rapid hypothermia. An acute dose of 3 to 4 units will increase blood pressure. It will cause irritation to gastric mucosae followed by erosion that heals after 3 to 4 weeks. At least 60% of chronic gastritis is reported with chronic alcoholics. Alcohol intake also affects sexual function. William Shakespeare wrote that alcohol "provides the desire but takes away the performance." Alcohol can be source of energy but cannot be used as food.

***Features of Alcohol Dependence*** People who drink alcohol occasionally can progress to drinking as a routine and then a habit. Constant drinking makes

**Figure 14-1.** Effect of excess of alcohol on car driving.

the individual feel guilty, unable to concentrate or discuss the problem, and lose control. Later, the individual may behave aggressively and many times promise to stop drinking but fail. Eventually, the drinker will avoid friends and family, and prefer to be alone. The individual may neglect work, and restrict food, and progress to moral deterioration. The chronic drinker cannot initiate any action; thinking becomes impaired. Eventually, the drinker will admit defeat and enter into a vicious circle that makes life miserable. Help must be given to the alcoholic drinker to overcome the problem.

***Alcohol Withdrawal Syndrome*** Withdrawal symptoms of alcohol include alcohol craving, tremor, irritability, nausea, sleep disturbance, tachycardia, hypertension, sweating, seizures 12 to 48 hours after last drinking, severe agitation, confusion, and dilated pupils.

***Fetal Alcohol Syndrome*** First identified around 1970, the term *fetal alcohol syndrome* (FAS) is often used to describe the damage some children suffer when mothers drink during pregnancy. The damage can be moderate to severe, causing clumsiness, behavioral problems, stunted growth, disfigurement, and mental retardation. Little is known about the thresholds of alcohol that cause FAS. It also is assumed that genetic factors may be associated, as not all mothers who drink have FAS babies.

The mind-bending effects of alcohol begin soon after it appears in the blood. Within minutes it enters the brain, numbing the nerve cells and slowing messages to the body. Chronic alcoholism affects speech, vision, posture, and judgment. In extreme cases, the person will die of respiratory failure. Alcoholism increases the risk of heart disease, cancer, and liver failure.

***Medicinal Uses of Alcohol*** Alcohol is seldom prescribed for medicinal use. The medicinal use of alcohol is essentially restricted to external use and a as vehicle for liquid preparations such as syrups and tinctures. It is frequently used as an antiseptic, as a rubefacient for sprains and joint pain, to reduce the body temperature as an alcohol sponge, and as an injection to relieve neurologic pain.

***Caution of Methyl Alcohol (Wood Alcohol)*** Methanol is commonly added to the rectified spirit, which makes it unfit to drink. Mixing methanol with alcoholic beverages results in methanol poisoning. Methanol is metabolized to formaldehyde and formic acid by aldehyde dehydrogenase. High blood levels (750 mg/dL) cause severe poisoning, which leads to blindness and even death.

Methanol poisoning symptoms include vomiting, headache, bradycardia, and hypotension. Presence of formic acid causes retinal damage. Alcohol is contraindicated in peptic ulcers, hyperacidity, reflux esophagitis, epilepsy, liver dysfunction, unstable personalities, and pregnant women.

***Food Value*** Alcohol is metabolized rapidly and does not require digestion. It yields approximately 7 caL/g, and this energy cannot be stored. Although it has a high caloric value, it cannot be used as a food because it does not contain nutrients. Individuals who consume large volumes usually suffer from nutritional deficiencies.

### Sports and Misuse of Drugs

Drugs are sometimes used to enhance the performance of athletes in prestigious sport events. The efficacy of such drugs is not documented. Detection of these drugs (e.g., anabolic steroids) is a difficult task. For events such as boxing, rowing, weightlifting, and wrestling, anabolic agents, such as clenbuterol, methandienone, nandrolone, stanozolol, and testosterone, are commonly used. These drugs are taken along with a high-protein diet and exercise, which increases body weight but not necessarily strength. High doses cause liver damage if the drugs are taken for a long duration. Growth hormones, such as somatotropine and corticotropin, are combined with anabolic steroids. Athletes who compete in bicycling and marathon running prefer to increase the oxygen-carrying capacity of the blood with erythropoietin. Sports — which require a steady hand in case of pistol or rifle shooting competitions — use $\beta_2$ receptor-blocking agents. For minor injuries during athletic training, nonsteroidal antiinflammatory drugs (NSAIDs) and corticosteroids are permitted to suppress inflammatory symptoms. The route of administration is restricted, and strong opioids are not allowed. Prescription of such drugs to athletes may disqualify them for participation in some sport competitions.

### Drugs Used for Torture, Interrogation, and Judicial Execution

Drugs, such as suxamethonium hallucinogens, thiopentone, nemoleptics, amphetamines, apomorphine, and cyclophosphamide, have been used for torture, interrogation, and judicial execution.[15a] These drugs have been used to hurt, frighten, and confuse the victim to reveal the truth in case of murder or theft.

## REFERENCES

1. D.W. Maurer, and V.H. Vogel, *Narcotics and Narcotic Addiction*. Charles C. Thomas Springfield, 1962
1a. T.R. Kosten, and L.E. Hollister, Drugs of Abuse, in *Basic and Clinical Pharmacology*, Bertram G. Katzung, 8th ed., The McGraw-Hill Companies, Inc., 537, 2001.
1b. M.R. Picciotto, Common aspects of the action of nicotine and other drugs of abuse. *Drug Alcohol Depend.* **51**:165, 1998.
1c. N.L. Benowiz, Pharmacology of nicotine: Addiction and therapeutics. *Ann. Rev. Pharmacol.* **36**:597–613. (General review article, including information on potential therapeutic uses of nicotine other than reduction of smoking.)
1d. H.P. Rang, M.M. Dale, and J.M. Ritter, Drug dependence and drug abuse, in *Pharmacology*, H.P. Rang, et al., ed. 4th ed., Churchill Livingstone, 618–623, 1999.
2. R. Peto, A.D. Lopez, J. Boreham, M. Thun, C. Heath, R. Doll, Mortality from smoking worldwide. *Br. Med. Bull.* **52**:12–21, 1996.
3. Royal College of Physicians Report, *Smoking or Health*, Pitman Medical Publishing, Tunbridge Wells, UK, 1977.
4. D.J.K. Balfour, and K.O. Fagerstrom, Pharmacology of nicotine and its therapeutic use in smoking cessation and neurodegenerative disorders. *Pharmacol. Ther.* **72**:51–81, 1996.
5. R. Maldonado, et al., Absence of opiate rewarding effects in mice lacking dopamine $D_2$ receptors. *Nature* **388**:586–589, 1997.
6. R.G. Pertwee, Pharmacology of cannabinoid $CB_1$ and $CB_2$ receptors. *Pharmcol. Ther.* **74**:129–180, 1997.
7. W.L. Dewey, Cannabinoid pharmacology. *Pharmacol. Rev.* **38**:151–178, 1986.
8. M.E. Abood, and B.R. Martin, Molecular neurobiology of the cannabinoid receptor. *Int. Rev. Neurobiol.* **39**:197–221, 1996.
9. M.E. Abood, and B.R. Martin, Neurobiology of marijuana abuse. *Trends Pharmacol. Sci.* **13**:201–206, 1992.
10. G. Zernig, K. Fabisch, H. Fabisch, Pharmacotherapy of alcohol dependence. *Trends Pharmacol. Sci.* **18**:229–231, 1997.
11. M.A. Shuckit, *Drug and Alcohol Abuse*, Plenum press, New York, 1995.
12. H.J. Little, Mechanisms that may underlie the behavioral effects of ethanol. *Prog. Neurobiol.* **36**:223–225, 1991.
13. J.W. Hanson, A.P. Streissguth, D.W. Smith, The effects of moderate alcohol consumption during pregnancy of fetal growth and morphogenesis. *J. Pediatr.* **92**:457–460, 1978.

14. J.C. Crabbe, J.K. Belknap, K.J. Buck, Genetic animal models of alcohol and drug abuse. *Science* **264**:1715–1723, 1994.
15. M.E. Charness, and R.P. Simon, D.A. Greenberg, Ethanol and the nervous system. *N. Engl. J. Med.* **321**:442–454, 1989.
15a. D.R. Laurence, P.N. Bennett, and M.J. Brown, Poisoning, overdose, antidotes, in *Clinical Pharmacology*, D.R. Laurence, et al., ed. 8th ed., Churchill Livingstone, 149, 1997.

CHAPTER 15

# Drugs and Pharmaceuticals: Safety Considerations

## INTRODUCTION

Humans and animals are prone to a variety of diseases and infections. Humans suffer due to known (identifiable) and unknown (not well diagnosed) diseases and ailments, making it imperative to introduce a number of newer drugs. Often, these conditions require multiple-drug therapy for an effective cure. In this context, the safety and efficacy of therapy is extremely important.

Drugs are known to possess physical and chemical incompatibilities. During manufacture of drugs and dosage forms, care should be taken to contain ill effects. Also, complete care is necessary in handling and disposal, especially with potent drugs and the associated problems that can result from their improper storage and packaging in pharmacies and health care industries. This chapter addresses therapeutic incompatibilities, such as drug interactions and associated toxicities, accidental or deliberate poisoning, and treatment of adverse effects in the health care industry. Understanding drug interactions is essential because it is associated with the potentiation or alteration of therapeutic effects vis-à-vis the background knowledge of professionals in factories, storehouses, and clinical industries.

During multiple-drug therapy, two medicines of different classes are mixed and administered to the patient. Under such conditions, it is possible that some drugs either lose their activity or (rarely) may have an enhanced effect. The change may occur in the form of physical, chemical, or therapeutic manifestation. There is every possibility that such combinations may affect the safety, efficacy, and appearance of medicine—either in the form of discoloration or precipitations. This change is known as *incompatibility*. In this instance, the pharmacist/doctor/nurse must use pharmaceutical knowledge and pharmacologic background to help determine an appropriate line of action.

Most prescriptions are official, and all details are available (e.g., list of ingredients, quantities, forms it must be dispensed). Despite these details, there are serious incompatibilities reported that may undergo physicochemical or therapeutic

*Industrial Guide to Chemical and Drug Safety*, By T.S.S. Dikshith and Prakash V. Diwan
ISBN 0-471-23698-5 

interactions. The important physicochemical incompatibilities of medicines are given in the *British Pharmacopeial Codex and Extra Pharmacopeia.*

There are many reports of interactions of plastic bags, plastic syringes, and aluminium needles with medicines where additives/materials of these packagings either react or leach out into the formulations. This information will help the manufacturer to exercise caution in the packaging and storage of drugs meant for human consumption. Handling toxic compounds, such as anticancer agents that are irritants to the skin and mucous membranes, requires proper protective clothing and gloves. There are drugs that antagonize the activity of a potent compound when mixed. The knowledge of likely drug incompatibilities is essential to persons handling the drugs. Industrial pharmacists or workers should be well aware of the possible physical and physicochemical incompatibilities due to loss of active ingredient potency or production of toxic components. Workers should also know therapeutic drug interactions, as some agents potentiate or antagonize the action of potent drugs. This may manifest as serious toxic effects. Here, we discuss examples of accidental or deliberate poisoning and related drug treatments, leading to adverse effects or abuse.

## ANTIMICROBIAL AGENTS

### Antibacterials

***Aminosalycylic Acid*** Dose forms should be freshly prepared, since aqueous solutions are unstable. Aminosalicylate acid interacts with other salicylates, such as probenecid and procaine.

***Ampicillin Sodium*** Ampicillin sodium stability depends on its concentration, pH, temperature, and type of vehicle. Ampicillin sodium has stability problems with glucose, fructose, sodium bicarbonate, hetastarch, and lactate. Reconstituted solutions of ampicillin sodium should be administered immediately.

Ampicillin sodium and aminoglycosides are not compatible. Interactions are found with other antibacterials, particularly those in high concentrations.

***Aztreonam*** Incompatibility of aztreonam has been reported with metronidazole, nafcillin, and vancomycin.[1]

***Benzylpenicillin*** Salts of benzylpenicillin are hydrolyzed in aqueous solutions, particularly in acids, alkalis, and concentrated solutions. Instability is minimum (about pH 6.8); using citrated buffers can enhance its stability.

Benzylpenicillin is incompatible with rubber products and metal ions. Stability is affected by alcohols, surfactants, oxidizing and reducing agents, macrogols and other hydroxy compounds, glycerol, glycols, some paraffins and ointment bases, preservatives such as chlorocresol or thiomersal, blood and blood products, and antibacterials such as amphotericin, cephalosporins, and vancomycin. Injections of benzylpenicillins and aminoglycosides should be administered separately.

Benzylpenicillin interacts with probenecid, some antibacterials (as mentioned above), and anticoagulants.

***Cefoperazone Sodium*** Admixture of cefoperazone sodium and aminoglycosides leads to inactivation of these drugs. Cefoperazone sodium is not compatible with other drugs that include diltiazem, pentamidine, perphenazine, pethidine, and promethazine.[2]

***Ceftazidime*** Ceftazidime is not compatible with aminoglycosides, vancomycin, and pentamidine; visual incompatibilities or inactivation have been reported.[3]

***Ceftriaxone Sodium*** Ceftriaxone sodium is not compatible with aminoglycosides, labetalol, vancomycin, fluconazole, amsacrine, and pentamidine.[4]

***Cephalothin Sodium*** Cephalothin sodium precipitates if formulated below pH 5 and is incompatible with aminoglycosides. When administered with aminoglycosides or diuretics, it may cause kidney problems.

***Chloramphenicol Palmitate*** Chloramphenicol has concentration-dependent incompatibility with many drugs. Chloramphenicol has severe toxic effects, and overdose may be treated by hemoperfusion of charcoal. It interacts with coumarin anticoagulants, some oral hypoglycemics, antiepileptic drugs, and nutrients such as iron and vitamin $B_{12}$.

***Chlortetracycline*** Preparations of chlortetracycline are not compatible with alkaline dose forms and drugs that are unstable with low pH. Drugs of this class interact with drugs that are exclusively metabolized and cleared by liver and inorganic cations.

***Ciprofloxacin*** Ciprofloxacin is not compatible with a range of antibacterials, and an infusion must not be mixed with drugs that are unstable with low pH.[5]

***Clindamycin*** Solutions of clindamycin salts are incompatible with alkaline doses or other drugs that are unstable at low pH. Clindamycin phosphate is not compatible with some rubber packing materials. pH-dependent degradation was reported with clindamycin, and phosphate salt was found to be most stable. Therapy with clindamycin should be stopped if diarrhea or colitis are found. The patient should be treated with other antibiotics. Clindamycin potentiates the actions of neuromuscular-blocking agents, opioids, and the antagonistic action of parasympathomimetics.

***Dapsone*** Supportive treatment for the adverse effects of dapsone may be initiated with stomach wash and activated charcoal. Methylene blue could be given to treat methemoglobinemia, but this is not effective in patients with glucose-6-phosphate dehydrogenase deficiency. Infusion of human erythrocytes can be given to treat hemolysis. Dapsone interacts with probenecid, trimethoprim, and rifampicin, and may reverse the antiinflammatory action of clofazimine.

***Doxycycline Hydrochloride*** Doxycycline hydrochloride is incompatible with alkaline dose forms or drugs that are unstable at low pH.

***Erythromycin*** Erythromycin is incompatible with preparations that are highly acidic or alkaline in nature. Pancreatitis disorders are reported with erythromycin overdose. Erythromycin and other macrolides interact with other drugs and are detailed elsewhere.[6]

***Gentamicin Sulfate*** Aminoglycoside antibiotics are incompatible with other antibiotics, such as penicillins and cephalosporins. Among aminoglycosides, amikacin is more stable, but tobramycin is rapidly inactivated. Other antibiotics that rapidly inactivate aminoglycosides are ampicillin, benzylpenicillin, carbenicillin, and antipseudomonal penicillins. Gentamicin is incompatible with frusemide, heparin, drugs with alkaline pH, or drugs that are unstable at acidic pH. Hence, aminoglycoside antibiotics such as gentamicin should not be mixed with the infusion containing beta lactams and other antibiotics. If required, they must be administered separately.[7]

The gentamicin sulfate solution containing a concentration of 10 to 40 mg/mL should not be stored in plastic and glass syringes, since it produces a brown precipitate.

***Isoniazid*** Isoniazid is incompatible with sugars. Isoniazid overdose may be severe to fatal, and treatment is symptomatic and supportive, including stomach wash for control of convulsions and treating metabolic acidosis. Administration of pyridoxine and hemodialysis may be needed. Isoniazid interacts with carbamazepine, phenytoin, diazepam, triazolam, chlorzoxazone, theophylline, ethosuximide, enflurane, cycloserine, and warfarin.

***Lincomycin Hydrochloride*** Lincomycin hydrochloride is incompatible with drugs unstable at low pH. Adverse effects of lincomycin hydrochloride can be treated with hemodialysis, peritoneal dialysis, and administration of calcium salts.

***Mezlocillin Sodium*** Mezlocillin sodium is not compatible with ciprofloxacin, metronidazole, aminoglycosides, and tetracyclines.

***Nafcillin Sodium*** Nafcillin sodium is not compatible with other antibacterials, including aminoglycosides, acidic drugs, and alkaline drugs.

***Nalidixic Acid*** Nalidixic acid should not be given with nitrofurantoin or melphalan. It interacts with cyclosporin, probenecid, and warfarin.

***Neomycin*** Neomycin interferes with the absorption of digoxin, methotrexate, vitamins, some penicillins, acarbose, and oral contraceptives.

***Nitrofurantoin*** Nitrofurantoin interacts with carbonic anhydrase inhibitors and probenecid.

***Polymyxin B*** Preparations that are highly acidic or alkaline in nature and other antibacterials are incompatible with polymyxin B sulfate. Polymyxin interacts with neuromuscular blockers and nephrotoxic drugs.

***Streptomycin Sulfate*** Streptomycin sulfate should be protected from moisture and is incompatible in the presence of acids and alkalis. Professionals who handle streptomycin should wear masks and rubber gloves, since streptomycin may cause severe dermatitis.

***Sulfadiazine Sodium*** Solutions of sulfadiazine sodium are alkaline, and incompatibility may reasonably be expected with acidic drugs or preparations unstable at high pH. Sulfadiazine interacts with paraaminobenzoic acid derivatives.

***Sparfloxacin*** Sparfloxacin should not be administered with antihistamines, cisapride, erythromycin, pentamidine, phenothiazines, and tricyclic antidepressants.

***Tetracycline*** Tetracycline injections have an acid pH. Incompatibility may reasonably be expected with alkaline preparations or with drugs unstable at low pH. Care should be taken when administering tetracyclines, since chelation takes place with metal ions. Tetracyclines interact with inorganic metal ions. They should not be used with drugs that cause hepatotoxicity and nephrotoxicity (e.g., digoxin, theophylline, ergot alkaloids, methotrexate, oral contraceptives, and penicillins).

***Trimethoprim*** Trimethoprim lactate injections are incompatible with other antibacterials.

***Vancomycin Hydrochloride*** Solutions of vancomycin hydrochloride have an acid pH and incompatibility may reasonably be expected with alkaline preparations or drugs unstable at low pH. Vancomycin interacts with aminoglycosides, loop diuretics, general anesthetics, and neuromuscular blockers.

### Antifungals

***Amphotericin*** Generally, amphotericin should not be mixed with other drugs. Change in pH and disruption of the colloidal suspension are the major reasons for precipitation and, hence, incompatibility. Any formulations containing isotonic sodium chloride should be avoided as they can cause precipitation of amphotericin.[8]

Stability of amphotericin in lipid emulsions was moderate,[9,10] but some researchers have reported instability of amphotericin with lipid emulsion.[11,12] Vigorous mixing of amphotericin and a lipid emulsion mixture has a good effect on the stability of the formulation.

Aspirin, paracetamol, and hydrocortisone are used to control febrile reactions of amphotericin. Patients with a history of adverse effects with amphotericin should be prophylactically treated with antipyretics and hydrocortisone. Antiemetics and pethidine also are used for the treatment of adverse effects of amphotericin. With sodium supplements and hydration therapy, damage to the kidney can be reduced. If conventional amphotericin is not well tolerated by the patient, colloidal carriers can be used as alternative options. Administration of amphotericin with a nephrotoxic drug, such as cyclosporin, may further increase toxicity. Diuretics and anticancer drugs should be avoided with amphotericin.

***Fluconazole*** Fluconazole may increase the plasma concentration of phenytoin, sulfonylurea (hypoglycemic), cyclosporin, zidovudine, cisapride, and terfenadine. Coadministration of fluconazole with rifampicin results in the reduced plasma concentration of fluconazole. Griseofulvin may reduce the plasma concentration of salicylates, coumarin, anticoagulants, and oral contraceptives.

***Miconazole*** Storage of miconazole in polyvinylchloride (PVC) containers is unsafe.[13] Miconazole is stable to dry-heat sterilization at 160 °C for 90 minutes when used with castor oil or arachis oil.[14]

### Antiprotozoals

Alcohol should be avoided during furazolidone and metronidazole therapy. Toxic psychosis has been seen in patients under treatment with furazolidone. Metronidazole should be used with caution when coadministering with warfarin, phenytoin, lithium, omeprazole, and phenobarbital. The coadministration of pentamidine with drugs toxic to the kidney should be avoided.

The concentrated form of metronidazole hydrochloride reacts with aluminium needles used for drug administration. Metronidazole hydrochloride should not be mixed with other drugs, including antibiotics, during intravenous infusions.[15]

Pentamidine isethionate solution is incompatible and precipitates immediately with cephalosporin and cephalomycin injections when admixed with 5% glucose infusions.[16]

### Antivirals

Coadministration of acyclovir and cidofovir with other nephrotoxic drugs may cause renal impairment. Simultaneous therapy of zidovudine and acyclovir is safe, and it has no adverse effects except fatigue. Indinavir sulfate is contraindicated with drugs having a narrow therapeutic window such as cisapride or terfenadine. Caution should be exercised when using antiviral drugs with antibacterials, analgesics, antiarrhythmics, antidiabetics, ergot alkaloids, hormonal contraceptives, calcium channel blockers, and theophylline.[17]

Use of trimethoprim in high doses should be avoided when coadministering lamivudine. Coadministration of stavudine with drugs causing peripheral neuropathy (e.g., metronidazole) and pancreatitis (e.g., pentamidine) should be avoided. Care should be exercised with simultaneous administration of nephrotoxic and myelosuppressive drugs. Zidovudine interacts with nonsteroidal antiinflammatory drugs (NSAIDs), antibacterials, antifungals, antivirals, and paracetamol.

Foscarnet sodium should not be mixed with any intravenous infusion drug, since it interacts with glucose- and calcium-containing solutions and sodium, ganciclovir, certain antibacterials, amphotericin, pentamidine isethionate, and trimetrexate.[18] Care must be taken to avoid inhalation and exposure of the skin to zalcitabine powder.

## AUTOCOIDS: DRUGS USED FOR INFLAMMATORY-RELATED DISEASES

### Analgesic, Antipyretic, and Antiinflammatory Drugs

***Nonsteroidal Antiinflammatory Drugs*** The intake of these drugs in an excessive dose, either accidentally or deliberately, leads to adverse effects. Overdose with NSAIDs may lead to acute poisoning, and common side effects include nausea and vomiting, headache, drowsiness, blurred vision, and dizziness. Rare serious toxicities include seizures, hypotension, apnea, coma, and renal failure, although usually after ingestion of substantial quantities. Overdose of mefenamic acid may cause seizures. Supportive treatment of overdose can be performed by gastric lavage, multiple doses of activated charcoal and forced diuresis, hemodialysis, and (rarely) hemoperfusion.[19]

Many drugs are taken simultaneously, and these combinations may lead to interactions. Thus, the simultaneous use of more than one NSAID should be avoided because of the increased risk of adverse effects. The concomitant administration of NSAIDs with either corticosteroids or alcohol, or bisphosphonates or oxpentifylline, may lead to increased incidents of gastrointestinal (GI) disorders. Generally, NSAIDs interact with and enhance the effects of oral anticoagulants, as well as increase plasma concentrations of lithium, methotrexate, and cardiac glycosides. There will be increased nephrotoxicity and hyperkalemia given along with angiotensin-converting enzyme (ACE) inhibitors, cyclosporin, ticrolimus or diuretics, and potassium-sparing diuretics. Reduction of the antihypertensive effects of ACE inhibitors, beta-blockers, and diuretics may be found. Other effects include convulsions with the use of quinolones; phenytoin and sulfonylurea antidiabetics and their effects may be enhanced with moclobemide. There is a risk of hemotoxicity with concomitant use of zidovudine, and NSAIDs must be avoided along with mifepristone.[20,21] Acetylcholine chloride, acetylcholine, and carbacol ophthalmic preparations are ineffective in patients being treated with topical NSAIDs.

***Drug Dependence, Withdrawal, and Treatment of Opioid Dependence*** Drug dependence and self- or cross-tolerance develop with the use of opioid analgesics. Abrupt termination of therapy leads to severe withdrawal symptoms that include yawning, mydriasis, lacrimation, rhinorrhea, sneezing, muscle tremor, weakness, sweating, anxiety, irritability, disturbed sleep or insomnia, restlessness, anorexia, nausea, vomiting, loss of weight, diarrhea, dehydration, leukocytosis, bone pain, abdominal and muscle cramps, gooseflesh, vasomotor disturbances, and increases in heart rate, respiratory rate, blood pressure, and temperature. Opioid antagonists, such as naloxone, or an agonist/antagonist, such as pentazocine, could be used to treat withdrawal symptoms.[22,23] Other drugs for such therapy include methadone, diamorphine, dihydrocodeine, levomethadyl acetate, buprenorphine, clonidine, and naltrexone.[24–28] In some instances, a previously tolerated dose may prove fatal, since tolerance diminishes rapidly.[29]

Counseling and other psychosocial services play a role in withdrawal therapy.[30] Supportive drugs for withdrawal therapy include diphenoxylate with atropine, promethazine, propranolol, thioridazine, benzodiazepines, and chlormethiazole — the use of which all require care and caution. Gerada et al.[31] have discussed problems associated with the management of pregnant patients with opioid dependence.

Supportive therapy for acute opioid poisoning may be performed first by stomach wash, and then laxative and intensive supportive therapy to treat respiratory failure and shock. Further, naloxone may be given to treat severe respiratory depression and coma. However, intravenous infusions are preferred if the toxicity is the result of a long-acting opioid. The treatment of adverse effects itself may cause withdrawal symptoms.[32]

Drugs and chemicals are known to cause activated interaction. The depressant action of opioid drugs is enhanced by drugs acting on the central nervous system (CNS) such as alcohol, anesthetics, anxiolytics, hypnotics, tricyclic antidepressants, and antipsychotics. Concomitant administration of opioid analgesics and monoamine oxidase inhibitors (MAOIs) should be avoided, or extra care should be taken if such a therapy is inevitable. Fatal reactions are reported when treated along with selegiline. Interactions also are reported with cyclizine, cimetidine, mexiletine, cisapride, metoclopramide, or domperidone.

***Alfentanil Hydrochloride*** Alfentanil hydrochloride can cause drug dependence similar to that of opioid analgesics. Since this may cause irritation and related adverse effects, it is important to avoid skin contact and the inhalation of alfentanil hydrochloride particles to contain adverse effects.

***Amidopyrine*** Amidopyrine is a potential carcinogen, particularly for patients who are heavy smokers.[33]

***Aspirin*** Aspirin is the most commonly used over-the-counter (OTC) drug. It is known to cause GI bleeding and results in ulcers when taken before food. In view of this, aspirin and related salicylates should be used cautiously, particularly in children younger than 12 years of age, as it may lead to Reye's syndrome. Buprenorphine is another drug that can cause drug dependence and naloxone may not reverse the actions.[34]

***Butorphanol Tartrate*** With this drug, dependence, withdrawal, and interactions are similar to that of opioids. Butorphanol has less potential to produce dependence when compared with morphine, but it is a drug of abuse. Naloxone acts as an antagonist and can be used for the treatment of overdose.[35]

***Codeine Phosphate*** The presence of aspirin along with codeine, even at a low moisture level, leads to acetylation of codeine phosphate in solid dose forms and is incompatible.[36] Codeine sulfate solutions are more stable than phosphate salts.[37] Drug dependence and withdrawal resemble that of opioid analgesics. Overdose causes acute intoxication in children, as accidental or deliberate ingestion of cough preparations containing codeine.[38]

Persons who drive or who are engaged in skilled work should avoid using codeine phosphate and alcohol together since they are incompatible and lead to seizures and adverse effects.[39]

***Dextropropoxyphene*** Dextropropoxyphene, a drug of abuse, and any misuse or overdose of which causes intoxication if taken along with paracetamol or alcohol. The individual develops serious CNS depression leading to death.[40–42] Adverse effects can be treated by gastric lavage and administration of activated charcoal and naloxone.[43]

Many drugs such as ritonavir and dezocine interact, leading to adverse drug action, and hence require care and caution during treatment.

***Diamorphine*** Diamorphine is relatively unstable in aqueous solutions, and minimum decomposition was observed at pH 4. Preparations should be used within 4 weeks when kept at room temperature, but degradation products also have analgesic activity.[45,46] Diamorphine is used in the management of opioid dependence; this is also a drug of abuse and the overdose is fatal.[47,48] Interactions are similar to those of opioid analgesics. Withdrawal symptoms of opioid dependence can be treated with diamorphine and methadone.[49]

***Diamorphine Hydrochloride*** Diamorphine interacts with mineral acids and alkalis and with chlorocresol. Concentration-dependent precipitations were observed in mixtures of diamorphine hydrochloride with cyclizine and haloperidol, and color changes were observed with mixtures containing metoclopramide and diamorphine.[44]

***Diclofenac*** Diclofenac should not be administered intravenously to patients already receiving NSAIDs or anticoagulants, including low-dose heparin. Concomitant use of diclofenac with triamterene or cyclosporin affects kidney function.[50]

***Diflunisal*** Overdose (15 g) of diflunisal leads to poisoning, which could be fatal.[51] Treatment can be given by gastric lavage and supportive care. Interactions of diflunisal with indomethacin, paracetamol, antacids, benzodiazepines, and probenecid have been reported.[5] Concomitant use with indomethacin should be avoided, as this could cause fatal GI complications.

**Dihydrocodeine** Dihydrocodeine is drug of abuse, and overdose may lead to kidney and liver disorders, and anaphylactoid reaction and can be treated with naloxone.[52,53]

**Etorphine Hydrochloride** Maximum care must be exercised when handling etorphine hydrochloride. Spillage on any part of the body must be washed immediately to avoid fatal reactions. When the drug is administered accidentally, a reversing agent should be given. Dependence and withdrawal symptoms are similar to those of opioid analgesics.

**Fentanyl** Care should be taken to avoid skin contact and inhalation of fentanyl citrate to prevent adverse effects. Incompatibility has been reported with other drugs such as thiopentone sodium, methohexitone sodium, and fluorouracil.[54] This drug should not be administered along with alkaline drugs.

Fentanyl citrate is stable if stored at room temperature up to 48 hours under normal conditions. Fentanyl should not be formulated with alkaline drugs and stored in PVC containers for product stability reasons. Formulations containing fentanyl and bupivacaine were found to be adsorbed in PVC containers. The formulation containing adrenaline or fentanyl citrate becomes degraded.[55–57]

Fentanyl and its analogs are drugs of abuse. Overdose and dependence may lead to respiratory complications and death.[58] Adverse effects can be treated with drugs such as naloxone, atropine, and neuromuscular blockers. In general, interactions are similar to those of opioid analgesics.

Ophthalmic preparations containing acetylcholine chloride and carbacol were found infective in patients receiving topical NSAIDs.

**Hydromorphone Hydrochloride** Formulations containing hydromorphone with either minocycline hydrochloride or tetracycline hydrochloride were found incompatible and manifest as a color change from pale yellow to light green. Concentration-dependent incompatibilities are reported in formulations containing hydromorphone hydrochloride with dexamethasone sodium or phosphate,[59] and fluorouracil.[60] Visual incompatibility, such as haziness or precipitation, developed 4 hours after mixing thiopentone sodium and hydromorphone hydrochloride.[61] Dependence, withdrawal, and interactions are similar to those of opioid analgesics.

**Ibuprofen** Overdose of OTC ibuprofen manifests as a syndrome of coma, hyperkalemia with cardiac arrhythmias, metabolic acidosis, pyrexia, and respiratory and kidney failure. The treatment of adverse effects includes intubation, mechanical ventilation, fluid resuscitation, gastric lavage, and administration of activated charcoal.[62] Interactions are similar to those of NSAIDs. Moclobemide enhances the effects of ibuprofen.

**Indomethacin** Indomethacin sodium injections must be reconstituted either with preservative-free sodium chloride or with water for injection. Addition of glucose or reconstitution of pH below 6 must be avoided, as indomethacin precipitates under these conditions. Physical incompatibilities of indomethacin have been reported with tolazoline hydrochloride, glucose injection, calcium gluconate, dobutamine, dopamine, cimetidine, gentamicin sulfate, and tobramycin sulfate.[63–65] Indomethacin-reconstituted solution stored at 2° to 6 °C was found stable for 14 days in glass vial or polypropylene syringe.[66]

In general, interactions are similar to those of NSAIDs. Indomethacin interacts with probenecid and aminoglycoside antibiotics. Concomitant administration of indomethacin and diflunisal should be avoided because of fatal GI complications. Similarly, concurrent administration of indomethacin and haloperidol produces severe drowsiness and confusion.

***Ketorolac*** Interactions are similar to those of NSAIDs. Ketorolac is contraindicated in patients receiving anticoagulant therapy. Ketorolac should not be administered with probenecid or other NSAIDs.

***Mefenamic Acid*** Mefenamic acid overdose produces CNS toxicity, such as convulsions and coma.[67]

***Methadone Hydrochloride*** Visual incompatibilities of methadone hydrochloride were observed with solutions of aminophylline, ammonium chloride, amylobarbitone sodium, chlorothiazide sodium, heparin sodium, nitrofurantoin sodium, novobiocin, pentobarbitone sodium, phenobarbitone sodium, phenytoin sodium, quinalbarbitone sodium, sodium bicarbonate, sodium iodide, sulfadiazine sodium, sulfafurazole diethanolamine, or thiopentone sodium.

Although methadone is used for the management of opioid dependence, symptoms of overdose are similar to those of morphine poisoning.

Interactions are similar to those of opioid analgesics. Methadone has interactions with enzyme inducers. Withdrawal symptoms of methadone are observed upon concurrent administration with phenobarbitone, phenytoin, and rifampicin. Methadone toxicity increases if administered with cimetidine and fluvoxamine.

***Methyl Salicylate*** Dose forms, such as liniments or ointments containing methyl salicylate, should not be dispensed in polystyrene-type containers, as it is not stable. Accidental and deliberate injection causes severe and rapid salicylate poisoning because of its highly concentrated form and rapid absorption.[68]

Excessive topical applications of methyl salicylates have interactions with anticoagulants such as warfarin.[69]

***Morphine Sulfate*** Morphine sulfate precipitates in alkaline media and drugs that are incompatible with it, including aminophylline, sodium salts of barbiturates, and phenytoin. Precipitate was found after 2 hours when morphine sulfate was formulated with acyclovir sodium.[70] Incompatibilities also are reported with chlorpromazine hydrochloride injections containing chlorocresol. Admixture of morphine sulfate (1 mg/mL), doxorubicin in doxorubicin hydrochloride liposomal injection (0.4 mg/mL) in 5% dextrose,[71] admixture of morphine sulfate (1 mg/mL), fluorouracil (1–16 mg/mL) in 5% dextrose or 0.9% sodium chloride,[72] admixture of frusemide with morphine sulfate solutions, admixture with pethidine hydrochloride, admixture with prochlorperazine edisylate due to phenol preservative are incompatible. Incompatibility with heparin sodium was found if a morphine sulfate concentration is more than 5 mg/mL and can be prevented using 0.9% sodium chloride instead of water. Visual incompatibility was reported when promethazine hydrochloride (12.5 mg) was mixed with solutions containing morphine sulfate (8 mg). A color change from yellow to light green has been reported when solutions of minocycline hydrochloride or tetracycline hydrochloride were mixed with morphine sulfate (5%) in glucose injection.

Morphine sulfate (2 mg/mL) containing 0.9% sodium chloride showed better stability in polypropylene syringes, but 15% of its potency is lost when 0.1% sodium metabisulfite was incorporated.[73] These preparations showed incompatibility when stored in glass syringes.[74] Dependence, withdrawal, and interactions are similar to those of opioid analgesics.

***Nefopam Hydrochloride*** Overdose can be fatal; symptoms include euphoria, hallucinations, convulsions, and CNS and cardiovascular toxicity. Temporary discoloration of the urine may be found, and overdose can be cured with routine supportive treatment.[75]

***Oxaprozin*** Concomitant administration of oxaprozin with opioid analgesics gives false results for the concentration of benzodiazepines. Phenytoin and correct method should be chosen for the analysis of these drugs.[76–78]

***Paracetamol*** Overdose with paracetamol, whether accidental or deliberate, shows common effects such as vomiting, lethargy, sweating, and abdominal pain. The poisoning may lead to hypoglycemia, cerebral edema, liver and kidney damage, and death. Liver damage may be particularly seen in alcoholics or patients receiving enzyme-inducing drugs.

Supportive treatment is essential, and stomach wash and multiple doses of charcoal may be used. Acetylcysteine or methionine can be given to prevent absorption of charcoal. Antidote treatment must be started after suspected paracetamol ingestion and continued depending on blood levels of paracetamol. Appropriate care must be taken in patients receiving enzyme-inducing drugs.[79,80]

Concomitant administration of paracetamol with other hepatotoxic drugs or drugs acting on liver microsomal enzymes enhances paracetamol toxicity. Other drugs that interact with paracetamol are metoclopramide, probenecid, and cholestyramine.[81]

***Pentozocine Lactate*** Injections of pentozocine lactate are incompatible with sodium bicarbonate, barbiturates, diazepam, chlordiazepoxide, glycopyrronium bromide, and nafcillin sodium. Dependence, withdrawal, and treatment of adverse effects are generally similar to those of opioid analgesics.

***Pethidine Hydrochloride*** Pethidine hydrochloride solutions are not compatible with barbiturates, aminophylline, morphine sulfate, sulfadiazine sodium, methicilline sodium, nitrofurantoin, phenytoin sodium, heparin sodium, sodium iodide, diethanolamine, sulfafurazol, acyclovir sodium, liposomal doxorubicin hydrochloride, frusemide, imipenem, and idarubicin. Visual incompatibilities are reported with mixtures of pethidine hydrochloride and cefoperazone sodium or mezlocillin sodium, nafcillin sodium and minocycline hydrochloride or tetracycline hydrochloride in 5% glucose injection.[82,83]

Dependence, withdrawal, and treatment of adverse effects are similar to those of opioid analgesics, and synthetic analogs are abused.

Pethidine and MAOIs show fatal reactions when coadministered. Phenothiozine, pethidine, and retonavir should be avoided.

***Piroxicam*** Piroxicam and ritonavir should not be administered concurrently due to potential toxicity.

***Remifentanil Hydrochloride*** Remifentanil hydrochloride should not be mixed with lactated Ringer's injection, 5% glucose injection, or blood products. Solutions of remifentanil hydrochloride are incompatible with chlorpromazine hydrochloride, remifentanil, cefoperazone sodium, and amphotericin.[84]

***Sodium Aurothiomalate*** The treatment of adverse effects is symptomatic. Therapy is withdrawn and a chelating agent such as dimercaprol may be used.

Simultaneous administration of gold compounds with other nephrotoxic, hepatotoxic, or myelosuppressive and penicillamine drugs must be avoided because of associated toxicities. Metal toxicity can be treated with dimercaprol, and drugs must be withdrawn.

***Sufentanil Citrate*** Extreme care should be exercised to avoid skin contact and inhalation of sufentanil citrate. Dependence, withdrawal, and adverse effects are similar to those of opioid analgesics.

***Sulindac*** Careful monitoring of blood pressure is required when treating with antihypertensive and sulindac.[19]

### Antigout Drugs

Aspirin, salicylates, and thiazide diuretics should not be used with allopurinol. The dose of mercaptopurine should be reduced one-third or one-fourth when used with allopurinol. Acute poisoning of colchicine should be treated with gastric lavage and activated charcoal administration. Supportive maintenance measures for blood pressure and respiration should be provided. Probenecid is used by athletes to inhibit the urinary excretion of banned anabolic steroids.[85]

### Antihistamines

Although sedative antihistamines do not potentiate the effect of alcohol, they should be avoided in excess quantity. Overdose of astemizole can be treated with gastric lavage and supportive measures.[86] Coadministration of astemizole and terfenadine with antiarrhythmics, antipsychotics, cisapride, and diuretics should be avoided. Chlorpheniramine maleate has been found to be incompatible with phenobarbitone sodium, kanamycin sulfate, and calcium chloride. Cyclizines have been used alone or with opioids in tablets or in injectable form for euphoric effects. Cyproheptadine has shown dependence in long-term use. Diphenhydramine is reported to be incompatible with amphotericin, cephalothin sodium, and hydrocortisone sodium succinate. Diphenhydramine and pheniramine maleate are sometimes used as drugs of abuse. Studies have shown that promethazine is adsorbed onto glass, plastic containers, and infusion systems.[87]

### Bronchodilators and Antiasthma Drugs

Long-term use of caffeine may lead to tolerance. Headache is a main withdrawal symptom of caffeine. Use of caffeine in large amounts is banned in athletics. Clenbuterol is used as an illicit drug to increase weight gain in animals. Clenbuterol also is used by athletes for its anabolic effects. Overdose of salbutamol may cause tachycardia, tremor, and CNS stimulation. Regular inhalation of $\beta_2$-agonists such as salbutamol may cause tolerance in patients. Salbutamol interacts with antidepressants, cardiac glycosides, corticosteroids, diuretics, xanthine, and muscle relaxants, hence care should be taken when administering with these categories. Overdose of terbutaline has been reported on topical application.[91] Treatment of theophylline overdose includes gastric lavage, administration of activated charcoal, and symptomatic and supportive treatment. Use of beta-blockers, hemodialysis, and hemoperfusion are also advised for the treatment of theophylline overdose. Coadministration of theophylline and xanthine is not advisable.[88–91]

***Aminophylline*** Aminophylline should not be mixed with acidic drugs, as it becomes precipitated if the pH of the final solution falls below pH 8. Mixture with glucose may increase the pH above 10 where proteins, such as insulin and erythromycin, are unstable. Some drugs that are incompatible with aminophylline include amiodarone, benzylpenicillin potassium, cisatracurium, ceftazidime, ceftriaxone, dobutamine, tetracycline hydrochloride, verapamil hydrochloride, warfarin sodium, and vitamin B and C injection. Alcohol-free theophylline should be stored in amber-colored containers; maximum care must be taken to protect it against exposure to light.[92–94]

## CARDIOVASCULAR DRUGS

Adverse effects are reversible on withdrawal of ACE inhibitor therapy. Volume expansion with intravenous isotonic sodium chloride and treatment with angiotensin amide are other measures used for the treatment of adverse effects. Overdose of ACE inhibitors causes hypotension, which can be generally controlled by supportive treatment. ACE inhibitors should be used with caution with antacids, antidiabetics, cytokines, digoxin, interferons, diuretics, NSAIDs, and probenecid. Adrenaline overdose is treated with supportive measures, alpha-adrenoceptor blockers, and beta-blockers.

### Beta-Blockers

Overdose of beta-blockers should be treated with gastric lavage and activated charcoal. Hypotension is controlled by fluid administration, glucagons, and sympathomimetics. Bradycardia is treated with sympathomimetics and atropine. Beta-blockers have shown interactions with NSAIDs, oral contraceptives, antihypertensives, calcium channel blockers, anticoagulants, antibacterials, antimalarials, antacids, antiarrhythmics, and antidepressants. Clonidine may enhance the psychoactive potential of morphine. Clonidine also is used for the treatment of alcohol withdrawal. Treatment of acute digitalis poisoning consists of gastric lavage, activated charcoal, hemodialysis, forced diuresis, atropine, and maintenance of electrolyte balance. Severe digoxin overdose may be treated with digoxin-specific antibody fragment. Cardiac glycosides should be used with caution when combining with ACE inhibitors, calcium channel blockers, benzodiazepines, antibacterials, antifungals, antimalarials, antineoplastics, antiarrhythmics, diuretics, NSAIDs, and sympathomimetics. Adverse effects of diltiazem (calcium channel blocker) should be treated with gastric lavage, activated charcoal, plasma expanders, calcium gluconates, dopamine, and atropine. Diltiazem interacts with antidepressants, antiepileptics, beta-blockers, digoxin, theophylline, lithium, benzodiazepines, and cyclosporin. Misuse of diltiazem by body builders has been reported.[95,96]

### Glyceryl Trinitrate

Glyceryl trinitrate overdose should be treated with the patient's head lowered. Other measures include respiration maintenance, use of plasma expanders, and electrolyte balance. Withdrawal of heparin treatment or dose reduction should be performed with the overdose of heparin. Protamine sulfate may be used to reduce severe bleeding. Heparin should be used with caution with glyceryl trinitrate, aprotinine, alcohol, tobacco, and ACE inhibitors. Nifedipine should be used with care when coadministering with immunosuppressants, magnesium salts, tobacco, digoxin, antineoplastics, calcium channel blockers, antihistamines, antifungals, antiepileptics, antiarrhythmics, and alcohol.

### Quinidine

Quinidine overdose should be treated with symptomatic and supportive therapy. Quinidine should be used with caution when coadministering with calcium channel blockers, diuretics, beta-blockers, antibacterials, antifungals, and antiarrhythmics.[97]

Allergic reaction due to streptokinase overdose should be treated with corticosteroids and histamines. Severe hemorrhage requires discontinuation of streptokinase. Packed red blood cells are preferable for blood-replacement therapy, and volume expansion is advisable. Streptokinase may be used with caution with heparin, allopurinol, sex hormones, sulfonamides, tetracyclines, and dextran.

### Verapamil

Treatment for verapamil overdose includes gastric lavage, administration of activated charcoal, symptomatic and supportive cardiovascular treatment, and calcium gluconate injection. Verapamil should be used carefully with calcium salts, beta-blockers, antineoplastics, antiepileptics, antibacterials, anxiolytics, and antiarrhythmics.

### Warfarin

Warfarin should be stopped in the case of excessive bleeding; after that, phytomenadione, factor II, VII, IX, and X should be given. Warfarin should be used carefully with vitamins, vaccines, tobacco, sex hormones, prostaglandins, lipid-regulatory drugs, glucagons, ginseng, diuretics, anxiolytics, sedatives, hypnotics, antipsychotics, antiplatelets, antivirals, antiprotozoals, antihistamines, antibacterials, antineoplastics, antidepressants, antidiabetics, and antifungals.

Decrease in concentration of solutions of amiodarone hydrochloride is found when stored in flexible PVC bags and glass or rigid PVC bottles, and administered with PVC sets. This may be due to the presence of plasticizer, di-2-ethylhexylphthalate.[98]

### Amiodarone

Amiodarone injection is incompatible with aminophylline, flucloxacillin, heparin, sodium bicarbonate, and sodium chloride solutions.[99,100]

### Amrinone Lactate

Amrinone lactate is physically and chemically incompatible with alkaline solutions. Pharmaceuticals that are incompatible include digoxin, potassium chloride, procainamide hydrochloride, propranolol hydrochloride, verapamil hydrochloride, sodium chloride, and glucose solutions containing frusemide.[101]

Aqueous solutions of captopril are prone to oxidative degradation; this increases with an increase in pH above 4. There are reports of variations in stability of captopril solutions, depending on the formulation. Addition of sodium ascorbate increases stability of the aqueous solution made of captopril powder.

### Diltiazem Hydrochloride

Diltiazem hydrochloride degrades to desacetyldiltiazem, and this was enhanced at acidic pH and elevated temperatures.[102]

### Digitoxin

Digitoxin potentially adsorbs onto glass, plastic, and inline intravenous filters containing cellulose ester membranes, some aqueous solutions such as 5% glucose, or 0.9% sodium chloride. The loss in potency may be minimized by pretreatment of the filter with a polymer coating. Such losses in potency were not found with 30% alcohol solutions.[103,104]

### Dobutamine

Dobutamine is unstable with alkaline preparations and is incompatible with sodium bicarbonate 5% and alkaline drugs such as aminophylline, frusemide, thiopentone sodium, heparin, bumetanide, calcium gluconate, insulin, diazepam, and phenytoin.[105,106]

### Dopamine

Dopamine is degraded by alkaline preparation and is incompatible with sodium bicarbonate solutions, furosemide, ampicillin, amphotericin, gentamicin sulfate, cephalothin sodium, oxacillin sodium, and thiopentone sodium. Percussion or excessive heat may cause an explosion with undiluted erythrityl tetranitrate. Care must be exercised to avoid contact with skin, eyes, and mucous membranes when handling ethacrynic acid. Sodium ethacrynate solutions are incompatible and less stable at very high pH, below pH 5, and at higher temperatures.

### Flecainide

Liquid oral formulations of flecainide must be freshly reconstituted prior to administration, since storage of such preparations causes crystallization of the drug, which may lead to accidental administration of a toxic dose.[107]

### Frusemide

Frusemide is physically incompatible with diltiazem hydrochloride, dobutamine hydrochloride, dopamine hydrochloride, labetalol hydrochloride, midazolam hydrochloride, milrinone lactate, nicardipine hydrochloride, parenteral nutrient solutions, cisatracurium besylate, and vecuronium bromide.

### Glyceryl Trinitrate

Percussion or excessive heat may cause an explosion of undiluted glyceryl trinitrate; hence care must be exercised when handling. Glyceryl trinitrate is incompatible with phenytoin and alteplase. Loss of glyceryl trinitrate from a solution occurs due to adsorption or absorption with plastics and inline filters; this could be minimized by using polyolefin or polyethylene.[108,109]

## Heparin

Heparin and its salts are incompatible with many drugs including alteplase, amikacin sulfate, amiodarone hydrochloride, ampicillin sodium, aprotinin, benzylpenicillin potassium or sodium, cephalothin sodium, ciprofloxacin lactate, cytarabine, dacarbazine, daunorubicin hydrochloride, diazepam, dobutamine hydrochloride, doxorubicin hydrochloride, droperidol, erythromycin lactobionate, gentamicin sulfate, haloperidol lactate, hyaluronidase, hydrocortisone sodium succinate, kanamycin sulfate, methicillin sodium, netilmicin sulfate, some opioid analgesics, oxytetracycline hydrochloride, some phenothiazines, polymyxin B sulfate, streptomycin sulfate, tetracycline hydrochloride, tobramycin sulfate, vancomycin hydrochloride, vinblastine sulfate, cisatracurium besylate, labetalol hydrochloride, nicardipine hydrochloride, cefmetazole, sodium ions, and fat emulsion.[110–112]

## Hydralazine

Hydralazine solutions reacts with metals, hence injections must be prepared using nonmetal filters and doses must be withdrawn prior to administration.[113,114] Percussion or excessive heat may cause an explosion of undiluted isosorbide dinitrate and must be handled carefully.

## Isosorbide Dinitrate

Isosorbide dinitrate adsorbs to PVC plastic intravenous infusion sets from solution. Adsorption and loss of potency could be prevented if stored in glass bottles or polyethylene-, nylon-, or polypropylene-laminated bags.[115,116]

Patients who are receiving ketanserin must be cautious when driving and operating machinery, as it causes drowsiness.

## Labetalol Hydrochloride

Labetalol hydrochloride is incompatible with 5% sodium bicarbonate injection. White precipitate is formed immediately in 5% glucose solutions of frusemide (10 mg/mL), insulin (100 mL), and thiopentone.[117,118]

## Mannitol

Mannitol should neither be mixed with whole blood for transfusion nor be administered through the same sets by which blood is being infused. Visual incompatibilities are found when methyldopate hydrochloride is mixed with methohexitone sodium, amphotericin, tetracyclines, or sulfadiazine.[119]

## Nicardipine Hydrochloride

Nicardipine hydrochloride is incompatible with sodium bicarbonate and lactated Ringer's solutions. Nifedipine solutions must be prepared in the dark or under light of above 400 nm, since it has potential degradation when exposed to light. Nimodipine is incompatible with plastics and PVC containers. Noradrenaline acid tartrate is generally incompatible with alkaline preparations and some drugs, including barbiturates, chlorpheniramine, chlorothiazide, nitrofurantoin, novobiocin, phenytoin, sodium bicarbonate, sodium iodide, streptomycin, and insulin.[120]

### Procainamide

The stability of procainamide in glucose infusions may be improved by the addition of sodium bicarbonate. Patients receiving sodium chloride infusions of procainamide are prone to the risk of heart failure due to sodium load.[121,122] Quinidine gluconate is incompatible with intravenous infusion sets made of PVC due to drug loss by adsorption.[123]

### Nitroprusside

Only freshly prepared nitroprusside should be administered, and the container should be covered with lightproof materials, since they are decomposed upon exposure to light. Solutions may be discolored due to the presence of organic and inorganic substances and must not be used. Preservatives, such as sodium bisulfate and hydroxy benzoates, reduce the stability of nitroprusside.[124–126]

### Trimetaphan Camsylate

Trimetaphan camsylate is incompatible with alkaline preparations, bromides, iodides, gallamine triethiodide, and thiopentone sodium. Loss of urokinase is reported when diluted with glucose solutions in PVC intravenous infusion sets.[127]

### Verapamil Hydrochloride

Verapamil hydrochloride is incompatible with alkaline preparations and drugs such as aminophylline, nafcillin sodium, and sodium bicarbonate.[128,129]

### Warfarin Sodium

Warfarin sodium may be adsorbed to PVC and intravenous infusion sets but may be minimized with glass containers or polyethylene-lined containers. Warfarin sodium is incompatible with solutions of adrenaline hydrochloride, amikacin sulfate, metaraminol tartrate, oxytocin, promazine hydrochloride, tetracycline hydrochloride, aminophylline, bretylium tosylate, ceftazidime, cimetidine hydrochloride, ciprofloxacin lactate, dobutamine hydrochloride, esmolol hydrochloride, gentamicin sulfate, labetalol hydrochloride, metronidazole hydrochloride, and vancomycin hydrochloride.[130,131]

## DRUGS ACTING ON THE NERVOUS SYSTEM AND RELATED DYSFUNCTIONS

### Antidepressants

Treatment with antidepressants will affect a person's ability to drive or operate machinery. Among the tricyclic antidepressants, amitriptyline and doxepin impair skills compared with imipramine and nortriptyline. Fluoxetine and dothiepin also show similar effects, such as affecting ability to work. The United Kingdom's Medical Commission on Accident Prevention has recommended that patients on long-term psychotropic medication are unsuitable drivers of heavy vehicles or public transport services.[132,133]

***Amitriptyline*** This drug is unstable upon autoclaving at 115 °C to 116 °C in the presence of oxygen. As degradation is accelerated with metal ions, it could be prevented with disodium edetate.

***Tricyclic Antidepressants*** Tricyclic antidepressant overdose increases suicidal tendency and may be fatal.[134]

***Bupropion Hydrochloride*** Bupropion interacts with levodopa, nicotine, alcohol, and drugs that affect hepatic metabolic enzymes. It lowers the seizure threshold and potentiates the seizure threshold–lowering effect of other antidepressants.[135]

Fluoxetine is a potential drug of abuse.[136] Overdose with sertraline causes suicidal tendencies, whereas citalopram causes fatal reactions such as cardiac dysfunction.[137] The adverse effects could be treated with stomach wash, administration of activated charcoal, dialysis, and hemoperfusion.

Selective serotonin reuptake inhibitors (SSRIs), such as fluoxetine, interact with drugs including clarithromycin, warfarin, phenelzine, benzotropine, chlorpromazine, diazepam, and cyproheptadine. Cigarette smokers metabolize SSRIs faster.

***Lithium Carbonate*** Common symptoms of lithium toxicity are nausea, vomiting, and diarrhea followed by tremor, increased muscle tone, and rigidity. Acute kidney problems and nephrogenic diabetes insipidus may be possible hazards. In serious cases, coma and convulsions can be observed along with toxicity.

Lithium severely affects a person's ability to drive or operate industrial machinery.[138]

***Mianserin Hydrochloride*** An acute overdose of mianserin is most hazardous in patients due to multiple drug overdose. The main feature of overdose is mild coma without chance of deep coma or convulsions. Mianserin may potentiate the effects of CNS depressants such as alcohol, antipsychotics, and anxiolytics.

***Moclobemide*** Moclobemide overdose causes adverse effects such as agitation, aggression, and behavioral changes.[139]

Tyramine-rich foods, sympathomimetics, dextromethorphan, anorectics, and other antidepressants should be avoided when treating with moclobemide. It also interacts with opioid analgesics, serotonin agonists, NSAIDs, levodopa, and cimetidine.[140,141]

Overdoses of MAOIs (e.g., phenylzine and tranylcypromile) cause side effects such as muscles spasms, sweating, and increasing body temperature, and these effects may be fatal. Overdose can be started with stomach wash, and supportive therapy can be performed to treat CNS effects, hyperpyrexia, and cardiovascular effects. Sometimes, overdose can be observed after a long period, and patients should be carefully monitored. Various drugs given to treat overdose are discussed elsewhere.

MAOIs affect psychomotor performance; driving and operating machinery is not advised. Abrupt termination of antidepressants, including MAOIs, causes severe adverse effects. Symptoms may be treated by gradually withdrawing these drugs.[142]

MAOIs interact with foods containing tyramine, such as smoked food, cheese, and meat. Such food must be avoided during MAOI therapy. Protein-containing foods can be eaten only if they are fresh; alcohol-containing beverages must be avoided.[143]

MAOIs interact with sympathomimetics, barbiturates, hypoglycemics, antimuscarinics, alcohol, antihypertensives, and antidepressants. Care must be exercised during concomitant administration. Since adverse effects may be seen after a long period, patients must be monitored carefully, even after therapy.[144]

***Reboxetine*** Reboxetine should not be given, even after termination of MAOI therapy. Care must be exercised when treating with antihypertensive drugs, antiarrhythmics, cyclosporin, antipsychotics, tricyclics, fluvoxamine, antidepressants, azole antifungals, and macrolide antibacterials.

***Trazodone Hydrochloride*** Trazodone overdose causes severe toxic effects. These effects are severe if taken along with benzodiazepines or alcohol. Trazodone interacts with MAOIs, cardiovascular drugs, CNS depressants, and antiepileptics.

***Tryptophan*** Tryptophan interacts with MAOIs affecting phenothiazines and benzodiazepines.

***Venlafaxine Hydrochloride*** Venlafaxine must be avoided with MAOIs and cimetidine.

***Viloxazine Hydrochloride*** Viloxazine hydrochloride interacts with MAOIs, cardiovascular drugs, CNS drugs, and antiepileptics.

### Antimigraine Drugs

Acute or chronic poisoning with ergot alkaloids may be referred to as *ergotism*. The most important side effects are extreme thirst, severe CNS effects, circulatory disturbances including hypotension and cardiovascular complications, and possibly fatal reactions. There are reports of ergotamine dependence.[145] Treatment is symptomatic and may be performed by stomach wash and activated charcoal treatment. Treatment must be performed with drugs such as sodium nitroprusside and heparin to maintain proper circulation and prevent thrombosis.[146]

Ergot alkaloids interact with beta-blockers, antimigraine drugs, antibacterials, and glycerol nitrate. Concomitant administration with erythromycin causes ergotism.[147,148]

***Sumatriptan Succinate*** Sumatriptan is a drug of abuse, and it interacts with ergotamine, serotonin, rizatriptan, antidepressants, and antipsychotics.

### Antimuscarinics

Hydroxybenzoate preservatives inactivate atropine sulfate in physical mixtures. Atropine can be used for treating acetylcholine poisoning. Organophosphorous pesticides are treated by use of diazepam, phenothiazines, and physostigmine.[149]

Atropine interacts with antipsychotics, antihistamines, phenothiazines, and antidepressant drugs. Benzhexol, orphenadrine, and oxybutynin hydrochloride are drugs of potential abuse; poisoning and withdrawal symptoms are reported.[150]

### Anxiolytics, Sedatives, Hypnotics, and Antipsychotics

Barbiturate overdose may be treated with gastric lavage and oral administration of activated charcoal. Supportive therapy of cardiovascular, respiratory, and renal function also should be provided. Coadministration of alcohol and barbiturates may increase the sedative effect of chloral hydrate. Long-term use of barbiturates leads to dependence. Sudden discontinuation of an antipsychotic drug may cause withdrawal symptoms such as nausea, vomiting, anorexia, diarrhea, rhinorrhea, sweating, insomnia, restlessness, and vertigo.[151]

***Phenothiazine*** Phenothiazine is known to interact with alcohol, antacids, antiarrhythmics, antimalarials, antidepressants, anticoagulants, antidiabetics, beta-blockers, antihistamines, and antivirals. Its use with the above categories should be monitored carefully. Treatment of diazepam overdose starts with gastric lavage and symptomatic supportive therapy. Flumazenil also can be used. Benzodiazepines are used for the management of opioid and alcohol withdrawal symptoms. Haloperidol must be used carefully in patients taking lithium medication. Barbiturate dependence may develop after continuous use, even with a regular regimen. It should not be stopped abruptly but barbiturate dose should taper with time. Withdrawal symptoms are anxiety, headache, dizziness, irritability, tremors, nausea, vomiting, abdominal cramps, insomnia, distortion in perception, muscle twitching, and tachycardia. Temazepam in liquid form is used as an illicit drug.[152]

Care should be exercised when handling alprazolam to avoid skin contact and inhalation. Amylobarbitone is incompatible with many drugs, particularly acids and acidic salts. Chloral hydrate is not compatible with alkalis, alkaline earths, alkali carbonates, soluble barbiturates, borax, tannin, iodides, oxidizing agents, permanganates, and alcohol. Chlordiazepoxide is incompatible with benzquinamide hydrochloride and pentazocine lactate.

***Chlormethiazole Edisylate*** Chlormethiazole edisylate may penetrate or be adsorbed onto plastics and may soften plastic sets of intravenous infusions. Adverse reactions such as thrombophlebitis, fever, and headache were reported in children receiving infusions of this drug, presumably due to reaction of components of infusions sets. For intravenous infusions, the use of Teflon intravenous cannulas or motor-driven glass syringes may be preferred over plastic or glass sets containing silastic cannulae.[153,154]

***Chlorpromazine*** Chlorpromazine hydrochloride injection precipitates if the pH increases after dilution with other components. It is also incompatible with a number of drugs such as carbamazepine suspension, aminophylline, amphotericin, aztreonam, some barbiturates, chloramphenicol sodium succinate, chlorothiazide sodium, cimetidine sulfate, some penicillins, ranitidine hydrochloride, and remifentanil. Health care professionals handling chlorpromazine must exercise proper care, since it causes contact sensitization. Loss of chlorpromazine hydrochloride due to sorption was reported with infusion sets made of plastic (cellulose proprionate burette with PVC tubing) or glass syringe through elastic tubing.[154] This may be prevented if a glass syringe with polyethylene tubing is used to administer infusions.

***Diazepam*** Diazepam should not be administered with some plastics/PVC/volume control chambers of cellulose proprionate. Administered sets containing glass, polyolefin, polypropylene, and polyethylene may be used for such infusions. Diazepam is incompatible with many drugs, hence it should not be mixed with infusions containing other drugs.[155,156]

***Droperidol*** Droperidol has pH-dependent incompatibility with many drugs. It is not compatible with nafcillin sodium, fluorouracil, folic acid, frusemide, heparin, methotrexate, or barbiturates. Droperidol should not be stored in PVC bags containing intravenous infusions, although some solutions appear to be stable.[157]

***Haloperidol*** A concentrated solution of haloperidol (5 mg/mL) is incompatible with heparin sodium (in sodium chloride or glucose solutions), sodium nitroprusside (in glucose solutions), cefmetazole sodium, diphenhydramine, and sargramostim. Solutions of haloperidol precipitate if diluted with a 0.9% sodium chloride solution.[158,159] Photodegradation of haloperidol could be prevented using benzyl alcohol and vanillin.[160]

***Lorazepam*** Lorazepam is incompatible with sargramostim and aztreonam.[161] Concentration and pH-dependent incompatibility of lorazepam was reported in mixtures of glucose and sodium chloride solutions.[162] Lorazepam has better stability in polyolefin or glass compared with PVC or polypropylene containers.[163]

***Midazolam Hydrochloride*** Midazolam hydrochloride is incompatible with a number of drugs, and precipitation is observed, presumably due to a pH of 5 or more with dimenhydrinate, pentobarbitone sodium, perphenazine, frusemide, thiopentone, prochlorperazine edisylate, ranitidine hydrochloride, and various intravenous fluids.[164] Solutions of midazolam hydrochloride should not be combined with Hartmann's solution, as the potency of midazolam may be affected.[165]

***Paraldehyde*** Care must be exercised when using paraldehyde. It must not be administered if is brownish in color or gives off a sharp odor of acetic acid. Degradation is much more pronounced once the containers are opened. Polypropylene or glass syringes with natural rubber-tipped plastic plungers are acceptable only for the immediate administration or measurement of paraldehyde doses.

***Prochlorperazine Edisylate*** Prochlorperazine edisylate is not compatible with sodium chloride solutions containing methyl hydroxybenzoate and propyl hydroxybenzoate as preservatives, but is compatible with solutions containing benzyl alcohol. Prochlorperazine edisylate salts are incompatible with a number of drugs such as aminophylline, amphotericin, ampicillin sodium, some barbiturates, benzylpenicillin salts, calcium gluconate, cefmetazole sodium, cephalothin sodium, chloramphenicol sodium succinate, chlorothiazide sodium, chloramphenicol, morphine sulfate containing phenol, magnesium trisilicate mixture, sodium succinate, chlorothiazide sodium, dimenhydrinate, heparin sodium, hydrocortisone sodium succinate, midazolam hydrochloride, and some sulfonamides.[166]

***Promazine Hydrochloride*** Promazine hydrochloride injection should not be mixed with other injections, since it is incompatible with a number of drugs such as aminophylline, some barbiturates, benzylpenicillin potassium, chloramphenicol sodium succinate, chlortetracycline, chlorothiazide sodium, dimenhydrinate, heparin, hydrocortisone sodium succinate, naficillin sodium, phenytoin sodium, prednisolone sodium phosphate, and sodium bicarbonate. Promazine hydrochloride should be administered by intravenous sets comprising a glass syringe with polyethylene tubing, since it adsorbs to plastic or glass with elastic tubing. Concentration-dependent photodegradation of promazine was seen with sodium chloride or glucose solutions.[167]

Solution of thioridazine hydrochloride is incompatible with carbamazepine suspensions. Cessation of therapy or drug reduction with troopiclone may cause dependence and withdrawal symptoms. Patients taking this drug should avoid operating machinery or industrial work and driving.[168]

## DRUGS USED FOR GASTROINTESTINAL TRACT DISORDERS

Care should be taken when administering aluminium salts such as aluminium hydroxide in patients with renal dysfunction and bone diseases. Patients with renal dysfunction receiving aluminium salt therapy should avoid citrate salt preparations such as effervescent and dispersible tablets or granules. Gastric emptying and pH are important factors that affect the bioavailability of antacids. Antacids should be taken carefully when coadministering with ethambutol, isoniazid, nitrofurantoin, quinolones, tetracyclines, indomethacin, theophylline, ranitidine, and corticosteroids.[169]

### Bismuth Compounds

Bismuth compounds are used as an antidiarrheal. Topical applications are used in skin disorders. Overdose may cause acute bismuth intoxication but gastric lavage, purgation, use of chelating agents, 2,3-dimercapto-1-propane sulfonic acid, and hemodialysis are steps to be taken.[170–172]

### Cimetidine

Cimetidine overdose should be treated with gastric lavage, emesis induction as an early measure, and should be followed by supportive and symptomatic treatment. Cimetidine interacts with antimuscarinics, antacids, sucralfate, antiarrhythmics, antiepileptics, antidiabetics, and alcohol.[173]

### Diphenoxylate

Diphenoxylate in high doses may produce dependence if coadministered with atropine. Diphenoxylate combined with thioridazine is associated with opioid withdrawal. Famotidine should be taken with care when coadministered with antacids, theophylline, and probenecid. Kaolin should not be administered with any other oral drug. Loperamide overdose should be treated with naloxone hydrochloride.[174,175]

### Metoclopramide

Incompatibilities of metoclopramide depend on drug concentration, pH, and temperature. It is incompatible with cephalosporins, chloramphenicol, sodium bicarbonate, doxorubicin, cisplatin, and cyclophosphamide. Caution should be exercised with simultaneous administration of metoclopramide with lithium, sympathomimetics, antidepressants, bromocriptine, and carbamazepine. Omperazole interacts with tolbutamide, clarithromycin, and phenytoin. Coadministration of rantidine and cisapride increases the plasma concentration of rantidine. Abuse of senna laxative has been reported and may cause hepatitis.[176–178]

### Sucralfates

Simultaneous administration of sucralfates with other nonantacid medication should be given after an interval of at least 2 hours. Sucralfates interact with rantidine, cimetidine, digoxin, quinidine, theophylline, and warfarin. Sulfasalazine should be administered with caution with antibacterials and antineoplastics.[179]

## DRUGS USED FOR PARASITIC DISEASES

### Anthelmintics

***Albendazole*** Albendazole interacts with dexamethasone.[180]

***Antimony Potassium Tartrate*** This drug is not compatible with acids, alkalis, albumin, soap, salts of heavy metals, or tannins. Poisoning with antimony compounds resembles arsenic poisoning. Treatment is similar; dimercaprol may be of use in treatment.

***Mebendazole*** Accidental mebendazole poisoning in infants is associated with convulsions, respiratory arrest, and tachyarrhythmia.[181] If administered concomitantly, mebendazole interacts with phenytoin, carbamazepine, and cimetidine.

***Piperazine*** Concomitant administration of piperazine and pyrantel should be avoided, since therapy becomes ineffective. Also, adverse effects of phenothiazines may be potentiated if the drug is taken with piperazine.

***Praziquantel*** Interactions of this drug are reported with carbamazepine, phenytoin, dexamethasone, and chloroquine.[182]

### Antimalarials

Sorption has been reported during membrane filtration of solutions of chloroquine sulfate, amodiaquine hydrochloride, mefloquine hydrochloride, or quinine sulfate. Care should be exercised when handling these drugs.[183] In the case of chloroquine toxicity, symptomatic supportive treatment, maintenance of respiration, and control of cardiovascular disturbance should be initially attended along with adrenaline and diazepam administration.[184] Chloroquine should not be given with halofantrine, omiodarone, or antacids. Halofantrine should not be given with drugs having the potential to induce cardiac arrhythmias. Mefloquine hydrochloride is reported to show photodegradation in water.[185] Mefloquine overdose shows dose-dependent cardiac, hepatic, and neurologic symptoms.[186] Primaquine should not be used with drugs causing hemolysis.[187] Pyrimethamine may cause toxicity in infants. Coadministration of pyrimethamine with other antagonists of folate may cause bone marrow depression.

Although the *British Pharmacopeia* (BP) directs that proguanil hydrochloride should be protected from light, studies suggest that it is a very stable compound with only small amounts of its major decomposition product (4-chloroaniline) formed during thermal and photochemical stress.[188]

## HORMONES AND HORMONE-REGULATING DRUGS

### Antidiabetics

***Acarbose*** Acarbose interacts with other hypoglycemics and drugs that act on the GI tract. Neomycin and cholestyramine interact with acarbose.

***Biguanides*** This class of antidiabetics may cause acute poisoning with adverse effects such as acidosis and may be treated by supportive therapy. These drugs have therapeutic interactions with other antidiabetic drugs, alcohol, drugs that affect kidney function, and cimetidine.

Antidiabetic drugs affect the ability to drive and operate machinery. Constant monitoring of blood glucose levels is required to check the patient's fitness for undertaking such jobs. Driving is not permitted when hypoglycemic awareness has been lost. If it is inevitable, patients can take short-acting antidiabetics when doing these jobs.

***Guargum*** Drugs that are given orally should be administered 1 hour prior to guargum, since it affects the absorption of many drugs from the GI tract.

***Insulin*** Insulin may be stable for 1 month if stored at a temperature of 25 °C, but care should be exercised not to expose it to heat or light. However, for better stability, storing at refrigerator temperature (2° to 8 °C) is recommended. Use of carrier proteins such as albumin can be formulated along with insulin preparations, since they adsorb to containers.

Adverse effects that result from insulin treatment include hypoglycemia and hypoglycemic coma and can be treated by administering glucose through parenteral routes. In conscious patients, oral sugar-based drinks can be given. Insulin-dependent diabetics must carry sugar along with them. Insulin interacts with alcohol, steroids, NSAIDs, tetracyclines, antidepressants, tricyclic antidepressants, some cardiovascular drugs, oral contraceptives, cyclophosphamide, isoniazid, mebendazole, and many other drugs. Care should be exercised with concomitant administration.

***Sulfonylureas*** In acute poisoning with sulfonylureas, the stomach should be washed and treated with activated charcoal, and hypoglycemia must be treated. Sulfonylureas interact with oral contraceptives, thiazide diuretics, corticosteroids, adrenaline, chlorpromazine, ACE inhibitors, some NSAIDs, antihistamines, anticoagulants, MAOIs, antidepressants, and many other drugs. Care must be exercised when treating with sulfonylureas.

***Troglitazone*** Troglitazone interacts with sulfonylureas, some oral contraceptives, and cholestyramines.

***Carbamazepine*** Carbamazepine suspension is incompatible with thioridazine hydrochloride solution, since it produces visual incompatibility in the form of a rubbery orange mass. Carbamazepine should be diluted before nasogastric administration.[189] Carbamazepine may be stored in silica gel sachets and must be protected from humidity for effective stability.[190]

Overdose of carbamazepine may be treated by gastric lavage, administration of activated charcoal, supportive respiratory function, and monitoring of cardiovascular function and electrolyte balance. Hemoperfusion may be performed if required.[191] Carbamazepine interacts with phenytoin, some antibacterials, sex hormones, barbiturates, anticoagulants, alcohol, antidepressants, antiepileptics, antihelmintics, antifungals, antiprotozoals, antipsychotics, antihistamines, benzodiazepines, and corticosteroids.[192]

***Clonazepam*** Clonazepam interacts with carbamazepine, benzodiazepine, alcohol, and phenytoin.

***Ethosuximide*** Ethosuximide interacts with isoniazid, phenytoin, phenobarbitone, carbamazepine, valproic acid, antipsychotics, and antidepressants.[193]

***Phenobarbitone Sodium*** Phenobarbitone sodium decomposes in aqueous solutions. Barbiturate poisoning may be treated with stomach wash and administration of activated charcoal. Monitoring respiratory, cardiovascular, and renal functions, hemodialysis, charcoal administration, forced diuresis, symptomatic and supportive therapy, and peritoneal dialysis may be performed.

***Phenytoin Sodium*** Phenytoin sodium is incompatible with many drugs when mixed with intravenous infusions. Solutions of phenytoin are only stable at a pH greater than 10.[194,195]

Phenytoin enhances the hypotensive activity of dopamine and the cardiac depressant activity of lignocaine. Care must be taken when treating.[196]

***Primidone*** Some problems associated with kidney functions are reported with primidone overdose. Primidone interacts with barbiturates and other antiepileptics.[197]

***Valproate*** Care must be exercised when administering valproate with aspirin, warfarin, or other hepatotoxic drugs.

***Vigabatrin*** This drug interferes in diagnostic tests of patients undergoing metabolic disorders, since it changes the urinary excretion of amino acids.[198]

### Bone-Modulating Drugs

Carrier proteins should be incorporated in calcitonin formulations for intravenous infusions, since this protein may be adsorbed to administration sets. Care must be exercised for dose adjustment when mixing with cardiac glycosides.

## MISCELLANEOUS

### Anticancer Agents and Immunosuppressants

Supportive care, administration of antiemetics, corticosteroids, and nutritional supplements may be used to treat the adverse effects of anticancer drugs. Scalp tourniquets and an ice-pack method have been used to reduce the antineoplastic concentration in the scalp. Infusion of blood products and colony-stimulating factors may be given for the treatment of adverse effects.[199] Azothioprine doses should be reduced when coadministered with allopurinol. Pulmonary toxicity has been reported with bleomycin treatment.[200]

Calcium channel blockers and prostaglandins have been used to reduce the nephrotoxic effects of cyclosporin during kidney transplantation.[201] Interleukin-2 should be taken cautiously when coadministered with NSAIDs.[202] Coadministration of tamoxifen with allopurinol and anticoagulants should be monitored carefully.[203]

***Amsacrine*** Amsacrine is incompatible with sodium chloride mixtures and glucose-containing injections with acyclovir sodium, ganciclovir sodium, aztreonam, amphotericin, cimetidine hydrochloride, ceftazidime, frusemide, ceftriaxone sodium, heparin sodium, methylprednisolone sodium succinate, metoclopramide hydrochloride, and some antineoplastics.[204] Asparaginase is incompatible with many drugs and rubber, and should be stored at 2° to 8 °C for stability.

***Bleomycin*** Bleomycin is incompatible and loses its potency if it is administered with solutions of benzylpenicillin sodium, carbenicillin, cephazolin or cephalothin sodium, hydrocortisone sodium succinate, mitomycin, methotrexate, nafcillin sodium, aminophylline, ascorbic acid, terbutaline, divalent and trivalent cations (especially copper), compounds containing sulfhydryl groups, and precipitation by hydrophobic anions, essential amino acids, riboflavine, dexamethasone, and frusemide.

Bleomycin sulfate should not be diluted in 5% glucose-infusion solutions. Loss of activity of bleomycin was reported with different packaging materials.[205] Patients under treatment with bleomycin must avoid driving.[206]

***Carboplatin*** Carboplatin is incompatible with aluminum and should not be mixed with 0.9% sodium chloride, since it degrades to cisplatin, thereby increasing risk of associated toxicity.[207]

To prevent the liquefication and decomposition of carmustine, it must be stored at 2° to 8 °C. Carmustine is incompatible with some packaging materials. Dilute solutions undergo rapid degradation.

***Cisplatin*** Cisplatin is incompatible and should not be mixed with infusions of bisulfite preservatives, sodium bicarbonate, fluorouracil, mannitol and potassium chloride, thiotepa, and aluminum containers. Cisplatin is unstable in sodium chloride solutions and concentrated forms.[208,209]

***Cyclophosphamide*** Cyclophosphamide or ifosfamide may be destroyed using alkaline hydrolysis in the presence of dimethylformamide or refluxing with hydrochloric acid, neutralizing, or reaction with sodium thiosulfate. The last method is ineffective for the disposal of ifosfamide and should not be used to degrade ifosfamide.[210] Cyclophosphamide is rapidly absorbed through intact skin and continues even after washing. Care must be exercised in formulation and clinical management by professionals such as pharmacists and nurses.

***Cyclosporin*** Cyclosporin should not be administered through PVC tubing or stored in PVC containers, since it leaches some carcinogenic materials.[211] Cyclosporin was less stable in dilute NaCl solutions and glucose and requires extensive mixing.[212] The bioavailability of cyclosporin appears to increase if taken with food. Better bioavailability is reported with some microemulsions.

***Dacarbazine*** Dacarbazine is incompatible with hydrocortisone and sodium succinate. Concentrated forms of dacarbazine have visual incompatibility with heparin.[213] Dacarbazine is photodegradable and must not be exposed to sunlight.[214]

Care should be exercised when handling anticancer drugs such as busulfan, dactinomycin, daunorubicin, doxorubicin, mustine hydrochloride, methotrexate, and fluorouracil to avoid contact with skin and mucous membranes. Dactinomycin adsorbs cellulose ester filters. The infusion of these drugs should be faster, as they may be adsorbed to packaging materials.[215] Daunorubicin and doxorubicin are incompatible with heparin sodium and dexamethasone sodium phosphate, and both drugs are incompatible with aluminum materials. Liposomal daunorubicin should not be mixed with saline solution, as the formulation is prone to aggregation. Both these drugs interact with many drugs. The physicochemical incompatibilities of doxorubicin with allopurinol, cefepime, and ganciclovir may be prevented by liposomal formulation. Even such formulations are incompatible with a number of drug solutions including amphotericin B, docetaxel, gallium nitrate, hydroxyzine hydrochloride, metoclopramide hydrochloride, miconazole, mitoxantrone hydrochloride, opioids, paclitaxel, sodium bicarbonate, and some antibacterials. These drugs are relatively stable in mixtures having an acidic pH. Adsorption was minimum with polymethylene and polypropylene containers.

Precipitation occurs when doxorubicin is combined with diazepam or hydrocortisone sodium succinate and frusemide or heparin sodium. Visual incompatibility forms a purple color when doxorubicin is mixed with aminophylline.[216] Doxorubicin is not photodegradable but is sensitive to light at low concentrations.

Sulfuric acid and potassium permanganate are used to dispose of doxorubicin or daunorubicin wastes. Doxorubicin shows mutagenicity only at high concentrations. Care must be exercised by wearing protective clothing when handling urine and feces after administration of doxorubicin, etoposides, fluorouracil, methotrexate, mitoxantrone, mustine, procarbazine, melphalan, mercaptopurine, vinblastine, vincristine, and hydroxyurea.[217]

Fluorouracil was relatively stable in PVC containers compared with glass, but this depends on concentration and storage temperature. Fluorouracil solutions are incompatible with synthetic elastomers containing packaging and infusion materials.[218,219]

***Ifosfamide*** Parenteral administration of fluorouracil requires caution, as injection near the sciatic nerve may result in pain and nerve trauma. People undergoing therapy should avoid driving or operating machinery. Ifosfamide has stability problems in aqueous solutions, and the degraded products may cause severe toxicity in biological fluids.[220] Ifosfamide is incompatible with preservatives such as benzyl alcohol and should not be mixed with water for injections containing preservatives, since incompatibility manifests in the form of turbidity.[221] Ifosfamide penetrates into latex and PVC gloves. Hence, suitable gloves should be used when handling ifosfamide, or gloves should be changed every 2 hours.[222]

Aldesleukin adsorbs to infusion devices and is incompatible with a loss of its potency when mixed with ganciclovir sodium, lorazepam, pentamidine isethionate, prochlorperazine edisylate, and promethazine hydrochloride. Suitable analytical methods should be selected for testing.[223]

Lomustine waste can be disposed of by reacting with hydrobromic acid in glacial acetic acid. This method cannot be used for disposing of carmustine, semustine, or PCNU wastes. Accidental overdose of lomustine may be fatal.[224]

***Melphalan*** Melphalan is relatively stable at 20 °C in sodium chloride infusion. Increased temperature or decreased concentration of chloride ions may cause faster degradation.[225]

The mixture of methotrexate sodium with cytarabine, fluorouracil, hydrocortisone sodium, and prednisolone sodium phosphate produces precipitate upon storage, although it is not observed immediately. A concentration-dependent photodegradation was reported for methotrexate and considered high in the presence of bicarbonate ions and unprotected polybutadiene tubings.[226] Methotrexate wastes may be disposed of by oxidation with potassium permanganate and sulfuric acid or by oxidation with aqueous alkaline hypochlorite.

Unused injections of mustine and equipment used for its preparation and administration may be neutralized with sodium thiosulfate and sodium bicarbonate. Further, the equipment used when manufacturing may be cleaned using solutions of sodium carbonate and sodium hydroxide in a mixture of methylated spirit and water.

Paclitaxel injection containing alcohol and polyethoxylated castor oil may deplete diethylhexyl phthalate from plastic sets.[227]

***Plicamycin*** Plicamycin is adsorbed to some inline filters of intravenous infusion sets more commonly with glucose solutions than with sodium chloride. Care must be taken to avoid administration of plicamycin with solutions containing divalent cations and trace elements, since this drug has complexation property.

Streptozen waste may be disposed of by reacting with hydrobromic acid in glacial acetic acid or by oxidation with a solution of potassium permanganate in sulfuric acid. Thiotepa powder (1 mg/mL) in 5% glucose solution is incompatible if it is combined with solutions of cisplatin or minocycline hydrochloride.[228]

***Vinblastine Sulfate*** Vinblastine sulfate loses its potency if stored along with 5% glucose solution in intravenous sets. This is maximum with cellulose propionate sets and minimum with methacrylate butadiene styrene materials. Solutions of this drug were more stable in polybutadiene tubing compared with PVC tubing.

### Blood Products, Plasma Expanders, and Hemostatics

Aprotinin administration during surgery should follow heparin treatment to avoid any risk. Patients under the treatment of antifibrinolytic therapy should be treated cautiously when coadministering drugs affecting hemostasis and estrogens.[229] Accumulation of albumin may occur in patients with impaired renal function if large volumes of albumin solutions are administered.

Care must be exercised to include 0.9% sodium chloride or 0.5% glucose when diluting albumin solutions. Accidental dilutions with water makes it hypoosmolar, and such a dose may cause severe hemolysis and renal failure.[230,231]

***Aprotinin*** Aprotinin should not be mixed with corticosteroids, heparin, nutrient-containing amino acids or fat emulsions, and tetracyclines. Care must be exercised when diluting solutions of colony-stimulating factors, as they become adsorbed to glass or plastic materials, and the preparation must contain a carrier protein such as albumin to avoid the losses.

Plasma volume expanders such as dextrans may be incompatible at acidic pH.[60,70] Care must be exercised with dilutions. Crystals that are formed upon storage may be redissolved by gentle warming. Care must be exercised when preparing doses of recombinant human erythropoietin solution for neonates to maintain the necessary amount of carrier proteins in the formulation.[232]

Gelatin plasma expanders should not be mixed with vancomycin. Hetastarch is incompatible with many injectable antibiotics and critical care drugs.

Platelets may be administered for transfusions, although ABO incompatibility manifests as acute hemolysis with high antibody titers, and appropriate medical screening may be required.

### Corticosteroids

The treatment of adverse effects of corticosteroids includes symptomatic management and dose tapering on slow withdrawal of drug. Sudden withdrawal of corticosteroids may cause adrenocortical insufficiency and may lead to death. Gradual withdrawal of corticosteroids is advisable, and supplementary corticosteroid therapy should be provided for the patient with a history of corticosteroid withdrawal.

### General Anesthetics

General anesthetics are administered by intravenous, inhalation, or intramuscular routes. Adverse effects of general anesthetics are hypersensitivity, involuntary muscle movements, bronchospasm, cardiac arrhythmias, and respiratory depression. General anesthetics should be used with caution when treating with patients taking corticosteroids, aspirin, estrogens, MAOIs, oral anticoagulants, and lithium. General anesthetics interact with sympathomimetics, ACE inhibitors, antihypertensives, MAOIs, antipsychotics, beta-blockers, and tricyclic antidepressants.[233,234]

***Halothane*** Illicit use of halothane has been performed by either ingestion or injection. Occupational exposure of nitrous oxide can cause serious toxicity, such as bone marrow and neurologic impairment. Effective ventilation should be provided to control the nitrous oxide pollution in the area of its use. Nitrous oxide has been reported to affect the fertility of male and female workers.[235–238]

## Antidotes

Hemoperfusion with activated charcoal may show some adverse effects. Activated charcoal should not be given when other antidotes (e.g., methionine) are used. Activated charcoal reduces the potential of many orally administered drugs and may reduce the action of emetics.

***Amifostine*** Amifostine is incompatible with many drugs such as acyclovir sodium, amphotericin, cefoperazone sodium, hydroxyzine hydrochloride, miconazole, minocycline hydrochloride, and prochlorpherazine edisylate.[239] Care should be exercised when handling amyl nitrate, since it is highly flammable. Volatile nitrites, such as "poppers," are abused and fatal adverse effects are reported.[240,241]

***Calcium Polystyrene Sulfonate*** Calcium polystyrene sulfonate may affect the oral absorption of tetracyclines.

***Desferrioxamine*** Desferrioxamine is incompatible with heparin. This is used in chronic and acute poisoning by iron and other metal ions, such as aluminium load. Edetic acid and its salts chelate metal ions and drugs having such ions. Naloxone hydrochloride is incompatible with bisulfate, metabisulfite, high molecular weight anions, and alkaline preparations. Care should be exercised when treating patients with naloxone and naltrexone, as they may cause withdrawal symptoms.

Penicillamine along with metals, antacids, and food should not be given since it has metal-chelating properties. Probenecid interacts with this drug. Sodium calciumedetate should not be given orally when treating lead poisoning. Sodium cellulose phosphate should be given carefully, as it interacts with magnesium and calcium, and additives used in pharmaceutical formulations.

Corticosteroids interact with barbiturates, carbamazepine, phenytoin, primidone, frusemide, thiazide, NSAIDs, antidiabetics, and antihypertensive drugs. Benzquinamide hydrochloride is incompatible with chlordiazepoxide, diazepam, and some barbiturates. Care should be exercised when handling bisacodyl to avoid contact with skin and mucosal membranes.

## REFERENCES

1. L.A. Trissel, et al., Compatibilities and stability of aztreonam and vancomycin hydrochloride. *Am. J. Health Sys. Pharmacol.* **52**:2560–2564, 1995.
2. A.A. Gayed, et al., Visual compatibility of diltiazem injection with various diluents and medications during simulated Y-site injection. *Am. J. Health Sys. Pharmacol.* **52**:516–520, 1995.
3. J.D. Lewis, and A. El-Gendy, Cephalosporin-pentamidine isethionate incompatibilities. *Am. J. Health Sys. Pharmacol.* **53**:1461–1462, 1996.
4. D. Pritts, and D. Hancock, Incompatibility of ceftriaxone with vancomycin. *Am. J. Hosp. Pharmacol.* **48**:77, 1991.
5. L.K. Jim, Physical and chemical compatibility of intravenous ciprofloxacin with other drugs. *Ann. Pharmacother.* **27**:704–707, 1993.
6. T.M. Berger, et al., Acute pancreatitis in a 12-year-old girl after an erythromycin overdose. *Pediatric* **90**:634–636, 1992.
7. R.J. Tindula, et al., Aminoglycoside inactivation by penicillins and cephalosporins and its impact on drug-level monitoring. *Drug Intell. Clin. Pharmacol.* **17**:906–908, 1983.
8. E. Spina, et al., Clinically significant pharmocokinetic drug interactions with carbamazepine: An update. *Clin. Pharmacokinet.* **31**:198–214, 1996.
9. M. Giaccone, Effect of enzyme inducing anticonvulsants on ethosuximide pharmacokinetics in epileptic patients. *Br. J. Clin. Pharmacol.* **41**:575–579, 1996.
10. K.I. Akinwande, and D.M. Keehn, Dissolution of phenytoin precipitate with sodium bicarbonate in an occluded central venous access device. *Ann. Pharmacother.* **29**:707–709, 1995.
11. C.S.T. Tse, and R. Abdullah, Dissolving phenytoin precipitate in central venous access device. *Ann. Intern. Med.* **128**:1049, 1998.
12. R.L. Nation, et al., Pharmacokinetic drug interactions with phenytoin. *Clin. Pharmacokinet.* **18**:37–60, 131–150, 1990.
13. R.W. Fincham, and D.D. Schottelium, Primidone: Interactions with other drugs, in Levy R.H., et al., eds, *Antiepileptic Drugs*, (4th ed., New York, Raven Press, 467–475, 1995.
14. V.E. Shih, and A. Tenanbaum, Aminoaciduria due to vinyl-gaba administration. *New Engl. J. Med.* **323**:1353, 1990.
15. G.S. Ogawa, et al., Dispensing pin problems. *Am. J. Hosp. Pharmacol.* **42**:1042–1045, 1985.
16. L.A. Trissel, et al., Compatibility of doxorubicin hydrochloride liposome injection with selected other drugs during simulated Y-site administration. *Am. J. Health Sys. Pharmacol.* **54**:2708–2713, 1997.
17. J. Harris, and L.J. Dodds, Handling waste from patients receiving cytotoxic drugs. *Pharm. J.* **235**:289–291, 1985.
18. V. Corbrion, et al., Precipitation of fluorouracil in elastomeric infusers with a polyisoprene and in polypropylene syringes with elastomeric joint. *Am. J. Health Sys. Pharmacol.* **54**:1845–1848, 1997.

19. S.C. Smolinske, et al., Toxic effects of nonsteroidal anti-inflammatory drugs in overdose: An overview of recent evidence on clinical effects and dose-response relationship. *Drug Safety* **5**:252–274, 1990.
20. J.R.B.J. Brouwers, and P.A.G.M. de Smet, Pharmacokinetic-pharmacodynamic drug interactions with nonsteroidal anti-inflammatory drugs. *Clin. Pharmacokinet.* **27**:462–485, 1994.
21. A. Bishnoi, et al., Effect of commonly prescribed nonsteroidal anti-inflammatory drugs on thyroid hormone measurements. *Am. J. Med.* **96**:235–238, 1994.
22. DOH, *Drug Misuse and Dependence: Guidelines on Clinical Management*, London: HMSO, 1991.
23. N.A. Seivewright, and J. Greenwood, What is important in drug misuse treatment? *Lancet* **347**:373–376, 1996.
24. M. Yaster, et al., The management of opioid and benzodiazepine dependence in infants, children, and adolescents. *Pediatrics* **98**:135–140, 1996.
25. D.S. Charney, et al., The combined use of clonidines and naltrexone as a rapid, safe, and effective treatment of abrupt withdrawal from methadone. *Am. J. Psychiatry* **143**:831–837, 1986.
26. C. Brewer, et al., Opioid withdrawal and naltrexone induction in 48–72 hours with minimal drop-out, using a modification of the naltrexone-clonidine technique. *Br. J. Psychiatry* **153**:340–343, 1988.
27. M. Farrell, et al., Methadone maintenance treatment in opiate dependence: A review. *Br. Med. J.* **309**:997–1001, 1994.
28. J.P. Gonzalez, and R.N. Brogden, Naltrexone: A review of its pharmacodynamic and pharmacokinetic properties and therapeutic efficacy in the management of opioid dependence. *Drugs* **35**:192–213, 1988.
29. K.P. Martindale, *The Complete Drug Reference*, 32nd ed., London, U.K., 1999.
30. A.T. McLellan, et al., The effects of psychosocial services in substance abuse treatment. *JAMA* **269**:1953–1959, 1993.
31. C. Gerada, et al., Management of the pregnant opiate user. *Br. J. Hosp. Med.* **43**:138–141, 1990.
32. J. Henry, and G. Volans, ABC of poisoning. Analgesics: Opioids. *Br. Med. J.* **289**:990–993, 1984.
33. E. Boyland, and S.A. Walker, Catalysis of the reaction of aminopyrine and nitrite by thiocyanate. *Arzeimittelforschung* **24**:1181–1184, 1974.
34. F.R. Christensen, and L.W. Anderson, Adverse reaction to extradural buprenorphine. *Br. J. Anaesth.* **54**:476, 1982.
35. J.M. Wagner, and S. Cohen, Fibrous myopathy from butorphanol injections. *J. Rheumatol.* **18**:1934–1935, 1991.
36. R.N. Galante, et al., Solid state acetylation of codeine phosphate by aspirin. *J. Pharmacol. Sci.* **68**:494–498, 1979.
37. M.F. Powell, Enhanced stability of codeine sulphate: Effect of pH, buffer, and temperature on the degradation of codeine in aqueous solutions. *J. Pharmacol. Sci.* **75**:901–903, 1986.
38. K.E. von Muhldendahl, et al., Codeine intoxication in childhood. *Lancet* **ii**:303–305, 1976.

39. M. Linnoila, and S. Hakkinen, Effects of diazepam and codeine, alone and in combination with alcohol, on simulated driving. *Clin. Pharmacol. Ther.* **15**:368–373, 1974.
40. R.J. Young, Dextropropoxyphene overdosage: Pharmacological considerations and clinical management. *Drugs* **26**:70–79, 1983.
41. P.S. Madsen, et al., Acute propoxyphene self-poisoning in 222 consecutive patients. *Acta Anaesthesiol. Scand.* **28**:661–665, 1984.
42. A.T. Proudfoot, Clinical features and management of distalgesic overdose. *Hum. Toxicol.* **3**(suppl.):85S–94S, 1984.
43. F.S. Tennant, Complications of propoxyphene abuse. *Arch. Intern. Med.* **132**:191–194, 1973.
44. J.S. McEwan, and G.H. Macmorran, The compatibility of some bactericides. *Pharmacol. J.* **158**:260–262, 1947.
45. E.A. Davey, and J.B. Murray, Hydrolysis of diamorphine in aqueous solutions. *Pharmacol. J.* **203**:737, 1969.
46. E.A. Davey, and J.B. Murray, Determination of diamorphine in presence of its degradation products using gas liquid chromatography. *Pharmacol. J.* **207**:167, 1971.
47. P. Kintz, et al., Toxicological data after heroin overdose. *Hum. Toxicol.* **8**:487–489, 1989.
48. A. Stewart, et al., Body packing—a case report and review of the literature. *Postgrad. Med. J.* **66**:659–661, 1990.
49. A.H. Ghodse, et al., Comparison of oral preparation of heroin and methadone to stabilize opiate misusers as inpatients. *Br. Med. J.* **300**:719–720, 1990.
50. J.P. Branthwaite, and A. Nicholls, Cyclosporin and diclofenac interaction in rheumatoid arthritis. *Lancet* **337**:252, 1991.
51. B. Levine, et al., Diflunisal related fatality: A case report. *Forensic Sci. Int.* **35**:45–50, 1987.
52. M.Z. Panos, et al., Use of naloxone in opioid-induced anaphylactoid reaction. *Br. J. Anaesth.* **61**:371, 1988.
53. H. Swadi, et al., Misuse of dihydrocodeine tartrate (DF 118) among opiate addicts. *Br. Med. J.* **300**:1313, 1990.
54. Q.A. Xu, et al., Rapid loss of fentanyl citrate admixed with fluorouracil in PVC containers. *Ann. Pharmacother.* **31**:297–302, 1997.
55. S.R. Kowlski, and G.K. Gourlay, Stability of fentanyl citrate in glass and plastic containers and in a patient-controlled delivery system. *Am. J. Hosp, Pharm.* **47**:1584–1587, 1990.
56. Y.-H. Tu, et al., Stability of fentanyl citrate and bupivacaine bydrochloride in portable pump reservoirs. *Am. J. Hosp. Pharm.* **47**:2037–2040, 1990.
57. P.J. Dawson, et al., Stability of fentanyl, bupivacaine and adrenaline solutions for extradural infusion. *Br. J. Anaesth.* **68**:414–417, 1992.
58. J.F. Buchanan, and C.R. Brown, 'Designer drugs': A problem in clinical toxicity. *Med. Toxicol.* **3**:1–17, 1988.
59. S.E. Walker, et al., Compatibility of dexamethasone sodium phosphate with hydromorphone hydrochloride or diphenhydramine hydrachloride. *Am. J. Hosp. Pharmacol.* **48**:2161–2166, 1991.

60. M.F. Chiu, and M.L. Schwartz, Visual compatibility of injectable drugs used in the intensive care unit. *Am. J. Health Sys. Pharmacol.* **54**:64–65, 1997.
61. Q.A. Xu, et al., Stability of compatibility of fluorouracil with morphine sulfate and hydromorphone hydrochloride. *Ann. Pharmacother.* **30**:756–761, 1996.
62. G.B. Zuckerman, and C.C. Uy, Shock metabolic acidosis, and coma following ibuprofen overdose in a child. *Ann. Pharmacother.* **29**:869–871, 1995.
63. E.D. Marquardt, Visual compatibility of tolazoline hydrochloride with various medications during simulated Y-site injection. *Am. J. Hosp. Pharmacol.* **47**:1802–1803, 1990.
64. D.Y. Ishisaka, et al., Visual compatibility of indomethacin sodium trihydrate with drugs given to neonates by continuous infusion. *Am. J. Hosp. Pharmacol.* **48**:2442–2443, 1991.
65. D.F. Thompson, and N.R. Heflin, Incompatibility of injectable indomethacin with gentamicin or tobramycin sulfate. *Am. J. Hosp. Pharmacol.* **49**:836–838, 1992.
66. S.E. Walker, et al., Stability of reconstituted indomethacin sodium trihydrate in original vials and polypropylene syringes. *Am. J. Health Sys. Pharmacol.* **55**:154–158, 1998.
67. H. Court, and G.N. Volans, Poisoning after overdose with non-steroidal anti-inflammatory drugs. *Adv. Drug React. Acute Poison. Rev.* **3**:1–21, 1984.
68. T.Y.K. Chan, Potential dangers from topical preparations containing methyl salicylate. *Hum. Exp. Toxicol.* **15**:747–750, 1996.
69. T.S. Tam, et al., Warfarin interactions with Chinese traditional medicines: Danshen and methyl salicylate medicated oil. *Aust. N. Z. J. Med.* **25**:258, 1995.
70. C.B. Pugh, et al., Visual compatibility of morphine sulphate and meperidine hydrochloride with other injectable drugs during simulated Y-site injection. *Am. J. Hosp. Pharmacol.* **48**:123–125, 1991.
71. L.A. Trissel, et al., Compatibility of doxorubicin hydrochloride liposome injection with selected other drugs during simulated Y-site administration. *Am. J. Health Sys. Pharmacol.* **54**:2703–2713, 1997.
72. Q.A. Xu, et al., Stability and compatibility of fluorouracil with morphine sulphate and hydromorphone hydrochloride. *Ann. Pharmacother.* **30**:756–761, 1996.
73. P.F. Grassby, The stability of morphine sulphate in 0.9% sodium chloride stored in plastic syringes. *Pharmacol. J.* **248**:HS24–HS25, 1991.
74. P.F. Grassby, and L. Hutchings, Factors affecting the physical and chemical stability of morphine sulphate solutions stored in syringes. *Int. J. Pharm. Pract.* **2**:39–43, 1993.
75. D.M. Piercy, et al., Death due to overdose of nefopam. *Br. Med. J.* **283**:1508–1509, 1981.
76. M. Pulini, False-positive benzodiazepine urine test due to oxaprozin. *JAMA* **273**:1905, 1995.
77. H. Raphan, and M.H. Adams, False positive benzodiazepine urine test due to oxaprozin. *JAMA* **273**:1905–1906, 1995.
78. T. Patel, et al., Assay interaction between oxaprozin and phenytoin. *Ann. Pharmacother.* **31**:254, 1997.
79. G.G. Collee, and G.C. Hanson, The management of acute poisoning. *Br. J. Anaesth.* **70**:562–573, 1993.

80. L.F. Prescott, Paracetamol overdose, in *Paracetamol (acetaminophen): A critical bibliographic review*, Taylor & Francis, London, UK, 401–473, 1996.
81. F. Kamali, The effect of probenecid on paracetamol metabolism and pharmacokinetics. *Eur. J. Clin. Pharmacol.* **45**:551–553, 1993.
82. L.A. Tressel, et al., Compatibility of doxorubicin hydrochloride liposome injection with selected other drugs during simulated Y-site administration. *Am. J. Health Sys. Pharmacol.* **54**:2708–2713, 1997.
83. WHO, WHO expert committee on specifications for pharmaceutical preparations: Thirty-first report. *WHO Tech. Rep. Ser.* **790**: 1990.
84. L.A. Trissel, et al., Compatibility of remifentanil hydrochloride with selected drugs during simulated Y-site administration. *Am. J. Health Sys. Pharmacol.* **54**:2192–2196, 1997.
85. Anonymous, Does probenecid mask steroid use? *Pharmacol. J.* **239**:299, 1997.
86. K. Laine, et al., The effect of activated charcoal on the absorption and elimination of astemizole. *Hum. Exp. Toxicol.* **13**:502–505, 1994.
87. E.A. Kowaluk, et al., Interactions between drugs and intravenous delivery systems. *Am. J. Hosp. Pharmacol.* **39**:460–467, 1982.
88. M. van Dusseldorp, and M.B. Katan, Headache caused by caffeine withdrawal among moderate coffee drinkers switched from ordinary to decaffeinated coffee: A 12 week double blind trial. *Br. Med. J.* **300**:1558–1559, 1990.
89. J. Dawson, Beta agonists put meat in the limelight again. *Br. Med. J.* **301**:1238–1239, 1990.
90. C. Spann, and M.E. Winter, Effect of clenbuterol on athletic performance. *Ann. Pharmacother.* **29**:75–77, 1995.
91. G.J. Ingrams, and F.B. Morgan, Transcutaneous overdose of terbutaline. *Br. Med. J.* **307**:484, 1993.
92. C.E. Johnson, et al., Compatibility of aminophylline and verapamil in intravenous admixtures. *Am. J. Hosp. Pharmacol.* **46**:97–100, 1989.
93. S.M. Bahal, et al., Visual compatibility of warfarin sodium injection with selected medications and solutions. *Am. J. Health Sys. Pharmacol.* **54**:2599–2600, 1997.
94. M.A. Parrish, et al., Stability of ceftriaxone sodium and aminophylline or theophylline in intravenous admixtures. *Am. J. Hosp. Pharmacol.* **51**:92–94, 1994.
95. M. Dick, et al., Digitalis intoxication recognition and management. *J. Clin. Pharmacol.* **31**:444–447, 1991.
96. H. Richards, et al., Use of diltiazem in sport. *Br. Med. J.* **307**:940, 1993.
97. S.Y. Kim, and N.L. Benowitz, Poisoning due to class IA antiarrhythmic drugs quinidine, procainamide and disopyramide. *Drug Safety* **5**:393–420, 1990.
98. S.J. Weir, et al., Sorption of amiodarone to polyvinyl chloride infusion bags and administration sets. *Am. J. Hosp. Pharmacol.* **42**:2679–2683, 1985.
99. A. Taylor, and R. Lewis, Amiodarone and injectable drug incompatibility. *Pharmacol. J.* **248**:533, 1992.
100. J.M. Korth-Bradley, Incompatibility of amiodarone hydrochloride and sodium bicarbonate injections. *Am. J. Health Sys. Pharmacol.* **52**:2340, 1995.
101. C.M. Riley, and P. Junkin, Stability of amrinone and digoxin, procainamide hydrochloride, propranolol hydrochloride, sodium bicarbonate, potassium chloride,

or verapamil hydrochloride in intravenous admixtures. *Am. J. Hosp. Pharmacol.* **48**:1245–1252, 1991.

102. G. Caille, et al., Stability study of diltiazem and two of its metabolites using a high performance liquid chromatographic method. *Biopharm. Drug Dispos.* **10**:107–114, 1989.
103. M. Kanke, et al., Binding of selected drugs to a "treated" inline filter. *Am. J. Hosp. Pharmacol.* **40**:1323–1328, 1983.
104. L. Molin, et al., Solubility, partition, and adsorption of digitalis glycosides. *Acta Pharmacol. Suec.* **20**:129–144, 1983.
105. S.K. Yamashita, et al., Compatibility of selected critical care drugs during simulated Y-site administration. *Am. J. Health Sys. Pharmacol.* **53**:1048–1051, 1996.
106. M.F. Chiu, and M.L. Schwartz, Visual compatibility of injectable drugs used in the intensive care unit. *Am. J. Health Sys. Pharmacol.* **54**:64–65, 1997.
107. A.G. Stuart, et al., Is there a genetic factor in flecainide toxicity? *Br. Med. J.* **298**:117–118, 1989.
108. M.S. Roberts, et al., The availability of nitroglycerin from parenteral solutions. *J. Pharm. Pharmacol.* **32**:237–244, 1980.
109. D.M. Baaske, et al., Nitroglycerin compatibility with intravenous fluid filters, containers, and administration sets. *Am. J. Hosp. Pharmacol.* **37**:201–205, 1980.
110. S.R. Hutching, et al., Compatibility of cefmetazole sodium with commonly used drugs during Y-site delivery. *Am. J. Health Sys. Pharmacol.* **53**:2185–2188, 1996.
111. W. Anderson, and J.E. Harthill, The anticoagulant activity of heparine in dextrose solutions. *J. Pharm. Pharmacol.* **34**:90–96, 1982.
112. A. Wright, and J. Hecker, Long term stability of heparin in dextrose saline intravenous fluids. *Int. J. Pharmacol. Pract.* **3**:253–255, 1995.
113. G. Enderlin, Discoloration of hydralazine injection. *Am. J. Hosp. Pharmacol.* **41**:634, 1984.
114. V. Das Gupta, et al., Stability of hydralazine hydrochloride in aqueous vehicles. *J. Clin. Hosp. Pharmacol.* **11**:215–223, 1986.
115. E.A. Kowaluk, et al., Drug loss in polyolefin infusion systems. *Am. J. Hosp. Pharmacol.* **40**:118–119, 1983.
116. H.J. Martens, et al., Sorption of various drugs in polyvinyl chloride, glass, and polyethylene-lined infusion containers. *Am. J. Hosp. Pharmacol.* **47**:369–373, 1990.
117. P.-H.C. Yuen, et al., Compatibility and stability of labetalol bydrochloride in commonly used intravenous solutions. *Am. J. Hosp. Pharmacol.* **40**:1007–1009, 1983.
118. A.S. Alam, Identification of labetalol hydrochloride precipitate. *Am. J. Hosp. Pharmacol.* **41**:74, 1984.
119. B.B. Riley, Incompatibilities in intravenous solutions. *J. Hosp. Pharmacol.* **28**:228–240, 1970.
120. S.K. Yamshita, et al., Compatibility of selected critical care drugs during simulated Y-site administration. *Am. J. Health Sys. Pharmacol.* **53**:1048–1051, 1996.
121. A. Sianipar, et al., Chemical incompatibility between procainamide bydrochloride and glucose following intravenous admixture. *J. Pharm. Pharmacol.* **46**:951–955, 1994.
122. J.I. Metras, et al., Stability of procainamide hydrochloride in an extemporaneously compounded oral liquid. *Am. J. Hosp. Pharmacol.* **49**:1720–1724, 1992.

123. D. Darbar, et al., Loss of quinidine gluconate injection in a polyvinyl chloride infusion system. *Am. J. Health Sys. Pharmacol.* **53**:655–658, 1996.
124. S.W. Davidson, and D. Lyall, Sodium nitroprusside stability in light protective administration sets. *Pharmacol. J.* **239**:599–601, 1987.
125. D. Lyall, Sodium nitroprusside stability. *Pharmacol. J.* **240**:5, 1988.
126. A.F. Asker, and R. Gragg, Dimethyl sulfoxide as a photoprotective agent for sodium nitroprusside solutions. *Drug Dev. Inv. Pharmacol.* **9**:837–848, 1983.
127. J.P. Patel, et al., Activity of urokinase diluted in 0.9% sodium chloride injection or 5% dextrose injection and stored in glass or plastic syringes. *Am. J. Hosp. Pharmacol.* **48**:1511–1514, 1991.
128. C.E. Johnson, et al., Compatibility of aminophylline and verapamil in intravenous admixtures. *Am. J. Hosp. Pharmacol.* **46**:97–100, 1989.
129. R. Tucker, and J.F. Gentile, Precipitation of verapamil in an intravenous line. *Ann. Intern. Med.* **101**:880, 1984.
130. H.J. Martens, et al., Sorption of various drugs in polyvinyl chloride, glass, and polyethylene-lined infusion containers. *Am. J. Hosp. Pharmacol.* **47**:369–373, 1990.
131. S.M. Bahal, et al., Visual compatibility of warfarin sodium injection with selected medications and solutions. *Am. J. Health Sys. Pharmacol.* **54**:2599–2600, 1997.
132. J.G. Ramaekers, et al., A comparative study of acute and subchronic effects of dothiepin, fluoxetine and placebo on psychomotor and actual driving performance. *Br. J. Clin. Pharmacol.* **39**:397–404, 1995.
133. F. Barbone, et al., Association of road-traffic accidents with benzodiazepine use. *Lancet* **352**:1331–1336, 1998.
134. Anonymous, Antidepressants have been associated with a higher suicide rate. *WHO Drug Inf.* **7**:18–20, 1993.
135. A.P. Popli, et al., Bupropion and anticonvulsants drug interaction. *Am. Clin. Psychiatry.* **75**:99–102, 1995.
136. M. Ostrom, et al., Fatal overdose with citalopram. *Lancet* **348**:339–340, 1996.
137. L.A. Pagliaro, and A.M. Pagliaro, Fluoxetine abuse by an intravenous drug user. *Am. J. Psychiatry.* **150**:1898, 1993.
138. A. Cremona, Mad drivers: Psychiatric illness and driving performance. *Br. J. Hosp. Med.* **35**:193–195, 1986.
139. P.G. Myrenfors, et al., Moclobemide overdose. *J. Intern. Med.* **233**:113–115, 1993.
140. M.P. Schoerlin, et al., Cimetidine alters the disposition kinetics of the monoamine oxidase inhibitors moclobemide. *Clin. Pharmacol. Ther.* **49**:32–38, 1991.
141. P.K. Gillman, Serotoxin syndrome with moclobemide and pethidine. *Med. J. Aust.* **162**:554, 1995.
142. S.C. Dilasaver, Withdrawal phenomena associated with antidepressant and antipsychotic agents. *Drug Safety* **10**:103–114, 1994.
143. S.B. Lipman, and K. Nash, Monoamine oxidase inhibitor update: Potential adverse food and drug interactions. *Drug Safety* **5**:195–204, 1990.
144. M.C. Livingston, and H.M. Livingston, Monoamine oxidase inhibitors: An update on drug interactions. *Drug Safety* **14**:219–227, 1996.
145. J.R. Saper, Ergotamine dependency—a review. *Headache* **27**:435–438, 1987.

146. R. Horstmann, et al., Kritische extremitatenischamie durch ergotismus: Behandlung mit intraarterieller prostaglandin-E1-infusion. *Ditsch. Med. Wochenschr.* **118**:1067–1071, 1993.

147. R. Ghali, et al., Erythromycin associated ergotamine intoxication: Arteriographic and electrophysiologic analysis of a rare cause of severe ischemia of the lower extremities and associated in ischemic neuropathy. *Ann. Vasc. Surg.* **7**:291–296, 1993.

148. F.J. Caballero-Granado, et al., Ergotism related to concurrent administration of ergotamine tartrate and ritonavir in an AIDS patient. *Antimicrob. Agents Chemother.* **41**:1207, 1997.

149. S. Afzaal, et al., High dose atropine in organophosphorous poisoning. *Postgrad. Med. J.* **66**:70–71, 1991.

150. WHO, WHO expert committee on drug dependence: Twenty-ninth report. *WHO Tech. Rep. Ser.* **856**: 1995.

151. S.C. Dilsaver, Withdrawal phenomena associated with antidepressant and antipsychotic agents. *Drug Safety* **10**:103–114, 1994.

152. M. Farrell, and J. Strang, Misuse of temazepam. *Br. Med. J.* **297**:1402, 1988.

153. E.A. Kowaluk, et al., Dynamics of chlormethiazole edisylate interaction with plastic infusion systems. *J. Pharm. Sci.* **73**:43–47, 1984.

154. M.G. Lee, Sorption of four drugs to polyvinyl chloride and polybutadiene intravenous administration sets. *Am. J. Hosp. Pharmacol.* **43**:1945–1950, 1986.

155. E.A. Kowaluk, et al., Interactions between drugs and intravenous delivery systems. *Am. J. Hosp. Pharmacol.* **39**:460–467, 1982.

156. H.J. Martens, et al., Sorption of various drugs in polyvinyl chloride, glass, and polyethylene lined infusion containers. *Am. J. Hosp. Pharmacol.* **47**:369–373, 1990.

157. E.L. Jeglun, et al., Nafcillin sodium incompatibility with acidic solutions. *Am. J. Hosp. Pharmacol.* **38**:462,464, 1981.

158. S.R. Hutchings, et al., Compatibility of cefmetazole sodium with commonly used drugs during Y-site delivery. *Am. J. Health Sys. Pharmacol.* **53**:2185–2188, 1996.

159. I.A. Ukhun, Compatibility of haloperidol and diphenhydramine in a hypodermic syringe. *Ann. Pharmacother.* **29**:1168–1169, 1995.

160. K. Thoma, R. Klimek, Photostabilisation of drugs in dosage forms without protection from packaging materials. *Int. J. Pharm.* **67**:169–175, 1991.

161. L.A. Trissel, J.F. Martinez, Compatibility of aztreonam with selected drugs during simulated Y-site administration. *Am. J. Health Sys. Pharmacol.* **52**:1086–1090, 1995.

162. J.I. Boullata, et al., Precipitation of lorazepam infusion. *Ann. Pharmacother.* **30**:1037–1038, 1996.

163. L.A. Trissel, S.D. Pearson, Storage of lorazepam in three injectable solutions in polyvinyl chloride and polyolefin bags. *Am. J. Hosp. Pharmacol.* **51**:368–372, 1994.

164. L.A. Trissel, et al., Compatibility of parenteral nutrient solutions with selected drugs during simulated Y-site administration. *Am. J. Health Sys. Pharmacol.* **54**:1295–1300, 1997.

165. Y.V. Pramar, et al., Stability of midazolam hydrochloride in syringes and intravenous fluids. *Am. J. Health Sys. Pharmacol.* **54**:913–915, 1997.

166. D.E.L. Zuber, Compatibility of morphine sulfate injection and prochlorperazine edisylate injection. *Am. J. Hosp. Pharmacol.* **44**:67, 1987.

167. I.R. Tebbett, et al., Stability of promazine as intravenous infusion. *Pharmacol. J.* **237**:172–174, 1986.
168. S. Sikdar, Physical dependence on zopiclone. *Br. Med. J.* **317**:146, 1998.
169. J. Main, and M.K. Ward, Potentiation of aluminium absorption by effervescent analgesic tablets in a haemodialysis patient. *Br. Med. J.* **304**:1686, 1992.
170. M. Hudson, et al., Reversible toxicity in poisoning with colloidal bismuth subcitrate. *Br. Med. J.* **299**:159, 1989.
171. F. Huwez, et al., Acute renal failure after overdose of colloidal bismuth subcitrate. *Lancet* **340**:1298, 1992.
172. M.A. Vernace, et al., Chronic salicylates toxicity due to consumption of over the counter bismuth subsalicylate. *Am. J. Med.* **97**:308–309, 1994.
173. T.J. Meredith, and G.N. Volans, Management of cimetidine overdose. *Lancet* **ii**:1367, 1979.
174. DOH, *Drug Misuse and Dependence: Guidelines on Clinical Management*, London, HMSO, 1991.
175. N.A. Minton, and P.G.D. Smith, Loperamide toxicity in a child after a single dose. *Br. Med. J.* **294**:1383, 1987.
176. T. Anderson, et al., Pharmacokinetics of various single intravenous and oral doses of omeprazole. *Eur. J. Clin. Pharmacol.* **39**:195–197, 1990.
177. S.R. Smith, and M.J. Kendall, Ranitidine versus cimetidine: A comparison of their potential to cause clinically important drug interactions. *Clin. Pharmacokinet.* **15**:44–56, 1988.
178. U. Beuers, et al., Hepatitis after chronic abuse of senna. *Lancet* **337**:372–373, 1991.
179. C.L. Szumlanski, and R.M. Weinshilboum, Sulphasalazine inhibition of thiopurine methyltransferase: Possible mechanism for interaction with 6-mercaptopurine and azathioprine. *Br. J. Clin. Pharmacol.* **39**:456–459, 1995.
180. M. Homeida, et al., Pharmacokinetic interactions between praziquantel and albendazole in Sudanese men. *Ann. Trop. Med. Parasitol.* **88**:551–559, 1994.
181. S. El Kalla, and N.S. Menon, Mebendazole poisoning in infancy. *Ann. Trop. Paediatr.* **10**:313–314, 1990.
182. D.I. Quinn, R.O. Day, Drug interactions of clinical importance: An updated guide. *Drug Safety* **12**:393–452, 1995.
183. J.K. Baird, and C. Lambros, Effect of membrane filtration of antimalarial drug solutions on in vitro activity against *Plasmodium falciparum. Bull WHO* **62**:439–444, 1984.
184. A. Jagger, et al., Clinical features and management of poisoning due to antimalarial drugs. *Med. Toxicol.* **2**:242–273, 1987.
185. H.H. Tonnesen, and A.-L. Grislingaas, Photochemical stability of biologically active compounds II: Photochemical decomposition of mefloquine in water. *Int. J. Pharm.* **60**:157–162, 1990.
186. A. Bourgeade, et al., Intoxication accidentelle a la mefloquine. *Presse. Med.* **19**:1903, 1990.
187. G. Edwards, et al., Interactions among primaquine, malaria infection and other antimalarials in Thai subjects. *Br. J. Clin. Pharmacol.* **35**:193–198, 1993.
188. R.B. Taylor, et al., A chemical stability study of proguanil hydrochloride. *Int. J. Pharm.* **60**:185–190, 1990.

189. A.L. Clark-Schmidt, et al., Loss of carbamazepine suspension through nasogastric feeding tubes. *Am. J. Hosp. Pharmacol.* **47**:2034–2037, 1990.

190. J.T. Wang, et al., Effects of humidity and temperature on in vitro dissolution of carbamazepine tablets. *J. Pharmacol. Sci.* **82**:1002–1005, 1993.

191. L. Durelli, et al., Carbamazepine toxicity and poisoning: Incidence, clinical features and management. *Med. Toxicol. Adv. Drug Exp.* **4**:95–107, 1989.

192. E. Spina, et al., Clinically significant pharmocokinetic drug interactions with carbamazepine: An update. *Clin. Pharmacokinet.* **31**:198–214, 1996.

193. M. Giaccone, Effect of enzyme inducing anticonvulsants on ethosuximide pharmacokinetics in epileptic patients. *Br. J. Clin. Pharmacol.* **41**:575–579, 1996.

194. K.I. Akinwande, and D.M. Keehn, Dissolution of phenytoin precipitate with sodium bicarbonate in an occluded central venous access device. *Ann. Pharmacother.* **29**:707–709, 1995.

195. C.S.T. Tse, and R. Abdullah, Dissolving phenytoin precipitate in central venous access device. *Ann. Intern. Med.* **128**:1049, 1998.

196. R.L. Nation, et al., Pharmacokinetic drug interactions with phenytoin. *Clin. Pharmacokinet.* **18**:37–60, 131–150, 1990.

197. R.W. Fincham, and D.D. Schottelium, Primidone: Interactions with other drugs, in Levy R.H., et al., eds, *Antiepileptic Drugs*, 4th ed., New York, Raven Press, 467–475, 1995.

198. V.E. Shih, A. Tenanbaum, Aminoaciduria due to vinyl-gaba administration. *N. Engl. J. Med.* **323**:1353, 1990.

199. A.M. Gianni, et al., Granulocyte macrophage colony stimulating factor to harvest circulating haemopoietic stem cells for autotransplantation. *Lancet* **ii**:580–584, 1989.

200. W.M. Bennett, et al., Fatal pulmonary bleomycin toxicity in cisplatin induced acute renal failure. *Cancer Treat. Rep.* **64**:921–924, 1980.

201. M. Moran, et al., Prevention of acute graft rejection by the prostaglandin E1 analogue misoprostol in renal transplant recipients treated with cyclosporine and prednisone. *N. Engl. J. Med.* **322**:1183–1188, 1990.

202. J.A. Sosman, et al., Repetitive weekly cycles of interleukin-2 II: Clinical and immunologic effects of dose, schedule, and addition of indomethacin. *J. Natl. Cancer Inst.* **80**:1451–1461, 1988.

203. R. Lodwick, et al., Life threatening interaction between tamoxifen and warfarin. *Br. Med. J.* **295**:1141, 1987.

204. L.A. Trissel, et al., Visual compatibility of amsacrine with selected drugs during simulated Y-site injection. *Am. J. Hosp. Pharmacol.* **47**:2525–2528, 1990.

205. G.V. Stajich, et al., In vitro evaluation of bleomycin-induced cell lethality from plastic and glass containers. *DICP Ann. Pharmacother.* **25**:14–16, 1991.

206. C.L. Zanetti, Scuba diving and bleomycin therapy. *JAMA* **264**:2869, 1990.

207. M.A. Allsopp, et al., The degradation of carboplatin in aqueous solutions containing chloride or other selected nucleophiles. *Int. J. Pharm.* **69**:197–210, 1991.

208. C.F. Stewart, R.A. Fleming, Compatibility of cisplatin and fluorouracil in 0.9% sodium chloride injection. *Am. J. Pharmacol.* **47**:1373–1377, 1990.

209. L.A. Trissel, and J.F. Martinez, Compatibility of thiotepa (lyophilized) with selected drugs during simulated Y-site administration. *Am. J. Health Sys. Pharmacol.* **53**:1041–1045, 1996.

210. M. Casternaro, et al., eds., *Laboratory decontamination and destruction of carcinogens in laboratory wastes: Some antineoplastic agents*, IARC Scientific Publications 73, Lyon:WHO/International Agency for Research on Cancer, 1985.
211. S.D. Pearson, and T.A. Trissel, Leaching of diethylhexyl phthalate from polyvinyl chloride containers by selected drugs and formulation components. *Am. J. Hosp. Pharmacol.* **50**:1405–1409, 1993.
212. H.L. McLeod, et al., Stability of cyclosporin in dextrose 5%, NaCl 0.9% dextrose/amino acid solution, and lipid emulsion. *Ann. Pharmacother.* **26**:172–175, 1992.
213. R.W. Nelson, et al., Visual incompatibility of dacarbazine and heparin. *Am. J. Hosp. Pharmacol.* **44**:2028, 1987.
214. B. Kirk, The evaluation of a light protecting giving set. *Int. Ther. Clin. Monit.* **8**:78–86, 1987.
215. P.F. D'Arcy, Reactions and interactions in handling anticancer drugs. *Drug Intell. Clin. Pharm.*, **17**:532–538, 1983.
216. L.A. Trissel, et al., Compatibility of doxorubicin hydrochloride liposome injection with selected other drugs during simulated Y-site administration. *Am. J. Health Sys. Pharmacol.* **54**:2708–2713, 1997.
217. J. Harris, L.J. Dodds, Handling waste from patients receiving cytotoxic drugs. *Pharmacol. J.* **235**:289–291, 1985.
218. V. Corbrion, et al., Precipitation of fluorouracil in elastomeric infusers with a polyisoprene and in polypropylene syringes with elastomeric joint. *Am. J. Health Sys. Pharmacol.* **54**:1845–1848, 1997.
219. M.C. Allwood, Fluorouracil precipitate. *Am. J. Health Sys. Pharmacol.* **55**:1315–1316, 1998.
220. A. Kupfer, et al., Intramolecular rearrangement of ifosfamide in aqueous solutions. *Lancet* **335**:1461, 1990.
221. R.J. Behme, et al., Incompatibility of ifosfamide with benzyl-alcohol-preserved bacteriostic water for injection. *Am. J. Hosp. Pharmacol.* **45**:627–628, 1988.
222. S.A. Corlett, et al., Permeation of ifosfamide through gloves and cadaver skin. *Pharmacol. J.* **247**:R39, 1991.
223. S. Alex, et al., Compatibility and activity of aldesleukin (recombinant interleukin-2) in presence of selected drugs during simulated Y-site administration: Evaluation of three methods. *Am. J. Health Sys. Pharmacol.* **52**:2423–2426, 1995.
224. K.C. Trent, et al., Multiorgan failure associated with lomustine oversode. *Ann. Pharmacother.* **29**:384–386, 1995.
225. F. Pinguet, et al., Effect of sodium chloride concentration and temperature on melphalan stability during storage and use. *Am. J. Hosp. Pharmacol.* **51**:2701–2704, 1994.
226. J.C. McElnay, et al., Stability of methotrexate and vinblastine in burette administration sets. *Int. J. Pharm.* **47**:239–247, 1998.
227. D.J. Mazzo, et al., Compatibility of docetaxel and paclitaxel in intravenous solutions with polyvinyl chloride infusion materials. *Am. J. Health Sys. Pharmacol.* **54**:566–569, 1997.

228. L.A. Trissel, J.F. Martinez, Compatibility of thiopeta (lyophilized) with selected drugs during simulated Y-site administration. *Am. J. Health Sys. Pharmacol.* **53**:1041–1045, 1996.

229. B.J. Hunt, and J.M. Murkin, Heparin resistance after aprotinin. *Lancet* **341**:126, 1993.

230. D.R. Steinmuller, A dangerous error in the dilution of 25% albumin. *N. Engl. J. Med.* **338**:1226, 1998.

231. L.R. Pierce, et al., Hemolysis and renal failure associated with use of sterile water for injection to dilute 25% human albumin solution. *Am. J. Health Sys. Pharmacol.* **55**:1057, 1062, 1070, 1998.

232. R.K. Ohls, and R.D. Christensen, Stability of human recombinant epoetin alfa in commonly used neonatal intravenous solutions. *Ann. Pharmacother.* **30**:466–468, 1996.

233. C.J. Davies, et al., Delayed adverse reactions to drugs used in anaesthesia. *Adv. Drug React. Bull.* 647–650, April 1995.

234. E.S. Ransom, and R.A. Mueller, Safety consideration in the use of drug combinations during general anaesthesia. *Drug Safety* **16**:88–103, 1997.

235. P. Berman, and M. Tattersal, Self-poisoning with intravenous halothane. *Lancet* **i**:340, 1982.

236. B. Sweeney, et al., Toxicity of bone marrow in dentists exposed to nitrous oxide. *Br. Med. J.* **291**:567–569, 1985.

237. A.S. Rowland, et al., Reduced fertility among women employed as dental assistants exposed to high levels of nitrous oxide. *N. Engl. J. Med.* **327**:993–997, 1992.

238. J.B. Brodsky, Nitrous oxide and fertility. *N. Engl. J. Med.* **328**:284–285, 1993.

239. L.A. Trissel, and J.F. Martinez, Compatibility of amifostine with selected drugs during simulated Y-site administration. *Am. J. Health Sys. Pharmacol.* **52**:2208–2212, 1995.

240. B. Lockwood, Poppers: volatile nitrite inhalants. *Pharmacol. J.* **257**:154–155, 1996.

241. Anonymous, Treatment of acute drug abuse reactions. *Med. Lett. Drugs Ther.* **29**:83–86, 1987.

CHAPTER 16

# Target Organ Toxicity

## INTRODUCTION

Different occupations in chemical and drug manufacturing industries make workers susceptible to adverse effects. The nature of physicochemical substances and the type of adverse effects in workers are similarly large. The term *adverse* is defined as a large spectrum of undesired effects. Some adverse effects are large, and others are not. Side effects of some drugs are not only undesirable but also deleterious. The exposure of a worker may be direct or indirect and through a single or all three routes (oral, dermal, inhalation). Also, the contact substance may be a solid chemical, liquid, toxic gas, or mist with suspended particles. (In other words, substances include drugs, metals and their dusts, organic and inorganic dusts of quartz, asbestos, cement, glass fibers, minerals, liquids, and concentrates such as different kinds of organic solvents, soaps and detergents, acids, alkalis, and gases such as oxides of nitrogen, carbon dioxide, fumes, and vapors of metallic substances.)

Exposure to chemicals and drugs could be alone or in combination, resulting in interaction and potentiation of chemical/drug toxicity. The resulting toxicologic effects may cause pathologic lesions, which over time may lead to serious health effects and death. Further, depending on the exposure period and concentration of toxic substances, people may develop systemic toxicity and suffer from adverse effects. The list includes phenols and phenolic compounds, organic solvents, halogenated hydrocarbons, metals such as antimony, arsenic, beryllium, cadmium, lead, manganese, mercury, phosphorus, and selenium.

This chapter discusses specific organ systems that contact a variety of substances in the workplace environment (or due to drug abuse and addiction), leading to health hazards. Important organs discussed here include the skin, the respiratory system, the cardiovascular system, the renal system, and the central nervous system (CNS). In the United States and other countries of the world, the application of phenol is on the increase in the manufacture of aromatic compounds, explosives, fertilizers, paints, rubber products, wood preservatives, synthetic resins, drugs and pharmaceutical products, plastics, leather tanning and

*Industrial Guide to Chemical and Drug Safety*, By T.S.S. Dikshith and Prakash V. Diwan
ISBN 0-471-23698-5 

a host of other products. Phenols are known to cause different pathologic lesions in animals and humans (e.g., skin irritation, skin corrosion, damage of the kidney tubules, nodular pneumonia, congestion, and liver damage). Continued exposure to phenol has caused emphysema, edema, bronchopneumonia, and petechial hemorrhages in the pleura of humans. Thus, we discuss here the impact of chemical and drug abuse, and the pathomorphologic lesions normally observed in these organs vis-à-vis human safety.

## SKIN

Skin is the first organ that is exposed to a variety of physical and chemical reagents. Mammalian skin is the organ of body defense (i.e., a "barrier") from chemical injury. Physical and chemical forces lower temperature, humidity, and repeated reaction with chemicals and drugs (e.g., detergents, soaps, organic solvents), leading to the impairment of this barrier through water loss and skin dryness.

Absorption of chemicals through the skin depends on a number of factors, the most important of which are concentration, duration of contact, solubility, physical condition of the skin, and part of the body exposed. The skin is composed of multiple, lipid-containing layers. The outermost layer, the stratum corneum, functions as the primary protective barrier. For percutaneous absorption to occur, the chemical must pass through this tough layer and reach the epidermis. As with all lipid-soluble materials, once a chemical gains access to the dermis, rapid and complete absorption is usually ensured. Differences in skin structure and condition also affect the degree to which chemicals can be absorbed. In general, toxicants cross thin skin much easier than thick skin. Roughly, the ease of absorption follows the following schedule: scrotum > face > forearm > palm > foot. When the skin is damaged by abrasion or chemical irritation, the penetration of chemicals increases. Wet, saturated skin (as often occurs during prolonged glove use) is four to five times more susceptible to chemical absorption. Various organic solvents also may alter chemical absorption. Carrier solvents, such as dimethyl sulfoxide (DMSO), can facilitate movement of other chemicals through the skin.

Many chemicals are absorbed through intact skin, mucous membranes, or eyes, either by direct contact or (to a more limited extent) contact with its vapor. This is based on the irritation potential of chemicals with extensive industrial use and significant skin absorption under work conditions.

This chapter is not an exhaustive list of all chemicals with skin-penetrating abilities. Consequently, any exclusion of a chemical from this list does not necessarily imply an inability for percutaneous absorption. When in doubt, the user should check the chemical label, material safety data sheet, manufacture date, and other details to achieve chemical safety (Table 16-1).

Many skin diseases among occupational workers have been identified as a result of prolonged contact with irritant chemicals.[1–5] In certain instances, alkylating agents, acids (sulfuric, nitric, hydrochloric, and hydrofluoric acid), and

**TABLE 16-1 Industrial Chemicals and Their Categorization as Skin Irritants**

| Industrial Chemicals | Appearance | Irritancy Classification* |
|---|---|---|
| Aluminium nitrate | White crystals | EI |
| O-aminophenol | Beige crystals | SII |
| P-aminophenol | White powder | SII |
| Ammonium ferrioxalate | Green crystals | VI |
| Ammonium isothionate | White crystals | SII |
| Anhydrous ianilin | Thick yellow emulsion | NI |
| Caprylyl cystinic acid | White powder | SVI |
| Chloroprophylmethylpiperazine | Colorless liquid | VI |
| 4-Chloro-5-sulphamoylphthalimide (20% w/w in toluene) | Brown powder | SVI |
| Copper nitrate | Blue crystals | EI |
| Cyclopentanone | Colorless liquid | SVI |
| Cyamepromazine | Yellow powder | I |
| 1,6-Dibromohexane | Colorless liquid | I |
| Diethoxybenzene | Colorless liquid | I |
| Dimethylsulfate | Colorless liquid | EI |
| 4-Dimethylsulfamido-2-amino-2-nitrodiphenylsulfide | Yellow powder | SII |
| 4-Dimethylsulfamido-2-farmamido-2-nitrodiphenylsulfide | Yellow powder | SII |
| 3-Dimethylsulfamido-phenothiazine ethanol (90%) | Brown-yellow powder | SII |
| Phenothiazine ethanol (90%) | Colorless liquid | SVI |
| Gamma-glycidoxypropyl-trimethoxysilane | Colorless liquid | SII |
| Iminodibenzyl | Beige powder | SII |
| Iminophenacylthiazolidine | Light beige powder | I |
| Mepyramine | Beige powder | SII |
| Mineral oil (sterile codex) | Colorless viscous liquid | SII |
| Orthovanilline | Yellow chips | SVI |
| Oxalic acid | White powder | EI |
| Oxomemazine | White powder | I |
| 1-Palmitoyl-4-Polmitoyl oxyproline | Beige powder | VI |
| Palmitoylcystinic acid | White powder | SII |
| P-Phenetidine | Light brown liquid | VI |
| Polysiloxane resin | Light yellow viscous liquid | SII |
| Polysiloxane resin (30%) w/w in toluene | Light yellow viscous liquid | VI |
| Promethazine | White solid lump | EI |
| Reticulable silica paste in polysiloxane | Grayish paste | SVI |
| Silica paste in polysiloxane | Pasty emulsion | I |
| Sodium fluosilicate | White powder | EI |
| Toluene | Colorless liquid | I |
| White spirit (dilutine 5) | Colorless liquid | I |

*I — Irritant; SII — Slightly Irritant; SVI — Severely Irritant; VI — Very Irritant; EI — Extremely Irritant.
*Source*: Guillot JP. et al. 1982.[20]

strong alkalis damage the skin instantaneously and cause skin burn of different degrees. This exposure also results in necrosis and ulceration within a short time. Birmingham et al.[5] demonstrated that workers associated with the mining industry develop punched-out ulcers on the skin and skin damage when in contact with chemicals like arsenic trioxide, calcium arsenate, calcium nitrate, and slaked lime. Also, spraymen engaged in pest control operations without using proper protective measures become exposed to the action of pesticides. Laboratory studies indicated that repeated exposure to pesticides causes adverse effects in the skin of animals (Figs. 16-1–16-3).[5a]

Potent drugs do more harm than good, as observed in cases of skin diseases. The drugs can cause immediate or delayed hypersensitive reaction, which sometimes leads to anaphylactic shock and death. Skin absorption from foot to palm varies significantly. Absorption on the face, forearm, and scrotum is relatively faster. Absorption is enhanced if the skin is damaged by burn or inflammation. In occlusive dressing, absorption increases 10-fold; at the same time, occlusive dressing for a large area causes serious systemic toxicity. The transdermal delivery system is used for the administration of drugs through the skin. Drug formulations are available as creams, ointments, pastes, dusting powders, and gels. The systemic or topical use of drugs may cause skin lesions.

Scleroderma is a progressive systemic sclerosis with a multisystem connective tissue disease. A number of industrial chemicals have been implicated as causative factors in human scleroderma. Industrial chemicals (e.g., toluene, benzene, xylene, aromatic mixers or white spirit, vinyl chloride, trichloroethylene, perchloroethylene, naphtha-n-hexane, epoxy resins,

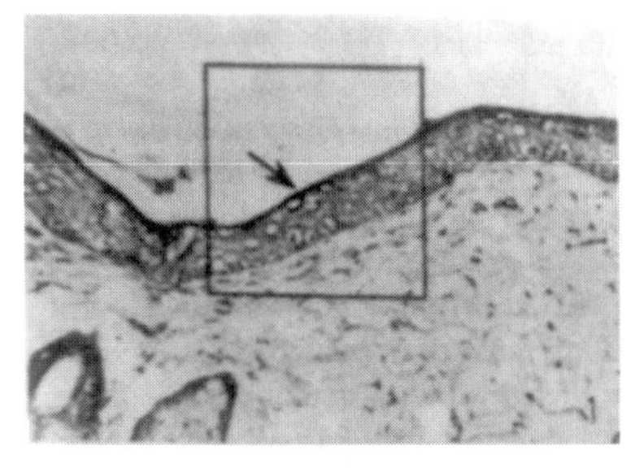

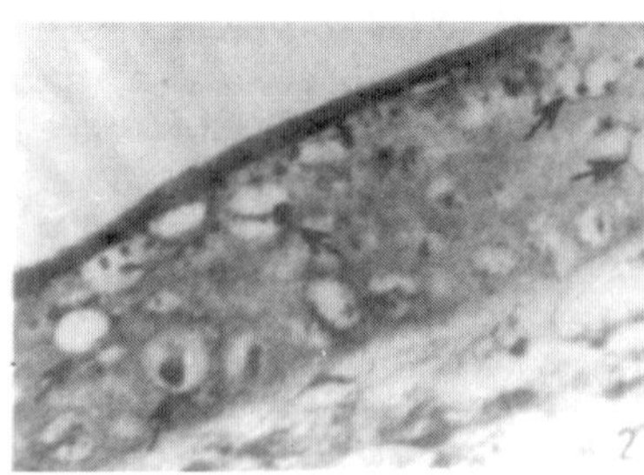

**Figure 16-1.** Rat skin showing normal structure of the epidermal cell layer.

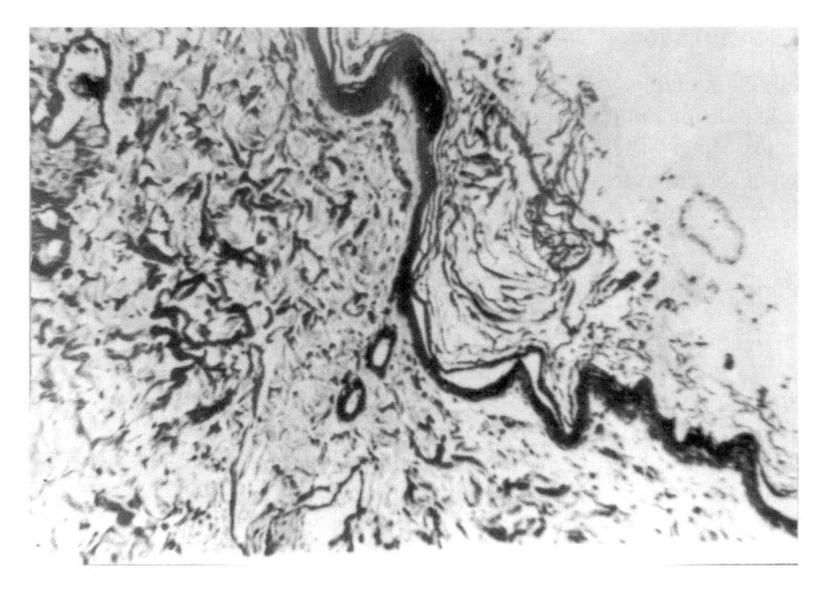

**Figure 16-2.** Section of rat skin exposed to toxic chemicals and hyperkeratinization.

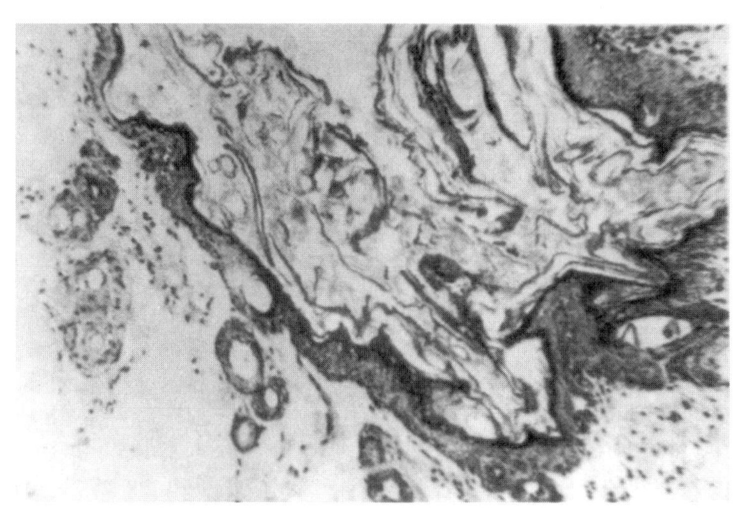

**Figure 16-3.** Effect of toxic chemicals on rat skin. Hyperkeratinization and thickening of the dermal layer.

urea, formaldehyde, metaphenylenediamine, and drugs such as bleomycin, carbidopa, L-5-hydroxytryptopha, cocaine, pentazocine, diethyl proprion, fenfluramine hydrochloride), have been implicated in this skin disorder. A majority of females in Afro-Caribbean countries have been found to suffer from this skin disease. Diffuse chlorosis is known to affect the peripheral vasculature, skin, gastrointestinal (GI) tract, heart, and muscular system.[6]

### Skin Sensitization

Industrial operations and chemical processing provide a suitable atmosphere for the induction of immunologic reactions as seen with skin sensitization and the respiratory tract. The antigen antibody reaction with a complex cascade-type system of serum proteins results in a variety of sensitizing reactions of the worker and disturbs his or her normal health.

### Skin and Phototoxicity

Generally, regulatory agencies around the world demand the application of animal study data before any drug or industrial chemical is made available to the public. Several antimicrobial and antibacterial compounds exert phototoxic effects on animals and humans. In fact, phototoxicity reactions of the skin are on the increase and have become an important problem in human dermatology. Reports have shown that many chemicals and drugs induce unwanted side effects on the skin.[7] Fluoroquinolones are known to cause phototoxicity, as measured by the development of erythema and edema in laboratory mice and guinea pigs.[8–10]

Workers in different situations become exposed to photosensitizing chemicals as well. Phototoxicity is an ill effect of skin in reaction to ultraviolet (UV) light. The most common type of phototoxity is skin damage by sunburn. Workers associated with operation and processing activities of electric furnaces, foundries, glassblowing, steel, welders, photocure inks, and printers have developed phototoxicity effects on the body. This is common among workers associated with coal tar industry who contact dusts, vapors, chemicals such as anthracene, phenanthrene, and acridine. The most common photoreceptive chemicals and interaction with sunlight result in hyperpigmentation. Workers who use different drugs quickly succumb to photosensitization effects of the chemical environment. Photosensitizers may be antibacterial agents (e.g., sulfonamides, hexachlorophenes, dibromosalicylanilide), tranquilizers, and coal tar constituents (e.g., pitch, acridine, creosote, chlorinated hydrocarbons).[11–14]

Workers also are affected by a variety of radiations, specifically UV. This radiation consists of UVA, UVB, and UVC, which induces certain skin lesions (Table 16-2). To protect the skin from the injuries from UVA and UVB radiation effects, certain drugs are suggested such as aminobenzoic acid, aminobenzoates, cinnamates, salicylates, and camphors for UVB protection, and benzophenones or dibenzoylmethanes for UVA protection.

### Primary Skin Irritants

Information on the potential of chemicals to cause acute skin irritation becomes important. This assumes great significance to establishing procedures and alternative methods for the safe handling, packing, and transportation of chemicals as well as for general safety-assessment purposes.[14–17]

Skin irritation is a common problem with chronic exposure to liquid aliphatic hydrocarbons used as industrial solvents. These chemicals extract fats from the

**TABLE 16-2 Ultraviolet Radiation and Skin Changes**

| Ultraviolet Radiation | Wavelength Range | Effect on Skin |
|---|---|---|
| UVA | 320–400 nm | Skin aging and likely skin cancer |
| UVB | 290–320 nm | Sunburns, tanning, chronic skin cancer and skin aging |
| UVC | 200–290 nm | Skin injury at high altitude |

skin, causing painful drying and cracking, called *chronic eczematoid dermatitis*. This eventually leads to severe itching and inflammatory reaction. The organic and inorganic alkalis damage keratin on the skin. These chemicals disturb lipid components of the skin and damage the protein and keratin layers. Some substances (e.g., arsenic, coal tar, methylcholanthrene, other substances) damage skin, lead to abnormal growth, and result in skin cancer. Schwartz et al.[18,19] compiled a number of chemicals that pose as primary skin irritants in industry workplaces. These may be broadly grouped into acids, alkalis, solvents, and metal salts.[18,19]

Drug-specific skin rashes are common and are categorized under different classes such as acne, purpura, urticaria, eczema, and other pathomorphologic disorders. The formation of acne in human skin can be triggered by misuse of certain drugs (e.g., corticosteroids, androgens). Skin erythema multiforme is caused by a variety of drugs (e.g., NSAIDs, sulfonamides, barbiturates, phenytoin). Allergic vasculitis is caused by drugs such as sulfonamides, NSAIDs, thiazides, chlorpropamide, phenytoin, penicillin, and retinoids. Purpura is caused by thiazides, sulfonamides, sulfonylureas, phenylbutazone, and quinine. In contrast, skin eczema is caused by drugs such as penicillins and phenothiazines. Urticaria and angioedema are caused by penicillins, enalapril, gold, and NSAIDs, for instance aspirin. Psoriasis may be aggravated by lithium and antimalarial drugs. Pruritus is associated with rash after the consumption of oral contraceptives, phenothiazines, and rifampicin. Hair loss occurs in people under medication of drugs such as cytotoxic anticancer drugs, acitretin, oral contraceptives, heparin, androgenic steroids (women), sodium valproate, and gold. Skin pigmentation occurs after an intake of oral contraceptives (in photosensitive distribution), phenothiazines, heavy metals, amiodarone, chloroquine (pigmentations of nails and palate, depigmentation of the hair), and minocycline.

## Ocular Irritation

Irrespective of work conditions in different situations (e.g., chemical factories, heavy industries, metal/package industries), eye protection becomes an important aspect of human safety. Present-day industries use a variety of substances that are mildly irritant, moderately irritant, and extremely irritant to human eyes. Further, society uses a variety of cosmetic products that often come into contact with the *conjunctiva*, *iris*, and *cornea* of the eye, eventually leading to ocular irritation or eye damage. To identify the primary ocular irritation or ocular corrosivity potential of chemicals, different protocols are recommended. These include the official method of testing cosmetics and toiletries in France, the OECD, and organizations like Association Française de Normalization (AFNOR). These protocols and ratings are important for the evaluation of chemicals and the interpretation of their effects on the human eye. In 1982, AFNOR proposed certain ratings for the development of ocular irritancy caused by chemicals based on the acute ocular irritation index (AOI), the individual ocular irritation index (IOI), and the mean ocular irritation index (MOI) (Table 16-3).[20]

A number of industrial chemicals were screened using AFNOR protocols with rabbits as the test systems. Information on the appearance and irritancy potential of different chemicals (and the classification thereof) offers valuable clues to the worker to avoid skin injury (Table 16-4). Certain drugs normally used by humans can cause adverse effects on eyes. The types of ocular disturbance are blurring of vision and diplopia, impairment of visual acuity, yellow vision (xanthopsia), corneal opacities, and lenticular opacities and drugs producing or precipitating the formation of cataract, which have been implicated with the use of candidate drugs.

For instance, blurring of vision and diplopia are caused by the use of imipramine, iproniazid, chlorpromazine, thioridazine, and promethazine. Impairment of visual acuity is caused by chlorpropamide, tolbutamide, alcohol, chlorpromazine, phenylbutazone, indomethacin, chloroquine, sulfonamides, ethambutol, chloramphenicol, isonex, clioquinol, quinine, streptomycin, and paraaminosalicylate. Yellow vision (xanthopsia) has been traced to the use of sulfonamides, streptomycin, methaqualone, barbiturates, chlorothiazide,

**TABLE 16-3 AFNOR Scale of Ocular Irritancy and the Interpretation***

| Acute Ocular Irritancy | Day 7 Individual Ocular Irritancy | Mean Ocular Irritancy Index | Result | Irritancy Class |
|---|---|---|---|---|
| 0–5 | — | 0–5 after 48 hours | Nonirritant | NI |
| 5–15 | — | Less than 5 after 48 hours | Slightly irritant | SII |
| 15–30 | — | Less than 5 after 4 days | Irritant | I |
| 30–60 | 30 in all 6 rabbits | 20 after 7 days | | |
| | 15 in at least 4/6 | | Very irritant | VI |
| 60–80 | 40 after 7 days | 60 in all 6 rabbits | | |
| | | 30 in at least 4/6 | Severely irritant | SVI |
| 80–110 | — | — | Extremely irritant | EI |

*Theoretic maximum score from a total of: cornea (80); iris (10); conjunctiva (20).
*Source*: Guillot JP. et al. 1982.[20]

**TABLE 16-4 Industrial Chemicals Causing Respiratory Diseases**

| Agents | Source of Exposure | Type of Lesion |
|---|---|---|
| Asbestos | Mining, construction, ship-building, manufacture of asbestos-containing materials | Asbestosis, lung cancer |
| Aluminum dust | Manufacture of aluminum products, fireworks, ceramics, paints, electrical goods, abrasives | Fibrosis |
| Aluminum | Manufacture of abrasives, smelting | Fibrosis initiated from short exposures |
| Ammonia | Ammonia production, manufacture of fertilizers, chemical production, explosives | Irritation |
| Arsenic $(Pb_3AsO_4)_2$ | Manufacture of pesticides, pigments, glass, alloys | Lung cancer, bronchitis, laryngitis |
| Beryllium | Ore extraction, manufacture of alloys, ceramics | Dyspnea, interstitial granuloma, fibrosis, corpulmonae, chronic disease |
| Boron | Chemical process | Acute neurotoxicity |
| Cadmium oxide (fume dust) | Welding, manufacture of electrical equipment, alloys, pigments, smelting | Emphysema |
| Carbides of tungsten, titanium, tantalium | Manufacture of cutting edges on tools | Pulmonary fibrosis |
| Chlorine | Manufacture of pulp and paper, plastics, chlorinated chemicals | Irritation |
| Chromium (IV) | Production of chromium compounds, paint pigments, reduction of chromite ore | Lung cancer |
| Coal dust | Coal mining | Pulmonary fibrosis |
| Coke oven emissions | Coke production | Lung cancer (9 times greater than other steel workers) |
| Hydrogen fluoride | Manufacture of chemicals, photographic film, solvents, plastics | Irritation, edema |

(*continued overleaf*)

**TABLE 16-4** (*continued*)

| Agents | Source of Exposure | Type of Lesion |
|---|---|---|
| Iron oxides | Welding, foundry work, steel manufacture, hematite mining, jewelry making | Diffuse fibrosis |
| Kaolin | Pottery making | Fibrosis |
| Manganese | Chemical and metal industries | |
| Nickel | Nickel or extraction, nickel smelting, electroplating, fossil fuel | Nasal cancer, lung cancer, acute pulmonary edema |
| Osmium tetraoxide | Chemical and metal industry | |
| Oxides of nitrogen | Welding, silo filling, explosives manufacture | Emphysema |
| Ozone | Welding, bleaching flour, deodorizing | Emphysema |
| Phosgene | Production of plastics, pesticides, chemicals | Edema |
| Perchloroethylene | Dry cleaning, metal degreasing, grain fumigating | Edema |
| Silica | Mining, stone cutting, construction, farming, quarrying | Silicosis (fibrosis) |
| Sulfur dioxide | Manufacture of chemicals, refrigeration, bleaching, fumigation | |
| Talc | Rubber industry, cosmetics | Fibrosis, pleural sclerosis |
| Tin | Mining, tin processing | |
| Toluene 2,4-diisocyanate | Plastics manufacture | Decrease of pulmonary function |
| Vanadium | Steel manufacture | Irritation |
| Xylene | Manufacture of resins, paints, varnishes, other chemicals, general solvent for adhesives | Edema |

furosemide, oral contraceptives, ethambutol, and hypervitaminosis A. Corneal opacities are observed in the use of chlorpromazine, indomethacin, gold, chloroquine, hydroxychloroquine, mepacrine, and vitamin D. Lenticular opacities or drugs producing or precipitating cataract have been identified with the use of steroids, anticholinesterases, busulfan, chlorambucil, and chlorpromazine.

More chemicals also are known to induce primary skin irritation in animals and humans. The following are some selected skin irritants commonly used in laboratory and industrial operations.

### Selected List of Primary Skin Irritants

Essentially, all primary skin irritants include acids, alkalis, metals, salts, and solvents. Among organic acids one may include acetic acid, acrylic acid, carbolic acid, chloroacetic acid, formic acid, lactic acid, oxalic acid, and salicylic acid. Among inorganic acids one may list arsenious acid, chromic acid, hydrochloric acid, hydrofluoric acid, nitric acid, phosphoric acid, and sulfuric acid. Alkalis include butylamines, ethylamines, ethanolamines, methylamines, propylamines, and triethanolamine. One also may include ammonium carbonate, ammonium hydroxide, calcium carbonate, calcium cyanamide, calcium hydroxide, calcium oxide, potassium carbonate, potassium hydroxide, sodium carbonate (soda ash), sodium hydroxide (caustic soda), and sodium silicate.

Many metal salts are well-known primary skin irritants. These substances include antimony trioxide, arsenic trioxide, chromium and alkaline chromates, cobalt sulfate, nickel sulfate, mercury chloride, and zinc chloride. In addition to the above industrial chemicals, several solvents are known to act as primary skin irritants among workers, such as, carbon tetrachloride ($CCl_4$), chloroform, ethylene dichloride, epichlorohydrin, ethylene chlorohydrin, perchloroethylene, and trichloroethylene, in addition to cool tar solvents such as naphtha, toluene, and xylene.

## THE RESPIRATORY SYSTEM

### Respiratory Irritants

Much is known about the induction of drug toxicity in animals and humans. A variety of drugs have been used for the treatment of human health disorders. Occupational workers, young, adult men, women, and pregnant women often become victims when exposed to chemicals and drugs in workplaces. Many drugs and related chemicals are known to disturb and induce respiratory diseases in such exposed workers. Therefore, it is important to monitor and properly manage workers' health to achieve safety.

Many chemicals (e.g., bromine, chlorine, chlorine oxides, cyanogen bromide, cyanogen chloride, dimethyl sulfate ozone, sulfur chlorides, phosphorus trichloride, phosphorus pentachloride) induce an irritant effect on the upper respiratory tract and lung tissue. Other chemicals (e.g., arsenic trichloride, nitrogen dioxide, nitrogen trioxide, phosgene) disturb and damage the terminal passages and air sacs of the lung.[21–24] Irritant materials are corrosive or vesicant in character and known to cause inflammatory reaction on the moist surface.

Different irritant chemicals affect parts of the mammalian respiratory system differently. The upper respiratory tract of the mammalian lung is affected by chemicals such as aldehydes, ammonia, alkaline dust, chromic acid, sulfur dioxide, sulfur trioxide, hydrogen fluoride, hydrogen chloride, and ethylene oxide). It is now known that many fluorine-containing organic compounds, such as chlorofluorocarbons (CFCs), have played a major role in the asthmatic deaths of workers associated with aerosol propellant processing with CFC aerosol. Table 16-4 provides a list of industrial chemicals that are known to cause respiratory disorders in workers. Extensive use of a variety of drugs in the form of antibiotics, antivirals, antigoutes, anticancers, antihypertensives, steroids, sedatives, and lipid-lowering drugs to contain ailments and disease has become a common feature in modern society. However, use of these drugs has been implicated in causing disturbances to the normal function of the respiratory system. The list includes several drugs of different categories (Table 16-5).

**TABLE 16-5 Drugs with Possible Side Effects on the Respiratory System**

| | |
|---|---|
| Acetylsalicylic acid | Colchicine |
| Acrylate | Cotrimoxazole |
| Acyclovir | Cromoglycate |
| Adenosine and derivatives | Cyclophosphamide |
| Adrenaline | Cyclosporin |
| Albumin | Cytarabine (Cytosine arabinoside) |
| Albuterol | Cytokines |
| Allopurinol | Danazol |
| Aminoglycoside antibiotics | Dapsone |
| Amiodarone | Daunorubicin |
| Amitriptyline | Dexamethasone |
| Amphotericin B | Dexfenfluramine |
| Ampicillin | Dextropropoxyphene |
| Amrinone | Dextran |
| Atenolol | Diclofenac |
| Azapropazone | Dihydroergocristine |
| Azathioprine | Dihydroergocryptine |
| Azithromycin | Dihydroergotamine |
| Barbiturates | Diltiazem |
| Bleomycin | Dimethyl sulfoxide |
| Bromocriptine | Diphenylhydantoin |
| Busulfan | Doxorubicin |
| Baptopril | Enalapril |
| Carbamazepine | Epinephrine |
| CCNU | Ergometrine |
| Cephalexin | Ergotamine |
| Cephalosporins | Ergots |
| Chlorambucil | Erythromycin |
| Chloroquine | Ethambutol |
| Chlorpromazine | Etoposide |
| Chlorpropamide | Fenfluramine/dexfenfluramine |
| Ciprofloxacin | Fenoprofen |
| Clofibrate | Fibrinolytics |
| Clomiphene | Floxuridine |
| Clonidine | 5-fluorouracil |

**TABLE 16-5** *(continued)*

| | |
|---|---|
| Fluoxetine | Metoclopramide |
| Flurbiprofen | Metoprolol |
| Fosinopril | Metronidazole |
| Fotemustine | Miconazole |
| Furazolidone | Midazolam |
| Glibenclamide | Minocycline |
| Gliclazide | Minoxidil |
| G(M)-CSF | Mitomycin C |
| Gold salts (aurothiopropanosulfonate) | Morphine |
| Gonadotropin | Nalidixic acid |
| Haloperidol | Naloxone |
| Heparin | Naproxen |
| Heroin | Nitric oxide (NO) |
| Hydralazine | Nitrofurantoin |
| Hydrochlorothiazide | Nitroglycerin |
| Hydrocortisone | Nitrosoureas |
| Hydroxyurea | Estrogens |
| Ibuprofen | Oxprenolol |
| Ifosfamide | Oxyphenbutazone |
| Imipramine | Paclitaxel |
| Immunoglobulins | Para-(4)-aminosalicylic acid |
| Indinavir | Paracetamol |
| Indomethacin | Paraffin |
| Insulin | Parenteral nutrition |
| Interferon-alfa | Penicillamine |
| Interferon-beta | Penicillins |
| Interferon-gamma | Pergolide |
| Interleukin-2 | Phenylbutazone |
| Isoniazid | Phenytoin |
| Ketamine | Piroxicam |
| Ketorolac | Polyethylene glycol |
| Levodopa | Procainamide |
| Lidocaine | Procarbazine |
| Lipids | Propylene glycol |
| Lisinopril | Propoxyphene |
| Medroxyprogesterone | Propranolol |
| Mefloquine | Pyrimethamine |
| Melphalan | Quinidine |
| Mephenytoin | Radiations |
| Mercaptopurine | Retinoic acid |
| Mesalamine | Rifampicin |
| Metformin | Roxythromycin |
| Methadone | Salbutamol |
| Methotrexate | Simavastin |
| Methyldopa | Steroids |
| Methylprednisolone | Streptokinase |
| Methysergide | Streptomycin |

*(continued overleaf)*

**TABLE 16-5** *(continued)*

| | |
|---|---|
| Sulfasalazine | L-Tryptophan |
| Sulindac | D-Tubocurarine |
| Tacrolimus | Urokinase |
| Tamoxifen | Valproic acid |
| Terbutaline | Vancomycin |
| Tetracycline | Vasopressin |
| Thiopental | Vinblastine |
| Triazolam | Vinorelbine |
| Trimethoprim | Vitamin D |
| Troglitazone | Warfarin |

Inert dusts (e.g., free crystalline silica, which exists as quartz, tridymite, cristobalite, and coesite) has a different biologic effect on the mammalian respiratory system by accumulating in the alveolar regions of the lung. Moolgavkar et al.[24,25] observed that the main determinant of carcinogenicity of synthetic fibers in the mammalian lung has been ascribed to fiber burden in the lung. Based on the available experimental and epidemiologic data, the International Agency for Research on Cancer classified artificial fibers as possible human carcinogens.[26] Several metals, such as aluminum oxide during abrasive production, fumes from aluminum soldering flux, inhalation of salts of chromium, cobalt, nickel, and platinum, also cause inhalation allergy, bronchial asthma, and rhinitis.

## ASTHMA

Asthma is physiologically characterized by increased responsiveness of the trachea and bronchi to chemicals and drugs as stimuli. This results in narrowing of the airways and causes changes in the normal function of the respiratory system. The pathologic features appear in the form of airway smooth muscle contraction, mucosal thickening from edema, cellular infiltration, and inspissations in the airway lumen of abnormally thick, viscid plugs of mucus. Drugs likely to precipitate bronchial asthma in humans include allergens, insulin, penicillin, antisera, and antibiotics such as cephalothin, erythromycin, and streptomycin. Also, drugs such as opiates (heroin, methadone), diuretics (hydrochlorthiazide), antineoplastic agents (bleomycin, cyclophosphamide, vinblastine), beta-blockers (propranolol), and beta-receptor agonists (adrenaline, terbutaline, ritodrine) are known for their precipitating action with pulmonary edema. 1,2-dibromo-3-chloropropane (DBCP) is extensively used as a soil fumigant to control soil nematodes. Exposure to DBCP has caused bronchial dilation, emphysema, and alveola distention (Figs. 16-4 and 16-5). McConnell, et al. conducted a study with 3,535 children in southern California with no history of asthma. In communities with high ozone concentrations, the relative risk of children developing asthma was associated with air pollution and outdoor exercise.[27]

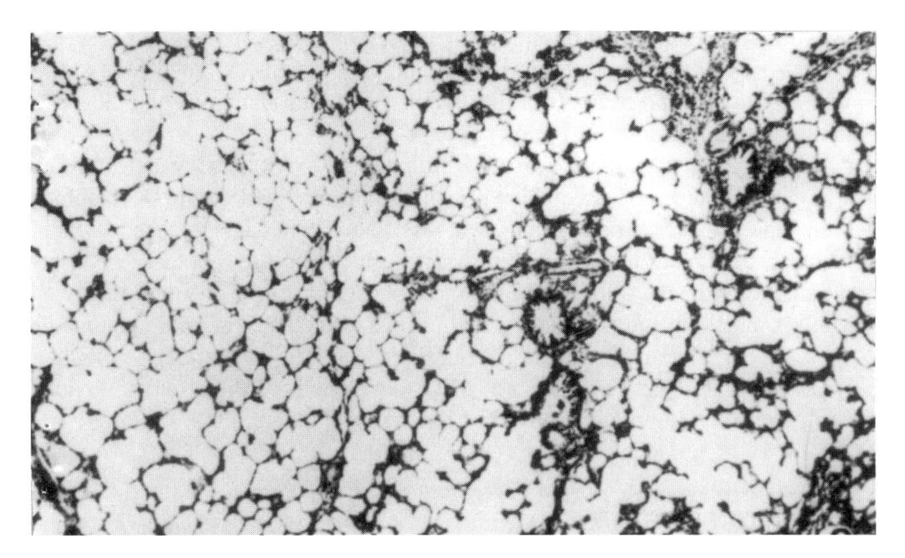

**Figure 16-4.** Section of rat lung unexposed to toxic chemicals.

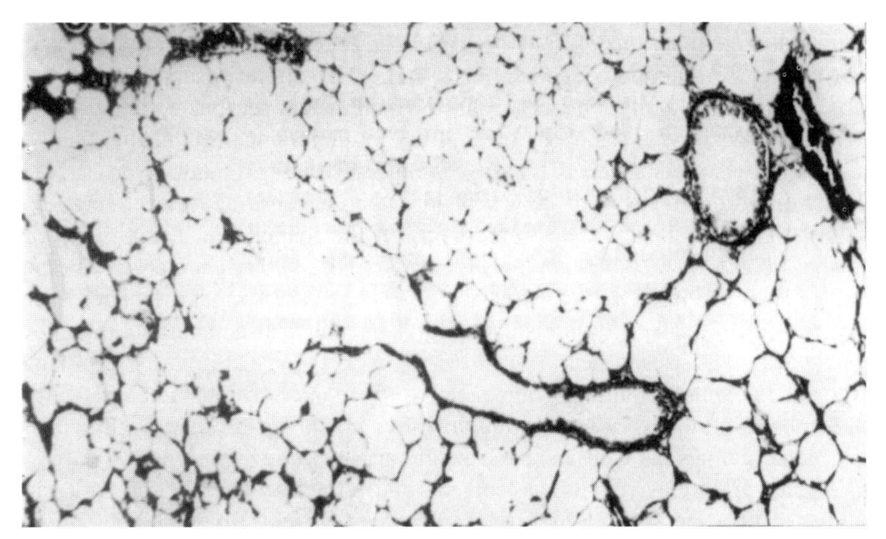

**Figure 16-5.** Section of rat lung exposed to toxic chemicals. Note the bronchial dilation and emphysematous alveolar distention.

One of the worst industrial accidents occurred in Bhopal, India, on December 2 and 3, 1984. It was due to the leakage of methyl isocyanate (MIC) released from the Union Carbide pesticide manufacturing plant. More than 3,000 people who resided in areas adjacent to the manufacturing plant died within a few hours after exposure to MIC. Death was attributed to severe pulmonary toxicity, followed by cardiac arrest (as supported by autopsy cases). The lungs of the victims were 2 to 3 times heavier than those of normal people. The lesions included edema, substantial destruction and necrosis of the alveolar wall, desquamative and ulcerative bronchiolitis, and infiltration by macrophage.[28]

All isocyanates are known to cause pulmonary toxicity. Isocyanates are the most common causes of occupational asthma and have led to the development of immediate or late asthma among workers. Isocyanates have caused bronchitis, rhinitis, conjunctivitis, chronic obstructive lung disease, contact sensitivity, dermatitis, allergic alveolitis, and immunologic hemorrhagic pneumonitis.[29]

MIC is used in industry as a chemical intermediate in the manufacture of polyurethane foams, plastics, adhesives, coating materials, pesticides, and paints.[30] Humans exposed to MIC developed cardiovascular toxicity, genotoxicity, reproductive toxicity, and biochemical changes.[31–34]

## THE NERVOUS SYSTEM

The mammalian nervous system is very complex. It consists of the CNS and the peripheral system. This system controls a variety of important functions of the body (e.g., movement, vision, hearing, sensory, thought, emotions, autonomic, neuroendocrine functions). The term *neurotoxicity* refers to any adverse effect on the structure or function of the nervous system related to exposure to a chemical substance.[35] As with other target organs, the mammalian nervous system becomes delicate or sensitive to the abuse or ill effects of continuous chemical toxicity. It has been already discussed that many organic solvents, aromatic hydrocarbons, aliphatic hydrocarbons, and alicyclic hydrocarbons cause adverse effects (e.g., loss of righting reflexes, CNS depression) on the mammalian nervous system. Inhaled vapors of these compounds/gas cross the alveolar-capillary membrane and are absorbed into the bloodstream. Reports have shown that prolonged exposure to toluene results in cerebellar encephalopathy and dementia. Continuous glue-sniffing has caused hearing loss among workers.[36–41] Similarly, prolonged exposure to paints containing heavy metals and organic solvents has caused structural changes in the CNS with symptoms of dementia, syndromes, cognitive and neurologic deficits, as well as neurasthemic and neurotic syndromes.[42]

Ketones are extensively used as solvents and extracts in a number of worldwide industries. The most widely used ketones are acetone, methylethyl ketone, methylisobutyl ketone, and cyclohexanone. The interaction of methylethyl ketone with other chemicals, such as n-hexane, potentiates neurotoxicity in mammals and demonstrates solvent-induced neuropathy. High concentrations of methylethyl ketone in workplace air caused depression, narcosis, and numbness of the extremities.[43]

The CNS is the target organ disturbed by all organophosphorus pesticides in different formulations used for the control of crop pests or household pests. It has been observed that exposure to organophosphorous pesticides has caused many neurologic effects. Animals exposed to organophosphorous pesticides have shown pathomorphologic effects in the cerebellum, particularly in the Purkinje cells (Figs. 16-6 to 16-8).[5a] For details, refer to Chapters 4 and 5. Based on the mandate from the Toxic Substances Control Act of the United States, the United States Environmental Protection Agency (USEPA) brought out a series of guidelines for the evaluation of neurotoxicity among workers and the general public.[44,45] These include the following:

- Functional observational battery
- Motor activity
- Neuropathology
- Neurotoxic esterase (NTE) neurotox assay
- Schedule-controlled operant behavior
- Acute delayed neurotoxicity of organophosphorus substances
- Subchronic delayed neurotoxicity of organophosphorus substances
- Peripheral nerve function

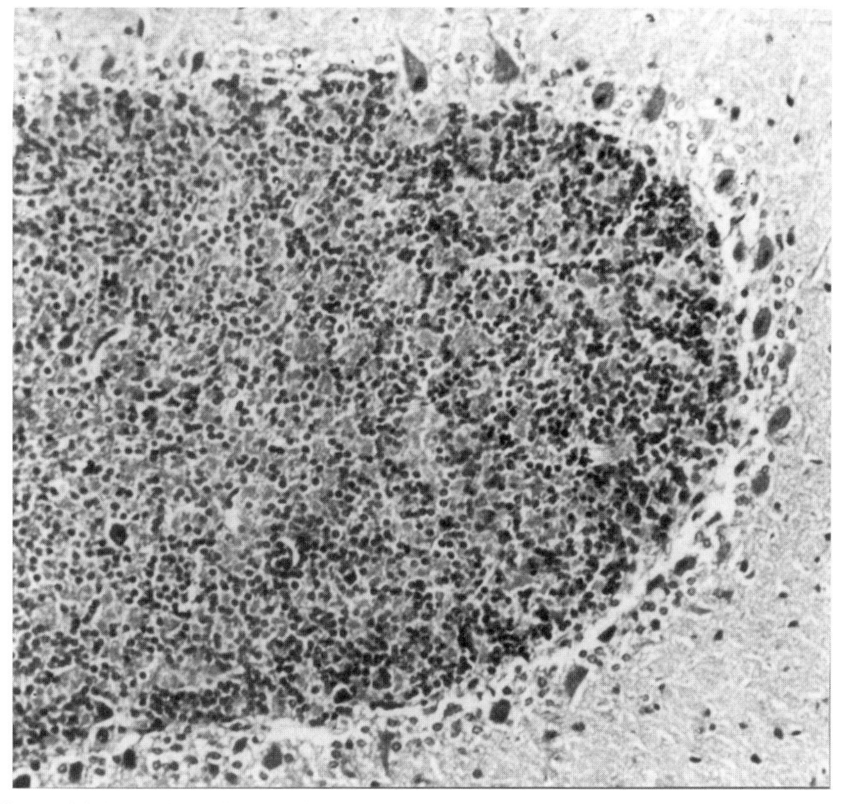

**Figure 16-6.** Section of rat brain and cerebellum. Note the presence of Purkinje cells lining the granular and molecular layer and the Purkinje cells with dendrites.

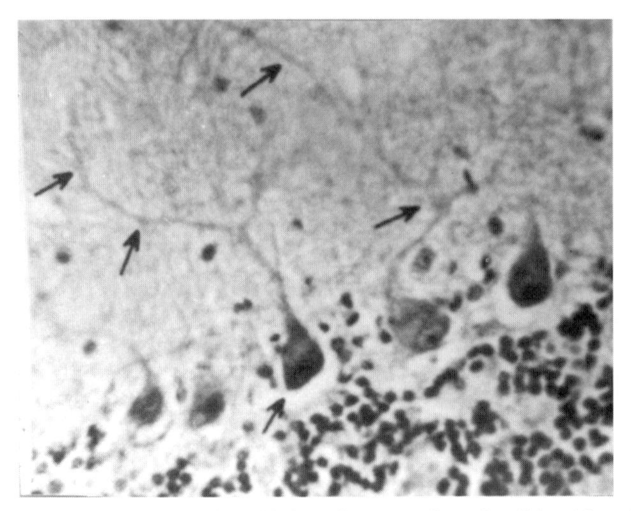

**Figure 16-7.** An enlarged view of Purkinje cells (pear-shaped cells) with a thick nucleus and branched dendrites.

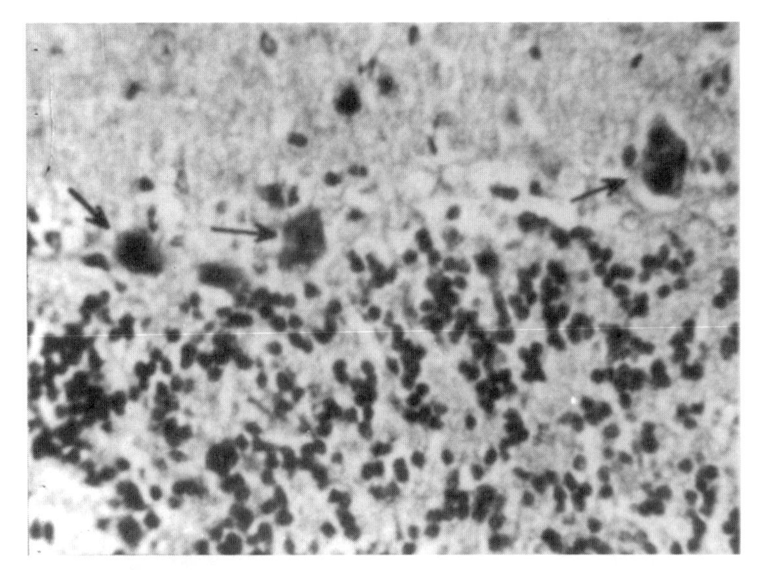

**Figure 16-8.** Section of rat brain and cerebellum zone after exposure to toxic chemicals. Note the damaged Purkinje cells and the loss of dendrites.

Subsequently, the above test guidelines were revised in March 1991 with important changes.[35] First and significantly, the application of the above guidelines was extended to neurotoxicity testing under the *Federal Insecticide, Fungicide, and Rodenticide Act (FIFRA)*. Second, functional observational battery, motor activity, and neuropathology guidelines were folded into one guideline with the title "neurotoxicity screening battery." Third, the developmental neurotoxicity screen (earlier a part of the standard test and requirement for diethyleneglycolbutyl ether and diethyleneglycolbutyl ether acetate) was revised and named "developmental neurotoxicity study." Finally, a new guideline was introduced as an appendix to the neurotoxicity screening battery as a new title, "guideline for assaying glial fibrillary acidic protein." On the basis of these guidelines and to comply with safety regulations, it became necessary to evaluate all substances used by the public and workers to achieve safety. More information on the neurotoxic potential of industrial and environmental chemicals and their adverse effects is available in the literature.[38–43,46–48]

## HEPATOTOXICTY

The mammalian liver is a vital organ associated with body functions such as detoxification and activation of toxic substances that enter the system. Adverse effects of chemicals on the liver have been studied extensively. The liver has been identified as one of the target organs affected by halogenated hydrocarbons, particularly polychlorinated biphenyls and pesticides.[49,50] Experimental animals exposed to acute and chronic PCBs showed liver hypertrophy, fatty degeneration, and central atrophy. Workers exposed to chlorobiphenyl and chloronaphthalene fumes suffered severe hepatoxic effects leading to death[51] (Figs. 16-9 to 16-13).

Several organochlorine insecticides also are known to cause injury to the mammalian liver in addition to the CNS. These substances are DDT, aldrin, dieldrin, chlordane, heptachlor, and other related compounds that trigger formation of lesions in the mammalian liver. Liver damage included increased organ weight, enlarged liver cells, and proliferation of the smooth endoplasmic reticulum, with no or minimal evidence of necrotic lesions. Subchronic and chronic treatments of laboratory rodents and dogs with aldrin demonstrated liver lesions, liver weight increase, and induction of tumors, particularly in mice.[51] Laboratory animals exposed to pesticides have shown pathomorphologic changes in the liver (Figs. 16-14 and 16-15). Dietary feeding of heptachlor to rats induced liver changes, as indicated by the increase in microsomal enzymes, reduction in liver glycogen, increased serum glucose, and leukocyte counts.[52] Prolonged exposure to petroleum and its products often results in petroleum toxicity, as characterized by the formation of oil droplets in the liver and other tissues.[53] Ketones, such as cyclohexanone, also are known to cause hepatocellular neoplasms (adenomas and carcinomas) in experimental rats. Female mice demonstrated a high incidence of malignant lymphoma.[54] Cellular and histopathologic changes in animal livers studied experimentally after chronic exposures to toxicants may be listed as follows:

- Cellular hypertrophy
- Cellular margination
- Fatty degeneration

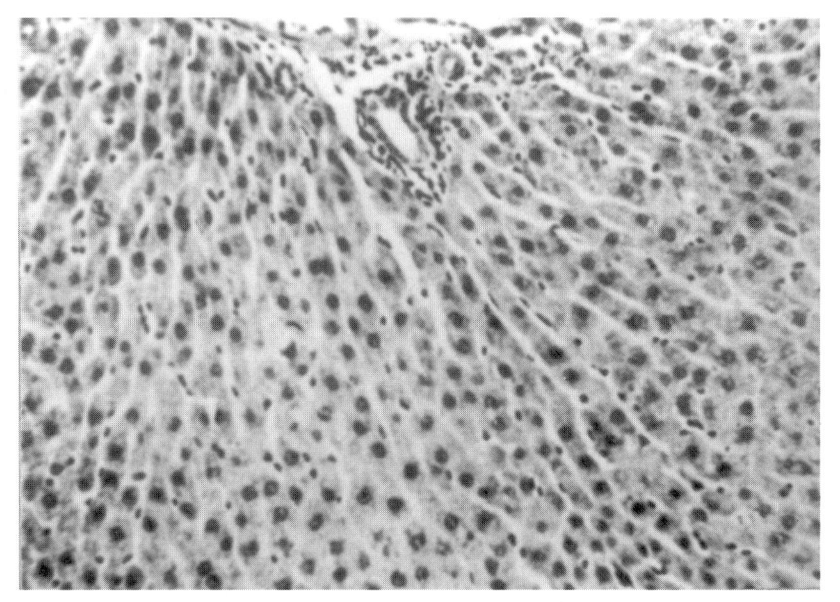

**Figure 16-9.** Section of rat liver unexposed to toxic chemicals. Note the normal structure of hepatocytes along with sinusoidal space.

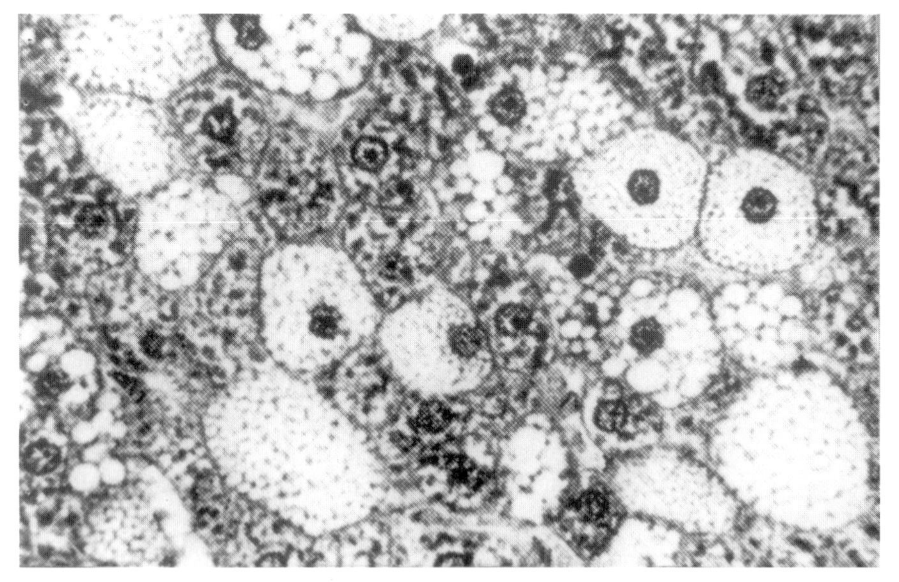

**Figure 16-10.** Section of rat liver exposed to toxic chemicals. Note the microdroplet fatty degeneration (vacuolation) in the centrilobular area of liver. Also note the presence of hyaline bodies (Fig. 16-11).

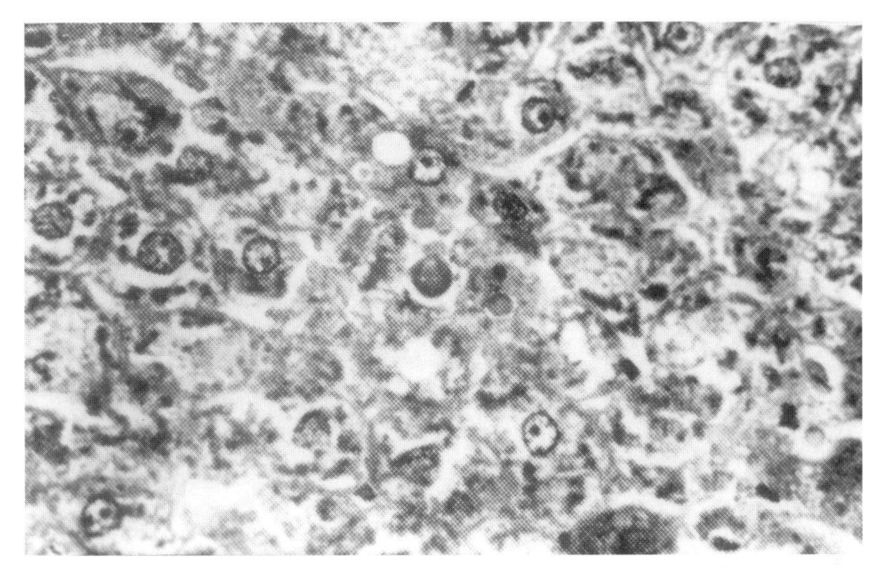

**Figure 16-11.** Section of rat liver exposed to toxic chemicals. Note the microdroplet fatty degeneration (vacuolation) in the centrilobular area of liver. Also note the presence of hyaline bodies (Fig. 16-11).

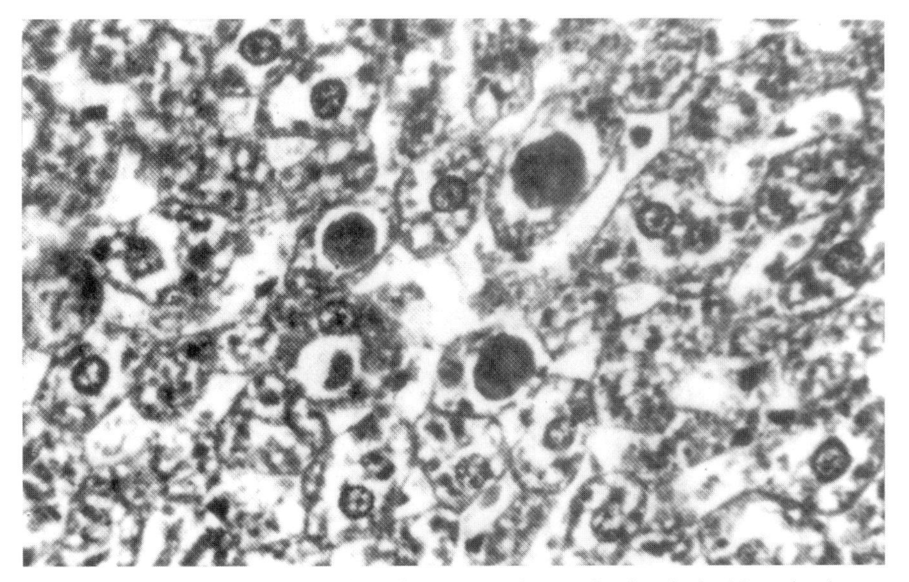

**Figure 16-12.** Another section of rat liver exposed to toxic chemicals. Note the degenerative changes.

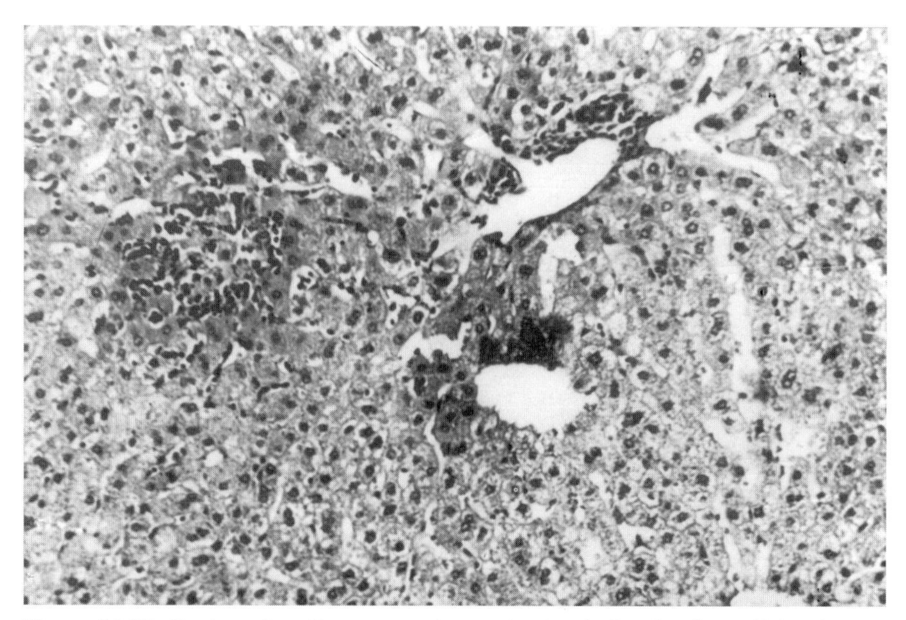

**Figure 16-13.** Section of rat liver exposed to toxic chemicals, showing cellular damage and necrosis.

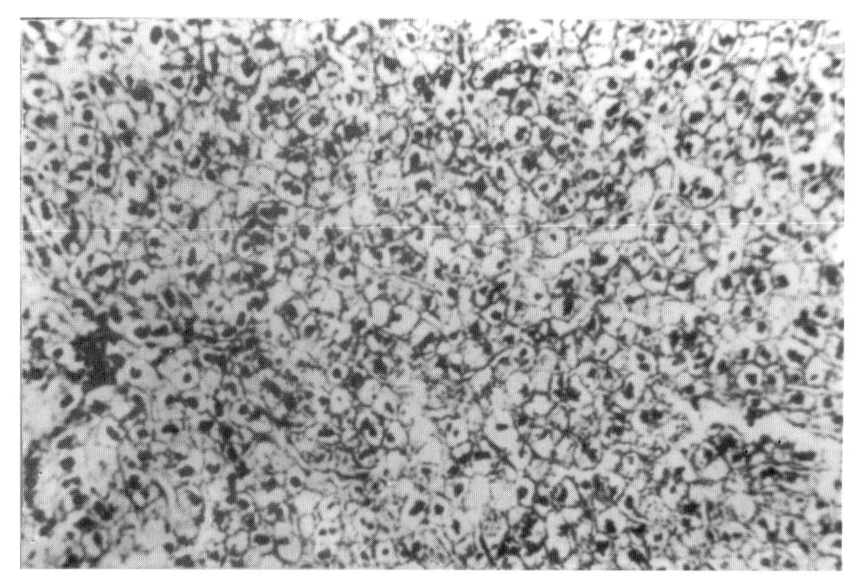

**Figure 16-14.** Another section of rat liver exposed to toxic chemicals showing cellular damage.

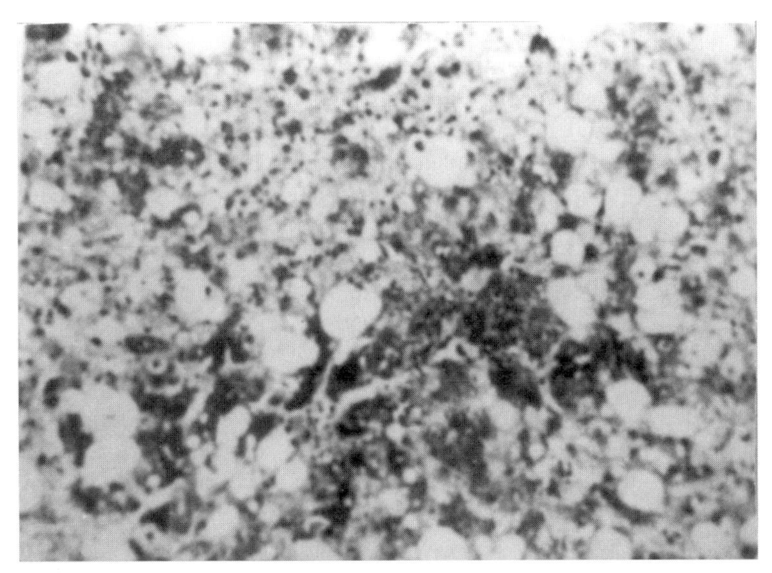

**Figure 16-15.** Another section of rat liver exposed to toxic chemicals, showing cellular damage.

- Increased basophilia
- Increased fat granules
- Cytoplasmic vacuolation
- Necrosis
- Atypical mitochondria
- Scattered, swollen, degenerated hepatocytes
- Enlargement of nuclei
- Hypertrophy of nucleoli

Alcohol abuse is the most common cause of liver damage in many populations around the world, specifically in the United Kingdom. In fact, more than two-thirds of cases of cirrhosis have been treated because of severe alcoholism in this region. Liver damage is produced by general mechanisms of cell injury, which may be due to the interaction of hepatocytes with reactive metabolite(s) of chemicals and drugs formed via $P_{450}$ enzymes. Drugs that cause liver damage are paracetamol, isoniazid, halothane, and methotrexate.[55,56]

## NEPHROTOXICITY

Global changes and extensive industrialization during the twentieth century made animals and humans become exposed to a variety of chemicals. Exposure to chemical forms of mercury (e.g., elemental mercury vapor [$Hg^{\circ}$], inorganic mercury [$Hg^{+}$] and mercuric [$Hg^{2+}$], organic mercury [$R{-}Hg^{+}$ or $R{-}Hg{-}R$])

has become well established in many industrial operations. The risk of mercury exposure and the subsequent adverse effects has caused concern because all forms of mercury have caused nephritic effects or kidney damage in humans.[57] With inorganic mercury and mercuric salts, the mammalian kidney is most prominently affected. Animal and human kidneys are primary target organs for the accumulation of inorganic mercury, whereas organic forms of mercury (e.g., methyl mercury) have extensive, diffuse systemic distribution and affect other target organs, such as hemopoietic and CNS.[58] Inorganic mercury accumulates primarily in the renal cortex, and outer strip of the outer medulla.[59,60] It has been demonstrated by histochemical autoradiography and tubular microdissection in rats and rabbits that inorganic mercury accumulates primarily along the segments of the proximal tubule.[61,62]

Several heavy metals, particularly lead, are known to cause major adverse effects to the mammalian kidney, resulting in kidney function impairment. Adverse effects to the mammalian kidney caused by lead include lesions on the proximal tubule and Henle's loop, and the presence of lead inclusion bodies. The metal also is known to cause aminoaciduria, phosphaturia, glycosuria, and renal tubular acidosis. Workers associated with lead-smelting industries also have shown kidney cancer.

Mercuric chloride ($HgCl_2$)–induced nephrotoxic effect is a classic example of industrial chemical toxicity. Mercuric chloride exposure disturbs the membrane phospholipid composition and function and eventually leads to proximal tubular necrosis and renal dysfunction.[63] Chronic mercuric poisoning results in proximal tubular and glomerular damage, in addition to ultrastructural lesions with the loss of brush border in proximal tubules. In a toxicity study, the National Toxicology Program evaluated the incidence of kidney lesions in male and female laboratory rats (F334, F344, and B6C3F) exposed to mercuric chloride for 2 years. Kidney lesions included adenoma, adenocarcinoma, hyperplasia, mild, moderate, and marked-type nephropathy, the incidence of which were of different frequencies[64] (Figs. 16-16 and 16-17).

Nephrotoxicity is caused by drugs that principally affect the renal hemodynamics of the patient depended on vasodilator prostaglandin biosynthesis or angiotensin converting enzyme (ACE) mediated vasoconstriction drugs causing nephrotoxicity include NSAIDs (fenprofen), ACE inhibitors (captopril, and cyclosporin).

## FETOTOXICITY

Chemical safety and worker health should not be ignored. This is particularly important wherever pregnant women are working in different chemical industries. Many chemicals are known to cause adverse effects to both working women (mothers) and the growing fetus (Table 16-6). Knowledge of chemicals and their adverse effects should provide guidelines to protect the health and safety of these workers.

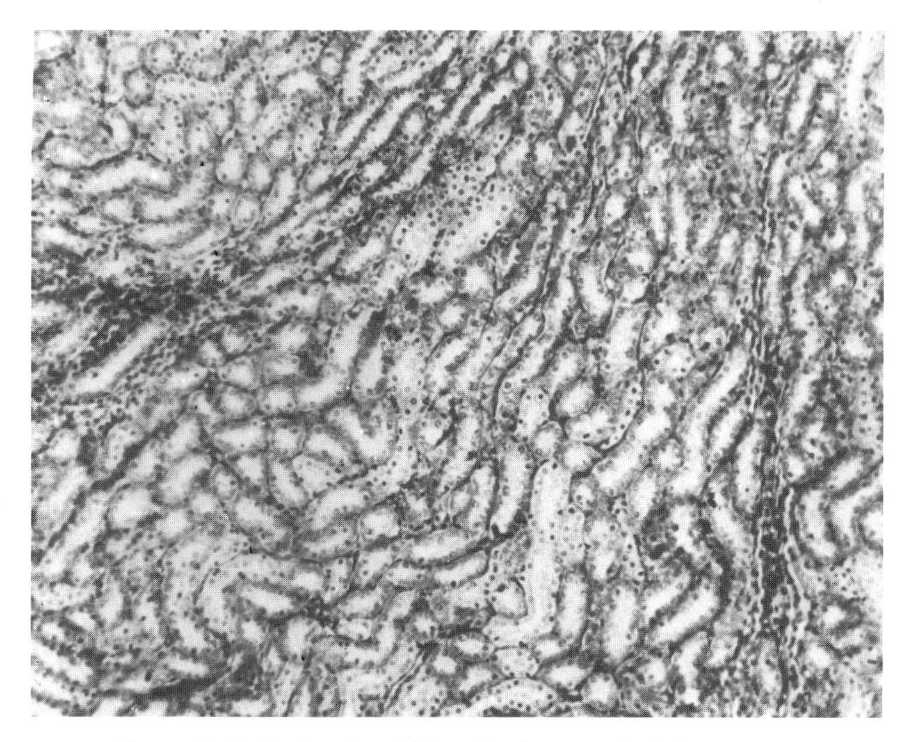

**Figure 16-16.** Section of rat kidney showing normal tubular structure.

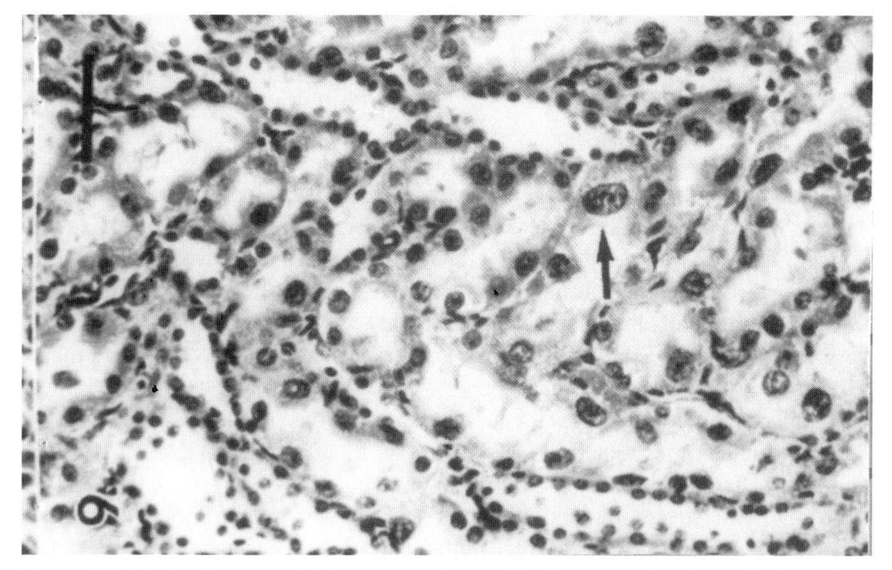

**Figure 16-17.** Section of rat kidney exposed to toxic chemicals, showing tubular cells in the outer medulla. Also note the appearance of giant epithelial cells on the margin of the tubules.

**TABLE 16-6 Adverse Effects of Drugs During Pregnancy**

| Chemicals/Drugs | Adverse Health Effects |
|---|---|
| Ammonium chloride | Acidosis |
| Antibacterials (streptomycin, tetracyclins) | Deposition in bones, discoloration of teeth, nerve damage (8th cranial nerve) |
| Antidiabetics (tolbutamide, chloropropamide, phenformin, insulin) | Thrombocytopenia, prolonged hypoglycemia, lactic acidosis, shock |
| Antihistamines | Infertility |
| Antithyroid drugs | Hypothyroidism |
| Barbiturates | Coagulation defects |
| Chloroquine | Death |
| Coumarin | Hemorrhage, death |
| Diazepam | Hypothermia |
| Diphenylhydantoin | Withdrawal syndrome |
| Erythromycin | Liver damage |
| Heroin, morphine | Withdrawal syndrome |
| Magnesium sulfate | Central depression, neuromuscular block |
| Nitrofurantoin | Hemolysis |
| Novobiocin | Hyperbilirubinemia |
| Phenobarbital | Neonatal bleeding, death |
| Phenothiazines | Hyperbilirubinemia, depression, hypothermia |
| Primidone | Withdrawal syndrome |
| Reserpine | Nasal congestion, lethargy, respiratory |
| Sulfonamides | Anemia |

*Source*: Grover JK. 1985[70].

Several studies indicate that different methods cause adverse effects to embryonic and fetal tissues and eventually lead to the development of teratogenic effects. Metals are omnipresent in the living environment. A variety of anthropogenic activities (e.g., smelting metallic ore, industrial and metal fabrication, commercial application, burning of fossil fuels) have caused adverse effects to the developing fetus. In fact, notorious elements, such as cadmium, lead, and mercury, have been associated with injury and malformation to the growing embryo and fetus of animals and humans.[65]

Similarly, high-dose arsenic causes malformations in animals, such as eye defects, renal agenesis, and gonadal agenesis in experimental golden hamsters and rats.[66] High doses of cadmium and mercury have caused malformation, abortions, and severe fetotoxicity. Holt and Webb[67] demonstrated that exposure to even minute amounts of cadmium during the gestation period caused inhibition of embryonic DNA and protein synthesis. However, extensive studies by Cohen and Roe[68] have not confirmed that lead causes teratogenic effects in animals.

## THE CARDIOVASCULAR SYSTEM

The cardiovascular system, like other body systems, is important for human health. Industrial chemicals (e.g., heavy metals, organic solvents, ketones, drugs and pharmaceuticals) often cause adverse effects on the mammalian cardiovascular system. Propranolol, a known beta-blocker, and quinidine, procainamide, lignocaine, mexiletine, and verapamil — all well-known antiarrhythmic agents — are known to cause adverse effects to the cardiovascular system in humans. Some chemical agents such as nicotine, coffee or tea, antiinflammatory agents, antianginal drugs (nifedipine), oral contraceptives (estrogen-containing pills), and oxytocin all trigger myocardial ischemia.[69]

Although mild to moderate alcohol consumption has beneficial effects on the mammalian system, high consumption of alcohol for a prolonged period has resulted in alcoholic heart muscle disease (AHMD). AHMD is characterized by dilated cardiomyopathy with concomitant ventricular dysfunction and histopathologic lesions.[70,71]

Exposure to allylamine in industries synthesizing pharmaceuticals and other commercial products is a known occurrence. Allylamines are known to cause adverse effects, especially to the liver, kidney, heart and/or blood vascular systems in experimental humans. It has been reported that exposure to methyl-, ethyl-, heptyl-, and allylamines results in severe pathologic lesions of the above-mentioned vital organs in animals and humans.[72] High doses of allylamines are always associated with the induction of fatal cardiovascular injury.[72,73]

## REFERENCES


1. A.M. Kligman, The biology of the stratum corneum, in *The Epidermis*, W. Montagna, and W.C. Lobitz, Jr., eds., Academic Press, New York, pp. 387–433, 1964.
2. I.H. Blank, The skin as an organ of protection against the external environment, in *Dermatology in General Medicine*, 2nd ed., T.B. Fitzpatrick et al., eds., McGraw-Hill, New York, 1979.
3. R.M. Adams, *Occupational Dermatology*, Grune & Stratton, New York, 1983.
4. H.I. Maibach, *Occupational and Industrial Dermatology*, 2nd ed., Year Book Medical Publishers, Chicago, 1987.
5. D.J. Birmingham, M.M. Key, D.A. Holaday, and V.B. Perone, An outbreak of arsenical dermatoses in a mining community. *Arch. Dermatol.* **91**:457–465, 1964.

5a. T.S.S. Dikshith, *Safety Evaluation of Environmental Chemicals*, New Age International Publishers (formerly Wiley Eastern Ltd.), New Delhi, India, 1996.

6. A. Silman, and M.C. Hochberg, Occupational and environmental influences on scleroderma. *Rheum. Dis. Clin. North Am.* **22**(4):737–749, 1996.
7. C.A. Elmets, Cutaneous phototoxicity, in *Clinical Photomedicine*, H.W. Lim, and N.H. Soter, eds., Marcel Dekker, New York, pp. 207–226, 1993.
8. K. Shimoda, M. Nomura, and M. Kato, Effect of antioxidants, antiinflammatory drugs, and histamine antagonists on sparfloxacin induced phototoxicity in mice. *Fundam. Appl. Toxicol.* **31**:133–140, 1996.

9. H. Kusajima, and R. Ishida, *Photodecomposition of Gatifloxacin (AM-1155) and its related drugs, and their phototoxic potency in guinea pigs, in 20th International Congress of Chemotherapy*, Sydney, Abstract No. 5318, 1997.
10. J. Ferguson, McEwen, K. Gohler, and A. Mignot, *A double blind placebo and positive controlled, randomized study to investigate the phototoxic potential of Gatifloxacin, a new fluoroquinolone antibiotic, in 38th Interscience Conference on Antimicrobial Agents and Chemoterapy*, San Diego, Abstract No. A. 78, 1998.
11. J. Epstein, Adverse cutaneous reactions to the sun, in *Year Book of Dermatology*, F.D. Malkinson and R.W. Pearson, eds., Year Book Medical Publishers, Chicago, 1971.
12. E. Emmett, and J.R. Kaminski, Allergic contact dermatitis from acrylates in ultraviolet cured inks. *J. Occup. Med.* **19**:113, 1977.
13. V. De Leo, and L.C. Harber, Contact photodermatitis, Chapter 25 in *Contact Dermatitis*, A.A. Fisher, ed., Lea & Febiger, Philadelphia, 1986.
14. European Centre for Ecotoxicology and Toxicology of Chemicals, *Skin Irritation and Corrosion: Reference Chemicals Data Bank*, *ECETOC* Technical Report No. 66, p. 247, Brussels, Belgium, 1995.
15. OECD, *Revised Proposal for the Harmonization of Hazard Classification Based on Skin Irritation/Corrosion*, ENV/MC/CHEM/HCL (98), **4**, OECD, Paris, France, 1998.
16. P.A. Botham, L.K. Earl, J.H. Fentem, R. Roguet, and J.J.M. van de Sandt, *Alternative Methods for Skin Irritation Testing: The Current Status*, ECVAM Skin Irritation Task Force Report 1, ATLA, **26**, 195–211, 1998.
17. J.H. Fentem, D. Briggs, C. Chesne, G.R. Elliott, J.W. Harbell, J.R. Haylings, P. Portes, R. Roguet, J.J.M. van de Sandt, and P.A. Botham, A prevalidation study on in vitro tests for acute skin irritation: Results and evaluation by the management team. *Toxicol. Invitro.* **15**:57–93, 2001.
18. L. Schwartz, L. Tulipman, and D.J. Birmingham, *Occupational Diseases of the Skin*, 3rd ed., Lea & Febiger, Philadelphia, 1957.
19. M. Lippmann, Review: Asbestos exposure indices. *Environ. Res.* **46**(1):86, 1988.
20. J.P. Guillot, J.F. Gonnet, C. Clement, L. Caillard, and R. Trauhaut, Evaluation of the ocular irritation potential of 56 compounds. *Fed. Chem. Toxicol.* **20**:573–582, 1982.
21. W.R. Parkes, *Occupational Lung Disorders*, Butterworths, London, 1982.
22. T.F. Hatch, and P. Gross, *Pulmonary Deposition and Retention of Inhaled Aerosols*, Academic Press, New York, 1964.
23. J.T. Boyd et al., Cancer of the lung on iron ore (hematite) miners. *Br. J. Ind. Med.* **27**:97, 1970.
24. S.H. Moolgavkar, E.G. Luebeck, J. Turim, and R.C. Brown, Lung cancer risk associated with exposure to man-made fibers. *Drug Chem. Toxicol.* **23**:223–242, 2000.
25. S.H. Moolgavkar, J. Turim, R.C. Brown, and E.G. Luebeck, Long man-made fibers and lung cancer risk. *Reg. Toxicol. Pharmacol.* **33**:138–146, 2001.
26. International Agency for Research on Cancer, *Man-Made Mineral Fibers and Radon*, International Agency for Research on Cancer, Lyon, France, Vol. 43, 1988.
27. R. McConnell, K. Berhane, F. Gilliand, S.J. London, T. Islam, W.J. Gauderman, E. Avol, H.G. Margolis, and J.M. Peters, Asthma in exercising children exposed to ozone: A cohort study. *Lancet* **359**:386–391, 2002.

28. United States Environmental Protection Agency, *Pesticide Assessment Guidelines, Subdivision F. Hazard Evaluation: Human and Domestic Animals, Addendum 10, Neurotoxicity., Series 81, 82, and 83, Health Effects Division*, Office of Pesticides Programs, Publication PB 91–154617, 1991.
29. C. Marwick, Bhopal tragedy's repercussions may reach American physicians. *JAMA* **253**:2001–2013, 1985.
30. M.H. Karol, Comparison of clinical and experimental data from an animal model to pulmonary immunologic sensitivity. *Ann. Allergy* **66**:485–489, 1991.
31. X. Baur, Isocyanates. *Clin. Exp. Allergy* **21**(suppl. 1):241–246, 1990.
32. Indian Council of Medical Research, *Health Effects of the Bhopal Gas Tragedy*, Report pp. 1–16, Indian Council of Medical Research, New Delhi, India, 1986.
33. D.R. Varma, Epidemiological and experimental studies on the effects of methyl isocyanates on the course of pregnancy. *Environ. Health Perspect.* **72**:153–157, 1987.
34. D.R. Varma, Pregnancy complications in Bhopal women exposed to methyl isocyanate vapor. *J. Envir. Sci. Health* **A26**:1437–1447, 1991.
35. D.R. Varma, and I. Guest, The Bhopal accident and methyl isocyanate toxicity. *J. Toxicol. Environ. Health* **40**:513–529, 1993.
36. World Health Organization, Target organ toxicity, eye, ear, and other special senses. *Environ. Health Perspect.* **44**, 1–27, World Health Organization Geneva, Switzerland, 1982.
37. World Health Organization, Principles and Methods for the Assessment of Neurotoxicity Associated with the Exposure to Chemicals. *Environ. Health Criteria,* **60** Office of Publications, World Health Organization, Geneva, Switzerland, 1982.
38. J.L. O'Donoghue, *Neurotoxicity of Industrial and Commercial Chemicals*, J.L. O'Donoghue, ed., Vol. 2, CRC Press, Boca Raton, FL, pp. 169–177, 1985.
39. Z. Annau, *Neurobehavioral Toxicology*, Johns Hopkins University Press, Baltimore, MD, 1986.
40. B.L. Johnson, *Advances in Neurobehavioral Toxicology: Applications in Environmental and Occupational Health*, Lewis Publisher, Chelsea, MI, 1990.
41. European Chemical Industry Ecology and Toxicology Centre, *Interpretation and Evaluation of the Neurotoxic Effect of Chemicals in Animals*, Special Report No. 6, Brussels, Belgium, 1993.
42. M. Van Bose, and M. Zaudig, Neuropsychiatric disorders caused by organic solvents. *Psychiatr. Prac.* **18**:25–29, 1991.
43. Y. Takeuchi, Y. Ono, N. Hisanaga, M. Iwata, M. Aoyama, J. Kitoh, and Y. Suriura, An experimental study of the combined effects of n-Hexane and methyl ethyl ketone. *Brit. J. Ind. Med.* **40**:199–203, 1983.
44. United States Environmental Protection Agency, *Toxic Substances Control Act Test Guidelines, Health Effects Testing Guidelines, Subpart G, Neurotoxicity, Fed. Reg.*, Vol. 50, No. 188, pp. 39458–39470, 1985.
45. United States Environmental Protection Agency, *Revision of TSCA Test guidelines, Fed. Reg.*, Vol. 51, No. 9, p. 1542, 1986.
46. United States Environmental Protection Agency, *Diethylene Glycol Butyl Ether and Diethylene Glycol Butyl Ether Acetate, Test Standards and Requirements, Fed. Reg.*, Vol. 53, No. 38, pp. 5932–5953, 1988.

47. R. Gilioli, M.G. Cassitto, and V. Foa, *Neurobehavioural Methods in Occupational Health*, Pergamon Press, Oxford, UK, 1983.
48. J.P.J. Maurissen, and J.L. Mattsson, Neurotoxicology: An orientation, in *Patty's Industrial Hygiene and Toxicology*, 3rd ed., Vol. 3, Part B, L.J. Cralley, L.V. Cralley, and J.S. Bus, eds., John Wiley & Sons, New York, pp. 231–254, 1995.
49. ATSDR, Toxicological Profile for Selected PCBs (Aroclor-1260, 1224, 1248, 1242, 1232, 1221, and 1016), NTIS Report prepared by Syracuse Research Corp. for the Agency for Toxic Substances and Disease Registry, United States Public Health Service and EPA; ATSDR/TP-88/21, PB 89–22403, June, 1989.
50. International Agency for Research on Cancer, *Polychlorinated Biphenyls and Polybrominated Biphenyls*, Vol. **18**: October, 1978.
51. H. Von Wedel, W.A. Holla, and J. Denton, *Rubber Age, 53, 419, 1943, in Documentation of the Threshold Limit Values and Biological Exposure Indices*, 5th ed., American Conference of Governmental Industrial Hygienists, Inc., 1990.
52. A.G. Smith, in *Handbook of Pesticide Toxicology*, Vol. 2, W.J. Hayes and E.R. Laws, Jr., eds., Academic Press, New York, 1991.
53. Agency for Toxic Substances and Disease Registry, Toxicological Profile for Heptachlor/Heptachlor Epoxide, NTIS Report, PB 89–194492, 1989.
54. R. Snyder, ed., *Ethel Browing's Toxicity and Metabolism of Industrial Solvents*, 2nd ed., Vol. 1, Hydrocarbons, Elsevier, Amsterdam, 1987.
55. J.B. Saunders, J.R.F. Walters, P. Davies, and A. Paton, A 20 year prospective study of cirrhosis. *Br. Med. J.* **282**:263–266, 1981.
56. D.I.N. Sherman, and R. Williams, Liver damage mechanisms and management. *Br. Med. Bull.* **50**(1):124–138, 1994. In *Alcohol and Alcohol Problems*, G. Edwards and T.J. Peters, eds., Churchill Livingstone, London, UK, 1994.
57. W.F. Fitzgerald, and T.W. Clarkson, Mercury and monomethyl mercury: Present and future concerns. *Environ. Health Perspect.* **96**:159–166, 1991.
58. N. Ballatori, Mechanisms of metal transport across liver cell plasma membranes. *Drug Metab. Rev.* **23**:83–132, 1991.
59. R.K. Zalups and M.G. Cherian, Renal metallothionein metabolism after a reduction of renal mass. I Effect of unilateral nephrectomy and compensatory renal growth on basal and metal induced renal metallothionein metabolism. *Toxicology* **71**:83–102, 1992.
60. R.K. Zalups, Early aspects of the intrarenal distribution of mercury after the intravenous administration of mercuric chloride. *Toxicology* **79**:215–228, 1993.
61. P.M. Rodier, B. Kates, and R. Simons, Mercury localization in mouse kidney over time: Autography versus silver staining. *Toxicol. Appl. Pharmacol.* **257**:235–245, 1988.
62. R.K. Zalups, Autometallographic localization of inorganic mercury in the kidneys of the rats: Effect of unilateral nephrectomy and compensatory renal growth. *Exp. Molec. Pathol.* **54**:10–21, 1991.
63. W. Lijinsky, and R.M. Kovatch, Chronic toxicity study of cyclohexanone in rats and mice. *J. Natl. Cancer Inst.* **77**:941–949, 1986.
64. V.H. Ferm, The teratogenic effects of metals on mammalian embryos. *Adv. Teratol.* **6**:51–75, 1972.

65. R.P. Beliles, The Metals, in *Patty's Industrial Hygiene and Toxicology, Toxicology*, Vol. II, Part C, Chapter 27, pp. 1879–2352, 4th ed., G.D. Clayton and F.E. Clayton, eds., John Wiley & Sons, New York, 1981.
66. A.R. Beaudoin, Teratogenicity of sodium arsenate in rats. *Teratology* **10**:153–158, 1974.
67. D. Holt, and M. Webb, The toxicity and teratogenecity of mercuric mercury in the pregnant rat. *Arch. Toxicol.* **58**:243–248, 1986.
68. A.J. Cohen, and F.J.C. Roe, Review of lead toxicology relevant to the safety assessment of lead arsenate as a hair coloring. *Food Chem. Toxicol.* **29**:485–507, 1991.
69. National Toxicology Program, Bioassay for Mercuric Chloride, 1990.
70. J.K. Grover, *Classification of Drugs and Their Adverse Reactions*, CBS Publishers and Distributors, Delhi, India, 1985.
71. World Health Organization, Safety evaluation of certain food additives and contaminants, International Program on Chemical Safety, WHO Food Additive Series, **46**, 2001, Geneva, Switzerland.
72. V.R. Preedy, and P.J. Richardson, Ethanol induced cardiovascular disease. *Br. Med. Bull.* **50**(1):152–163, 1994.
73. R.R. Beard, and J.T. Noe, Aliphatic and alicyclic amines, Chapter 44, in *Patty's Industrial Hygiene and Toxicology*, 3rd ed., Vol. 2B, G.D. Clayton and F.E. Clayton, eds., John Wiley & Sons, New York, pp. 3135–3173, 1981.

# CHAPTER 17

# Disposal of Hazardous Chemicals

## INTRODUCTION

Human activities are associated with the use and disposal of a variety of chemicals and chemical products. This is the situation for a householder, a laboratory student, and also the industry worker. Many materials have properties that make them hazardous. They can create physical (fire, explosion) or health hazards (toxicity, chemical burns). However, there are many ways to work with chemicals which can both reduce the probability of an accident and reduce the consequences should an accident occur. Risk minimization depends on safe practices, appropriate engineering controls for chemical containment, the proper use of personnel protective equipment, use of the least amount of material necessary, and substitution of a less-hazardous chemical for a more hazardous one. Before beginning any chemical processing or operation, ask "What would happen if . . .?" The answer to this question requires understanding of the hazards associated with chemicals, the equipment, and the procedure involved. The hazardous properties of the material and its intended use will dictate the precautions to be taken.

Disposal of hazardous waste is dangerous and expensive, even when the contents of the waste are identified. Fortunately, most chemical waste produced in a laboratory or work area is identifiable. When the contents of a reagent bottle, reaction flask, or gas cylinder are not identified, the process of disposal is more dangerous, expensive, and difficult. Without mitigating information, all unknown materials must be treated as if they were potentially lethal and hazardous. In all cases, chemical unknowns cannot be disposed of until a general profile of the unknown has been generated. Even then, the disposal cost is a premium. Additionally, there is a constant threat of personal injury or death to individuals who handle these potentially dangerous materials. No price tag can be attached to an avoidable personal injury.

Another important distinction to be known is the difference between hazard and risk. The two terms are sometimes used as synonyms. In fact, the term *hazard* is a much more complex concept because it includes conditions of use. The hazard presented by a chemical has two components: (1) the inherent capacity

*Industrial Guide to Chemical and Drug Safety*, By T.S.S. Dikshith and Prakash V. Diwan
ISBN 0-471-23698-5 

of the candidate chemical to do harm by virtue of its toxicity, flammability, explosiveness, or corrosiveness; and (2) the ease with which the chemical can come into contact with any person or other object of concern. The two components determine risk (the likelihood or probability that a chemical will cause harm). Thus, an extremely toxic chemical, such as strychnine, cannot cause poisoning if it is in a sealed container and does not contact the handler. It is important to remember that a chemical that is not highly toxic can also become lethal if a large amount is ingested.

Chemical safety is inherently linked to other safety issues including laboratory procedures, personal protectives, equipment, electrical safety, and hazardous waste disposal.

Not all chemicals are considered as hazardous. Examples of nonhazardous chemicals include buffers, sugars, starches, agar, and naturally occurring amino chemicals. The following sections provide general guidelines for chemical safety.

1. Primary metallurgical manufacturing industries associated with metals such as zinc, lead, copper, aluminium, and steel
2. Paper, pulp, and newsprint
3. Pesticides
4. Refineries
5. Fertilizers
6. Paints
7. Dyes
8. Leather tanning
9. Rayon
10. Sodium/potassium cyanide
11. Basic drugs
12. Foundry
13. Storage batteries (lead acid type)
14. Acids/ alkalies
15. Plastics
16. Rubber (synthetic)
17. Cement
18. Asbestos
19. Fermentation industry
20. Electroplating industry

## LABORATORY MATERIALS: EXPLOSIVE CHEMICALS

An explosive chemical or mixture of chemicals is one that can undergo violent or explosive decomposition under appropriate conditions of reaction or initiation.

Some chemicals and combinations used in laboratories are known to be explosive. Laboratory manipulations with known explosive chemicals or reagent combinations should be performed only by trained personnel who are thoroughly familiar with the hazards involved and the precautions that must be taken. The worker should know the procedures for destroying or disposing of potentially explosive materials. Any laboratory procedure that results in an unexpected explosion should be investigated to ascertain the probable cause, and a laboratory safety rule established to prevent recurrence. Circumstances of an unexpected explosion should be brought to the attention of workers, team members, management, and the concerned public to help observe caution under similar work conditions.

Explosive materials must be disposed of in a way that protects all personnel from consequences that might occur during handling. Potentially explosive material must not be disposed of in landfills, even in a laboratory pack. Small quantities of commercial explosives can be incinerated after reducing the explosive potential by dilution with a flammable solvent or solid such as sawdust. Small containers of diluted explosive should be fed into an incinerator one at a time. The incinerator operator should be fully aware of the nature of the materials being handled. Also any potentially explosive material(s) should only be transported on public roads with specialized handling equipment and adequate protection.

The option for disposal of potentially explosive materials is to have it detonated under carefully controlled conditions. Some laboratories and industries may have personnel trained in explosive handling, and they may be able to remove and detonate the material on their site where no damage will result. Alternatively, some contract waste disposal firms have the capability to remove and dispose of explosive material. It also is possible to make arrangements with a local squad who handles explosives (or even a fire department) to collect, remove, and detonate the material under safe conditions. In all situations, the chemist should provide the disposal expert with whatever information is available on the hazards of the chemical(s).

Small quantities of explosive laboratory chemicals can be destroyed following known methods. By adopting standard methods, it is possible to destroy or reduce the dangerous nature of laboratory chemicals and check their reaction. Hydrocarbons (e.g., alkanes, alkenes, alkynes, arenas) burn well and can be disposed of by incineration or as fuel supplants. Also, many hydrocarbons commonly used in chemical laboratories may be easily ignited. Some cyclic compounds such as alkanes and cyclohexane may form explosive peroxides. Personnel trained in handling explosives should destroy these compounds using detonation. Many poly(nitro) aromatic compounds are explosive, and their disposal requires the services of an expert.

Picric acid (2,4,6-trinitrophenol) is normally sold in a damp condition, containing 10% to 15% water; in this state, it is relatively safe to handle. However, dry picric acid may explode on initiation by friction, shock, or sudden heating. Moreover, picric acid forms salt on contact with metals, and heavy metal picrates are highly sensitive to detonation by friction, shock, or heat. It is possible that some older reported explosions of picric acid were initiated by detonation of a minute quantity of metal picrate found in the threads of a metal-capped container. Although picric acid is now sold in plastic-capped containers, it is possible for material in such containers to dry out after repeated opening, and thus become hazardous. If picric acid in a plastic-capped container appears to have dried out, the bottle can be immersed upside down in water for a few hours to allow water to wet the threads. The bottle can then be uncapped and filled with water. The water-filled bottle should be allowed to stand a few days to ensure complete wetting of the contents. Gram quantities of picric acid can be destroyed by reduction with tin and hydrochloric acid. Larger quantities of the chemical require a commercial waste disposal service or the local squad on explosives or fire department. A metal-capped container of picric acid should be handled only by a trained expert or a squad trained in explosives.

Organic peroxides and hydroperoxides, including peroxide-containing solvents, can be treated. However, any solvent capable of forming peroxides from which a solid has crystallized should be handled by personnel trained in handling explosives. Most diazonium salts are not explosive and can be converted into disposable material by the coupling procedure. However, some diazonium salts are explosive when dry and should be carefully moistened before any manipulation is attempted.

Chloric (VII) acid also can be disposed of following known methods by trained workers. Ammonium chlorate (VII), heavy metal chlorates (VII), and chlorates (VII) of organic cations, pyridinium chlorate (VII), and 2,4,6-trimethylpyrylium chlorate (VII) are however, inherently explosive and should be disposed of as explosives. Alkyl chlorates (VII) also are explosive, and their formation should be avoided. If one is formed inadvertently, it should never be isolated, and the mixture in which it exists should be disposed of as an explosive. It must be cautioned again that these operations should be handled by trained workers under supervision.

Sodium azide and other heavy metal azides are too dangerous to be handled by this procedure and should be treated as explosives. Sodium amide and potassium amide can be destroyed by standard procedures. Many other compounds in which nitrogen is linked to a metal should be disposed of as potential explosives.

## LABORATORIES AND WASTE MANAGEMENT SYSTEMS

Waste management systems for single-laboratory operations, for a small college, and for a large university or industrial complex must have substantial differences in detail, but each system should have certain basic characteristics in common. Essentially, four are important in any laboratory waste-management program or system. These are as follows:

- Commitment of the laboratory director or chief executive to the principles and practice of good waste management.
- A waste management plan.
- Assigned responsibility for the waste management system.

- Policies and practices directed to reducing the volume of waste generated in the laboratory.

It is important that personnel at all levels — department heads, supervisors, academic faculty — exhibit a sincere and open interest in the waste management plan, and each one supports the other continuously. It is not sufficient to support the plan at its outset and to assume that it will then operate. Success depends on the participation and cooperation of the laboratory workers, who will be conditioned by their perception of management commitment. Any program that must be perceived and has only nominal support will come to be ignored by laboratory personnel.

## WASTE MANAGEMENT PLAN

It is the responsibility of laboratory management to see that a plan is developed for the handling of surplus, waste, and hazardous waste chemicals. The plan should be tailored to laboratory operations and should conform to all pertinent legal regulations. Written policies and procedures should be prepared to cover all phases of waste handling, from generation to ultimate safe and environmentally acceptable disposal. Although the necessary documents should be organized and written by the individuals who will be responsible for their implementation, much can be gained by having laboratory personnel participate in the planning process.

The documents should describe all aspects of the system for the particular laboratory and should spell out responsibilities and specific procedures to be performed by laboratory personnel, supervisors, management, and waste management organizations. The waste management plan should be reviewed at regular intervals to ensure that it covers changes that may have occurred in laboratory operations.

## RESPONSIBILITY FOR THE WASTE MANAGEMENT SYSTEM

Any successful waste management system requires a team effort from laboratory managers, supervisors, personnel, stockroom personnel, the local waste management organization, and the local safety and health organization. The responsibilities of these groups should be clearly set forth in the waste management plan. Although small laboratories may not have organizational subdivisions that correspond to each of these functions, the functions do exist, and the people who perform them should assume the responsibilities assigned by the waste management plan.

Responsibility for implementing the system must be specifically assigned. It is important that the responsible individual has enough knowledge of chemistry and laboratory work to understand the problems faced by laboratory personnel in their part of the system and to make a chemically sound judgment about waste handling situations. If a waste manager does not have a broad chemical background, a consultant with appropriate chemical knowledge should be employed. Laboratory personnel are in the best position to identify any chemicals that might pose unusual hazards. They are responsible for putting waste in proper performance under the plan. The responsibilities of stockroom personnel will depend on the size and complexity of the laboratory. Their activities may include dating the chemical labels whose use is time-limited, monitoring stocks of such chemicals, and operating an exchange clearinghouse for unneeded chemicals. Although these responsibilities in the operation of the waste management system are made clear in the plan, they should be reinforced by training and refresher sessions to all categories of workers.

## GENERATION OF WASTE: REDUCTION IN VOLUMES

Policies and practices for reducing the waste volume generated in the laboratory and for avoiding special disposal problems should be an integral part of the waste management plan.

The planning of every experiment should include 1) consideration of the hazardous properties of the starting materials; 2) the disposal of leftover starting materials; and 3) the disposal of the products and byproducts that will be generated. The worker and the management should ask the following before any processing or chemical operations:

- Are chemicals being acquired in proper quantities? Can any material be recovered for reuse?
- Will the experiment generate any chemical that should be destroyed by a special laboratory procedure? If so, what procedure or specific areas in laboratory?
- Can any unusual disposal problem be anticipated? If so, inform the waste management organization beforehand.

## STORAGE AND SHELF LIFE OF REAGENTS

Indefinite and uncontrolled accumulation of excess reagents, (i.e., holding of chemicals) often creates storage problems and safety hazards. These problems can be alleviated and purchase costs saved by having an excess chemicals storage room to which laboratory workers can go for unused chemicals instead of ordering new material. Chemicals that deteriorate with time can pose difficult disposal problems if allowed to accumulate in the laboratory. The waste management system should provide for periodic searching of chemicals and reagents. For instance, reagents that react readily with oxygen or water and deteriorate when stored for long times after the original container has been opened. Unused or deteriorated samples of water-reactive chemicals and pyrophoric chemicals should not be allowed to remain in the laboratory. Severe hazards can be created by peroxide-forming chemicals that have not been dated after opening the original container or that have exceeded the storage time limit after opening.

## REAGENT LABELING AND MAINTENANCE

All chemicals and reagents should be labeled. Deterioration of labels on reagent bottles or containers is a common occurrence, and more so on old reagent containers. If the reagent or the container has not deteriorated, the container should be relabeled if its identity is certain. However, if a reagent container label has disappeared or become illegible and the chemical is not known, it should be discarded. The method for proper disposal can usually be determined by simple laboratory tests.

## AVOIDING ORPHAN REACTION MIXTURES

Laboratory glassware containing reaction mixtures of unknown nature (and sometimes of unknown origin) can pose difficult disposal problems. Such materials occur frequently in research laboratories, particularly in those that have a high rate of personnel turnover. Simple laboratory tests may provide enough information for safe disposal. The waste management system should provide a procedure designed to prevent the occurrence of such orphan wastes.

Laboratories should require that any reaction mixture stored in glassware be labeled with its chemical composition, its date of preparation or formation, the name of the laboratory worker responsible, and a notebook reference. This procedure can provide the information necessary to guide disposal of the mixture if the responsible laboratory worker is no longer available. It should be recognized, however, that such a procedure must be enforced. It cannot guarantee that a departing worker will not leave behind unlabeled mixtures.

## THE WASTE MANAGEMENT SYSTEM

The waste management system for any laboratory should be tailored to the:

- Volume and variety of wastes generated.
- Number and locations of generating sources within the laboratory.
- Options chosen for waste disposal. These parameters should be determined as the system is being created.

### Initial Survey

The first two of these parameters require a survey of the entire laboratory by the waste system manager. For small laboratories, a brief survey, which includes discussion with the individual(s) responsible for each segment of the operation, may suffice. For large, diverse departments it may be desirable to precede the survey with a simple questionnaire. It is almost inevitable that such a survey will uncover waste problems that must be solved at the outset, (e.g., caches of deteriorated or old chemicals, forgotten chemicals unknown origin or identity). The system must arrange for their safe disposal and should be designed to prevent such accumulations in the future.

The survey should include all units of the institution. Some laboratories will find units that are using chemicals and generating hazardous waste without the awareness of any hazard. This situation is particularly likely in academic institutions, where chemicals are used routinely in many areas, such as the biology, geology, electrical engineering, art, and physics departments, and in hospitals where workers may have little or no training in chemistry.

The survey should be designed to reveal 1) the volume and types of nonhazardous waste being generated; 2) the types of hazardous wastes being generated; and 3) in what locations these are generated. This information will determine many characteristics of the waste management system, such as procedures for classification, segregation, and collection, as well as methods to be chosen for hazardous waste disposal.

## WASTE MANAGEMENT MANUAL

The waste disposal plan should be summarized in a manual that is understandable by and available to all laboratory personnel. New laboratory workers should be given a copy at the time of their arrival or during an early induction program. The manual can be made more readable by putting long lists in the appendixes. Examples are:

- A glossary of terms and abbreviations.
- Listings of chemicals such as: incompatible chemicals; potentially explosive chemicals and reagent combinations; water-reactive chemicals; pyrophoric chemicals; and peroxide-forming chemicals, including time limits on retention after opening the original container.

Some common laboratory chemicals react violently with water, and they should always be stored and handled so that they do not contact liquid water or vapor. They must not be disposed of in landfills, even in a lab pack, because of their characteristic reactivities (Table 17-1).

## PEROXIDE-FORMING CHEMICALS

Many common laboratory chemicals can form peroxides in the presence of atmospheric oxygen. A single opening of a container to remove contents can introduce enough air for peroxide formation to occur. Some compounds form peroxides that are violently explosive in a concentrated solution or as solids. Accordingly, peroxide-containing liquids should never be evaporated to dryness. Peroxide formation also can occur in many polymerizable unsaturated compounds, and these

**TABLE 17-1 Water-Reactive Chemicals**

| |
|---|
| Alkali metals |
| Alkali metal hydrides |
| Alkali metal amides |
| Metal alkyls, such as lithium alkyls and aluminium alkyls |
| Grignard reagents |
| Halides of nonmetals, such as $BCl_3$, $BF_3$, $PCl_3$, $PCl_5$, $SiCl_4$, $S_2Cl_2$ |
| Inorganic acid halides, such $POCl_3$, $SOCl_2$, $SO_2Cl_2$ |
| Anhydrous metal halides, such $AlCl_3$, $TiCl_4$, $ZrCl_4$, $SnCl_4$ |
| Phosphorus (V) oxide |
| Calcium carbide |
| Organic acid halides and anhydrides of low molecular weight |

can initiate a "runaway," sometimes explosive, polymerization reaction. Testing procedures for peroxides (and their removal in small amounts from laboratory chemicals) is available in literature.[1–4]

Based on the structural characteristics of organic compounds, they peroxidize and form peroxides (Table 17-2). Although the tabulation of organic structures may seem to include a large fraction of common organic chemicals, they are listed in approximate order of decreasing hazard. Reports of serious incidents involving the last five organic structural types are extremely rare, but they should be known by laboratory workers. These workers should be aware that the compounds could

**TABLE 17-2 Types of Chemicals That May Form Peroxides**

| |
|---|
| Organic Structures |
| Ethers and acetyls with alpha hydrogen atoms |
| Olefins with allylic hydrogen atoms |
| Chloroolefins and fluoroolefins |
| Vinyl halides, esters, and ethers |
| Dienes |
| Vinylacetylenes with alpha-hydrogen atoms |
| Alkylacetylenes with alpha-hydrogen atoms |
| Alkylarenes that contain tertiary-hydrogen atoms |
| Alkanes and cycloalkanes that contain tertiary-hydrogen atoms |
| Acrylates and methacrylates |
| Secondary alcohols |
| Ketones that contain alpha-hydrogen atoms |
| Aldehydes |
| Ureas, amides, and lactams that have an H-atom linked to C attached to N |
| Inorganic Substances |
| Alkali metals, especially potassium, rubidium, and cesium |
| Metal amides |
| Organometallic compounds with a metal atom bonded to carbon |
| Metal alkoxides |

form peroxides which can influence the course of experiments in which they are used.

Common chemicals can become serious hazards because of peroxide formation. Suggested time limits are given for retention or testing of these compounds after opening the original container. Although some laboratories mark containers of such chemicals with the original date of receipt, it should be recognized that such dating does not take into account the unknown time span between original packaging and the date of receipt. The opening date the original container of a chemical that is likely to form peroxides should always be marked on the container with proper identity. Students and laboratory workers should understand correct labeling instructions and should affix samples date and details of peroxide-forming reagents.[1–4]

Many common laboratory chemicals form peroxides on exposure to air. They should be tested for the presence of peroxides before being distilled or used as solvents. The recommended retention times begin with the date of synthesis or of opening the original container (Table 17-3).

Many laboratory and industrial chemicals are hazardous. Some are potentially explosive, others are shock-sensitive compounds, and still others are incompatible.[5–6] The following appendixes (Appendix 1 to 4) provide a partial list of these chemical categories. It also is important that all laboratory workers and drug formulation units be aware of the details of hazardous chemicals that they handle regularly and all the consequences of thereof (Appendix 5). Safe working is the most important rule that all laboratory and occupational workers must follow. In fact, safety awareness becomes a part of each worker's habit when repeatedly discussed with skilled and unskilled workers. Management, supervisors, and group leaders should exercise regular formal safety and housekeeping inspections. In housekeeping management, several nonhazardous organic and inorganic chemicals are used. The worker and students should check these chemicals before use and disposal in a chemical laboratory (Appendix 6).

In conclusion, the obvious goal of any disposal policy in a chemical industry is to bring the number of unknowns as close to zero as possible by following the chemical hygiene plan and the hazard communication protocol. Labeling all chemicals contained glassware, and disposing of all old, outdated, or questionable chemicals are important steps in the management plan. Recycling unneeded chemical reagents, maintaining separate containers for different waste classes, and maintaining a log of all wastes placed into disposable containers goes a long way to curb chemical hazards in a working laboratory. There are three well-known procedures to identify the unknown in any chemical laboratory.

- Preparation and maintenance of the "unknown profile form," normally available from any government safety office.

The lab supervisor and the group lead should have adequate knowledge of the material, as well as the analytical test to identify the unknown chemicals. He or she should provide the physical description, appearance, odor, and quantity

**TABLE 17-3 Common Peroxides Forming Chemicals**

| | |
|---|---|
| Severe peroxide hazard on storage with exposure to air: Discard within 3 months | |
| Diisopropyl ether | Sodium amide (sodamide) |
| Divinylacetylene* | Vinylidene chloride (1,1-dichloroethylene)* |
| Potassium metal | Potassium amide |
| Peroxide hazard on concentration: Do not distill or evaporate without first testing for the presence of peroxides: Discard or test for peroxides after 6 months | |
| Acetaldehyde diethyl acetyl (acetyl) | Ethylene glycol dimethyl ether (glyme) |
| Cumene (isopropylbenzene) | Ethylene glycol ether acetates |
| Cyclohexane | Ethylene glycol monoethers (cellosolves) |
| Cyclopentene | Furan |
| Decalin (decahydronaphthalene) | Methylacetylene |
| Diacetylene | Methylcyclopentane |
| Dicyclopentadiene | Methyl isobutyl ketone |
| Diethyl ether (ether) | Tetrahydrofuran |
| Diethylene glycol dimethyl ether (diglyme) | Tetralin (tetrahydronaphthalene) |
| Dioxan/Dioxolan | Vinyl ethers* |
| Hazard of rapid polymerization initiated by internally formed peroxides* | |
| List A. Normal liquids: Discard or test for peroxides after 6 months† | |
| Chloroprene (2-chloro-1,3-butadiene)‡ | Vinyl acetate |
| Styrene | Vinylpyridine |
| List B. Normal gases: Discard after 12 months§ | |
| Butadiene‡ | Vinylacetylene‡ |
| Tetrafluoroethylene‡ | Vinylchloride |

*Monomers may be polymerized and should be stored with a polymerization inhibitor, from which the monomer can be separated by distillation just before use.

†Although common acrylic monomers, such as acrylonitrile, acrylic acid, ethyl acrylate, and methyl methacrylate, can form peroxides, they have not been reported to develop hazardous levels in normal use and storage.

‡The hazardous peroxide formation in these compounds is substantially greater when they are stored in the liquid phase. If stored in this form, without an inhibitor, they should be included in List A.

§Although air cannot enter a gas cylinder in which gases are stored under pressure, these gases are sometimes transferred from the original cylinder to another in the laboratory, and it is difficult to ensure that there is no residual air in the receiving cylinder. An inhibitor should be put into secondary cylinder before transfer. The supplier can suggest an appropriate inhibitor to be used. The hazard posed by these gases is much greater if there is a liquid phase in the secondary container. Even inhibited gases that have been put into a secondary container under conditions that create a liquid phase should be discarded within 12 months.

of the unknown. The specific chemicals presented as hazardous waste should be properly identified and indicated before submission for disposal.

- Always attach the sheet to the material submitted for disposal.
- Whenever uncertain, call the department safety office.
- Knowledge + Common Sense + Caution = Chemical Safety

## REFERENCES

1. H.L. Jackson, W.B. McCormack, C.S. Rondesvedt, K.C. Smeltz, and I.E. Viele, *Safety in the Chemical Laboratory*, Vol. 3, N.V. Steere, ed., American Chemical Society, Easton, PA, 1974.
2. G. Lunn, and E.G. Sansone, *Destruction of Hazardous Chemicals in the Laboratory*, Chichester, U.K.: John Wiley & Sons, 1990.
3. D.A. Pipetone, ed., *Safety Storage of Laboratory Chemicals*, 2nd ed., New York, John Wiley & Sons, 1991.
4. IUPAC-IPCS, *Chemical Safety Matters*, Cambridge University Press, Cambridge, U.K., 1992.
5. L. Bretherick, ed., *Hazards in the Chemical Laboratory*, 4th ed., Royal Society of Chemistry, London, U.K., 1986.
6. L. Bretherick, *Handbook of Reactive Chemical Hazards*, 4th ed., London, U.K., Butterworths, 1990.

## APPENDIX 17-1 SHOCK-SENSITIVE CHEMICALS

*Acetylenic compounds*, especially polyacetylenes, haloacetylenes, and heavy metal salts of acetylenes (copper, silver, and mercury salts are particularly sensitive)

*Acyl nitrates* (V)

*Alkyl nitrates* (V), particularly polyol nitrates (V) such as nitrocellulose and nitroglycerine

*Alkyl and acyl nitrites* (III)

*Alkyl chlorates* (VII)

*Amine metal oxosalts*: metal compounds with coordinated ammonia, hydrazine, or similar nitrogenous donors and ionic chlorate (VII), nitrate (V), manganate (VII), or their oxidizing group

*Azides*, including metal, nonmetal, and organic azides

*Chlorate* (III) salts of metals, such as $AgClO_2$ and $Hg(ClO_2)_2$

*Chlorate* (VII) salts. Most metal, nonmetal, amine, and organic cation chlorates (VII) can be detonated/undergo violent reaction in contact with combustible materials

*Diazo compounds*, such as $CH_2N_2$

*Diazonium salts*, when dry

*Fulminates* (silver fulminate, AgCNO, can form in the reaction mixture from the Tolens' test for aldehydes if it is allowed to stand for some time. This can be prevented by adding dilute nitric (V) acid to the test mixture as soon as the test has been completed.)

*Hydrogen peroxide* becomes increasingly treacherous as the concentration rises above 30%, forming explosive mixtures with organic materials and decomposing violently in the presence of traces of transition metals

*N-Halogen compounds* such as difluoroamino compounds and halogen azides

*N-Nitro compounds* such as N-nitromethylamine, nitrourea, nitroguanidine, and nitric amide

*Oxo salts of nitrogenous bases*: chlorates (VII), dichromates (VI), nitrates (V), iodates (V), chlorates (III), chlorates (V), and manganates (VII) of ammonia, amines, hydroxylamine, guanidine, etc.

*Peroxides and hydroperoxides*, organic

*Peroxides* (solid) that crystallize from or are left from evaporation of peroxidizable solvents

*Peroxides*, transition metal salts

*Picrates*, especially salts of transition and heavy metals, such as nickel, lead, mercury, copper, and zinc; picric acid is explosive but is less sensitive to shock or friction than its metal salts, and is relatively safe as a water wet paste

*Polynitroalkyl compounds*, such as tetranitromethane and dinitroacetonitrile

*Polynitroaromatic compounds*, especially polynitrohydrocarbons, phenols, and amines

## APPENDIX 17-2 POTENTIALLY EXPLOSIVE COMBINATION OF SOME COMMON REAGENTS

Acetone with chloroform in the presence of base
Acetylene with copper, silver, mercury, or their salts
Ammonia (including aqueous solutions) with $Cl_2$, $Br_2$, or $I_2$
Carbon disulfide with sodium azide
Chlorine with an alcohol
Chloroform or carbon tetrachloride with powdered aluminium or magnesium
Decolorizing carbon with an oxidizing agent
Diethyl ether with chlorine (including a chlorine atmosphere)
Dimethyl sulfoxide with an acyl halide, SOC12, or POC13 or with CrO3
Ethanol with calcium chlorate (I) or silver nitrate (V)
Nitric (V) acid with acetic anhydride or acetic acid
Picric acid with a heavy-metal salt, such as lead, mercury, or silver
Silver oxide with ammonia with ethanol
Sodium with a chlorinated hydrocarbon
Sodium chlorate (I) with an amine

## APPENDIX 17-3 CHEMICALS GENERALLY INCOMPATIBLE

| Acids and oxidizing agents | Bases, metals, and reducing agents |
|---|---|
| Chlorates (VI) | Ammonia, anhydrous and aqueous |
| Chromates (VI) | Carbon |
| Chromium (VI) oxide | Metals |
| Halogens/Halogenating agents | Metal hydrides |
| Hydrogen peroxide | Nitrates (III) |
| Manganates (VII) | Organic compounds in general |
| Nitric (V) acid/Nitrates (V) | Phosphorus |
| Peroxides | Silicon |
| Sulfates (VII) | Sulfur |

## APPENDIX 17-4 SPECIFIC CHEMICALS WITH INCOMPATIBLES

| Group I | Group II |
|---|---|
| Acetylene, monosubstituted acetylenes | Group IB and IIB metals and their salts |
| | Halogens/Halogenating agents |
| Ammonia, anhydrous and aqueous | Halogens/Halogenating agents |
| | Mercury |
| | Silver |
| Alkali and alkaline earth | Water |
| Carbides | Acids |
| Hydrides | Halogenated organic compounds |
| Hydroxides | Halogenating agents |
| Metals | Oxidizing agents |
| Oxides/peroxides | |
| Azides, inorganic | Acids |
| | Heavy metals and their salts |
| | Oxidizing agents |
| Cyanides, inorganic | Acids/Strong bases |
| Mercury and its amalgams | Acetylene |
| | Ammonia, anhydrous and aqueous |
| | Nitric (V) acid |
| | Sodium azide |
| Nitrates (V), inorganic | Acids |
| | Reducing agents |
| Nitric acid | Bases |
| | Chromic (VI) acid |
| | Chromates (VI) |
| | Metals |
| | Manganates (VII) |
| | Reducing agents |
| | Sulfides |
| | Sulfuric acid |
| Nitrates (III), inorganic | Acids |
| | Oxidizing agents |
| Organic compounds | Oxidizing agents |
| Organic acyl halides | Bases |
| | Organic hydroxy and amino compounds |
| Organic anhydrides | Bases |
| | Organic hydroxy and amino compounds |
| Organic halogen compounds | Group IA and IIA metals |
| | Aluminium |
| Organic nitro compounds | Strong bases |

(*continued overleaf*)

| Group I | Group II |
|---|---|
| Oxalic acid | Mercury and its salts |
| | Silver and its salts |
| Phosphorus | Oxidizing agents |
| | Oxygen |
| | Strong bases |
| Phosphorus (V) pentoxide | Alcohols |
| | Strong bases |
| | Water |
| Sulfides, inorganic | Acids |
| Sulfuric acid (concentrated) | Bases |
| | Potassium manganate (VII) |
| | Water |

## APPENDIX 17-5 QUICK GUIDE TO CHEMICAL TOXICITY AND SAFETY

| Chemicals | Toxic response and pathologic reasons |
|---|---|
| Acetaldehyde (acetic aldehyde:ethanol) | Eye and respiratory tract irritation, narcosis, bronchitis, and liver damage |
| Acetic anhydride (acetyl oxide; ethanoic anhydride) | Strong eye and upper respiratory tract irritation, and corrosive action |
| Acetone (dimethyl ketone; 2-propanone) | Slight eye, nose, and throat irritation, and narcosis |
| Acetonitrile (methyl cyanide) | Respiratory irritation and cyanide poisoning |
| Acrolein | Lacrimation and respiratory irritation |
| Ammonia | Eye irritation and pulmonary edema |
| Aniline (aminobenzene; phenylamine) | Cyanosis due to methemoglobinemia, slight narcosis, and respiratory paralysis |
| Benzene | Narcosis, leukemia, liver damage, and aplastic anemia |
| Benzidine | Abdominal pain, nausea, skin irritation, and cancer |
| Carbon tetrachloride (tetrachloromethane) | Headache, nausea, slight jaundice, loss of appetite, narcosis, liver and kidney damage, and GI disturbances |
| Chloroform (trichloromethane) | As for carbon tetrachloride |
| Cyanogen bromide | Abdominal pain, nausea, diarrhea, blurred vision, and pulmonary edema |
| Cytochalasin dioxane | Narcosis, liver and kidney damage, and cancer |
| Diethyl ether | Vomiting, eye irritation, and addictive |
| Formaldehyde (formalin) | Respiratory, skin and mucous membrane irritation, and pulmonary edema |
| Glutarol | Respiratory and mucous membrane irritation |
| Methanol (methyl alcohol) | Narcosis, mucous membrane irritation, and damage to the retina and optic nerve |

(*continued overleaf*)

| Chemicals | Toxic response and pathologic reasons |
|---|---|
| Mercury | Vomiting, diarrhea, headache, nausea, eye pain, CNS disturbance, swollen gums, and loose teeth |
| a-Naphthylamine | Cyanosis due to methemoglobinemia, slight narcosis, |
| b-Naphthylamine | anemia, reduced methemoglobinemia with cyanosis, |
| Nitrobenzene (nitrobenzol) | bladder irritation, and liver damage |
| Phenol | Abdominal pain, vomiting, diarrhea, skin irritation, eye pain, corrosive action, CNS disturbance, and coma |
| Pyridine | Liver and kidney damage, and neurotoxicity |
| Selenium | Burning skin, eye pain, cough, CNS disturbance, and teratogenesis |
| Tetrahydrofuran (diethyl oxide; tetramethyl oxide) | Narcosis, liver and kidney damage, and eye and respiratory irritation |
| Thallium | Abdominal pain, vomiting, nausea, diarrhea, neuropathy, visual problems, muscle weakness, and ataxia |
| O-Tolidine | Cancer |
| Toluene (methyl benzene; phenyl methane; toluol) | Narcosis, nonspecific neurologic impairment, and addiction possible |
| Trichloroethylene (ethinyl trichloride) | Narcosis, liver damage, and nonspecific neurologic impairment |

## APPENDIX 17-6 NONHAZARDOUS ORGANIC AND INORGANIC CHEMICALS SUITABLE FOR SANITARY SEWER DISPOSAL

A. Organic chemicals

Alcohols
- Alkanols with fewer than five carbon atoms
- t-Amyl alcohol
- Alkanediols with fewer than eight carbon atoms
- Glycerol
- Sugars and sugar alcohols
- Alkoxyalkanols with fewer than seven carbon atoms
- n-$C_4H_9OCH_2CH_2OCH_2CH_2OH$
- 2-Chloroethanol

Aldehydes
- Aliphatic aldehydes with fewer than five carbon atoms

Amides
- $RCONH_2$ and RCONHR with fewer than five carbon atoms
- $RCONR_2$ with fewer than 11 carbon atoms

Amines*
- Aliphatic amines with fewer than seven carbon atoms
- Aliphatic diamines with fewer than seven carbon atoms
- Benzylamine
- Pyridine

Carboxylic acids*
- Alkanoic acids with fewer than six carbon atoms
- Alkanedioic acids with fewer than six carbon atoms
- Hydroxyalkanoic acids with fewer than six carbon atoms
- Aminoalkanoic acids with fewer than seven carbon atoms
- Ammonium, sodium, and potassium salts of the above acid classes with fewer than 21 carbon atoms
- Chloroalkanedioic acids with fewer than four carbon atoms

Esters
- Esters with fewer than five carbon atoms
- Isopropyl acetate

Ethers
- Tetrahydrofuran
- Dioxolane
- Dioxane

(*continued overleaf*)

B. Inorganic chemicals

Ketones
  Ketones with fewer than six carbon atoms
Nitriles
  Acetonitrile
  Propionitrile
Sulfonic acids
Sodium or potassium salts of most are acceptable
B. Inorganic chemicals[†]

| Cations | Anions |
|---|---|
| Aluminium (III) | $BO_3^{3-}$, $B_4O_7^{2-}$ |
| Calcium (II) | $Br^-$ |
| Copper (II) | $CO_3^{2-}$ |
| Iron (II), (III) | $Cl^-$ |
| Hydrogen | $HSO_3^-$ |
| Lithium | $OH^-$ |
| Magnesium | $I^-$ |
| Sodium | $NO_3^-$ |
| NH (IV) | $PO_4^{3-}$ |
| Tin (II) | $SO_4^{2-}$ |
| Strontium | $SCN^-$ |
| Titanium (III), (IV) | |
| Zinc (II) | |
| Zirconium (II) | |

*Those amines and acids with a disagreeable odor, such as 1,4-butanediamine, di-methylamine, butyric acids, and valeric acids, should be neutralized and the resulting salt solutions flushed down the drain, diluted with at least 1,000 volumes of water.

[†]This list comprises water-soluble salts in which both the cation and anion have a low-toxic hazard. Any of these salts that are strongly acidic or basic should be neutralized before being flushed.

CHAPTER 18

# Good Laboratory Practice

## INTRODUCTION

The economic and social implications of new technologies are closely linked with new chemicals. They have offered us both benefits and dangers and/or adverse effects. To be specific, the application of number of chemicals, either for the control of crop pests or for the improvement of the quality of life, has become common practice and order of the day around the world. Industrial growth and human activities have lead to many health problems and human fatalities. A need, therefore, has arisen to regulate the use and disposal of chemicals. Many international organizations have been working for the regulation of toxic substances around the world. All the organizations have one common goal: human safety and protection of the environment. The present discussion is for the proper understanding of the principles of good laboratory practice (GLP) and its importance in achieving chemical safety. This is essential for all researchers, students, and workers for the generation of data of quality and integrity.

Generation of data of quality and integrity is the product of work conducted on the basis of a proper protocols. A team of qualified personnel and integrity in laboratory studies enhance the data quality. Human and environmental safety from the large-scale use and disposal of a variety of substances and chemicals largely depends on the qualified worker, whether in a laboratory study or in a product development or manufacturing industry. In the 1970s, false data was submitted to regulatory agencies for product registration without conducting required toxicologic investigations. Prolonged investigations of 25 animal safety studies by the United States Food and Drug Administration (USFDA) and the United States Environmental Protection Agency (USEPA) (at the cost of 16.6 million dollars), identified 620 new cases of fraud. A draft discussing GLP was published in November 1976, was held for comments until March 1977, and was made effective in June 1977. An enormous amount of time and money was spent to identify serious lapses of integrity in the workplace, personnel, and documentation. The FDA then made GLP to be effective from June 1977.[1,2] This regulation was strengthened by the inspection of laboratories to assess laboratory competency in

*Industrial Guide to Chemical and Drug Safety*, By T.S.S. Dikshith and Prakash V. Diwan
ISBN 0-471-23698-5 

the generation of quality data, per GLP. Toxicologic tests and toxicologists are responsible for the generation of quality data on different products. Many products (e.g., drugs, cosmetics, soaps and detergents, pesticides, food additives, child care and dairy products, medical devices, other agents) have been in use by us all in daily life; hence, we must have correct and satisfactory information about these products' toxicity and safety. The discussions here will emphasize how common sense and, good science would help a laboratory student, and an ordinary factory or field worker collect and document observations for the generation of quality data. Further, these principles help train students who will compete in the job market, to prepare them to conduct research according to the mandates of quality research.[1] For the past three decades, global agencies and regulatory bodies have repeatedly emphasized the importance of GLP in conducting toxicologic studies, and this information is available in literature.[3–5]

In recent years, economic cooperation among European and other countries formed the Organization for Economic Cooperation and Development (OECD) in Paris. Essentially, the purpose of the rules and regulations of these organizations reflect the same philosophy, GLP. In short, it now has become mandatory to observe all regulations, follow quality methods, establish credibility, and present documentary evidence of the work performed to get the required benefit. These regulations have stood the test of time to protect human health and the environment.[3–5] Agencies such as OECD and regulatory bodies such as USEPA and Federal Insecticide Fungicide Rodenticide Act (FIFRA) have continuously made demands for the generation of quality data based on GLP.[6–8]

## Good Laboratory Practice

GLP is concerned with organizational processes and how all types of studies in a laboratory/test house should be planned, performed, monitored, recorded, and reported. By adhering to GLP principles a laboratory ensures proper planning of studies and the provision of means to meaningful study conclusions. Studies performed according to GLP ensure the quality and integrity of data and allows its use by regulatory authorities in the risk assessment of chemicals. GLP standards are prescribed for toxicology studies on agricultural chemicals. Primarily, GLP is intended to ensure the quality and integrity of product data generated in a laboratory. A violation would occur if a set protocol is not followed. The USEPA has regulations and guidelines that suggest what studies are required and how they are to be performed. Today, GLP standards are recognized throughout the world.[9–11,16–18]

## International Organizations

Among international organizations, the Food and Agriculture Organization (FAO) and the World Health Organization (WHO) of the United Nations stand supreme. For proper management and chemical safety, many guidelines have been made. These guidelines are aimed at identifying possible adverse effects during the use of substances and determining the capability of industries and workers to observe set guidelines. These regulations have become necessary because of situations in some of the countries. As mentioned earlier, lack of adherence to set regulations ends in several consequences. Regulations are set to help workers, organizations, and the general public adhere to and practice goals of chemical safety. Toxicology addresses the manner and methods of handling toxic substances and safeguards for the health of animals and humans. The USFDA made it mandatory to inspect laboratories conducting toxicologic investigations.

### International Program on Chemical Safety

The International Program on Chemical Safety (IPCS) was established in 1980. IPCS is a joint program of three cooperating organizations, namely, International Labor Organization (ILO), United Nations Environment Program (UNEP), and WHO. IPCS implements activities related to chemical safety with an intersectoral coordination and a scientifically based program. WHO is the executing agency of IPCS, and it has two main roles: 1) to establish scientific basis for safe use of chemicals; and 2) to strengthen national capacities for chemical safety. Essentially, the areas of activity include (1) evaluation of chemical risks to human health and the environment; (2) methodologies for the evaluation of hazards and risks; (3) prevention and management of toxic exposures and chemical emergencies; and (4) development of the human resources required for the above areas.

Toxicologic evaluations of chemicals are based on accepted methods and guidelines. These guidelines are formulated by established scientific organizations and institutions, and implemented through government regulatory bodies. Organizations such as the WHO, USFDA, USEPA, and the Code of Federal Regulations (CFR) of the United States are taking measures to help workers and industry understand the benefits of GLP. In 1989, the USEPA expanded regulations to require compliance with GLP standards for testing of ecological effects, chemical fate, residue chemistry, and product performance. The USEPA has amended these regulations to ensure the quality and integrity of data submitted in conjunction with pesticide product registration or other marketing and research.[5]

The philosophy of GLP came into active force during the 1960s. In the course of product evaluation, false documentation of test results in a U.S. laboratory came to the attention of regulatory agencies. A prolonged investigation by the U.S. Federal Agencies confirmed fraud in the laboratory. Subsequently, regulatory bodies toned up the established rules, generated product toxicity data and made GLP mandatory. As a result of this and globalization of trade and product movements, every country has been moving toward the generation of quality data. These organizations have made tremendous efforts to incorporate and harmonize test methodologies adopted in different countries. Today, the OECD guidelines have offered a valuable tool to reach this goal; they consist of several tests to establish safety to human health and the environment.

The test guidelines are periodically updated to keep pace with progress in science. In addition, new test guidelines are developed and agreed upon, based on specific needs identified by OECD-member countries. The OECD-wide network of experts, derives the latest knowledge from scientists, academia, government, and industry to the guidelines. Broad acceptance and recognition of the guidelines as the international standard has been achieved through these networks. Since adoption in 1981, the guidelines have become the recognized reference tool for chemical testing and assessment of their potential hazards. The guidelines also have become an important component of in the acceptance of quality data.

### Harmonization

The process of harmonization seeks ways to increase burden-sharing between national regulatory agencies and reduce regulatory burden on industry. Harmonization is not a uniform standardization. In fact, harmonization attempts to understand differences between regulatory approaches and deliver more efficient and effective regulatory programs while maintaining national standards. Through OECD, the Intergovernmental Forum for Chemical Safety (IFCS) and other forums, governments, and industries have been seeking cooperative efforts on new chemical-approval processes to reduce study duplication. This benefits governments and industries in that it allows for better use of resources. In June 1999, the OECD endorsed a program to enhance harmonization of notification activities between member states. Activities such as exchange, and greater use of information, assessments, and common data submission formats are being pursued. The overall objective for the program is that "Notification and assessment of a new substance in one country be used to facilitate notification and assessment in other countries by 2005." Thus, the examination of the knowledge-based economy, the globalization challenge, and economic performance and competitiveness has given the world a lead.

It is important to know about the International Conference on Harmonization of Technical Requirements for Registration of Pharmaceuticals for Human Use (ICH). This is a unique project bringing together the regulatory authorities of Europe, Japan, and the United States to discuss scientific and technical aspects of product registration. The purpose was to achieve seamless interpretation and application of technical product guidelines to reduce duplication of testing during the development of new medicines. The goals are for a more economical use of resources, for more rapid development and availability of medicines, and for the maintainence of safeguards on quality, safety, and efficacy. In fact, these goals are well embodied in the Terms of Reference of ICH.

### Harmonization of Environmental Measurement

The Harmonization of Environmental Measurement (HEM) center was established in 1989 by UNEP with financial support from the German Ministry of the Environment (BMU) as part of the Global Environment Monitoring System (GEMS). The initiative to establish HEM was taken by the Economic Summit in Venice in 1987. This initiative was based on the work of the Secretariat of the Environment Experts of the Economic Summit (EEES). Currently, HEM is concentrating on the standardization of environmental monitoring and research programs to improve collection and management of data, and enhance the quality of environmental information worldwide.

The United Kingdom conducted the first compliance inspections with the principles of GLP, in facilities generating toxicity and other health and environmental safety data for submission to regulatory authorities. GLP inspectors in 1982 established a unit to monitor industrial chemical testing. Prior to this, most toxicology laboratories implemented GLP only to meet registration requirements set by the USFDA. In April 1983, another inspection was made by the Department of Health and Social Security, to monitor laboratories for compliance with GLP principles. This inspection led to establishment of the United Kingdom's GLP Monitoring Authority (UKGLPMA) and extended the range of facilities inspected to include those involved in safety testing of pharmaceuticals, agrochemicals, cosmetics, and food additives. In May 1986, it formally assumed responsibility for the monitoring of laboratories involved in industrial chemical testing, which had previously been monitored by the health and safety executive. The UKGLPMA is at present part of the United Kingdom's Department of Health. Many of the organizations are associated with studies on organic, inorganic, bioanalytical, and radioanalytical capabilities, including customized approaches and routine analyses using methods specified by the USFDA, ASTM, USEPA, and NIOSH. This has provided opportunities for staff members to develop validations of the test methods.

In contrast to chemicals, biomaterial and medical device testing constitutes an extremely diverse, heterogeneous category of evaluation. Because the use of these products normally entails direct or indirect contact with patients, there is an obligation for manufacturers to establish the product's safety before marketing. Medical device safety evaluation assesses the risk of adverse effects due to normal use and misuse. Since adverse effects could result from exposure to the materials from which a device is made, preclinical assessment is needed to minimize the potential hazard to the user, namely, the patient.

### Good Laboratory Practice Mind-Set

To develop any good practice, discipline and basic knowledge of the subject is essential; GLP is no different. The research student, supervisor, worker, or manager should develop interest in quality work. The organization should adopt direct and indirect methods of working the process. First, data have a different purpose and review process for GLP studies. The data generated from GLP studies are generally compiled by a sponsor (e.g., a pharmaceutical company). Subsequently, the data are submitted as part of a registration package to a federal agency. Federal authorities will review the data to determine adequacy and validity for the grant of product registration. The GLP mindset is possible by questioning the work and data generated. For instance, "How is it possible?" and "How do I know?" should be answered by documentary evidence and support for legal challenges, in the case that it occurs. Generation of data, along with reliable and valid documentation, is the GLP touchstone.

## Good Laboratory Practice Provisions

GLP has several provisions, and all are interlinked. The purpose of these provisions is to help generate data of quality and integrity. These may be listed in brief:

- General: (1) study purpose; (2) definitions; and (3) notification by the sponsor to the contracting laboratory.
- Personnel Organization: (1) personnel; (2) testing facility management; (3) study director; and (4) quality assurance unit (QAU).
- Facilities: (1) general; (2) animal care facilities; (3) animal supply facilities; (4) test and control substances facilities; (5) laboratory operation area; and (6) administrative facilities.
- Equipment: (1) equipment design and function; (2) equipment location; and (3) equipment control.
- Operation (of the laboratory activities): (1) standard operating procedure (SOP); (2) reagents and solutions; and (3) animal and quarantine care.
- Test and Control Substances: (1) characterization; (2) handling; and (3) mixtures and carriers.
- Protocol for the Conduct of the Study: (1) protocol; and (2) performance and/or
  conduct of the study.
- Reports and records: (1) final report; (2) storage/archive; retrieval of records and specimens; and (3) retention period of records and specimens.

It is important to understand the GLP provision *in toto*. However, present discussions are largely about the worker, who should know how to conduct studies in a laboratory/workplace to comply with GLP. Generation of quality data gives credibility to the results and report, to the laboratory or industry, and to the organization as a whole. Keeping this in view, emphasis has been given to the selected parts of GLP, which are as follows:

1. Organization, management, and testing facility
2. Study purpose
3. Study director
4. Quality assurance
5. Facilities: personnel, laboratory, animal, materials, and equipment
6. SOP
7. Protocol
8. Conduct of the study

9. Documentation and storage
10. Final report

## ORGANIZATION, MANAGEMENT, AND THE TESTING FACILITY

A research project in academia may have a number of coprincipal investigators (PI), and each one has variable responsibilities. In the case of a GLP project, one PI is designated as the study director. The study director is the individual responsible for the overall conduct of the nonclinical laboratory study and is the single point of study control.[6–9] The USFDA assigns a study director with the responsibilities as follows:

### Responsibilities of the Test Facility Management

- Each test facility management should ensure that it complies with GLP.
- The facility, at a minimum, should ensure that a statement exists identifying the individual(s) within a test facility who fulfill management responsibilities as defined by GLP.
- The facility should ensure that a sufficient number of qualified personnel, facilities, equipment, and materials are available for the timely and proper conduct of the study.
- The facility should ensure record(s) maintenance of the qualifications, training, experience, and job description for each professional and technical individual.
- The facility should ensure that personnel clearly understand the functions they are to perform and, where necessary, provide training for these functions.
- The facility should ensure that appropriate and technically valid SOPs are established and followed and approve all original and revised SOPs.
- The facility should ensure that there is a quality assurance (QA) program with designated personnel and that the QA responsibility is performed in accordance within GLP.
- The facility should ensure that for each study, an individual with appropriate qualifications, training, and experience is designated by management as the study director before the study is initiated. Replacement of a study director should be done according to established procedures and should be documented.
- The facility should ensure that in a multisite study (if needed), a PI is designated, who is appropriately trained, qualified, and experienced to supervise the delegated phase(s) of the study. Replacement of a PI should be done according to established procedures and should be documented.
- The facility should ensure the documented approval of the study plan by the study director.

- The facility should ensure that the study director has made the approved study plan available to QA personnel.
- The facility should ensure maintenance of a historic file of all SOPs.
- The facility should ensure that an individual is identified as responsible for archive(s) management.
- The facility should ensure the maintenance of a master schedule.
- The facility should ensure that test facility supplies meet requirements appropriate for use in a study.
- The facility should ensure that in a multisite study, clear lines of communication exist between the study director, PIs, the QA program, and the study personnel.
- The facility should ensure that test and reference items are appropriately characterized.
- The facility should establish procedures to ensure that computerized systems are suitable for their intended purpose and are validated, operated, and maintained in accordance with GLP.
- When a phase(s) of a study is conducted at a test site, test site management (if appointed) will have the responsibilities.

### Study Purpose

Each study should clearly indicate its purpose. It should describe the reasons for the conduct of such a study and benefits the study may add to existing knowledge.

### Study Director, Qualifications, and Responsibilities

The study director represents the central point of control with ultimate responsibility for the overall conduct of the study. This is the prime role of the study director, and all responsibilities as outlined in GLP principles stem from it.[6] The study director accomplishes this by coordinating inputs of management, scientific/technical staff, and the QA program.

***Qualifications*** Qualifications for a study director are dictated by the requirements of the individual study. Setting the criteria is the responsibility of management. Furthermore, management has the responsibility for selection, monitoring, and support of the study director, to ensure that studies are performed in compliance with GLP principles. The study director indicates compliance with GLP by his or her signature on the final study report. Minimal qualifications established by management for the position of study director should be documented in appropriate personnel records. In addition to a strong technical background, the role of study director requires an individual with strengths in communication, problem-solving, and managerial skills.

***Responsibilities*** The study director is usually the scientist responsible for study plan design and approval, as well as overseeing data collection, analysis, and reporting. The study director is responsible for drawing the final conclusions from the study. The study director also must coordinate with other key study scientists and their findings until completion of the project. Management must coordinate with the study director and provide all resources (e.g., personnel, equipment, and facilities).

***Interface with the Study*** The phrase, "responsibility for the overall conduct of the study and for its report" may be interpreted in a broad sense for those studies where the study director may be geographically remote from actual experimental work. With multiple levels of management, study personnel, and QA staff, it is critical that there are clear lines of authority and communication, and assigned responsibilities, so that the study director can effectively carry out GLP responsibilities. This should be documented in writing.

For studies that have delegated responsibilities to a PI, the study director will rely on that individual to ensure that relevant phase(s) of the study are conducted in accordance with the study plan, relevant SOP, and principles of GLP. The PI should contact the study director when event(s) occur that may affect the objectives defined in the study plan. All communications should be documented. Communication between the study director and the QA is required at different stages of the study. For instance; (1) to review study plans; (2) to review new and revised SOPs; (3) attendance of QA personnel at study initiation meetings and in resolving potential problems related to GLP; (4) by responding to inspection and audit reports promptly; (5) by indicating corrective action; and (6) by necessary liaison with QA staff, and scientific and technical personnel. In certain unforeseen conditions, change of study director becomes essential. Under such conditions, replacement of the study director should take immediate effect.

***Replacement*** Because the study director has responsibility for the overall conduct of a study according to the GLP principles, he or she must ascertain that every phase of a study fully complies with these principles, that the study plan is followed faithfully, and that all observations are fully documented. Theoretically, this responsibility can only be fulfilled if the study director is present during the whole study. This is not always feasible in practice, and there will be periods when replacement may be necessary. Although the circumstances under which a study director would be replaced are not defined in GLP principles, they should be addressed by the facility SOPs. These SOPs also should address procedures and documentation necessary to replace a study director.

The decision for replacement (or temporary delegation) is the responsibility of management. All decisions should be documented in writing. Two circumstances in which replacement might be considered are important only in longer-term studies, since the presence of a study director during a short study may be assumed. The first circumstance is termination of employment. In the event of termination of a study director, the need for replacement is obvious. In this case, one responsibility of the replacement study director is (with the assistance of QA personnel) to become familiarized with the GLP compliance in the study to date.

The replacement of a study director, and the reasons for it, must be documented and authorized by management. It also is recommended that the results of any interim GLP review should be documented in case deficiencies or deviations have been found.

The second circumstance is when a study director is temporarily absent because of holidays, scientific meeting, illness, or accident. An absence of short duration might not necessitate the formal replacement of the study director, if it is possible to communicate should problems or emergencies arise. If critical study phases are expected to fall in the period of absence, these phases may either be moved to a more suitable time (with study plan amendment, if necessary) or a replacement study director may be considered, either by formally nominating a replacement or by temporary delegation of responsibilities to competent staff. Should the unavailability of the study director be of longer duration, a replacement should be named rather than delegation to staff. The returning study director must ascertain whether deviations from GLP Principles have occurred, irrespective of whether he or she was formally replaced. Any deviations from GLP principles during the absence should be documented by the returning study director.

### Quality Assurance Unit

QA in analytical laboratories, irrespective of the discipline, is considered very casually. Many laboratories find it very difficult to effectively and systematically implement the necessary procedures of QA. In fact, laboratories with marginal budgets, or smaller research laboratories working without much competition, often do not have the necessary resources and incentives to engage in a comprehensive QA effort, whereas laboratories seeking accreditation attempt to follow QA requirements. Proper training and refresher courses are necessary for this important program. Neglected aspects of laboratory work also include keeping full, systematic records, and drafting and implementation of proper operating procedures and protocols.

The QAU must be a separate and independent unit of the facility/laboratory. It must be different from the personnel engaged in the study.[4] The role of the QAU is to ensure management that the facilities, equipment, personnel, methods, practices, records, and control are compliant with GLP regulations.[6,10,11] The QAU must maintain a master schedule of ongoing studies and protocols of those studies. The QAU inspects each study at intervals. Adequate efforts are made to ensure the integrity of the study, and keep records of those inspections, report the finding of those inspections to the study director and management, and review the final study report. Also, the QAU should ensure that such a report accurately describes the methods and SOP, and that the reported results accurately reflect the raw data of the study. The QAU also will prepare and sign a statement in the final report specifying only the dates that inspections were made and the dates that the findings were reported to management and the study director.

***Responsibilities of the Quality Assurance Personnel*** The responsibilities of QA personnel are many. The QA should:

- Ascertain that the study plan and SOP are available to personnel conducting the study.
- Ensure that the study plan and SOP are followed by periodic inspections of the test facility or by auditing the study in progress. Records of such procedures should be retained.
- In view of the high frequency and routine nature of some short-term studies, it has been recognized in the OECD Consensus Document on QA and GLP that each study need not be inspected individually by the QA during the experimental phase of the study. In such cases, a process-based inspection may cover each study-type.
- The frequency of inspections should be specified in the relevant SOP of the QA. Further, an SOP must be available describing the regularly conducted processes and inspections.
- Prepare and sign a statement, to be included with the final report, which specifies the dates inspections were made and the dates any findings were reported to management and the study director.
- It is important, whenever individual study-based inspections do not take place, that the QA statement must clearly describe the same. For instance, whether the inspection is related to process-based and when performed. The QA statement must also indicate that the final report was audited.

### Facilities

There are major facilities and closely related facilities required to perform a GLP study. In fact, a test facility should be of suitable size, construction, and location to meet these requirements. These characteristics are to minimize disturbances that would interfere with the study validity. The design of the test facility should provide an adequate degree of separation for activities, to ensure the proper conduct of each study. Further, management is responsible for providing personnel, laboratory, materials, equipment, test system, and other facilities for the GLP study. In brief, these may be grouped as: (1) personnel; (2) laboratory; (3) test system; (4) archive; and (5) disposal. The details of the facilities may be seen in the "International Organizations" heading of this chapter.

## STANDARD OPERATING PROCEDURE

A test facility should have written SOPs approved by test facility management. Performing studies based on written SOPs is strongly encouraged, since it helps to minimize systematic errors. By following an SOP at each step, the chance of inadvertent mistakes could be avoided. Therefore, each student or worker should be familiar with the SOPs to perform different studies. In fact, adherence to an SOP enhances the quality and integrity of the data generated.

A typical SOP will induced the following: (1) a title; (2) SOP number; (3) a stated purpose; and (4) an outline of the procedure. The lab supervisor or a senior researcher and the author should sign off on the SOP and indicate the date it is to be effective. An SOP should be written by the people actually performing the procedures who know best how a procedure is conducted. The written SOP should be practical and clear, and could be a training document for new staff learning the procedure. With the advancement of science and development of newer techniques, the procedures are modified. In such situations, the SOP should be revised and updated. The outdated SOP should be retained in a historic archive to reference how a procedure was performed previously. Published textbooks, analytical methods, articles, and manuals may be used as supplements to an SOP. Any deviations from a particular study-related SOP should be documented and acknowledged by the study director and PI.

Each unit of the test facility should have an immediately available current SOP. SOPs normally include the following:

- The receipt, identification, labeling, handling, sampling, and storage of the test and reference items.
- Use, maintenance, cleaning, and calibration of apparatus, materials, and reagents.
- Record keeping, reporting, storage, and retrieval.
- Coding of studies, data collection, preparation of reports, indexing systems, handling of data, including the use of computerized systems.
- Room preparation and environmental room conditions for the test system.
- Procedures for receipt, transfer, proper placement, characterization, identification, and care of the test system.
- Test system preparation, observations, and examinations, before, during, and at the conclusion of the study.
- Handling of test system individuals found dead during the study.
- Collection, identification, and handling of specimens, including necropsy and histopathology.
- Siting and placement of test systems in test cages or rooms.
- Operation of QA personnel in planning, scheduling, performing, documenting, and reporting inspections.

## PROTOCOL

Each GLP study shall have an approved written protocol. A protocol should contain several details, required during and after the study period (Table 18-1).

### Protocol Development and Amendment

A protocol differs from a grant proposal in that the protocol need not justify the project, since it is not intended to be a fundraising tool. Rather, it shall indicate the objectives of the study, identify the testing facility, indicate the responsible

**TABLE 18-1 Protocol Content**

1. A descriptive title and statement of the study purpose.
2. Identification of the test and control articles by name, chemical abstract number, and code number.
3. The sponsor name and the name and address of the testing facility at which the study is being conducted.
4. The number, body weight range, gender, supply source, species, strain, substrain, and age of the test system.
5. The procedure for identification of the test system.
6. A description of the experimental design, including the methods for the control of bias.
7. A description and identification of the diet used in the study as well as solvents.
8. Emulsifiers, or other materials used to solubilize or suspend the test or control articles before mixing with the carrier. The description shall include specifications for acceptable contaminant levels that are expected to be present in the dietary materials, and are known to be capable of interfering with the study purpose or conduct if present at levels greater than established by the specifications.
9. Each dosage level, expressed in milligrams per kilogram of body weight or other appropriate units, of the test or control article to be administered and the method and frequency of administration.
10. The type and frequency of tests, analyses, and measurements to be made.
11. The records to be maintained.
12. The approval date of the protocol by the sponsor and the dated signature of the study director.
13. Statement of the proposed statistical methods to be used. All changes in or revisions of an approved protocol and the reasons therefore shall be documented, signed by the study director, and dated. This should be maintained with the protocol.

parties, describe the test system, the experimental design, the observations to be made, the records to be maintained, and the statistics to be used. The protocol must be approved by the sponsor, dated, and signed by the study director. Any deviation from the protocol must be documented with justification given, signed, and dated by the study director. Different inputs required for a GLP study are included in Appendixes 1 through 11.

## PERFORMANCE OF THE STUDY

The study starts only after meeting the following prerequisites:

1. The study director must agree to a study plan, objective, and conduct.
2. The study director should take responsibility for the study by dated signature of the study plan.
3. The date and signature by the study director makes the plan an official document.

4. The study director should ensure that the study plan is also signed by the sponsor and management (as required by national programs).
5. The study director should ensure that copies of the study plan are supplied to all key personnel involved in the study and QA staff.
6. The management should be presented with clear needs of adequate resources, personnel, and materials to perform the study and should ensure that adequate test materials and systems are available.

## Study Plan

The study director has complete responsibility for the overall conduct of the study. The procedures identified in the study plan are fully followed and documented legibly with date. The study plan should contain, but not be limited to, the information listed in Table 18-2.

The study director should review study procedures and data at every stage of the project. Documentation of regular reviews, frequency, and compliance of

**TABLE 18-2 Study Plan**

1. Identification of the study, the test item and reference item (descriptive title).
2. Statement that reveals the study nature and purpose.
3. Identification of the test item by code or name.
4. The reference item to be used.
5. Name and address of the sponsor.
6. Name and address of any test facilities and test sites involved.
7. Name and address of the study director.
8. Name and address of the principal investigator(s) and the phase(s) of the study delegated by the study director and under the responsibility of the principal investigator(s).
9. The approval date of the study plan by signature of the study director.
10. The approval date of the study plan by signature of the test facility management.
11. The date and signature of the sponsor (as required by national regulation/country).
12. The proposed experimental starting and completion dates.
13. Test methods (e.g., OECD, any other established guideline).
14. The justification for selection of the test system (animal/plant/organism).
15. Characterization of the test system (the species, strain, substrain, source of supply, number, body weight range, gender, age, other pertinent information).
16. The route or method of administration of the test substance to the test system and the reason for its choice (oral, dermal, inhalation, injection, diet).
17. The dose levels or concentration(s), frequency, and duration of administration, application, or exposure period.
18. Detailed information on the experimental design, including a description of the chronologic study procedure, all methods, materials and conditions, type and frequency of analysis, measurements, observations, and examinations to be performed, and statistical methods to be used (if any).

SOPs by the group, with the support of computer-generated data avoids many pitfalls. All documentation should be accurate, complete, and honest.

***Study Plan Amendment*** A study plan amendment should be issued to document a planned change in study design *before* the event occurs. An amendment also may be issued as a result of unexpected occurrences during the study that will require significant action. Amendments should indicate the reason for the change and be sequentially numbered, dated, signed, and distributed to all recipients of the original study plan by the study director.

### Study Deviations

Studies are conducted to evaluate the adverse effects of chemicals and drugs, and are usually based on an earlier study plan. Deviations from study plans are not common. However, such deviations cannot be anticipated or totally ignored. In such a situation, there is a need for documentary evidence. Whereas an amendment is a planned change to the study plan, a deviation is an unplanned change that occurs during the study. Study information, such as a deviation from the study plan, should be noted in documentation. Such notes may be initiated by other personnel involved in the study but should be acknowledged by the study director who must approve any corrective action taken. The study director should consider whether to consult with other scientists to determine the impact of any such information on the study and should report (and discuss where necessary) these deviations in the final report.

### Documentation Requirements

Data collected for GLP studies must be recorded promptly, honestly, and legibly in ink. All data entries should be dated on the day of entry and signed or initialed by the person entering the data. There may be study files that contain numerous documents pertaining to the study. These files may contain telephone memos, correspondence, shipping labels, organizational charts for the unit, internal memos, and QA findings. A study worker may, for example, have files of equipment maintenance or calibration records, SOPs, environmental or weather data, training records, animal room logs, veterinary care reports, pathology reports, analytical lab reports, statistical analysis data, purchase orders, protocol amendments, protocol deviations, unusual occurrences, chain-of-custody records, manifests, and interview records. All documents must be archived at the end of the study. This will help the sponsor and regulatory agencies to reconstruct the study and understand exactly what transpired during the course of the investigation. This also provides a complete information about the procedures adopted, the quality control employed, the data integrity, the validity of the results, and the study conclusions. Needless to say, the organization and legibility of files will leave an impression on those reviewing the data. Thus, the GLP mindset can force a research team to be organized and systematic during the study and documentation.[10,12–14]

***Correction of Error*** Any study, no matter what precautions are taken, does have errors in date and documentation. Changes or corrections to the data should be made so as not to obscure the original entry. Documents should indicate clearly the reason for the change, dated and signed at the time of the change.[6–8] Changes may be circled with code numbers such as: (1) misspelling or carelessness; (2) instrument was misread; (3) misunderstood or heard incorrectly; (4) miscount; (5) recorded in wrong place; (6) transposed or out of proper sequence; (7) illegible writing; (8) wrong word or phrase for meaning; (9) previous error(s); or (10) other unlisted reason(s). Codes also could appear as **"WE"** for "wrong entry," or **"WD"** for "wrong date."

### Data Storage and Retrieval, Archiving, Raw Data Availability

It may take years for a company to compile and submit a regulatory package of completed studies in support of an application. Thus, data storage and retrieval are important. All raw data, documentation, records, protocols, specimens, and final reports generated as a result of a study should be retained. Specimens obtained from mutagenicity tests, specimens of soil, water, and plants, and wet specimens of blood, urine, feces, and biologic fluids, need not be retained after QA verification.[6–8]

Many companies use a commercial archive facility to handle this task. Thus, the records are retained as long as the sponsor holds any research or marketing permit for which the study is pertinent. The records will be archived at least 2 years after completion of the study, if it is not submitted to a regulatory agency. Further, records are maintained at least 5 years following the date on which the study results are submitted to the regulatory agency. Once the study is completed, it is the responsibility of the study director to ensure that the study plan, final report, raw data, and related material are transferred to archives in a timely manner. The final report should include a statement indicating where all specimens, raw data, study plan, final report, and other related documentation are stored. Once data are transferred to the archives, the responsibility for it lies with management.

### Final Report

After the completion of the study, a final report must be prepared. It should be produced as a detailed scientific document and contain the study purpose, methods and materials used, the summary, and the conclusions from the study. Any deviation(s) or otherwise with the QA statement should be part of the final report (Table 18-3). The study director should sign and date the final report only after full satisfaction that the findings are complete, true, accurate, comply with GLP, and are valid.

Today, society is flooded with innumerable chemicals both for good and adverse effects. Much has been discussed about adequate testing and evaluation in a reasonable time. Safeguarding human health and the environment against deleterious effects received appreciable significance after the Earth Summit held

**TABLE 18-3 Content of the Final Report**

1. Identification of the study, the test item, and reference item
2. A descriptive title
3. Identification of the test item by code or name
4. Identification of the reference item by name
5. Characterization of the test item including purity, stability, and homogeneity
6. Name and address of the sponsor
7. Name and address of any test facilities and test sites involved
8. Name and address of the study director
9. Name and address of the principal investigator(s) and details, if applicable
10. Name and address of scientists contributing to the final report
11. Experimental starting and completion dates
12. A quality assurance statement listing the details (types of inspections, dates, phase(s), results, reporting date to management and study director)
13. Description of methods and materials used
14. A summary of results
15. All information and data required by the study plan
16. A presentation of the results, including calculations and determinations of statistical significance
17. An evaluation and discussion of the results and conclusions where appropriate
18. The location(s) where the study plan, samples of test and reference items, specimens, raw data, and the final report are to be stored
19. Signature of principal investigators and scientists involved in the study and dates
20. Signature by the study director and dates indicating acceptance of responsibility for the validity of the data; the extent of compliance with good laboratory practice

*Note*: Any corrections and additions to a final report should be in the form of amendments. Amendments should clearly specify the reason for the corrections or additions and should be signed and dated by the study director.
Reformatting of the final report to comply with the submission requirements of a national registration or regulatory authority does not constitute a correction, addition, or amendment to the final report.

in Rio de Janeiro, Brazil, in June 1992. Since this summit, society has been well supported with the mechanisms of regulatory agencies for chemical control.

Although rules and regulations have been available for some time, nearly 20 years ago, a private company engaged in toxicologic investigations of chemicals submitted false data without performing any animal studies. Regulatory bodies did not respond quickly to the problem. Much later, the USFDA introduced the GLP system to improve the credibility and validity of studies. Why the lack of fast response? It happened because of the lack of qualified, trained, and honest people. We have come a long way now. Industrial growth (along with disasters) have increased, but with urgent attention to human safety. A qualified and trained worker should properly identify a hazardous chemical at the very outset. In fact, this is the critical first step in chemical safety and waste management programs. Failure to identify regulated hazardous waste will result in mismanagement and could lead to environmental damage and human fatalities, in addition to civil penalties. Chemical waste is considered hazardous or toxic if it contains certain chemicals that have a propensity to migrate into the environment.

The proper attitude and skill comes from proper guidance. In short, there is a need for manpower development with the right qualifications, experience, and training to contain the spreading of disasters. Chemical safety requires human force for proper function. It is essential to understand the degree of contamination and the consequences thereof to function with credibility. Therefore, basic knowledge about chemicals, toxicity profile, and adherence to GLP in all workplaces is a must. This group of specialists, toxicologists, must have a multidisciplinary approach to chemical safety. As indicated earlier, specializations in toxicology include analytical toxicology, aquatic toxicology, biochemical toxicology, clinical toxicology, ecotoxicology, environmental toxicology, epidemiology, genetic toxicology, immunotoxicology, and regulatory toxicology, to name a few. The qualification and experience of these workers comes only after understanding the basics of toxicology and the behavior of different chemicals with biologic systems. Therefore, workplaces in chemical industries and laboratories should have competent, trained workers. The type of training, depending on the nature of workplace, should progress from the manager to the worker. A worker in a life science laboratory should know the proper handling of a laboratory rat, correct identification of the group under treatment, its feeding with test chemical or otherwise, and the involvement during termination of the study. More information may be found in the literature.[10–12,18–19] Similarly, a laboratory analyst must know the physicochemical properties of a chemical, its use, equipment, importance of the log book, proper labeling of containers, and waste disposal[17] (Table 18-4).

Laboratory or industry analyst is an important person in a study process. He or she contributes much to the generation of quality data. Therefore, it is necessary that he or she has proper knowledge and skill for the correct documentation of the work done in a laboratory or field.

**TABLE 18-4 Basic Skills for a Laboratory Worker**

1. Maintaining a good laboratory notebook.
2. Designing an experiment and understanding experimental protocol development.
3. Writing a research proposal, including the use of Medline and Internet search results.
4. Working with different volumes to prepare solutions and the dilutions (e.g., μl to ml pipettes, graduated cylinder versus Erlenmeyer flask).
5. Working with solid reagents, to weigh reagents, prepare chemical solutions, and labeling.
6. Proper handling and use of a pH meter and adjusting solutions to the proper pH by acid/base titration.
7. Using a spectrophotometer and understanding the basics of spectrophotometry.
8. Using microliter pipettor and applying concepts of accuracy and precision.
9. Preparing aliquot reagents from stock bottles.
10. Preparing serial dilution from stock solutions.
11. Washing and sterilizing glassware using proper procedures.

A GLP-oriented mindset among workers and students substantially reduces possible chemical hazards. Manpower should include: (1) toxicologists; (2) scientists with different specialization; (3) laboratory assistant; (4) laboratory technician for different units; (5) managers and supervisors; and (6) inspectors and field assistants. Persons with proper knowledge, skills, and training are able to perform duties well organized and with valid documentation. Prudent practice in handling chemicals in the laboratory, factory, and field, and proper disposal after use, is the answer to the growing problem of chemical pollution. Any industry interested in broadening its activities must have manpower qualified and trained to generate quality data—and quality data comes from practicing valid procedures and providing honest documentation. Unless the worker understands (by training or guidance) the importance of the basics of toxicology when handling chemicals or waste, the goals of chemical safety cannot be achieved. Most of the problems in industry have been due to improper handling and negligence, rather than poor science. This could be resolved with proper education and training in GLP systems.

As indicated earlier, GLP is essentially governed by several important guidelines, which are all aimed at the generation of quality data. These guidelines must be well followed and understood by students and workers when quality data on a test chemical or drug must be generated. OECD documents these guidelines under different headings or topics (*See* Appendixes 18-1 through 18-11). These guidelines help facilitate a workers, researchers, students, trainees, and professionals in industries to properly document and preserve the integrity and quality of the study conducted.

## CONCLUSION

Proper basic knowledge in science in association with GLP data-collection standards helps to achieve research integrity and success. It is expected that all workers, organizations, and industries adhere to GLP standards for study documentation, sample handling, data management, and reporting. Only the use of GLP and manpower together can enrich chemical safety for the improvement of human health and safety worldwide.

## REFERENCES

1. C.E. Paget, ed., *Good Laboratory Practice*, MTP Press, Ltd., Lancaster, UK, 1979.
2. C.E. Paget, and R. Thomson, *Standard Operating Procedures in Toxicology*, MTP Press, Ltd., Lancaster, UK, 1979.
3. World Health Organization, *Principles and Methods for Evaluating Toxicity of Chemicals, Part I, Environmental Health Criteria No. 6*, WHO, Geneva, Switzerland, 1978.
4. Environmental Protection Agency, *Good Laboratory Practice Regulations as promulgated in the Code of Federal Regulations*, (FDA) 21 CFR Part 58, September 4, 1987; EPA: 40 CFR Part 160, August 17, 1989.

5. OSHA Federal Register, Good Laboratory Practice Standards; Federal Insecticide, Fungicide and Rodenticide Act (FIFRA). *Final Rule, Federal Register*, No. 54: 34052-74, CFR Title 40, August 17, 1989.
6. Organization for Economic Cooperation and Development, *OECD Series on Principles of Good Laboratory Practice and Compliance Monitoring*, No. 8: 1998, Paris, France.
7. Environmental Protection Agency, *Good Laboratory Standards for Health Effects*, United States Federal Register, Vol. 54: No. 158, 34067–34074, August 17, 1989.
8. Environmental Protection Agency, *Good Laboratory Standards for Health Effects*, United States Federal Register, Vol. 44: No. 91, 27334–27375, May 1991.
9. Good Laboratory Practice, *GLP Consensus document*, The Role and Responsibilities of the Study Director in GLP studies, 1999.
10. National Institute of Occupational Safety and Health, *Quality Assurance and Laboratory Operations Manual*, Cincinnati, OH, 1987.
11. T.A. Ratliff, *The Laboratory Quality Assurance System: A Manual of Quality Procedures and Forms*, Van Nostrand Reinhold, New York, 1993.
12. American Society for Testing and Materials, *Standard Guide for Laboratory Accreditation Systems*, E994-84, Philadelphia, PA, 1984.
13. International Organization for Standardization, *ISO 9000: Quality management and quality assurance standards—Guidelines for selection and use*, ISO, UK, 1987.
14. H.M. Kanare, *Writing the Laboratory Notebook, American Chemical Society*, Washington, DC, **145**, 1985.
15. International Organization for Standardization, *ISO Guide 25: General requirements for the competence of calibration and testing laboratories*, ISO, UK, 1990.
16. Japanese Ministry of Agriculture, Forestry and Fisheries. *Agricultural Chemicals Regulation Law*, (Provision of Paragraph 2 of Article 2 of the Law), No. 82: 1948, Japan.
17. British Association of Research Quality, UK, 1999.
18. National Research Council, *Prudent Practices in the Laboratory: Handling and Disposal of Chemicals*, National Research Council, Canada, 1995.
19. P. Ostrosky-Wegman, and M.E. Gonsebatt, Environmental toxicants in developing countries. *Environ. Health Perspect.* **104**(suppl. 3):599, 1996.

## APPENDIX 18-1 PROTOCOL CHECKLIST

Study Title:

Study Director Name:

Project Title:

| Particulars | Checked (Yes/No) | Remarks |
|---|---|---|

**1. General**

- Study number
- Study title
- Study description
- Study purpose
- Sponsor address
- Testing facility name

**2. Test/Reference Article**

- Test/Control article
- Test article name
- Test article preparation
- Test article supplier
- Test article purity
- Test article stability
- Control article name
- Control article preparation
- Control article supplier
- Control article stability

**3. Test System**

- Name of the species/strain
- Gender: Age: Body weight range:
- Animals source
- Justification of selection of the species/strain

- Number of animals
- Quarantine procedure
- Housing and environment: Room temperature: Humidity: Air changes:
- Animal feed and water
- Source of diet and water
- Method of identifying individual animals (code details)

**4. Compliance**

- GLP compliance
- Data retention
- Specimen retention
- Date of arrival of animals
- Acclimatization period
- Clinical laboratory tests date(s)
- Necroscopy date(s)
- Study start date: Study complete date: Final report:
- Study director's signature: Date:

## APPENDIX 18-2 ANIMALS DOSING AND OBSERVATIONS

Study Title:

Study Director name:

Particulars Check (Yes/No)

**1. Dosing of the Test Substance**

- Route and method of administration (exposure)
- Reasons for route choice
- Preparation of dose form
- Concentration of test substance in carrier (vehicle)
- Assay of dosage form
- Identity of vehicle (if used)
- Amount of test substance required
- Frequency of administration (exposure)
- Method to determine absorption, (if any)

**2. Experimental Design**

- Group designation (control/experimental)
- Dose levels
- Interim sacrifice (group/number/animals)
- Terminal sacrifice (number of animals/each group/gender)

**3. Observation–In Life**

- Observation type
- Frequency of observation
- Number of animals
- Specimen and records maintained

- Statistical analysis
- Laboratory analysis (hematology, blood chemistry, urine analysis)

**4. Observations/Necropsy/Remarks**

- Animal weight/organ weight
- Details of moribund animals/number/group
- Animal number/group number/gross pathology
- Animal number/group number/histopathology
- Specimens and records maintained
- Statistical analysis

**5. Protocol**

- Archive location and period
- Archive in charge
- Protocol amendment procedures
- Retention of raw data, records, and specimens
- Quality assurance
- References
- Protocol approval

## APPENDIX 18-3 FACILITY INSPECTION REPORT

Study location:

Study date:

Inspector:

| Particulars | Check (Yes/No) | Remarks |
|---|---|---|

**1. Personnel**

- Is an organizational chart available?
- Are personnel curriculum vitae available?
- Are job descriptions available?
- Are adequate sanitation facilities available?
- Is protective equipment available?
- Is documentation of personnel GLP training available?

**2. Quality Assurance Unit**

- Is there an established QAU?
- Does the organizational chart indicate independent QAU?
- Is a master schedule available?
- Is the master schedule adequate with contents (viz., test substance; initiation date; name of the sponsor; study director name)?
- Does the QAU maintain protocols and written audit reports in one location?
- Is a QAU SOP available?
- Is a documentation SOP available?

**3. Facilities**

- Does the facility maintain an archive?

- Does the archive appear adequately secure?
- Does the archive appear accessible and orderly?
- Is an archive SOP available?
- Does the archive SOP describe responsibility and access?
- Is test substance storage adequate and secure?
- Is test substance storage temperature monitored?
- Is a test substance handling SOP available?
- Is there adequate clean storage?
- Is there adequate sample (freezer) storage?
- Is sample storage temperature monitored?
- Is a sample storage SOP available?

**4. Equipment**

- Is a balance calibration SOP available?
- Is a calibration log maintained for each balance and major piece of equipment?
- Are standard weights certified?
- Are equipment logs maintained and available?
- Are reagents and solutions labeled correctly with expiration dates?

**5. Animal Facilities**

- Is sanitation and pest control documented?

- Is there adequate veterinary care?
- Are cage sizes adequate?
- Are there quarantine facilities?
- Are SOPs in proper place?
- Is the feed storage area clean and adequate?
- Are clean and dirty cages separated?
- Are records maintained for cage washer?
- Are there records for animal room preparation and room logs?

## APPENDIX 18-4 METHOD DEVELOPMENT AND VALIDATION (ANALYTICAL CHEMISTRY AUDIT FORM)

Project Title:

Project Number:

Study Director:

Date of Inspection:

| Name of the Auditor and Unit: | Checked (Yes/No) | Reviewed (Yes/No) | Remarks |
|---|---|---|---|
| • Name of the inspected item | | | |
| • What is the method source? | | | |
| • Is it sensitive, reproducible, and specific for study? | | | |
| • Was method developed for matrices tested? | | | |
| • Were alterations made to method? | | | |
| • Was there method validation? | | | |
| • How was validation conducted? | | | |
| • Was validation over expected concentration range? | | | |
| • Was method validated for metabolites required by protocol? | | | |
| • Was method validated for all matrices specified in protocol? | | | |
| • How were detection limit and quantification limit defined and determined? Calculated correctly? | | | |
| • Are method validation data available for audit? | | | |

## APPENDIX 18-5 STABILITY OF TEST SUBSTANCE IN SAMPLE MATRIX

| Particulars | Checked (Yes/No) | Reviewed (Yes/No) | Remarks |
|---|---|---|---|
| • Inspected item | | | |
| • Stability of the test substance | | | |
| • Protocol requires stability tests | | | |
| • All matrices used for stability tests | | | |
| • Levels of fortification used | | | |
| • Time frame tests matched or unmatched storage period | | | |
| • Same methods followed for analysis of fortified and study samples | | | |
| • Sample storage conditions adequate and comparable with stability test data | | | |
| • Sample stability tests data available for audit and agree with findings | | | |
| • SOPs are available in respective places | | | |

## APPENDIX 18-6 ANALYTICAL REFERENCE STANDARDS

| Checked Particulars | Reviewed (Yes/No) | Remarks (Yes/No) |
|---|---|---|
| • Name of the inspected item: | | |
| • Name the source for analytical reference standards: | | |
| • Indicate the date of characterization: | | |
| • Was analysis of reference standards done at the facility: | | |
| • Stability of the substance demonstrated: | | |
| • Characterization of the substance demonstrated as per GLP: | | |
| • Details of labeling and storage to avoid degradation of the substance done | | |

## APPENDIX 18-7 STANDARD REFERENCE SOLUTIONS AND INSTRUMENT CALIBRATION

| Particulars | Checked (Yes/No) | Reviewed (Yes/No) | Remarks |
|---|---|---|---|
| • SOP for preparation and handling of stock and working solutions? | | | |
| • Data regarding preparation of stock solutions available? | | | |
| • How were solutions stored? Containers OK? Any evaporation? | | | |
| • Proper labeling to avoid mix-up? | | | |
| • Were solutions standardized? Log available? | | | |
| • How often were fresh standards prepared? | | | |
| • Did more than one person use standards? | | | |
| • Was detector response to standards stable? | | | |
| • Was there a trend in deviation of response or was it random? | | | |
| • Were study personnel aware of changes in detector response? | | | |
| • Was cause documented and remedial action taken? | | | |
| • Was quantification made from calibration curves or from single standards? | | | |
| • Was response linear and, if not, how was it addressed in calculations? | | | |
| • How often was calibration curve prepared? | | | |

## APPENDIX 18-8 SAMPLE PREPARATION, EXTRACTION, AND CLEAN UP

| Particulars | Checked (Yes/No) | Reviewed (Yes/No) | Remarks |
|---|---|---|---|
| • Are SOPs available at the workplace? | | | |
| • Who prepared and extracted the sample? | | | |
| • Were worksheets or notebooks used properly with dates and signatures? | | | |
| • Were balances calibrated? If so, how often? Are logs available? | | | |
| • Did raw data document storage conditions? | | | |
| • How much time between receipt of samples and analysis? | | | |
| • How much time between beginning of sample preparation and final measurement? | | | |
| • If delays occurred, were reasonable explanations offered? | | | |
| • Were delays caused by too few techniques? | | | |
| • Were delays caused by instrument of facility problems? | | | |
| • Was study director aware of delays? | | | |
| • Was stability data available that covered the delay periods? | | | |

## APPENDIX 18-9 SAMPLE ANALYSIS

| Particulars | Checked (Yes/No) | Reviewed (Yes/No) | Remarks |
|---|---|---|---|
| • Were there changes in procedure from SOP? | | | |
| • Or protocol? Were they documented? | | | |
| • Were instruments calibrated? Documentation available? | | | |
| • Was automated data collection used? Hard copies? Written SOPs for ADC? | | | |
| • SOP regarding the need for reanalysis of samples? | | | |
| • SOP for the number of significant figures to be reported? | | | |
| • SOP for rounding technique? | | | |
| • Were analytical reference standards analyzed concurrently and at intervals? | | | |
| • Calculations made from linearity curves or a single reference point? | | | |
| • Is it possible to reconstruct study from raw data? | | | |

## APPENDIX 18-10 QUALITY CONTROL PRACTICES

| Particulars | Checked (Yes/No) | Reviewed (Yes/No) | Remarks |
|---|---|---|---|
| • Were quality control practices used? | | | |
| • Was there periodic analysis of replicate samples to determine reproducibility of analytical results? | | | |
| • Was there analysis of controls to determine potential interference from external sources? | | | |
| • Was there analysis of reagent or blanks to determine interference or contamination from reagents or glassware? | | | |
| • Was there analysis of control samples fortified with known quantities of analyte? | | | |
| • Were QSUM or other control charts used? | | | |
| • Did study protocol or SOPs call for QC procedures? | | | |

## APPENDIX 18-11 QUALITY ASSURANCE CHECKLIST

Study Title:

Study Director Name:

Signature: Date:

| Particulars | OK | Not OK |
|---|---|---|
| • GLP compliance confirmed by QA and sponsor | | |
| • Protocol reviewed by QA for completeness | | |
| • Study director identified | | |
| • Protocol signed by study director | | |
| • Testing facility management identified | | |
| • Personnel files in place and contain CVs, training records | | |
| • Organizational chart available showing independent QA | | |
| • Facility inspection by QA completed | | |
| • All personnel trained in GLPs | | |
| • All personnel familiar with study protocol | | |
| • SOPs for operation, calibration of all equipment in place | | |
| • Files prepared for room logs, pathology reports, shipping records, manifests, chain-of-custody, telephone logs, etc. | | |
| • Protocol deviation form prepared | | |
| • Error correction form in lab notebooks | | |
| • Critical events identified | | |
| • QA scheduled for in-life inspections | | |
| • Protocol amendments authorized by study director | | |

- LIMS systems validated according to GLPs
- Archival system and location identified
- QA audit of raw data
- QA audit of final report
- Final report signed by study director
- All data and specimens archived

CHAPTER 19

# Safety Evaluation: Methods and Procedures

## INTRODUCTION

Increased use of chemicals and drugs in industries and their impact on the health of animals, humans, and the environment has caused concern. In recent years, international regulatory agencies have made it mandatory to generate data on the toxicity profiles and any adverse effects caused by chemicals to achieve chemical safety. It has been well established that many chemicals, when misused, produce adverse effects. These effects range from acute intoxication to severe poisoning to death. In prolonged and chronic exposure, these chemicals, produce effects that may involve vital organs, such as the lung, liver, kidney, or heart. Mass intoxications can affect thousands of people worldwide. The fact that these disasters (and individual cases of accidental or occupational chemical poisoning) happen is only one aspect of the problem. Protocols and guidelines required for conducting toxicologic tests will be discussed here. As described in Chapter 2, "Principles of Toxicity and Safety," three important study phases must be remembered, namely acute toxicity, subacute toxicity, and chronic toxicity. Several publications are available on the safety evaluation of chemicals, and this chapter is based on some of them.[1–4]

### Acute Oral Toxicity

Acute oral toxicity is the study of adverse effects occurring shortly after oral administration of a single chemical dose or multiple doses given to an animal within 24 hours. In the evaluation of chemical safety, determination of acute oral toxicity becomes important, since it normally forms the first study step. Acute toxicity is important in establishing dose regimen for subchronic and other studies, and may provide initial information on the mode of chemical toxic action, as well as a basis of for classification and labeling.

Groups of animals are exposed to the test chemical orally administered by gavage in one graduated dose per group. Subsequently, animals that die during

*Industrial Guide to Chemical and Drug Safety*, By T.S.S. Dikshith and Prakash V. Diwan
ISBN 0-471-23698-5 

the test are necropsied; at the end of the test, the surviving animals are sacrificed and necropsied.

***Test System (Animal)*** Several mammalian species are used for acute toxicity studies, but the rat is the preferred rodent species. The body weight variation in test animals should not exceed 20% of the mean weight. At least 10 rodents (5 females, 5 males) should be used at each dose level, and females should be nulliparous and nonpregnant.

It is necessary to maintain the housing and feeding conditions for the test system both of treated and untreated (control) groups should be uniform. The temperature of the experimental room should be 22 °C (±3 °C) and the relative humidity 30% to 70%. Test animals may be caged individually or grouped by gender. In the latter case, the number of animals per group must be limited and should allow for proper observation of each animal. Wherever lighting is artificial, the sequence should be 12 hours light and 12 hours dark. The test animals should be given laboratory diets with an unlimited supply of drinking water.

The acute toxicity profile chemical and drugs should be tested with sufficient doses, at least in three doses. Doses should be spaced appropriately to produce test groups with a range of toxic effects and mortality rates. The dose should produce a dose-response curve and should permit an acceptable determination of the test chemical or drug $LD_{50}$ value. Before exposure to chemicals or drugs, healthy young adult animals should be acclimated to laboratory conditions for at least 5 days. Test animals are randomized and assigned to treatment groups. Whenever necessary, test chemicals are dissolved or suspended in a suitable vehicle. An aqueous solution should be used as far as possible, followed by a solution in oil, (e.g., peanut oil or corn oil) and if necessary, other vehicles. For nonaqueous vehicles, toxic characteristics of the vehicle should be known or determined before commencement of the study.

The maximum volume of liquid that can be administered orally at one time depends on the size of the test animal. For instance, in rodents, the volume should not exceed 1 mL/100 g body weight. In cases of aqueous solutions, the concentration must be adjusted to ensure a constant volume at all dose levels. If a dose of at least 5,000 mg/kg body weight using the above procedure produces no chemical or drug-related mortality, a full study using three dose levels may not be necessary.

All test animals should be starved prior to the administration of test chemicals. For instance, for the rat, food should be withheld overnight; for other rodents with higher metabolic rates, a shorter period of starvation is appropriate. Following the period of starvation, animals should be weighed and the test chemical administered in a single dose, by gavage using a stomach tube or a suitable intubation cannula. If a single dose is not possible, the dose may be given in smaller fractions over a period not exceeding 24 hours. After administration of the chemical, the food may be withheld further for 3 to 4 hours. After chemical or drug dosing, animals are provided with food and water *ad libitum.*

After the administration of the test chemical, observations are recorded systematically. Individual data sheets should be maintained for each animal. The observations should be made for a minimum period of 14 days. However, the duration of observations should not be fixed rigidly. It should be determined by toxic reactions, rate of onset, and duration of recovery, and may be extended when necessary. The time at which toxicity signs appear and disappear and the time of death are important indications, especially if there is a tendency for death to be delayed. A careful clinical examination should be made at least once every day. Additional observations of test animals should be made daily with appropriate actions taken to minimize loss of animals in the study, (e.g., necropsy or refrigeration of animals found dead, isolation or sacrifice of weak or moribund animals).

Cage-side observation of test animals should include changes in the skin and fur, eyes, mucous membranes, respiratory, circulatory, autonomic and central nervous system (CNS), somatomotor activity, and behavior pattern. Particular attention should be directed to observations of tremors, convulsions, salivation, diarrhea, lethargy, sleep, and coma (Table 19-1). The time of death should be recorded as precisely as possible. Individual weight should be determined shortly before the test chemical is administered, weekly thereafter, and at death. Changes in weight should be calculated and recorded when survival exceeds 1 day. At the end of the test, surviving gross necropsy of all animals should record the nature of toxic effects, if any. All gross pathologic changes should be recorded. Microscopic examination of organs, showing evidence of gross pathology in animals surviving 24 hours or more, also should be included as useful information.

The acute oral $LD_{50}$ value should always be considered in light of observed toxic effects as well as any chemical- or drug-induced necropsy findings. The $LD_{50}$ value is a relatively coarse measurement, useful only as a reference value for classification and labeling purposes, and for expression of the lethal potential of a test chemical by ingestion. The test system (animal species) should be mentioned in each assay of acute toxicity and arrive at the $LD_{50}$ value obtained. Evaluation should include the relationship (if any) between the exposure of the animal to the test chemical and the incidence and severity of all abnormalities, including behavioral and clinical abnormalities, gross lesions, body weight changes, mortality effects, and other toxic effects attributable to the test chemical.

The results of experiments may be summarized in a tabular form. They may include the following information:

- Nature of the test chemical as solid/liquid/aerosal.
- Nature of the vehicle as distilled water/edible oil (used for the administration of the test chemical to animals).
- Particulars of the experimental animals such as species/strain/gender/age/body weight/source.
- Experimental conditions: Room temperature/humidity/lighting/food and water.
- Test chemical doses used for the study and the number of animals in each group.
- Date of commencement and termination of the experiment.

**TABLE 19-1 Diverse Effects Observed in Animals on Experimentation**

| Observation |
|---|
| Bizarre physical position |
| Bizarre position |
| Exploratory behavior |
| Aggressiveness toward species |
| Inactivity |
| Convulsions, spontaneous |
| Dyspnea |
| Sedation |
| Nystagmus |
| Cyanosis |
| Abnormal excreta |
| Salivation |
| Nasal discharge |
| Piloerection |
| Phonation |
| **Physical examination** |
| Altered muscle tone |
| Catatonia |
| Muscle tremors |
| Aggressiveness toward experimenter |
| Coma |
| Convulsions to touch |
| Alterations in cardiac rate and rhythm |
| Paralysis |
| Change in papillary size |
| Sensitivity to pain |
| Skin lesions |
| Corneal opacity |
| Placing reflexes |
| Righting reflexes |
| Grasping reflexes |
| Pinnal reflexes |
| Death |

- Reference of the statistical method used for the determination of the $LD_{50}$ value and 95% confidence interval for $LD_{50}$.
- Description of the clinical signs and symptoms of toxicity shown by each group of animals after treatment.

- Number of deaths and survivors in each group at the time of conclusion of the experiment.
- Details of growth increase or decrease, gross pathology, microscopy, hematology, enzyme chemistry, and urinanalysis.
- Evaluations made with particular reference to the toxicity rating of the test chemical.
- Any additional information relevant to the study.

The Joint Meeting of OECD decided to delete and abolish the $LD_{50}$ Acute Toxicity Test—OECD Test guideline 401. In fact, this is a major step in animal welfare. The decisions made are as follows: During a meeting in November 2001, the OECD Joint Meeting of the Chemicals Committee and Working Party on Chemicals, Pesticides and Biotechnology agreed that the $LD_{50}$ Draize test for acute oral toxicity, known as OECD Test Guideline 401, was subjected to criticism. Because it was heavily criticized by the animal welfare protectorate, the Joint Meeting decided to abolish and delete Test Guideline 401 from the OECD manual of internationally accepted test guidelines.

The delegates agreed that the three alternative methods now available for this disputed animal test can provide sufficient information to replace Guideline 401. Subsequent to this agreement, OECD Council adopted the decision to delete Guideline 401 by December 17, 2002. 401 Acute Oral Toxicity (Deleted Guideline, date of deletion: December 20 2002) Thus, OECD Test Guideline 401 will be deleted and the Acute Oral $LD_{50}$ Test will not be conducted using large number of laboratory animals of different species".[5]

## Acute Dermal Toxicity

Acute dermal toxicity is the study of adverse effects occurring within a short time of dermal application of a single-dose test chemical. In evaluating the safety of a chemical, determination of acute dermal toxicity is useful when exposure by the dermal route is likely and more predominant. It provides information on health hazards likely to arise from short-term exposure by the dermal route. Data from an acute dermal toxicity study may serve as a basis for chemical classification and labeling. It is an initial step in establishing a dose regimen in subchronic (and other) studies, and may provide information on dermal absorption as well as a chemical's mode of toxic action.

The test chemical is applied to the skin of groups of animals in graduated doses, such as one dose per group. Subsequently, observations of toxic effects and deaths are made. Animals that die during the test are necropsied. At the conclusion of the test, surviving animals are sacrificed and necropsied as necessary.

For evaluation of the dermal toxicity of a chemical or drug, the adult rat, rabbit, or guinea pig have been considered useful species. However, use of other species requires a justification. Animals with the following weight ranges are useful and facilitate the toxicity test: rat 200 to 300 g; rabbits 2 to 3 kg; guinea pig 350 to 450 g. Equal numbers of each gender with healthy intact skin are required for each dose level. At least 10 animals (5 female, 5 male) should be used at each dose level. Females should be nulliparous and nonpregnant. In the case of larger animals, (e.g., dog and monkeys), a smaller number may be used without sacrificing the statistical requirement to determine $LD_{50}$. Housing and feeding conditions for experimental animals should be uniform and as described for acute oral toxicity.

It is important to use different doses of a chemical or drug to evaluate chemical safety vis-à-vis skin toxicity. Dose levels to be tested for acute dermal toxicity in animals should be sufficient in number, (e.g., three or more and spaced appropriately to produce test groups with a range of toxic effects and mortality rates). The data should permit an acceptable determination of $LD_{50}$.

Before administration of the test substance, selected healthy, young adult animals are acclimated to laboratory conditions for at least 5 days. Before the test, animals are randomized and assigned to treating groups. Approximately 24 hours before the test, the fur on the skin should be clipped or shaved from the dorsal area of the trunk of the test animal. Care must be taken to avoid abrading the skin, which could alter the permeability of the test chemical through the skin. Approximately 10% of the body surface area should be prepared for application of the test chemical.

Solid forms of chemicals and drugs may be pulverized to a fine powder. The test chemical should be moistened sufficiently with water or, where necessary, with a suitable vehicle to ensure good contact with the skin. When a vehicle is used, the influence of the vehicle on penetration by the test chemical should be considered. Chemicals in solid form are generally used undiluted. When a test at one dose level of at least 2,000 mg/kg body weight, using the procedures described for this study, produces no compound-related mortality, a full study using three dose levels may not be necessary.

The test chemical or drug should uniformly cover an area (approximately 10% of the total body surface area). With highly toxic chemicals, the surface area covered may be less, but the area should be covered with a thin, uniform film of the test chemical.

The test chemical should be held in contact with the skin with a porous gauze dressing and nonirritating tape throughout the 24-hour exposure period. The test site should be covered to retain the gauge dressing and test chemical and ensure that the animal cannot ingest the test substance. A restrainer may be used for this purpose, but complete immobilization is not recommended.

At the end of the exposure period, the residual test chemical should be removed, using (if necessary) water or an appropriate solvent. The observation period for acute dermal toxicity in the animals should be at least 14 days. For clinical examination and pathology, the parameters tested for acute oral toxicity may be followed.

The acute dermal $LD_{50}$ value of the test chemical should always be considered in conjunction with the observed toxic effects and the necropsy findings. The experimental animal species should be mentioned in which the $LD_{50}$ value was obtained. An evaluation should include the relationships (if any) between the animals' exposure to the test chemical and the incidence and severity of all abnormalities, including behavioral and clinical abnormalities, gross lesions, body weight changes, effects on mortality, and other toxic effects attributable to the test chemical. The results of the acute dermal toxicity study may be summarized in a tabular form, as in the case of acute oral toxicity study.

### Acute Dermal Irritation/Corrosion Test

The acute dermal irritation is the study of reversible inflammatory changes in the skin of test animals following the application of a test chemical. Acute dermal corrosion is the study of irreversible tissue damages in the skin following the application of a test chemical. In the evaluation of toxic characteristics of a chemical, determination of the irritant or corrosive effects on mammal skin is an important study step. Information derived from this test indicates the existence of hazards likely to arise from skin exposure to the test chemical.

The chemical to be tested is applied in a single dose to the skin of several animals, each animal serving as its own control. The degree of irritation is recorded and scored at specified intervals, and further described to provide a complete evaluation of the chemical toxicity. The study duration should be long enough to fully evaluate the reversibility (or irreversibility) of the effects observed in the animals. The test chemical should be prepared as indicated for acute dermal toxicity test. However, strongly acidic or alkaline chemicals, (e.g., with a demonstrated pH of 2 or less or 11.5 or greater) need not be tested for dermal irritation, because of their predictable corrosive properties.

For the acute dermal irritation/corrosion study, the albino rabbit is considered as the preferred species, although several mammalian species may be used. At least three healthy adult animals should be used. Additional animals may be required to clarify equivocal responses. All experimental animals should be individually caged. Feeding and maintenance of animals should be uniform as described for acute oral toxicity. Approximately 24 hours before the test, the fur from the dorsal area of the trunk should be clipped or shaved. Care should be taken to avoid abrading the skin. Only animals with healthy, intact skin should be used for the study.

A dose of 0.5 mL of liquid or 0.5 g of solid or semisolid test chemical is applied to the site. Separate animals are not required for an untreated control group, since adjacent areas of untreated skin serve as controls for the test. The test chemical should be applied to a small area (approximately 6 $cm^2$) of skin and covered with a gauze patch, held in place with nonirritating tape. In the case of liquids or some pastes, it may be necessary to apply the test chemical to the guaze patch and apply that to the skin. The patch should be loosely held in contact with the skin by means of a suitable occlusive dressing for the duration of the exposure period. The animals' access to the patch and resultant ingestion of the test chemical by licking or by inhalation should be prevented.

In the normal course, the exposure duration for acute dermal irritation is 4 hours. However, longer exposures may be indicated under certain conditions based on the expected pattern of human use and exposure. At the end of the exposure period, the residual test chemical should be removed, wherever practicable, using water or an appropriate solvent, without altering the existing response or the integrity of the epidermis.

The duration of the observation period should not be fixed rigidly but should be long enough to fully evaluate the reversibility or irreversibility of the effects observed. However, under normal conditions, it need not exceed 14 days after the application of the test chemical. Animals should be examined for signs of erythema and edema and the responses scored at 30 to 60 minutes, then at 24, 28, and 72 hours after patch removal. In a majority of instances, albino rabbits are more sensitive than humans to a variety of environmental and irritant chemicals.

Dermal irritation scores should be evaluated in conjunction with the nature and reversibility of the responses observed. Individual scores do not represent an absolute standard for the irritant properties of a chemical, and they should be viewed as reference values that are only meaningful when supported by a full description and evaluation of the observations. The use of an occlusive dressing is a severe test, and the results are relevant to likely human exposure conditions.

***Results*** Extrapolation of the results of dermal irritant/corrosivity from animals to humans is valid only to a limited degree. The findings of similar test results in other animal species may give more weight to extrapolation of animal studies to humans. The results of the experiment may be presented in a tabular form incorporating all the aspects of the study, and as with acute oral toxicity.

### Sensitization Studies (Skin and Mucous Membranes)

Sensitization is a phenomenon that manifests itself by local effects on skin or mucous membranes. It differs from primary irritation in that it also may involve tissues remote from the application site. These reactions represent allergy to a chemical that is not obtained on first contact but that requires a number of prior exposures for development. The number exposures required may vary greatly, depending on the sensitization potential of the agent and the individual's susceptibility. The sensitized tissue exhibits a greatly increased capacity to react to subsequent exposure of the offending chemical. Thus, subsequent exposures may produce severe reactions. Humans are so heterogenous that few chemicals have been incriminated as sensitizers at one time or another. It is this heterogenicity that makes it difficult to set a rigid criterion for experimental tests, whereby a chemical should or should not be considered acceptable based on its sensitizing properties. Such a criterion of acceptability must rest not only on the incidence of reaction (total numbers reacting in an experimental group) but also on the seriousness of the reactions.

***Guinea-Pig Technique*** Although the guinea pig intracutaneous technique may not be entirely adequate (borderline cases of sensitization may be missed), it has been found useful to screen out compounds that are severe sensitizers. White male guinea pigs weighing 300 to 500 g and subsisting on a commercial rabbit-pellet ration supplemented with greens (kale or lettuce) are identified, and hair is removed from back and flanks by close clipping. A 0.1% solution or suspension of the test material in physiologic saline is injected intercutaneously, using a 26-gauge hypodermic needle. Injections are made every other day or three times weekly, until a total of 10 have been made. The 10 sensitizing injections are made at random in the area of the back and upper flanks. The area measures 3 or 4 $cm^2$. The retest injection is made in an area just below the region or sites of sensitizing injections. The first injection consists of 0.05 mL, and the remaining nine injections consist of 0.1 mL each. Two weeks after the tenth injection, a retest injection is made, using 0.05 mL of a freshly prepared solution or suspension as before. Twenty-four hours following injections, readings of the diameter, height, and color of reactions are taken. A comparison of the reaction following the test is made with the average of the reading taken after each of the original 10 injections. If the value for the retest is substantially higher than for the average of the 10 original readings, the substance can be considered to produce sensitization. The degree of sensitization is proportional to the increase in the final reading compared with the average of the readings following the 10 original doses.

***Sensitization Technique in Humans*** The guinea pig intracutaneous technique should be used to screen all chemicals. In the case of chemicals that may be only mildly sensitizing, or where the guinea pig method is unable to furnish unequivocal results, humans appear to be the only feasible test subject. Human tests should employ preferably 200 individuals (100 male, 100 female, constituting as wide an age group as possible). The test chemical of 0.5 mL (or 0.5 g, if solid) is applied by patch to an area (randomized) on the arms or back. The area of erythema and edema is measured. Edema is estimated by the evaluation of the skin with respect to the contour of the unaffected skin. The subjects are given a day's rest and then given a second application. This procedure is repeated until a series of 10 consecutive exposures have been made with the test chemical. After this series of 10 applications, the subjects are given 10 to 45 days of rest, after which a retest dose is applied in the same manner as one of the original 10 exposures. A comparison of the reactions observed from the 10 sensitization doses with the reaction following the retest doses permits an estimation of the degree of sensitization (if any) obtained.

Repeated chemical patch applications to both humans and animals reveal that certain chemicals at a given concentration do not produce primary irritation, but may elicit severe skin reactions after a number of exposures. Nevertheless, these reactions may not be considered incidents of sensitization, because after 10 to 14 days of rest, the skin recovers its original resistance to the chemical injury. Such chemicals are neither primary irritants nor sensitizers.

For lack of a better term, such reactions have been called *skinfatigue*. This term is a misnomer in that it does not imply a condition in which the skin has lost significant normal physiologic function. This phenomenon consists of a suitable change in which the skin no longer exhibits its original refractories or resistance to the continued or repeated action of an agent. A number of chlorinated compounds, especially chlorinated phenols, are prone to produce this reaction. Compounds eliciting this reaction may be detected by the repeated-patch technique described above.

***Use Versus Patch Tests*** The use test is a method frequently employed to estimate the potential sensitizing ability of a product. Such a test is usually conducted by distributing the product to a large test panel, (e.g., the inhabitants of one or more large cities). Should the product be found acceptable to the general public, (e.g., tolerated without many serious complaints of injury), the manufacturer may feel justified in substituting this test procedure. The use test does not afford information on the product's acceptability and safety for use, and it is generally considered unsatisfactory since it lacks proper medical or other supervision of the test subject. Because of the lack of supervision, it is difficult to confer validity to the results, such as the extent of individual use, the nature, number, and seriousness of adverse reactions that the test chemical may produce.

***Photosensitization*** Following ingestion or topical application, some agents may produce ill effects in individuals subsequently exposed to light — especially sunlight. Among such agents, are sulfonamides and chlorolpromazine. In the cosmetic field, certain essential oils and lipstick dyes may be offenders. Test procedures consist of exposing treated subjects to a source of ultraviolet (UV) irradiation, which is generally a UV lamp. At first, the amount of irradiation that constitutes a suberythemal dose in subjects (albino rabbits) is determined. After determining the adjustments necessary to produce the desired UV dosage, six subjects treated with the test chemical and three controls (untreated) are exposed once daily until a total of 10 exposures (2 weeks) are made. The degree of photosensitization (if any) is determined by comparing local effects on the skin or other treated surfaces with the treated and control (untreated) subjects.

### Acute Eye Irritation Test

Eye irritation is the production of reversible changes in the eye following the application of a test chemical to the anterior surface of the eye. Eye corrosion is the production of irreversible tissue damage in the eye following application of a test chemical to the anterior surface of the eye.

In the evaluation of a chemical's toxic characteristics, determination of the irritant/corrosive effects on the eyes of mammals is an important parameter. Data obtained from this test indicate hazards likely to arise from exposure of the eyes and associated mucous membranes to the toxic chemical under test.

***Test System*** Although a variety of animals have been used for acute eye irritation tests, the adult albino rabbit is the preferred species. The animals should be individually caged under standard animal husbandry and feeding conditions as described for acute oral toxicity. The chemical is applied in a single dose to one of the eyes of all the adult albino rabbits or animals; the untreated eye serves as the control. The degree of chemical irritation or corrosion is evaluated and scored at specific intervals. The study duration should be long enough to fully evaluate the reversibility or irreversibility of the effects. Both eyes of each experimental animal selected for testing should be examined 24 hours before the test. Animals showing eye irritation, ocular defects or preexisting corneal injury should not be used.

Chemicals that are strongly acidic or alkaline in nature (i.e., with a pH of 2 or less 11.5 or greater) need not be tested because of their probable corrosive properties. Chemicals that demonstrate definite corrosion or severe irritation in a dermal study need not be further tested for eye irritation. It may be presumed that such chemicals produce similarly severe effects in the eyes.

The following dose levels are recommended for testing different chemicals: for liquids, a dose of 0.1 mL is used; for solids, pastes, and particulate chemicals, the amount used should have a volume of 0.1 mL or a weight of not more than 100 mg (the weight must always be recorded). If the material is solid or granular, it should be ground to a fine powder. The volume of particulates should be measured after gently compacting them, as by tapping the measuring container.

To test a chemical contained in a pressurized aerosol container, the eye should be held open and the test chemical administered in a single burst of about 1 second from a distance of 10 cm directly in front of the eye. The dose may be estimated by weighing the container before and after use. Care should be taken not to damage the eye. Pump sprays should not be used but instead, the liquid should be expelled and 0.1 mL collected and instillated into the eye as described for liquids.

The test chemical should be placed in the conjunctival sac of one eye of each animal after gently pulling the lower lid away from the eyeball. The lids are then gently held together for about 1 second to prevent loss of the material. The other eye, which remains untreated, serves as a control. If it is considered that the chemical could cause extreme pain, a local anaesthetic may be used prior to installation of the test chemical. The type and concentration of the local anesthetic should be carefully selected to ensure that no significant differences in reaction to the test chemical would result from its use. The control eye should be similarly anesthetized.

The eyes of the test animals should not be washed for 24 hours following installation of the test chemical. At 24 hours, a washout may be used if considered appropriate. For some chemicals shown irritating by this test, additional tests may be performed, using rabbits with eyes washed soon after installation of the chemical. In these cases, it is recommended that six rabbits be used. Four seconds after instillation of the test chemical, the eyes of three rabbits are washed and 30 seconds after installation, the eyes of the other three rabbits are washed. For both groups, the eyes are washed for 5 minutes using a volume and velocity of flow, which will not cause injury to the eyes. The eyes should be examined at 1,

24, 48 and 72 hours. If there is no evidence of irritation at 72 hours, the study may be terminated. Extended observations may be necessary if there is persistent corneal involvement or other ocular irritation to determine the progress of the lesions and their reversibility or irreversibility. In addition to the observations made in the cornea, iris, and conjunctiva, any other lesions should be recorded and reported. The grades of ocular reaction should be recorded at each examination.

Examination of reactions can be facilitated by use of a binocular loupe hand slit-lamp, biomicroscope, or other suitable device. After recording the observations at 24 hours, the eyes of any or all rabbits may be further examined with the aid of a fluorescent lamp. The duration of the observation period should not be fixed rigidly but should be long enough to fully evaluate the reversibility or irreversibility of the effects observed. This period, however, need not exceed 21 days after instillation of the test chemical.

Ocular irritation scores should be evaluated in conjunction with the nature and reversibility (or otherwise) of the response observed. The individual scores do not represent an absolute standard for the irritant properties of a chemical. They should be viewed as reference values and only meaningful when supported by a full description and evaluation of observations.

### Irritation of Mucous Membrane Test

In tests designed to evaluate the primary irritation of mucous membranes, the tissues of choice (wherever possible) should be the specific mucosal to which the chemical is applied. Differences in histology, absorptive capacity, and pH ranges of the mouth, eye, genitourinary, and rectal mucosa are great enough that results obtained on one mucosa may not apply for the other.

***Vaginal Preparation*** The rabbit has been found a suitable animal for most tests on mucous membranes, except the vaginal mucosa. The urethral meatus in the female rabbit is situated approximately halfway in the vaginal vault, thus test materials must be introduced in the upper half of the vaginal tract, or there is risk of the test material (especially if irritating) being voided by frequent urination. To obviate this difficulty whenever a rabbit is used as the animal, the test material must be introduced in the upper half of the vaginal vault. Another complicating factor in the use of rabbits in vaginal studies is that the probing or handling necessary to introduce the test chemical intravaginally frequently induces vascular changes in this organ, thus complicating the interpretation of resulting reactions. The dog is preferred species for vaginal tests, since the above difficulties observed in the rabbit are not encountered. The formulation is applied as per label directions, with some applications exceeding prescribed doses to demonstrate a margin of safety.

***Penile Mucosa*** In the case of the penile mucosa, the preparation is applied on the specific region and approximately 0.2 mL of the preparation is required. The erythema and edema are evaluated based on scores that total four severe injuries. If the preparation is found sufficiently irritating to cause necrosis and sloughing of the penile mucosa, the agent is reapplied in sufficient dilution so that no necrosis or sloughing results. Reactions are read at 1, 2, 24, and 48 hours after application. The study results may be presented in tabular form covering all aspects of the test.

### Acute Inhalation Toxicity

Acute inhalation toxicity is the study of adverse effects caused by a chemical following a single uninterrupted exposure by inhalation over a short period of time (i.e., 24 hours or less to a chemical capable of being inhaled). The $LC_{50}$ value is expressed as the weight of test chemical per standard volume of air (mg/L) or as parts per million (ppm).

In the evaluation of toxic characteristics of an inhalable environmental chemical (e.g., carbon monoxide, volatile chemical, or aerosol/particulate), determination of acute inhalation toxicity is an initial study step. It provides information on health hazards likely to arise from short-term exposure by inhalation. Data from an acute test help to establish a dose regimen in subchronic (and other) studies, and may provide additional information on a chemical's mode of toxic action.

Several groups of experimental animals are exposed to the test chemical in graduated concentrations for a defined period, one concentration used per group. Whenever a vehicle is used to generate an appropriate concentration of the chemical in the atmosphere, a vehicle control group should be used. Animals that die during the test are necropsied. At the conclusion of the test, surviving animals are sacrificed and necropsied as necessary.

***Test System*** Among mammals, the rat is the preferred species for acute inhalation toxicity tests. The weight variation in animals or between test groups should not exceed ±20% of the mean weight. At least 10 animals (5 females, 5 males) for each concentration level should be used. Females should be nulliparous and nonpregnant. For experimental animals, housing and feeding conditions should be uniform and according to the conditions already described for an acute oral toxicity study.

Healthy, young adult animals are acclimated to laboratory conditions for at least 5 days prior to the test, before they are randomized and assigned to groups. Wherever necessary, a suitable vehicle may be added to the test chemical to help generate appropriate concentration of the chemical in the atmosphere.

The animals should be tested with inhalation equipment designed to sustain a dynamic air flow of 12 to 15 air changes per hour, and to ensure an adequate oxygen content of 19% and an evenly distributed atmosphere. Whenever a chamber is used, its design should minimize crowding of test animals and maximize their exposure to the chemical. As a general rule, to ensure stability of the chamber atmosphere, the total volume of the test animals should not exceed 5% of the test chamber volume. Alternatively, oronasal, head only, or whole-body individual exposure chambers may be used.

Exposure concentrations or chemical test dosages should be sufficient in number (at least three) and should be spaced appropriately. This should produce a range of toxic effects and mortality rates in animals. It also should help to draw a concentration mortality curve and permit an acceptable determination of $LC_{50}$ values. In the case of potentially explosive test chemicals, care should be taken to avoid generation-explosive concentrations. However, for purposes of establishing suitable exposure concentration, a trial test is recommended.

The duration of exposure to the chemical should be at least 4 hours after the attainment of equilibrium of the chamber concentration. Other durations may be needed to meet specific requirements. If during a test at an exposure concentration of 5 mg/L (actual concentration of respirable chemical) for 4 hours (or where this is not possible due to physical or chemical properties of the test chemical), the maximum attainable concentration using the procedures described for this study produces no compound-related mortality, a full study using three dose levels may not be necessary.

Shortly before exposure to the test chemical, the animals are weighed and exposed to different test chemical concentrations in the designated chamber for 4 hours. The temperature at which the test is performed should be maintained at 22 °C (±2 °C). Ideally, the relative humidity should be maintained between 30% and 70%, but in certain cases (e.g., aerosols), even this may not be practicable. Food should be withheld during exposure. Water also may be withheld in certain cases. The observation period for acute inhalation toxicity in animals should be at least 14 days. For clinical examination and pathology, the parameters listed for acute oral toxicity may be followed. The following conditions should be monitored during the experiment using a standard inhalation chamber.

- The rate of air flow (preferably continuously).
- During exposure, the actual concentrations of the test chemical should be as constant as practicable.
- During the development of the generating system, particle size analysis should be performed to establish the stability of aerosol concentrations.

***Results*** The $LC_{50}$ value of the test chemical should be considered in conjunction with the observed toxic effects and necropsy findings. The animal in which the $LC_{50}$ value was obtained should be mentioned. An evaluation should include the relationship (if any) between exposure of animals to the test chemical and the incidence and severity of all abnormalities, including behavioral and clinical abnormalities, gross lesions, body weight changes, mortality, and other toxic effects.

Determination of $LC_{50}$ value provides an estimate of a chemical's relative toxicity by inhalation. Extrapolation of the $LC_{50}$ values and the results of acute toxicity studies from animals to humans is valid only to limited degrees. The results of acute inhalation toxicity tests may be summarized in tabular form. They should include all important aspects of the test, as in the case of acute oral toxicity studies.

## SHORT-TERM TOXICITY STUDIES

### Repeated-Dose, Oral Toxicity (14- and 28-Day Study)

In the evaluation of toxic characteristics of any environmental chemical, determination of oral toxicity using repeated doses may be performed after obtaining initial acute toxicity data. This provides information on possible adverse effects that may arise from repeated exposures to the test chemical over a limited period of time. Although there are major similarities in the 28-day and 14-day oral toxicity studies on rodents, the main difference lies in the time over which the dose is administered and the extent of the clinical and pathologic investigations that might be necessary for the shorter period of test.

***Principles of the Test Method*** Experimental animals will be administered graduated oral doses of the test chemical, daily for a period of 28 or 14 days. Throughout the period of exposure, the animals are observed for clinical signs of poisoning and mortality. Animals that die during the period of test are necropsied. The surviving animals are sacrificed and necropsied at the conclusion of the study.

***Description of the Test Procedure*** Healthy, young adult animals are acclimated to laboratory conditions for at least 5 days prior to commencement of the test. The animals are then randomized and divided into different groups. Depending on the requirements, the test chemical is administered to groups of animals by gavage, capsules, drinking water, or mixed with the diet. It is, however, important to remember that the manner of administration to different groups should always be identical during the entire period of experimentation. For example, if a vehicle or other additive is used to facilitate dosing, it should not influence absorption of the test chemical or trigger toxic effects in animals.

***Test System*** The housing and feeding conditions of animals used for repeated oral toxicity studies should be according to standard animal husbandry conditions already described for the acute oral toxicity study. It is necessary to use at least 10 animals (5 female, 5 male) for each of the test-dose regimem. Female animals should be nulliparous and nonpregnant. If interim sacrifices are planned, the number should be more and a satellite group of 10 animals (5 per gender) may be treated with a high-dose level of the chemical for 14 or 28 days. This should be associated with observations for reversibility, persistence, or delayed occurrence of toxicity symptoms for 14 days posttreatment.

For the study, a minimum of three dose levels and an appropriate control should be used. Except for the treatment of the test chemical, animals in the control group should be handled in an identical manner to the rest of the groups. The highest-dose level should result in toxic effects in animals but not cause fatalities of a large number of animals, which would interfere with meaningful evaluation of data. Similarly, the lowest dose of the chemical should not produce evidence of toxicity. In contrast, intermediate doses should produce minimal observable toxic effects. Whenever more than one intermediate dose is used, the dose levels of the test chemical should be so spaced as to have gradation of toxic effects. Care should be taken in the selection of doses, where the animal fatalities would be low, particularly in the control and intermediate dose groups, for purposes of correct evaluation of results.

It is important to ensure that whenever a chemical is administered in the diet, the quantity used does not interfere with the normal food intake of the animal. Further, when administered in diet, a constant dietary concentration (ppm) or a constant dose level in terms of the test animal's body weight may be used.

Whenever the chemical is administered by gavage, it should be given at a specific time of the day, and the dose level should be adjusted at a specific interval (e.g., every week or every fortnight). This method maintains a constant dose level in terms of animal body weight. When a repeated-dose study is used as a preliminary to a long-term study, a similar diet should be used.

If one dose level of at least 1,000 mg/kg body weight produces no observable toxic effects and if toxicity would not be expected based on the data from structurally related chemicals, a full study using three dose level may not be essential. After exposure to the test chemical, observations should be made for 14 or 28 days. Although the selection of this period is not rigid, it should be determined based on the toxicologic profile and related reactions characteristic of the test chemical.

Animals that comprise the satellite group for follow-up studies should be kept under observation for 14 days without treatment. This procedure detects persistence of toxic effects (if any), as well as manner of recovery. A careful clinical examination should be made every day. Action should be taken to minimize the loss of experimental animals, (e.g., identifying animals that are moribund, doing necropsy, refrigeration of animals that are found dead, sacrificing weak or moribund animals).

All animals should be dosed with the test chemical throughout the week, (i.e., for 7 days). This should continue for 14 or 28 days as scheduled. Signs of toxicity should be recorded when they become evident. This record should include, the time of onset, degree or intensity, and duration. Besides these, cage-side observations should be recorded, such as, changes in fur, eye or mucous membrane color respiration rate, circulation, autonomous and CNS, somatomotor activity, and behavioral patterns.

Measurements of food and water intake should be made every week, and animals should be weighed every week to record growth pattern. Regular observation is necessary to ensure that animals are not lost due to cannibalism, autolysins of tissues, or misplacement. At the end of the study, surviving animals in the nonsatellite treatment group are sacrificed.

***Clinical Examination*** A detailed clinical examination should include the following: hematology including hematocrit, hemoglobin concentration, erythrocyte count, total and differential leukocyte counts, measure of clotting potential, (e.g., clotting time, prothrombin time, thromboplastin time, platelet count). These evaluations should be performed at the end of the test period.

Similarly, clinical biochemistry determinations should be performed at the end of the test period. These should include blood parameters as well as liver and kidney function tests. The selection of specific biochemical tests are influenced by the chemical's mode of action. However, following are some of the most essential tests or determinations: calcium, phosphorus, sodium, potassium, fasting glucose (with period of fasting appropriate to the species), serum glutamic pyruvic transaminase (SGPT), serum glutamic oxaloacetic transaminase (SGOT), ornithine decarboxylase, gamma glutamyl transpeptidase, urea, nitrogen, albumin, blood creatine, total bilirubin, and total serum protein.

For adequate toxicologic evaluation, determinations should be made on lipids, hormones, acid/base balance, methemoglobin, and cholinesterase activity. In other words, additional clinical biochemistry may be resorted to wherever necessary to arrive at meaningful conclusions in association with the observed effects. Urinalysis, although not required on a routine basis, should be performed whenever there is an indication based on observed toxicity.

***Pathology*** Animals of different groups should be subjected to full gross necropsy, (e.g., examination of the external body surface, all orifices, the cranial, thoracic and abdominal cavities and their contents). It is also important to record the weight of liver, kidneys, testes, and adrenals as soon as possible and after dissection to avoid drying. All tissues and organs required for histopathologic examination should be preserved in a suitable medium, (e.g., neutral formal or bruin's fluid) to retain the cellular integrity. The most common tissues normally preserved are the liver, kidney, spleen, brain, gonads, heart, adrenals, lung, and other target organs that show gross lesions or changes in size.

The histologic examination should be performed on preserved organs tissues of the high-dose group and the control group. These examinations may be extended to other dose groups wherever necessary to investigate tissue damage as observed in the high-dose group. Animals of the satellite group should be examined histologically, with particular emphasis on organs and tissues that show adverse effects in other treatment groups.

The findings of a repeated-dose oral toxicity study should be considered in terms of observed toxic effects, necropsy, and histopathologic observations. The evaluation of the data will include the relationship between the test chemical dose and the presence or absence of abnormalities, the incidence and severity of such abnormalities (if any) behavioral and clinical changes, and gross lesions identified in target organs. In addition, it is important to record body weight changes, percent mortality, and any other general or specific toxic effects noticed among the animals. A properly conducted 14- or 28-day study will provide information on the effects of repeated-dose test chemicals. The study also can indicate the need for further long-term studies and provide useful clues for selecting dose levels whenever a long-term study is undertaken. Wherever the study demonstrates absence of any toxic effects of the test chemical in animals, further investigation to establish absorption and bioavailability should be carefully considered.

## Subchronic Oral Toxicity (90-Day Study)

The purpose of subchronic oral toxicity is to determine adverse effects of test chemicals after daily oral administration for 90 days to a group of animals. In evaluating the toxic characteristics of a test chemical, subchronic oral toxicity may be determined after obtaining initial information on the acute oral toxicity of the chemical. Subchronic oral toxicity studies provide information on possible adverse effects likely to arise from repeated oral exposures to the test chemical over time. The study also will provide information on the toxicity to target organs, the possibilities of accumulation of the chemical in the body, and an estimate of a no-effect level of exposure. All these are needed when selecting the dose schedule for performing chronic oral toxicity studies and establishing safety criteria for human exposure.

## Test System

Although a variety of species may be of use, the albino rat is preferred as the test animal for subchronic oral toxicity studies. Commonly used laboratory strains of young, healthy animals should be employed. Dosing of the test chemical should begin as soon as possible after weaning, ideally before the animals are 3 weeks old. Animals more than 8 weeks old are not suitable for the study.

The weight variation of animals at the commencement of the study should not exceed ±20% of the mean weight. It is also important that wherever a subchronic oral toxicity study is conducted as a preliminary to a long-term study, the same species and strain should be used in all studies. At least 20 animals (1 female, 10 male) should be used for each test dose. In view of the importance of the subchronic oral toxicity study, use of more animals would be advantageous. Females should be nulliparous and nonpregnant.

In the subchronic toxicity study, if interim sacrifices are planned, the number of animals should be increased so that there is no shortage of animals at the termination of the study; this ensures meaningful conclusions about the test chemical's toxicity. In addition, a satellite group of 20 animals (10 of each gender) may be treated with the high dose of the test chemical for 90 days. These animals should be observed for signs of adverse effects, (e.g., reversibility, persistence, delayed occurrence of toxic effects). The observation period should last for a posttreatment period of appropriate length not less than 28 days.

The housing and feeding conditions for animals should be according to the standard animal husbandry conditions as described earlier. Wherever group caging is followed, not more than five animals should be housed in a single cage. Healthy, young adult animals are acclimated to laboratory conditions for at least 5 days before commencement of the test. All animals are randomized and assigned to different groups. The test chemical may be administered to animals in different ways, (e.g., mixed with the diet, in capsules, by gavage, mixed in drinking water). However, it is important to remember that all animals should be dosed by the same method during the entire period of study. Whenever a vehicle or other additive is used to facilitate dosing, it should not interfere with absorption of the test chemical or produce adverse effects in animals.

In a subchronic oral toxicity study, the effect of the test chemical should be studied in at least three dose levels along with a control group. For treatment with a test chemical, control group animals should be handled in the same manner as those of treated groups. The highest-dose level of the test chemical should result in toxic effects but should not produce an incidence of fatality that might interfere with the statistical and meaningful evaluation. Similarly, the lowest dose of the test chemical should not produce any evidence of toxicity among the animals. Whenever one finds a usable estimation of human exposure the lowest test dose should exceed the selected dose. Intermediate dose levels of the test chemical should produce minimal observable adverse effects. If more than one intermediate dose is used, the level should be spaced so as to produce gradation toxic effects. It is important that in the low, intermediate, and control groups the incidence of fatalities should be as low as possible, permitting for a statically valid interpretation of test data.

For a chemical of low toxicity, it is important to ensure that when it is administered in the diet, the quantities involved do not interfere with normal nutrition of the animals. Whenever the chemical is mixed with the diet, either a constant dietary concentration (ppm) or a constant dose level in terms of the body weight (mg/kg) may be used.

When a test chemical is administered by gavage, the dose should be given at the same specified time each day. Furthermore, the test dose should be adjusted at regular intervals (e.g., weekly or biweekly), to maintain constant dose levels in terms of the changing body weight of the animal. Also, where a subchronic oral toxicity study is used as a preliminary to a long-term study, a similar dietary regimen should be used for both studies.

The observation period for the subchronic oral toxicity should be at least 90 days. Animals in a satellite group scheduled for follow-up observations should be kept for a further period of 28 days without treatment to detect recovery from, or persistence of, toxic effects. Careful clinical examination should be made at least once each day. Additional observations should be made and appropriate actions taken to minimize loss of animals; these actions include necropsy or refrigeration of those found dead, or isolation, or sacrifice of weak or moribund animals. The animals are dosed with the test chemical ideally 7 days per week for 90 days. However, based primarily on practical considerations, dosing by gavage or capsules 5 days per week is acceptable.

***Clinical Examination*** Signs of toxicity should be recorded when they are observed, including the time of onset, degree, and duration. Changes in skin and fur, eyes and mucous membranes, respiratory, circulatory, autonomic and CNS, somatomotor activity, and behavior pattern should be recorded. Measurements should be made weekly of food consumption when the test chemical is administered in the drinking water, and animals should be weighed weekly.

Regular observation is necessary to ensure that animals are not lost in the study due to cannibalism, autolysis of tissues, or misplacement. At the end of the

90-day period, all surviving animals are sacrificed. Any moribund animal should be removed and sacrificed when discovered.

An ophthalmologic examination should be performed, using an ophthalmoscope or equivalent suitable equipment, prior to administration and at the termination of the study, preferably in all animals but at least in the high dose and control groups. If changes in the eyes are detected, all animals should be examined. Hematology, including hematocrit, hemoglobin concentration, erythrocyte count, total and differential leukocyte count, and a measure of clotting potential (e.g., clotting time, prothrombin time, thromboplastin time, platelet count) should be investigated at the end of the test period.

Clinical biochemistry of blood should be performed out at the end of the test period, namely 90 days. The following are other test areas considered appropriate to all studies: electrolyte balance, carbohydrate metabolism, and liver and kidney function. The selection of specific tests will be influenced by observations on the chemical's mode of action. Suggested determinations are calcium, phosphorus, chloride, sodium, potassium, and fasting glucose (with period of testing appropriate to the species). SGPT, SGOT, ornithine decarboxylase, gamma glutamyl transpeptidase, urea nitrogen, albumin, blood creatinine, total bilirubin, and total serum protein measurements. Other determinations that may be necessary for an adequate toxicologic evaluation include lipid analysis, hormones, acid/base balance, methemoglobin, and cholinesterase activity. Additional clinical biochemistry may be employed whenever necessary to extend the investigation of observed effects. Urinalysis is not required on a routine basis, only when there is an indication based on expected or observed toxicity.

***Gross Necropsy*** All animals should be subjected to a full gross necropsy, including examination of the external body surface, all orifices, and the cranial, thoracic, and abdominal cavities and their contents. The liver, kidneys, adrenal, and testes must be weighed wet (i.e., as soon as possible after dissection) to avoid drying. The following organs and tissues should be preserved in a suitable medium for possible future histopathologic examinations: all gross lesions, brain including sections of medulla/pons, cerebellar cortex and pituitary, thyroid/parathyroid, thymus, trachea, lungs, heart, aorta, salivary glands, liver, spleen, kidneys, adrenals, pancreas, gonads, accessory genital organs, gallbladder (if present), esophagus, stomach, duodenum, jejunum, ileum, cecum, colon, rectum, urinary bladder, representative lymph node (female mammary gland*), thigh musculature*, peripheral nerve, eyes*, sternum with bone marrow*, femur including articular surface*, spinal cord at three levels: cervical, midthoracic, and lumbar*, and exorbital lachrymal glands*. The tissues with asterisks need only be examined if indicated by signs of toxicity or target organ involvement.

***Histopathology*** Full histopathology should be performed on organs and tissues of all test animals in the control and high-dose groups. All gross lesions should be examined. Target organs in other dose groups should be examined. Where rats are used, lungs in the low- and intermediate-dose groups should be subjected to histopathologic examination to determine evidence of infection, since this provides a convenient assessment of the state of health of animals. Further histopathologic examination may not be required on animals in these groups, but must always be performed in organs that show evidence of lesions in the high-dose group. When a satellite group is used, histopathology should be performed on tissues and organs identified as affecting other treated groups. Wherever one finds that histopathologic baseline data are inadequate, consideration should be given to the determination of hematological and clinical biochemistry parameters before commencement of treatment.

Subchronic oral toxicity study results should be evaluated in conjunction with preceding study findings and considered in terms of the toxic effects, necropsy, and histopathologic findings. Evaluation should include the relationship between the test chemical dose, the incidence and severity of abnormalities, including behavioral and clinical abnormalities, gross lesions, identified target organs, body weight change effects on mortality, and other general or specific toxic effects. A properly conducted subchronic test should provide satisfactory estimation of a noneffect level of the test chemical. In any study that demonstrates the absence of toxic effects, further investigation should be considered to establish absorption and bioavailability of the test chemical.

Subchronic and toxicity study results may be summarized in tabular form. In addition to the aspects indicated under acute oral toxicity, details of food and water intake, body growth rate, clinical symptoms of toxicity and mortality rate, after different test chemical doses may be included. Results also should include dose-related changes in hematology, clinical enzymes, and macroscopic and microscopic changes found in the vital organs of animals.

### Repeated-Dose Dermal Toxicity Test (21- and 28-Day Study)

In the evaluation of a chemical's toxic characteristics, the determination of subchronic dermal toxicity may be performed after initial information on toxicity has been obtained via acute dermal testing. The subchronic study provides information on health hazards likely to arise from repeated exposure by the dermal route over a limited period (21 or 28 days). The test chemical is applied daily to the skin in graduated doses to several groups of animals, one dose or group, for 21 or 28 days. During the period of application, animals are observed daily to detect signs of toxicity. Animals that die during the test are necropsied. At the conclusion of the test, the surviving animals are sacrificed and necropsied.

***Test System*** The adult rat, rabbit, or guinea pig have been considered useful for repeated dermal toxicity studies. Other species may be used if justified. The body weight for different species should range as follows: rats 200 to 300 g; rabbits 2 to 3 kg; guinea pig 350 to 450 gm.

At least 10 animals (5 male, 5 female) with healthy skin should be used at each dose level. Females should be nulliparous and nonpregnant. If interim sampling or sacrifices are planned, the number should be increased by the number scheduled to be sacrificed before completion of the study. In addition, a satellite group of 10 animals (5 per gender) may be treated with the high-dose level of the test chemical for 21 to 28 days. This group should be observed for reversibility, persistence, or delayed occurrence of toxic effects for 14 days posttreatment. Animals should be caged individually. Housing and feeding should be set up according to the standard conditions identified for acute oral toxicity.

The dose levels for repeated dermal toxicity studies in animals should have at least three concentrations, including an appropriate vehicle control. Except for treatment with the test chemical, the control group should be handled in a manner identical to the test group. The highest-dose level should not produce evidence of toxicity attributable to the test chemical.

Where there is usable estimation of human exposure, the lowest level should exceed this. Ideally, the intermediate-dose level(s) should produce minimal observable toxic effects. If more than one intermediate dose is used, the dose levels should be spaced to produce gradation of toxic effects. In the low and intermediate groups and in the controls, the incidence of fatalities should be low to permit a meaningful evaluation of the results. If application of the test chemical produces severe irritation, the concentration may be reduced. This reduction may result in a subsequent reduction in, or absence of, other toxic effects observed at the high-dose level. If the skin shows severe damage, it may be necessary to terminate the study and undertake a new study at lower concentrations.

Healthy, young adult animals are acclimated to laboratory conditions for at least 5 days prior to the test. Before the test, animals are randomized and assigned to treatment and control groups. Shortly before testing, fur is clipped from the dorsal area of the trunk. Repeat clipping or shaving is usually needed at weekly intervals. When clipping or shaving the fur, care must be taken to avoid abrading the skin, which could alter its permeability, unless a requirement for abraded skin is part of the test design. No less than 10% of the body surface area should be clear for application of the test chemical. The weight of the animal should be considered when deciding on the area to be cleared and the dimensions of the covering.

Solid chemicals may be pulverized to a fine powder. The test chemical should be moistened sufficiently with water or other suitable vehicle to ensure good contact with the skin. When a vehicle is used, the influence of the vehicle on skin penetration should be considered. Liquid chemicals are generally used without dilution.

If one dose level of at least 1,000 mg/kg body weight (expected human exposure may indicate the need for a higher-dose level), using the procedures described for this study, produces no observable toxic effects, and if toxicity would not be expected based on data from structurally related compounds, a full study using three dose levels may not be necessary.

The animals are treated with the test chemical ideally for at least 6 hours per day on a 7-day basis, for 21 or 28 days. However, based primarily on practical considerations, application on a 5-day-per-week basis is considered acceptable. Animals in a satellite group scheduled for follow-up observations should be kept for a further 14 days without treatment to detect recovery from, or persistence of, toxic effects.

The test chemical should be applied uniformly over an area approximately 10% of the total body surface area. With highly toxic chemicals, the surface area covered may be less. As much of the area as possible should be covered with a thin, uniform film of the chemical. Between applications, the test chemical is held in contact with the skin with a porous gauze dressing and nonirritating tape. The tape site should be further covered in a suitable manner to retain the gauze dressing and test substance, and ensure that the animals cannot ingest the test chemical. Restrainers may be used to prevent ingestion of the test chemical, but complete immobilization of the animal is not advisable.

A careful clinical examination of all animals should be made from Day 0 through Day 28. Additional observations should be made daily with appropriate actions taken to minimize loss of animals in the study, (e.g., by necropsy, by refrigeration of the animals found dead, by isolation or sacrifice of weak or moribund animals).

The signs of toxicity produced by the test chemical should be recorded. Observations should include the time of onset, degree, and duration of toxic signs and symptoms. Cage-side observations should include, but not be limited to, changes in skin and fur, eyes and mucous membranes, and respiratory, circulatory, autonomic, CNS, and somatomotor activity.

Measurements of food consumption should be made weekly, and the animals should be weighed weekly. Regular observation is necessary to ensure that animals are not lost from the study due to cannibalism, autolysis of tissues, or misplacement. At the end of the study, all survivors in the nonsatellite treatment groups are sacrificed. Moribund animals should be removed and sacrificed when discovered. The clinical examinations should be made on all animals at the end of test period. These examinations are discussed in detail below.

***Hematology*** This should include hematocrit, hemoglobin, erythrocyte count, total and differential leukocyte count, and a measure of clotting potential (e.g., clotting time, prothrombin time, thromboplastin time, platelet count).

***Clinical Biochemistry*** Blood parameters of liver and kidney functions are appropriate. The selection of specific tests will be influenced by observations on the chemical's mode of action. The following determinations are suggested: calcium, phosphorus, chloride, sodium, potassium, fasting glucose (with period of fasting appropriate to the species), SGPT, SGOT, ornithine decarboxylase, gamma glutamyl transpeptidase, urea, nitrogen, albumin, blood creatine, total bilirubin, and total serum protein measurements. Other determinations that may be necessary for an adequate toxicologic evaluation include analysis of lipids, hormones, acid/base balance, methemoglobin, and cholinesterase activity.

***Pathology*** All animals in the study should be subjected to a full gross necropsy that includes examination of the external surface of the body, all orifices, and the cranial, thoracic, and abdominal cavities and their contents. The liver, kidneys, adrenals, and tests must be weighed wet (as soon as possible after dissection) to avoid drying. The following organs and tissues should be preserved in suitable medium for future histopathologic examination: normal and treated skin, liver, kidney, and target organs, (i.e., those organs showing gross lesions or changes in size).

Histologic examination should be performed on preserved organs and tissues of the high-dose and control group. These examinations may be extended to animals of other dosage groups (if considered necessary) to investigate the changes. All observed results, quantitative and incidental, should be evaluated by an appropriate statistical method.

Evaluation should include the relationship between the test chemical dose and the incidence and severity of abnormalities, behavioral and clinical abnormalities, gross lesions, identified target organs, body weight changes, effect on mortality, and other general or specified toxic effects attributable to the test chemical.

A properly conducted 21- or 28-day study will provide information on the effects of repeated dermal application of a test chemical and can indicate the need for further studies (long-term). The study also can provide information on the selection of dose levels for long-term studies. The results of the repeated-dose dermal toxicity test may be presented in tabular form, including all salient aspects.

### Subchronic Dermal Toxicity (90-Day Study)

Subchronic dermal toxicity is the study of adverse effects occurring as a result of the repeated daily dermal application of a test chemical to animals for a part (not exceeding 10%) of the life span. In the evaluation of a chemical's toxic characteristics, the determination of subchronic dermal toxicity may be performed after initial information on toxicity has been obtained by acute testing. This study provides information on health hazards likely to arise from repeated exposure via the dermal route over a limited period of time.

***Test System*** The adult rat, rabbit, or guinea pig must be used for subchronic dermal toxicity test. The following body weight ranges for the test animal are recommended: rat 200 to 300 g; rabbits 2 to 3 kg; guinea pig 350 to 450 g. The remaining conditions for the selection of test animals may be the same as those in subchronic oral toxicity study. At least 20 animals (10 female, 10 male) with healthy skin should be used at each dose level. Females should be nulliparous and nonpregnant. If interim sacrifices are planned, the number should be increased by the number scheduled to be sacrificed before completion of the study. A satellite group of 20 animals (10 per gender) may be treated with the high-dose level of the test chemical for 90 days and observed for reversibility, persistence, or delayed occurrence of toxic effects for a post-treatment of appropriate length, not less than 28 days.

The housing and feeding of experimental animals should be according to standard animal husbandry conditions described for acute oral toxicity study.

Animals should be caged in groups by gender or individually. For group caging, no more than five animals should be housed per cage. Healthy, young adult animals are acclimated to laboratory conditions for at least 5 days prior to the test. Before the test, animals are randomized and assigned to treatment and control groups. Shortly before testing, fur is clipped from the dorsal area of the trunk. Shaving may be employed, but it should be performed approximately 24 hours before the test. Repeated clipping or shaving is usually needed at weekly intervals. When clipping or shaving the fur, care must be taken to avoid abrading the skin, which could alter the permeability of the test chemical. No less than 10% of the body surface area should be clear for application of the test chemical. The weight of the animal should be considered when deciding on the area to be cleared and dimensions of the covering. Solid test chemicals, may be pulverized into fine powder, if appropriate. Test chemicals should be moistened sufficiently with water or (when necessary) with a suitable vehicle to ensure good contact with the skin surface. When a vehicle is used, the influence of the vehicle on skin penetration by the test chemical should be considered. However, liquid test chemicals may be used.

At least three dose levels with a control and (where appropriate) a vehicle control should be used in a subchronic dermal toxicity study. Expect for treatment with the test chemical, the control group should be handled in a manner identical to the test group. The highest-dose level of the test chemical should result in toxic effects but not produce fatalities, which would prevent a meaningful evaluation of the results. The lowest-dose level of the test chemical should not produce evidence of toxicity. Where there is a usable estimation of human exposure, the lowest level should exceed this. Ideally, the intermediate-dose level(s) of the chemicals should produce minimal observable toxic effects. If more than one intermediate dose is used, the dose levels should be spaced to produce gradation of toxic effects. In the low and intermediate groups and in the controls, the incidence of fatalities should be low to permit a meaningful evaluation of results.

If application of the test chemical produces severe skin irritation, the concentration should be reduced, although this may result in a reduction in, or absence of, other toxic effects at the high-dose level. However, if the skin has been badly damaged early in the study, it may be necessary to terminate the study and undertake a new study at lower concentrations of the chemical.

If one dose level of at least 1,000 mg/kg body weight (expected human exposure may indicate the need for a higher dose level), using the procedures described for this study, produces no observable toxic effects and if toxicity would not be expected based on data from structurally related compounds, a full study using three dose levels may not be necessary. A careful clinical examination should be made daily with appropriate actions taken to minimize the loss of animals, (e.g., by necropsy, refrigeration of animals found dead, or isolation and sacrifice of weak or moribund animals).

Animals are treated with the test chemicals, ideally for at least 6 hours per day on a 7-day-per-week basis, for a period of 90 days. However, on the basis of practical considerations, application on a 5-day-per-week basis is considered acceptable. Animals in a satellite group scheduled for follow-up observations should be kept for at least 28 days without treatment to detect recovery from, or persistence of, toxic effects of the test chemical. The remaining procedure for subchronic dermal toxicity is similar to the one described for repeated dermal toxicity (28 days). Signs of toxicity should be recorded when they are observed during the period of experimentation. Signs include time of onset of toxicity, and the degree and duration of toxic symptoms, as described for subchronic oral toxicity study.

***Clinical Examination and Pathology*** The clinical examination of the test animals may be made as described for subchronic oral toxicity. These, include ophthalmologic examination, hematology, and clinical biochemistry. Gross necropsy and histopathology of all vital organs may be recorded as described for subchronic oral toxicity study. All observed results, quantitative and incidental, should be evaluated by an appropriate statistical method.

The findings of the subchronic dermal toxicity study should be evaluated in conjunction with previous study findings and considered in terms of the observed toxic effects, necropsy, and histopathologic findings. Evaluation will include the relationship between the test chemical dose and the incidence and severity of abnormalities, including behavioral and clinical abnormalities, gross lesions, identified target organs, body weight changes, effects on mortality, and any other general or specific toxic effects. A properly conducted subchronic dermal toxicity study should provide satisfactory estimation of a noneffect level of the test chemical. The results of the study may be presented in tabular form, as with subchronic oral toxicity test.

### Repeated Inhalation Toxicity (14- and 28-Day Study)

For evaluating the toxic characteristics of an inhalable material (e.g., gas, volatile chemical, or aerosol/particulate material), animals are subjected to repeated exposures after initial information on material toxicity has been obtained by acute testing. It provides information on health hazards likely to arise from repeated exposure via the inhalation route over a limited period of time. Hazards of inhaled chemicals are influenced by inherent toxicity and physical factors, such as volatility and particulate size.

There is similarity in the conduct of a 28-day or 14-day repeated-dose inhalation study. This similarity allows the use of one guideline to cover both test durations. The main differences, however, lie in the time over which dosing takes place and the extent of clinical and pathologic investigations, which might be appropriate for shorter test duration.

Several groups of animals are exposed daily to the test chemical for a defined period in graduated concentrations (one concentration per group) for a period of 28 days. When a vehicle is used to help generate an appropriate concentration of the test chemical in the atmosphere, a vehicle control group should be used.

During the exposure period, animals are observed daily to detect signs of toxicity. Animals that die during the test are necropsied, and animals that have survived at the conclusion of the test are sacrificed and necropsied.

***Test System*** To conduct a repeated-dose inhalation toxicity study, different species may be used. However, this guideline is intended primarily for use with rodents. When a rodent is required, the preferred species is the rat. Commonly used laboratory strains of young, healthy animals should be employed. At the commencement of the study, the weight variation of animals should not exceed ±20% of the mean body weight. When a repeated-dose inhalation study is conducted as preliminary to a long-term study, the same species should be used in both studies.

At least 10 animals (5 females, 5 males) should be used for each test group. Females should be nulliparous and nonpregnant. If interim animal sampling/sacrifice are planned, the number should be increased by the number scheduled to be sacrificed before the completion of the study. In addition, a satellite group of 10 animals (5 per gender) may be treated with a high concentration of test chemical for 28 or 14 days. The animals should be observed for reversibility, persistence, or delayed occurrence of toxic effects of the test chemical for 14 days post-treatment. The housing and feeding conditions (before and after exposure to chemicals) for the test animals should be uniform. The same standard conditions may be followed as described for acute oral toxicity.

Healthy, young adult animals are acclimated to laboratory conditions for at least 5 days prior to the test. Before the test, animals are randomized and assigned to the required groups. When necessary, a suitable vehicle may be added to the test chemical to help generate an appropriate concentration of the chemical in the atmosphere. If a vehicle is used, it should not influence absorption of the test chemical or produce toxic effects in test animals.

The particulars for the exposure of animals in the inhalation chamber are similar to those described for acute inhalation toxicity. A dynamic inhalation system with a suitable analytical concentration control system should be used. The rate of airflow should be adjusted to ensure that conditions throughout the equipment are essentially the same during the period of exposure.

For repeated inhalation toxicity, at least three concentrations of the test chemical with a control and (when appropriate) a vehicle control (corresponding to the concentration of vehicle at the highest exposure level) should be used. Except for exposure to the test chemical, the control group should be handled in a manner identical to the test group. The highest concentration of test chemical should result in toxic effects but not produce an incidence of fatalities, which would prevent a meaningful evaluation of results. The lowest concentration should not produce evidence of toxicity. When there is a usable estimation of human exposure, the lowest concentration of test chemical should exceed this.

Ideally, the intermediate concentration(s) of test chemical should produce minimal observable toxic effects. If more than one intermediate concentrations is used, the concentrations should be selected and spaced to produce gradation of toxic effects. In the low and intermediate groups and in the controls, the incidence of fatalities should be low to permit meaningful evaluation of the results. The duration of daily exposure to the test chemical should be 6 hours after equilibrium of the chamber concentrations. Other durations may be used to meet specific requirements. In the case of potentially explosive test chemicals, care should be taken to avoid generating explosive concentrations.

The animals are exposed to the test chemical ideally on 7-day basis for a period of 28 or 14 days. However, based primarily on practical consideration, exposure on a 5-day-per-week basis is considered acceptable. Animals in a satellite group scheduled for follow-up observations should be kept for a further 14 days without treatment to detect recovery from, or persistence of, toxic effects of the test chemical.

***Physical Measurement*** In a repeated-dose inhalation toxicity study, monitoring of the following conditions should be performed during the experiment:

- The rate of air flow should be monitored continuously.
- During the exposure period, actual concentrations of the test chemical should be maintained constant as far as practicable.
- During the development of the generating system, particle size analysis should be performed to establish the stability of aerosol concentrations.
- During the exposure period, analysis should be conducted as often as necessary to determine the consistency of particle size distribution.
- Temperature and humidity of the exposure chamber should be maintained constant as much as possible and monitored continuously.

The clinical examination of all animals should be made at least once each day. The parameters listed for acute oral toxicity may be followed. Observations should be made and recorded systematically, and individual records should be maintained for each animal, including the time of onset of toxicity, and the degree and duration of toxic effects. Cage-side observations should include, but not be limited to, changes in the skin and fur, eyes and mucous membranes, respiratory, circulatory, autonomic, and CNS, somatomotor activity, and behavioral pattern. Weekly measurements should be made of food consumption, and the animals should be weighed weekly. Regular observations of the animals are necessary in these studies. At the end of the study period, all survivors in nonsatellite treatment groups are sacrificed. Moribund animals should be removed and sacrificed when discovered.

***Hematology*** This observation should include hematocrit, hemoglobin, erythrocyte count, total and differential leukocyte count, a measure of clotting potential (e.g., clotting time, prothrombin time, thromboplastin time, platelet count) should be investigated at the end of the test period.

***Clinical Biochemistry*** The blood should be tested at the end of the study. The selection of specific tests depends on the observations on the mode of action of the test chemical. However, the following determinations are suggested: calcium, phosphorus, chloride, sodium, potassium, fasting glucose (with a period of fasting appropriate to the species), SGPT, SGOT, ornithine decarboxylase, gamma glutamyl transpeptidase, urea, nitrogen, albumin, blood creatine, total bilirubin, and total serum protein measurements. Other determinations, which may be necessary for an adequate toxicologic evaluation, include lipid analysis, hormones, acid/base balance, methemoglobin, and cholinesterase activity. Additional clinical biochemistry may be employed when necessary to extend the investigation of observed effects. Urinalysis is not required on a routine basis, only when there is an indication based on expected or observed toxicity. Histologic and clinical biochemistry parameters should be considered before commencing the dosing.

***Pathology*** All animals should be subjected to a full gross necropsy, which includes examination of the external body surface all orifices, and the cranial, thoracic, and abdominal cavities, and their contents. The liver, kidneys, adrenals, and tests must be weighed wet (as soon as possible after dissection) to avoid drying. The following organs and tissues should be preserved in a suitable medium for possible future histopathologic examination: lungs, which should be removed intact, weighed, and treated with a suitable fixative to ensure that structure is maintained (perfusion with the fixative is considered an effective procedure), liver, kidney, spleen, adrenals, heart, and other target organs, such as, those showing gross lesions or changes in size.

Histologic examination should be performed on the preserved organs and tissues of the high-dose group and control group. These examinations may be extended to animals to other concentrations or groups, if necessary, to investigate changes observed in the high-concentration group with emphasis on organs and tissues identified as affected. All observed results, quantitative and incidental, should be evaluated by an appropriate statistical method.

The findings of a repeated-dose inhalation study should be considered in terms of observed toxic effects, necropsy, and histopathologic findings. The evaluation shall include the relationship between the test chemical concentration, the duration of exposure, the incidence and severity of abnormalities, gross lesions, identified target organs, body weight changes, effects on mortality, and other general or specific toxic effects. A properly conducted 28- or 14-day study should provide information on the effects of repeated-inhalation exposure and indicate the need for further long-term studies. It also can provide information on the selection of test chemical concentrations for longer-term studies.

### Subchronic Inhalation Toxicity (90-Day Study)

Subchronic inhalation toxicity is the study of adverse effects that follow repeated daily exposure via inhalation to environmental chemicals for a part (not exceeding 10%) of the life span of the species.

In the evaluation of the toxic characteristics of a gas, volatile chemical, aerosol, or particulate matter, determination of subchronic inhalation toxicity may be performed after initial toxicity information has been obtained via acute testing. It provides information on health hazards likely to arise from repeated exposure by the inhalation route over a limited period of time. Hazards of inhaled chemicals are influenced by the inherent toxicity and by physical factors such as volatility and particle size. The housing and feeding conditions, before and after exposure to environmental chemicals, should be planned according to the standard animal husbandry conditions described for acute oral toxicity study.

***Test System*** At least 20 animals (10 female, 10 male) should be used for each test group. Females should be nulliparous and nonpregnant. If interim animal sacrifices are planned, the number should be increased by the number scheduled to be sacrificed before the completion of study. In addition, a satellite group of 20 animals (10 per gender) may be treated with the high-concentration level for 90 days and observed for reversibility, persistence, or delayed occurrence of toxic effects for a posttreatment of appropriate length, not less than 28 days.

Healthy, young adult animals — preferably rats — are acclimated to laboratory conditions for at least 5 days prior to the test. Before the test, animals are randomized and assigned to the required groups. When necessary, a suitable vehicle may be added to the test chemical in the atmosphere. If a vehicle is used, it should not influence the absorption of the test chemical or produce toxic effects in test animals.

The subchronic inhalation toxicity should be studied with at least three concentrations of the test chemical with a control (and where appropriate, a vehicle control corresponding to the concentration of the vehicle at the highest exposure level). The remaining details of exposure are similar to those described for acute oral and acute inhalation toxicity. The duration of daily exposure should be 6 hours after equilibrium of the chamber concentrations. Other durations may be used to meet specific requirements.

A careful clinical examination should be made at least once each day. Additional observations should be made daily with appropriate actions to minimize loss of animals to the study. This could be done through necropsy, refrigeration of animals found dead, or by isolation or sacrifice of weak or moribund animals. Further details of clinical examination, gross necropsy, histopathology, and statistical evaluation are similar to those described for subacute oral toxicity.

The findings of a subchronic inhalation toxicity study should be evaluated in conjunction with previous study findings and considered in terms of the observed toxic effects, necropsy, and histopathologic findings. The evaluation should include the relationship between the test chemical concentration, the duration of exposure, the incidence of severity of abnormalities, gross lesions, identified target organs, body weight changes, effects on mortality, and other general or specific toxic effects. A properly conducted 90-day subchronic inhalation toxicity study should provide a satisfactory estimation of a no-effect level of the test chemical.

The results of the subchronic inhalation toxicity study may be summarized in tabular form. These results may be based on the same lines as subchronic oral toxicity test in addition to experimental conditions used for the inhalation toxicity test.

## CHRONIC TOXICITY STUDIES

Objectives of the chronic toxicity study are to characterize the profile of a chemical in any mammalian species, following prolonged and repeated exposures, through oral, dermal, or inhalation route. In this respect, the chemical toxicity should be interpreted broadly to include any change from the normal. The duration of a chronic toxicity study for effects other than neoplasia is still widely debated. Chronic toxicity testing in animals requires exposure to the test chemical by appropriate route and at an appropriate dosage for much of the test animal's lifespan, if not for the entire life. Chronic toxicity tests assess potential toxicity from long-term exposures at low levels.

The application of guidelines should help to generate data on which to identify the majority of chronic effects and determine dose response relationships. Ideally, the design and conduct of chronic toxicity studies should allow for the detection of general toxicity including neurological, physiologic, biochemical, and hematologic effects and exposure-related morphologic (pathology) effects.

Prior to the initiation of chronic toxicity studies, there should be characterization of test chemical. Information on chemical identity and structure can sometimes be used in an analysis based on structure activity relationships to indicate biological or toxicologic activity. The physical and chemical characteristics of the test chemical provide important information for the selection of administration routes, study design, and handling and storage of the test chemical.

Composition of the test chemical, including major impurities, should be known prior to initiating the study. Relevant physicochemical properties, including stability of the test chemical, should be known prior to the initiation of a chronic toxicity study. The development of an analytical method for qualitative and quantitative determination of the test chemical (including major impurities when possible in the dosing medium and biologic material) should precede the initiation of long-term studies.

### Test System

Historically, it has been recommended that chronic toxicity testing be performed with two mammalian species, one a rodent and another a nonrodent. The rat has normally been the rodent of choice. In a majority of cases, rats and mice have been traditionally used. This is because of their lifespan, size, and cost which are well-suited for testing relatively large numbers of chemicals in a manageable time period. Further, years of chemicals testing have yielded extensive experience and information on the biologic characteristics of these animals.

Of the nonrodents, dogs and primates have been used due to their larger size, the ease in performing clinical and biochemical examinations on these species, and their general availability. It should be noted that availability of nonrodents for research purposes may be a problem internationally. The lack of results with a nonrodent may impose a serious reduction in test results of important effects that might be encountered in humans.

Such a dichotomy cannot be resolved in a generic manner. Although it may be advisable that chronic effect results be obtained from both a rodent and nonrodent, the selection of appropriate species for chronic toxicity tests should be based upon practical reasons and the results of previously conducted studies. In some cases, testing with a single species may provide sufficient data for assessing test chemical hazards. Strains of test animals should be well characterized for commonly found diseases and their resistance, and should be free from congenital defects.

For chronic toxicity studies, animals of both genders should be used. The study should be performed on young, healthy laboratory animals. For rodents, this should be during the rapid growth phase and as soon as possible after weaning and acclimatization. A sufficient number of animals should be used so that at the end of the study, enough animals in every group are available for thorough biologic evaluation. Adequate randomization procedures should be used for the proper allocation of test and control group animals. For rodents, each dose group and concurrent control group should contain at least 20 animals of each gender. For nonrodents, a minimum of 4 of each gender is recommended. If interim animal sacrifices are to be included, the number per group should be increased accordingly.

### Animal Care, Diet, and Water Supply

Stringent control of environment conditions and proper animal care techniques are mandatory for meaningful results. As part of such control, access to animal facilities should be monitored to prevent excessive traffic and other disturbances. Factors such as housing conditions, intercurrent disease, drug therapy, impurities in diet, air, water, and bedding can significantly influence the outcome of experiments.

The control of intercurrent infectious diseases or parasites is facilitated if rodents are bred and maintained in conditions free from specific pathogenic organisms. Bedding used in long-term studies should be sterilized. Animals should be housed in quiet, well-ventilated rooms, with controlled lighting, temperature, and humidity. Experiments should not be initiated until animals have been allowed a period of acclimation to the environment. Neither conditions nor animals from outside sources should be placed in tests without an adequate period of quarantine.

Housing of more than one animal species in a room should be avoided. Unless there is little chance of the inadvertent cross-exposure of animals to the test chemical, only one chemical should be tested in each room. The same consideration should be given to the housing of control animals in the same room as test animals. If control animals are housed in different quarters, additional problems in data evaluation may arise.

Cages, racks, and other equipment must be capable of regular, easy cleaning. The use of disinfectants and pesticides should be avoided, particularly where they may come in contact with the animals, since such biologically active compounds may affect the study results. More detailed information regarding animal husbandry management and procedures can be found in the scientific literature, animal welfare publications, and official documents of the GLP.

The diet should meet all nutritional requirements of the species tested and should be free from impurities that might influence the outcome of the test. Dietary contaminants and levels of various nutrients have been shown to alter physiologic processes of animals. Rodents should be fed and watered ad libitum with food replaced at least weekly. When a nonrodent such as the dog is used, it should be fed daily. At present, three types of diets are used: conventional (standard), synthetic, and various open-formula diets. Of these, the first two are more widely used in carcinogenicity bioassays. Whichever diet is chosen, suppliers must ascertain by periodic monitoring the nutrient quality and the contaminant level in the basal diet. The researchers should know the effect of the dietary regimen on metabolism and animal longevity.

Special attention should be paid to the dietary composition when the test material itself is a nutrient, (e.g., an industrially treated protein or starch, single-cell protein, irradiated food products). This is because some products are usually incorporated into the diet at levels as high as 20% to 60% at the expense of corresponding regular and normal nutrients. Examples are, modified versus unmodified starch and single-cell protein versus soybean meal.

Variations in the use pattern of industrial and agricultural chemicals throughout the world preclude standardization by international organizations such as OECD. Despite this fact, common dietary constituents, which are known to influence toxicity are antioxidants, unsaturated fatty acids, and selenium. These must be present in interfering concentrations. The potential impact of several common dietary contaminants on chronic toxicity assessment therefore, necessitates that special attention be given to their presence. In this respect, substances of concern include pesticide residues, chlorinated and polycyclic aromatic hydrocarbons, estrogens, heavy metals, nitrosamines, and mycotoxins.

In addition, the testing laboratory for both nutrients and unintentional contaminants, including carcinogens, may perform periodic analysis of the basal diet. The results of such analysis should be retained and included in the final report on each chemical. When the test chemical is administered in water or food, stability tests are essential. Properly conducted stability and homogeneity tests, prior to the chronic study, should be used to establish the frequency of diet preparation and monitoring required. When diets are sterilized, the effects of such procedures on the test chemical and dietary constituents should be known. Appropriate adjustments to nutrient levels should be performed. The effect of chemical sterilants, (e.g., ethylene oxide) on the bioassay should be ascertained.

During chronic studies, investigators should be aware of potential contaminants in the water used. Although water approved for human consumption is generally satisfactory, the investigator should ascertain the data available on the components in the water supply.

### Test Conditions

***Dose Levels and Frequency of Exposure*** In chronic toxicity studies, it is desirable to have a dose-response relationship as well as a no-observed-toxic-effect level (NOEL). Therefore, at least three dose levels should be used in addition to the concurrent control group. The highest-dose level should elicit some signs of toxicity without causing excessive lethality. The low dose should produce signs of toxicity. For a diet mixture, the highest concentration should not exceed 5% with the exception of nutrients.

Frequency of exposure is normally daily, but may vary according to the route chosen. If the chemical is administered in the drinking water or mixed in the diet, it should be continuously available. The frequency of administration may be adjusted according to the toxicokinetic profile of the test chemical.

***Controls*** A concurrent control group, identical in every respect to the exposed groups except for exposure to the test chemical, should be used. In special circumstances (e.g., in inhalation studies) involving aerosols or the use of an emulsifier of uncharacterized biologic activity in oral studies, a concurrent negative control group should be used. Whenever a negative control group is used, the animals of this group should be treated in the same manner as all the other test group, except that this control group should not be exposed to the test chemicals or any vehicles.

***Route of Administration*** There are three main routes of exposure for an environmental chemical to enter the body of an animal. These are the oral, dermal, or inhalation routes. The choice of administration route for a chemical depends on the physical and chemical characteristics of the test chemical and the form typifying exposure in humans. In general, the frequency of exposure may vary according to the administration route chosen and should be adjusted according to the toxicokinetic profile of the test chemical, if available.

### Oral Toxicity Studies

The oral route has been considered the third most common means through which several environmental chemicals enter the body of animals and humans. The gastrointestinal (GI) tract in animals may be regarded as a tube going through the body, starting at the mouth, through oral cavity, winding through the intestine, and ending at the anus. Despite the compactness of the system, the contents of the GI tract are essentially exterior to the body fluids. Because of this, any chemical that enters the GI tract could produce an effect only on the surface of the mucosal cells that line the tract, unless absorption of the chemical from the GI tract also takes place. A number of environmental chemicals, when administered orally, produce a systemic effect on the animal only after absorption has occurred from the oral cavity or the GI tract. The first site to which the orally administered chemicals are effectively transferred is the stomach, or the rumen, in animals that have such organs.

In the stomach, the chemicals contact other already-existing materials (stomach contents), secretions, (e.g., pepsin, gastric lipase, rennin), and hydrochloric acid. The effect of pH in the stomach, and the influence of pH on the ionization of weak organic acids and bases, plays a major role in the bioavailability of the chemical within the organism.

Following oral administration, absorption of chemicals from the gut necessarily involves translation of the chemical either to the lymphatic system or portal circulation. Those environmental chemicals, which enter the portal circulation, are transported directly to the liver.

The oral route is the preferred exposure, provided that the test chemical is not absorbed from the GI tract. Animals must receive the test chemical in their diet, dissolved in drinking water, or given by gavage or capsule for the length of time specified in the section on the study duration.

Ideally, daily dosing on a 7-day-per-week basis should be used because dosing in gavages or as capsules on a 5-day-per-week basis may permit recovery or toxicity withdrawal in the nondosing period, thus affecting the result and evaluation. However, based primarily on practical considerations, dosing on a 5-day-per-week basis is considered acceptable.

### Dermal Toxicity Studies

The most common and easiest manner of exposure to an environmental chemical is through accidental or intentional skin contact. To pass into the skin, the candidate chemical must either traverse the epidermal cells or enter through the follicles. The epidermal cells are the major pathway of potential to chemicals, since these constitute a larger surface area.

Cutaneous exposure may be selected to simulate the main route of human exposure and as a model for induction of skin lesions. During skin absorption, the chemical applied is transferred from the outer surface of the skin through the horny layer, the epidermis, the corneum, and into the systemic circulation. Absorption of chemicals through the skin is time-dependent, and this can be demonstrated by the application of occlusive bandages to prevent loss of the test material from the application site. Solid chemicals and chemicals soluble in secretions of the skin may dissolve in the secretions to a variable extent.

### Inhalation Toxicity Studies

The respiratory tract and lung are the main organ systems that interface with the external environment. An adult human inspires approximately 10,000 liters of air in 21,600 respirations every 24 hours. In other words, this amounts to delivering approximately 450 $cm^3$ of air in each breathing to the respiratory exchange region or specifically, to the alveolar region of the respiratory system.

Inhalation of environmental chemicals goes unnoticed and has become unavoidable (unless one uses a device) because of the large-scale contamination of the atmosphere. The actual and potential hazards associated with exposure to environmental chemicals through the respiratory tract have become evident wherever industrial working environments, pollution of the atmosphere, and high-density human populations are encountered.

It is interesting that the adult human lung has an enormous gas tissue interface, approximately 90 $m^2$, 70 $m^2$ alveolar space. This large surface, together with the blood capillary network surface of 140 $m^2$, with its continuous and profuse blood flow, offers an extremely rapid and efficient medium for the absorption of chemicals from the air, into the alveolar portion of the lungs, and into the bloodstream.

Although highly water-soluble chemicals, (e.g., soluble halogen salts, soluble chromates) may pass through the lung very rapidly and undetected, other industrial chemicals, due to their extreme insolubility in body fluids and rapid reactivity with lung constituents, remain in the lung for a prolonged period of time. This results in irritation, inflammation, edema, emphysema, granulomatosis, fibrosis, malignancy, and allergic sensitization. Several highly reactive industrial gases and vapors of low solubility are known to produce pulmonary edema and immediate irritation of the respiratory tract.

Inhalation studies are mostly associated with technical problems of greater complexity than the other type of assay. It is recognized, however, that intracheal instillation may constitute a valid alternative in specific situations.

Long-term respiratory exposures are usually patterned to projected industrial experience, giving the animal a daily exposure 6 hours after equilibrium of chamber concentrations, for 5 days a week (intermittent exposure) or 22 to 24 hours of environmental exposure per day, 7 days a week (continuous exposure), with 1 hour for feeding and maintaining the chambers. In both the cases, the animals are usually exposed to a fixed concentration of test materials. A major difference to consider between intermittent and continuous exposure is that in the former there is a period of 17 to 18 hours in which animals may recover from the effects of daily exposure, and an even longer recovery period during weekends.

The choice of intermittent or continuous respiratory exposure depends on the objectives of the study and on the human experience that is to be simulated. However, certain technical difficulties must be considered. For example, the advantage of continuous exposure for simulating environmental conditions may be offset by the necessity of watering and feeding during exposure and by the need for more complicated (and reliable) aerosol and vapor generation and monitoring techniques. Intermittent systems use simpler inhalation chambers because provision for food and water is not necessary.

***Exposure Chamber*** Animals should be tested in an inhalation chamber designed to sustain a dynamic flow of 12 to 15 air changes per hour to ensure an oxygen content of approximately 19% and an evenly distributed atmosphere. Control and exposure chambers should be identical in construction and design to ensure exposure conditions comparable in all respects, except for exposures to the test chemicals. Chambers should minimize the crowding of test animals to maximize their exposure to the test chemicals. As a general rule, the volume of the test animals should not exceed 5% of the chamber volume. Slight negative pressure inside the chamber is generally maintained to prevent leakage of the test chemical into the surrounding areas. In normal situations, the duration of animal exposure to a test chemical in a chronic toxicity study is 12 months and beyond.

***Observations*** A careful clinical examination of all animals should be made at least once each day. Additional observations should be made daily with appropriate actions taken to minimize loss of animals to the study, such as necropsy, refrigeration of those animals found dead, and isolation or sacrifice of weak or moribund animals. Careful observations should be made to detect onset and progression of toxic effects as well as to minimize loss due to diseases, autolysis, or cannibalism.

Clinical signs, including neurologic and ocular changes as well as mortality, should be recorded for all animals. Time of onset and progression of toxic conditions, including suspected tumors, should be recorded. Body weight should be recorded individually for all animals once a week during the first 13 weeks of the test period and at least once every 4 weeks thereafter. Food intake should be determined weekly during the first 13 weeks of the study and thereafter at approximately 3-month intervals, unless health status or body weight changes dictate otherwise.

***Hematologic Studies*** Hematologic study includes estimations of hemoglobin content, packed-cell volume, total erythrocytes, total leukocytes, platelets, or other measures of clotting potential. These should be performed on blood samples collected from all nonrodents, from 10 rats of both genders, from all groups at 3 months, 6 months, thereafter at approximately 6-month intervals, and at termination. If possible, these collections should be from the same rats of each interval. In additions, a pretest sample should be collected from nonrodents.

If clinical observations suggest a deterioration in health status of the animals during the study, a differential blood cell count of the affected animals should be performed. A differential blood cell count is performed on samples from animals in the highest-dosage group and the controls. Differential blood counts are performed for the next lower group only if there is a major discrepancy between the highest group and the control or if indicated from the pathologic examination.

***Urinalysis*** Urine samples should be collected for analysis from all nonrodents, from 10 rats, and genders of all groups, using preferably the same rats at the same intervals. The following determinations should be made, either on individual animals or on a pooled sample/gender/group for rodents: Appearance, volume, and density for individual animals, protein, glucose, ketones, occult blood (semiquantitatively), and microscopy of sediment (semiquantitatively).

***Clinical Chemistry*** Blood samples from all nonrodents, 10 rats, and genders of all groups should be drawn at approximately 6-month intervals and at study termination for clinical chemistry measurements. When possible, the samples should be drawn from the same rats at the same interval. In addition, a pretest sample should be collected from nonrodents. Serum is prepared from these samples and the following determinations are made: total protein concentrations and albumin concentrations. Liver and kidney function tests should include estimations of serum alkaline phosphate activity, SGPT activity, SGOT, gamma glutamyl transpeptidase, and ornithine decarboxylase estimation of carbohydrate metabolism, such as fasting blood urea nitrogen, also should be performed.

***Pathology*** The pathologic examination, which includes macroscopy as well as light microscopy, is often the cornerstone of the chronic toxicity study. These aspects, along with diagnosis, should command all necessary attention and should be described and reported in detail.

***Necropsy Procedures*** A well-performed gross necropsy may provide optimal information for light-microscopy studies and may, in certain cases, facilitate more restrictive microscopic examination. An inadequate gross necropsy cannot be replaced by microscopic examination, no matter how well performed. Gross necropsy should be performed under the guidance of a trained laboratory animal pathologist.

Complete gross examination should be completed in all animals, including those that died during the experiments or were killed in moribund conditions. Prior to sacrifice of all animals, blood samples should be collected for differential blood counts. All grossly visible lesions, tumors, or lesions suspect of being tumors should be preserved. An attempt should be made to correlate gross observations with microscopic findings.

All organs and tissues of test animals should be preserved for microscopic examination. Usually, the following organs and tissues are considered for microscopy: brain, (medulla/pons, cerebellar cortex, cerebral cortex) pituitary, thyroid glands, liver, spleen, kidneys, adrenals, esophagus, stomach, duodenum, jejunum, ileum, cecum, colon, rectum, uterus, urinary bladder, lymph nodes, mammary gland, skin musculature, peripheral nerve, spinal cord (cervical, thoracic lumber), sternum with bone marrow and femur (including joint), and eyes. Although, inflation of lungs and urinary bladder with a fixative is the optimal way to preserve these tissues, the inflation of the lungs in inhalation studies is essential for appropriate histopathologic examination.

In special studies, such as inhalation toxicity studies, the entire respiratory tract should be studied, including the nose, pharynx, and larynx. If other clinical examinations are performed, the information obtained from these procedures should be available before microscopic examination, because it may give significant guidance to the pathologist.

***Histopathology*** All grossly visible tumors and other lesions should be examined microscopically. In addition, the following procedures are recommended:

- Microscopic examination of all preserved organs and tissues with complete description of all found in the exposed animal: 1) all animals that die or were killed during the study; 2) all animals exposed to highest dose of the test chemical and the controls.
- Organs or tissues showing abnormalities caused or possibly caused by the test chemical are examined in the lower-dose groups.
- In case the result of the experiment gives evidence of substantial alteration of the animals' normal longevity or the induction of effects that might affect a toxic response, the next lower-dose level should be examined as described above.
- The incidence of lesions normally occurring in the strain of test animals, under the same laboratory conditions, historic control is indispensable for correctly assessing the significance of changes observed in them.

***Test Report*** The test report must include the following information:

- Identity of the laboratory where the test was performed.
- Dates of the test, indicating the commencement and termination.
- The individual responsible for the conduct and report of the study.
- The test report must include all information necessary to provide a complete and accurate description of the test procedures and an evaluation of the results. It should contain a summary of the data, an analysis of the data, and the conclusions drawn from the analysis. The summary must highlight data or observations and any deviations from control data which are indicative of toxic effects.

## REFERENCES

1. Food Safety Council, *A Proposed Food Safety Evaluation Process, Final Report of Board of Trustees*, The Nutrition Foundation, Inc., Washington DC, 1982.
2. Organization for Economic Cooperation and Development, *Testing guidelines prepared by the Organization of Economic and Cooperative Development of the United Nations*, Paris, France, 1992.
3. H.M. Peck, *An Appraisal of Drug Safety Evaluation in Animals and the Extrapolation of Results to Man*, Pennsylvania, 449–471, 1980.
4. T.S.S. Dikshith, *Safety Evaluation of Environmental Chemicals*, New Age International Publishers (formerly Wiley Eastern Ltd.), New Delhi, India, 1996.
5. Organization for Economic Cooperation and Development, *Joint meeting of the Chemicals Committee and Working Party on Chemicals*, OECD, Paris, November 2001.

CHAPTER 20

# Guidance for Laboratory Students and Occupational Workers

## INTRODUCTION

The toxicity of a chemical refers to its ability to damage an organ system (kidneys, liver), disrupt a biochemical process (e.g., the blood-forming process), or disturb an enzyme system at a site remote from the site of contact. Toxicity is a property of each chemical, which is determined by molecular structure. Any substance can be harmful to living things. Just as there are degrees of being harmful, there are also degrees of being safe. The biologic effects (beneficial, indifferent or toxic) of all chemicals are dependent on a number of factors.

For every chemical, there are conditions in which it can cause harm and, conversely, there are conditions in which it does not. A complex relationship exists between a biologically active chemical and the effect it produces. This relationship involves consideration of dose (the amount of a substance to which one is exposed), time (how often, and for how long during a specific time, the exposure occurs), the route of exposure (inhalation, ingestion, absorption through skin or eyes), and other factors such as gender, reproductive status, age, general health and nutrition, lifestyle factors, previous sensitization, genetic disposition, and exposure to other chemicals.

The most important factor is the dose-time relationship. The dose-time relationship forms the basis for distinguishing between two types of toxicity: acute and chronic. Acute toxicity of a chemical refers to its ability to inflict systemic damage as a result (in most cases) of a one-time exposure to relative large amounts of the chemical. In most cases, the exposure is sudden and results in an emergency situation.

Chronic toxicity refers to a chemical's ability to inflict systemic damage as a result of repeated exposures over a prolonged time period, to relatively low levels of the chemical. Some chemicals are extremely toxic and are known primarily as acute toxins (e.g., hydrogen cyanide); some are known primarily as chronic toxins (e.g., lead). Other chemicals, such as some chlorinated solvents, can cause either acute or chronic effects.

*Industrial Guide to Chemical and Drug Safety*, By T.S.S. Dikshith and Prakash V. Diwan
ISBN 0-471-23698-5 

Toxic effects of chemicals can range from mild and reversible (e.g., headache from inhaling petroleum naphtha vapors that disappears with fresh air) to serious and irreversible (e.g., liver or kidney damage from excessive exposures to chlorinated solvents). Toxic effects from chemical exposure depend on the severity of the exposures. Greater exposure and repeated exposure generally lead to more severe effects.

## GENERAL SAFETY GUIDELINES

Almost everyone works with or around chemicals and their products every day. Many of these materials have properties that make them hazardous; they can create physical (e.g., fire, explosion) or health hazards (e.g., toxicity, chemical burns). However, there are many ways to work with chemicals which can both reduce the probability of an accident and reduce the consequences to minimum levels, should an accident occur. Risk minimization depends on safe practices, appropriate engineering controls for chemical containment, the proper use of personnel protective equipment, the use of the least quantity of material necessary, and substitution of a less-hazardous chemical for the more hazardous one. Before beginning an operation, ask, "What would happen if...?" The answer to this question requires an understanding of the chemical hazards, equipment, and procedures involved. The hazardous properties of the material and intended use will dictate the precautions to be taken.

Another important distinction is the difference between *hazard* and *risk*. The two terms are sometimes used as synonyms. In fact, *hazard* is a much more complex concept because it includes conditions of use. The hazard presented by a chemical has two components: 1) its inherent capacity to do harm by virtue of its toxicity, flammability, explosiveness, or corrosiveness; and 2) the ease with which the chemical can come into contact with a person or other object of concern. Together, two components determine risk (i.e., the likelihood or probability that a chemical will cause harm). Thus, an extremely toxic chemical, such as strychnine, cannot cause poisoning if it is in a sealed container and does not contact the handler. In contrast, a chemical that is not highly toxic can be lethal if a large amount is ingested.

## CHEMICAL SAFETY GUIDELINES

Always follow these guidelines when working with chemicals in laboratory:

- Assume that any unfamiliar chemical is hazardous.
- Know all the hazards of the chemicals with which you work. For example, perchloric acid is a corrosive, an oxidizer, and a reactive. Benzene is an irritant that is also flammable, toxic, and carcinogenic.
- Consider any mixture to be at least as hazardous as its most hazardous component.

- Never use any substance that is not properly labeled.
- Follow all chemical safety instructions precisely.
- Minimize your exposure to any chemical, regardless of its hazard rating.
- Use personal protective equipment, as appropriate.
- Use common sense at all times.

The five prudent practices of chemical safety may be summed up under the following safety guidelines:

1. Treat all chemicals as if they were hazardous.
2. Minimize your exposure to any chemical.
3. Avoid repeated exposure to any chemical.
4. Never underestimate the potential hazard of any chemical or combination of chemicals.
5. Assume that a mixture or reaction product is more hazardous than any component or reactant.

In a clinical laboratory, each worker, student, and supervisor should know the following:

1. Workers know safety rules and follow them.
2. Adequate emergency equipment is available in full working order.
3. Training in the use of emergency equipment has been provided.
4. Information on special or unusual hazards in nonroutine work has been distributed to the laboratory workers.
5. Appropriate safety training has been given to individuals when they commence their work in the laboratory.

The laboratory workers should develop good personal safety habits:

1. Appropriate eye protection should be worn at all times.
2. Exposure to chemicals should be kept to a minimum.
3. Smoking, eating, drinking, and the application of cosmetics should be prohibited in areas where laboratory chemicals are present.

Advance planning is one of the best ways to avoid serious incidents. Before performing any chemical operation, the laboratory worker should consider, "What would happen if...?" and be prepared to take proper emergency actions.

Overfamiliarity with a particular laboratory operation may result in overlooking or underrating its hazards. This perception can lead to a false sense of security, which frequently results in carelessness. Every laboratory worker has a basic responsibility to plan and execute laboratory procedures in a safe manner.

## GENERAL PRINCIPLES

Every laboratory worker should observe the following rules:

1. Know the safety rules and procedures that apply to the work being performed. Determine the potential hazards (e.g., physical, chemical, biologic) and appropriate safety precautions before beginning any new operation.
2. Know the types of protective apparel available and use the proper type for each job.
3. Know the location of the safety equipment in the area, as well as how to use it and how to obtain additional help in an emergency. Be familiar with emergency and first aid procedures.
4. Be alert to unsafe conditions and actions, and call attention to them so that corrections can be made as soon as possible.
5. Prohibit smoking, the consumption of food or beverage, and the application of cosmetics in areas where laboratory chemicals are being used or stored.
6. Prevent hazards to the environment by following accepted waste disposal procedures. Chemical reactions may require traps or scrubbing devices to prevent the escape of toxic substances.
7. Be certain that all chemicals are correctly and clearly labeled. Combine reagents in the appropriate order.
8. Post warning signs when unusual hazards, such as radiation, laser operations, flammable materials, biologic hazards, or other special problems exist.
9. Remain out of the area of a fire or personal injury unless it is your responsibility to help meet the emergency. Curious bystanders may interfere with rescue and emergency personnel and endanger themselves.
10. Avoid distracting or startling any other worker. Practical jokes or horseplay cannot be tolerated in the laboratory.
11. Use equipment only for its designed purpose.
12. Position and clamp reaction apparatus carefully to permit manipulation without the need to move the apparatus until the reaction is completed.
13. Think, act, and encourage safety until it becomes a habit.
14. Participate in a continuous safety training program.

## HEALTH AND HYGIENE

Laboratory workers also must observe the following health practices meticulously:

1. Wear appropriate eye protection at all times.

2. Use protective apparel, including face shields, gloves, and other special clothing or footwear as needed.
3. Confine long hair and loose clothing when in the laboratory.
4. Do not use mouth suction to pipette chemicals or to start as siphon; a pipette bulb or an aspirator should be used.
5. Avoid exposure to gases, vapors, and aerosols. Use appropriate safety equipment whenever such exposure is likely.
6. Wash well before leaving the laboratory area. However, avoid the use of solvents for washing skin. They remove natural protective oils and may cause irritation or inflammation; moreover, they can facilitate absorption of chemicals through the skin.

## FOOD HANDLING

It has been observed often that a workbench becomes a lunch table in a chemical laboratory. This is a matter of concern; working places cannot be a place for food and drinks.

Contamination of food, drink, and smoking materials is a potential route for exposure to toxic substances. Food should not be stored, handled, or consumed in any laboratory area.

1. Well-defined areas should be established for storage and consumption of food and beverages. No food should be stored or consumed outside this area.
2. Areas where food is permitted should be prominently marked and a warning sign (e.g., Eating Area—No Chemicals) displayed. All chemicals, chemical equipment, and laboratory coats must be excluded from such areas.
3. Consumption of food or beverages and smoking must be excluded from areas where laboratory operations are performed.
4. Glassware or utensils that have been used for laboratory operations should never be used to prepare food or beverages. Laboratory refrigerators, ice chests, and cold rooms should not be used for food storage; separate equipment should be dedicated and prominently labeled for that use.

## HOUSEKEEPING

There is a definite relationship between safety performance and orderliness in the laboratory. When housekeeping standards fall, safety performance inevitably deteriorates. The work area should be kept clean, and chemicals and equipment should be properly labeled and stored.

1. Work areas should be kept clean and free from obstructions. Cleaning up should follow the completion of any operation or take place at the end of each day.

2. Wastes should be deposited in appropriate color-coded or labeled receptacles.
3. Spilled chemicals should be cleaned up immediately and disposed of properly. Disposal procedures should be established, and all laboratory personnel should be aware of them; the effects of other laboratory accidents also should be cleaned up promptly.
4. Unlabeled containers and chemical wastes should be disposed of promptly by appropriate procedures. Such materials, as well as chemicals that are no longer needed, should not be accumulated in the laboratory.
5. Floors should be cleaned regularly; accumulated dust, chromatography adsorbents, and various chemical residues may pose respiratory hazards.
6. Corridors, hallways, and stairways should not be used as storage areas, either for chemicals or equipment of any kind.
7. Access to exits, emergency equipment, and switches or control valves should never be blocked.
8. Equipment and chemicals should be stored properly.

## EQUIPMENT MAINTENANCE

Good equipment maintenance is important for safe, efficient operations. Equipment should be inspected and maintained regularly. Servicing schedules should be related to expected usage and the reliability of the equipment. Equipment awaiting repair or maintenance should be removed from service.

## GUARDING FOR SAFETY

All mechanical equipment should be furnished with guards that prevent access to electrical connections or moving parts. Each laboratory worker should inspect equipment before using it to ensure that guards are in place.

Careful design of guards is vital. An ineffective guard can be worse than none at all, as it can give rise to a false sense of security. Emergency shutoff devices may be needed, in addition to electrical and mechanical guarding.

## SHIELDING FOR SAFETY

Safety shielding should be used for any operation that is potentially explosive, such as:

1. An exothermic reaction being attempted for the first time.
2. A familiar reaction performed on a larger than usual scale.

3. Operations being performed under nonambient pressures. Shields must be placed so that all personnel in the area are protected from hazard.

## GLASSWARE

Accidents involving glassware are a major cause of laboratory injuries. These can be avoided or reduced by adhering to the following:

1. Careful handling and storage procedures should be used to avoid damaging glassware. Damaged items should be discarded or repaired by an experienced glassblower.
2. Glassblowing operations should not be attempted except at a properly equipped bench, with proper annealing facilities being available.
3. Vacuum-jacketed glass apparatus should be handled with extreme care to prevent implosion. Equipment such as Dewar flasks should be taped or shielded. Only glassware designed for vacuum work should be used for that purpose.
4. Thick gloves should be worn when picking up pieces of broken glass. Small fragments should be swept up with a brush into a dustpan.
5. Proper instruction should be provided in the use of glass equipment designed for special tasks which might present unusual risks. For example, stoppered separating funnels containing volatile solvents can develop considerable pressure.
6. It may be necessary to insert glass tubing into bungs or corks, or make rubber-to-glass hose connections. In these cases, tubing should be fire-polished and lubricated, and appropriate hand protection should be worn.

## FLAMMABILITY HAZARDS

Because flammable materials are widely used in laboratory operations, the following rules should be observed:

1. Store flammable materials correctly.
2. Never use an open flame to heat a flammable liquid or to perform distillation under reduced pressure.
3. In any laboratory procedure, an open flame should only be used when there is no better alternative. It should be extinguished once it is no longer required.
4. Before lighting a flame, remove all flammable substances from the immediate area. Check all containers of flammable materials in the area to ensure that they are tightly closed.
5. Notify other occupants of the laboratory before lighting a flame.

6. When volatile flammable materials may be present, use only intrinsically safe nonsparking electrical equipment, which is not liable to overheat.

## COLD TRAPS AND CRYOGENIC HAZARDS

Cryogenic materials and the surfaces they cool can cause severe "cold" burns if they are allowed to contact the skin. Thick gloves and a face shield may be needed when preparing or using some cold baths.

Neither liquid nitrogen nor liquid air should be used to cool a flammable mixture in the presence of air because both can condense oxygen from the air, resulting in a potentially explosive mixture. "Dry ice" (solid carbon dioxide) should be handled with dry leather gloves and should be added slowly to a liquid cooling bath to avoid foaming. Workers should avoid lowering their heads into a "dry ice" chest; a high concentration of carbon dioxide can inhibit the autonomic breathing system, which can lead to loss of consciousness or asphyxiation. Large quantities of liquid nitrogen also can cause asphyxiation in confined or poorly ventilated spaces.

## CLOSED SYSTEMS UNDER POSITIVE OR NEGATIVE PRESSURE

Reactions should never be performed in, or heat applied to, a closed-system unless it is designed and tested to withstand the expected pressure change. A pressurized apparatus should have an appropriate relief device.

## WASTE DISPOSAL PROCEDURES

Laboratory managers are responsible for establishing waste disposal procedures for routine and emergency situations, and for communicating these procedures to laboratory workers. Workers must follow them to avoid hazards or damage to the environment.

## WARNING SIGNS AND LABELS

Laboratory areas that have special unusual hazards should be posted with warning signs. Standard signs and symbols have been established for a number of special situations, such as radioactivity hazards, biologic hazards, fire hazards, and laser operations. Other signs should be posted to show the locations of safety showers, eyewash stations, fire extinguishers, and exits. Extinguishers should be sited carefully, and they should be clearly labeled with the type of fire for which they are intended. Waste containers should be color-coded or labeled indicating the type of waste that can be safely deposited in them. Safety and hazard signs in the laboratory should enable personnel to appropriately address an emergency situation.

When possible, labels on chemical containers should contain information concerning the hazard(s) associated with use of the chemical. Unlabeled bottles should not be allowed to accumulate. They should be handled with care when determining how they can be disposed of safely.

## UNATTENDED OPERATIONS

Laboratory operations are often performed continually or overnight with no one present. It is essential to plan for interruptions in utility services such as electricity, water, and gases. Operations should be designed to be safe, and plans should be made to avoid hazards in case of failure. Wherever possible, arrangements for periodic inspection of the operation should be made and, in all cases, the area should be properly lit and an appropriate sign placed on the door and the apparatus. Failure of availability of cooling water can have serious consequences. A variety of devices can be used that:

1. Automatically regulate water pressure to avoid surges that might rupture the supply lines; or
2. Monitor the water flow so that a failure will automatically result in the shut-down of electrical connections and water supply valves, and venting of any pressure build-up in the system.

## WORKING ALONE

It is prudent to avoid working in a laboratory alone. However, if this must be done, arrangements should be made between individuals working in separate laboratories outside of conventional hours to crosscheck periodically. Alternatively, security guards may be asked to check on a laboratory worker. A worker who is alone in a laboratory should not undertake experiments known to be hazardous.

Special rules may be necessary under unusual circumstances. The laboratory supervisor is responsible for determining whether the work requires special safety precautions, such as having two persons in the same room or in close proximity, during a particular operation.

## ACCIDENT REPORTING

Emergency telephone numbers to be used in the event of fire, accident, flood, or hazardous chemical spillage should be displayed prominently in each laboratory. In addition, the home telephone numbers of laboratory workers and supervisors should be displayed. The appropriate person must be notified promptly in the event of an accident or emergency.

Every laboratory should have an internal accident-reporting system. This includes provisions for investigating the cause of an injury as well as any potentially serious incident that does not result in injury. The primary aim of such investigations should be to make recommendations to improve safety, not to assign blame for an incident. Local legal regulations may require reporting procedures for accidents or injuries.

## EVACUATION

Not only should laboratory workers be aware of emergency telephone numbers, but they also should know the location and proper use of fire-fighting appliances, rescue equipment, and emergency exits. They also should be familiar with the prescribed procedure for evacuating the building and the location of the outside assembly point where personnel rollcalls will be checked.

## EVERYDAY HAZARDS

Laboratory workers should remember that injuries can and do occur outside the laboratory in other work areas. It is important that safety principles be practiced in offices, stairways, corridors, and similar places. Here, safety is largely a matter of common sense, but constant awareness of everyday hazards is vital.

A sound waste management system requires that laboratory wastes be properly identified, classified as to the types of hazards they present, and segregated to avoid chemical interactions. These steps are essential for the self-accumulation, transportation, and disposal of wastes. Moreover, in many countries, government regulations prescribe the classification and segregation of such wastes. The classification and segregation procedures of the laboratory should be identified clearly in the laboratory waste management plan and summarized in the waste management manual.

## IDENTIFICATION

The laboratory worker must decide if a material is no longer required and ought to be disposed of. Possibilities for recovering or recycling material should be considered. Laboratory operating guidelines should include information on the types of chemicals that can be recovered, recycled, or reused. Once a material is declared waste, those working with it should determine the degree of hazard it represents and provide sufficient information on its characteristics for correct disposal. Accordingly, they must be familiar with the hazardous characteristics by which wastes are classified as well as the procedures used by the laboratory for segregating and collecting wastes. The laboratory waste management manual should provide enough information for this purpose.

## CLASSIFICATION

Wastes should be classified according to the type and degree of hazard, if necessary, as prescribed by government regulation. The following guidelines should be helpful in classifying hazardous materials.

Acute hazardous wastes are substances that are fatal to humans in low doses or that have an oral $LD_{50}$ toxicity (in rats) of less than 50 mg/kg, an inhalation $LD_{50}$ toxicity (in rats) of less than 2 mg/L, a dermal $LD_{50}$ toxicity (in rabbits) of less than 200 mg/kg, or that are capable of causing serious, irreversible, or incapacitating illness.

Hazardous wastes are substances that meet any of the following criteria:

1. Flammability

   - Liquids, other than aqueous solutions containing less than 24% (v/v) alcohol, that have a flashpoint below 60 °C.
   - Materials other than liquids that are capable, under ambient conditions of temperature and pressure, of causing fire by friction, absorption of moisture, or spontaneous chemical changes and, which when ignited, burn so vigorously and persistently as to create a hazard.
   - Flammable compressed gases that form flammable mixtures at a concentration of 13% (v/v) or less in air or that have a flammable range in air wider than 12% (v/v), regardless of the lower limit.
   - Oxidizing agents such as chlorate (VII), manganate (VII), nitrate (V), or inorganic peroxides that readily yield oxygen to stimulate the combustion of organic matter.

2. Corrosivity (e.g., aqueous solutions that have a pH less than 2 or greater than 12.5).
3. Reactivity: This classification includes substances that react with water violently to produce gases or explosive mixtures; unstable or explosive substances; and substances that contain cyanide or sulfide or generate toxic gases when exposed to a pH in the range of 2 to 12.5.

Classification and segregation of wastes by such criteria are essential for their safe handling and disposal. If the waste is not a common chemical with known characteristics, enough information about it must be supplied to satisfy regulatory requirements and be certain that it can be handled and disposed of safely. For many wastes, only the principal components must be specified. However, if the waste contains a carcinogen or toxic metal, this information should be supplied. The information needed to characterize a waste also depends on the method of ultimate disposal.

## SEGREGATION

### Labeling

Classes of waste should be properly segregated for temporary accumulation and storage as well as for transportation and disposal. Accordingly, all wastes must be labeled properly before being removed. The label should contain sufficient information to ensure safe handling and disposal, including the initial of accumulation and chemical names of the principal components and any minor components that may be hazardous. The label also should indicate whether the waste is toxic, reactive, corrosive, metallic, flammable, an inhalation hazard, or lachrymatory.

Some laboratories, particularly research laboratories, use and synthesize many unusual chemicals that can become waste. In general, chemists involved with such work know qualitatively whether they are flammable, corrosive, or reactive. If large quantities of chemical waste with unknown hazards are being generated, analytical tests must be performed to determine such properties. However, for typically small quantities of laboratory chemicals, formal analysis is not warranted. Laboratory samples with unknown hazard characteristics, including orphan wastes, can be tested on a small scale in the laboratory for flammability with a small flame. The pH also should be determined (if the solution is aqueous), as should its reactivity with water or air. The presence of peroxides or other oxidizing compounds should be checked with potassium iodide. The one hazard characteristic that cannot be readily determined is toxicity, although the probability that a chemical is toxic can sometimes be inferred by analogy to closely related chemical structures. In the absence of a basis for judgment, a waste should be assumed to be toxic and labeled accordingly.

Waste generated in the laboratory can often be characterized from knowledge of the starting materials (e.g., hydrocarbon mixture, flammable laboratory solvents, chlorobenzene still bottoms). Professional expertise, common sense, judgment, and safety awareness of trained professionals performing chemical operations in the laboratory usually put them in a position to judge the type and degree of chemical hazard.

## DISPOSAL OF WASTES IN SEWER SYSTEMS

Limited quantities of some wastes can be disposed of in sanitary sewer systems but never in storm-sewer systems. A sanitary sewer is one that is connected directly to a waste-treatment plant, whereas a storm sewer usually discharges into a stream, river, or lake.

## ACCUMULATING WASTES FOR DISPOSAL

The first step in the disposal sequence usually involves accumulation or temporary storage of waste in or near the laboratory. Except when a single chemical is accumulated for recovery, waste accumulation generally involves several chemicals in a container. Only compatible chemicals should be placed in a container, whether packaged separately or mixed. In this context, "compatible" implies the absence of chemical interaction. Two common practices for accumulation of wastes are mixing compatible chemicals in a waste container and accumulating small containers of compatible wastes in a larger outer container (e.g., a lab pack). The method chosen and the scheme for segregating the wastes depend primarily on the intended mode of disposal.

If laboratory wastes are to be landfilled, the most common method of packaging is the lab pack. The waste management plan and manual should include specific directions for preparing lab packs as well as the assigned responsibility for preparing them. On the other hand, if a contract disposal service that prepares the lab packs is used, the manual should give directions for segregating and labeling wastes in accordance with the contractor's requirements. The principal consideration in segregation of chemicals for landfill disposal is compatibility. In addition, explosives and chemicals that present a reactivity hazard (expect for cyanide- and sulfide-bearing reactive wastes) must not be disposed into a landfill.

The method chosen for segregation and accumulation of wastes destined for incineration depends on the design of incinerator and its waste-feed mechanism, which vary widely. Some incinerators can handle only bulk liquid wastes, whereas others accept solid or packaged wastes such as fiber packs and glass or plastic bottles; a few even accept steel cans or drums.

Incinerators that accept only liquid wastes either blend them with fuels or incinerate directly. In either case, the disposer generally pumps the contents from the container. Small containers are less desirable than the standard 200-liter drum. Incinerators that accept solid waste generally incinerate without removing the waste from the container, avoiding the hazards of opening containers. Some facilities will accept a variety of containers, including individual bottles. Others prefer to accept wastes in fiber packs, which is a combustible version of the lab pack used in landfills.

Explosive compounds should not be put into fiber packs for incineration. Some operators will incinerate certain explosives by adding them to the incinerator feed in small quantities, preferably diluted with a flammable solvent or sawdust. Containers with more than 100 $cm^3$ of carbon disulfide produce an explosive hazard in incinerators and should be avoided.

If separate incinerator facilities for hazardous and nonhazardous waste are available, segregation of hazardous waste can be both cost-effective and environmentally more acceptable. Compared with incinerators for municipal waste, incinerators for hazardous waste require a more expensive design, more careful operation, higher costs for obtaining permits, and regulated treatment of ash and scrubber water. Consequently, incineration of nonhazardous waste in hazardous waste facilities is costly.

The proper sorting of wastes destined for incineration is essential. Halogenated wastes should be kept separate from nonhalogenated wastes, because the former must be burned in an incinerator with a scrubber that greatly reduces emissions of volatile halogen compounds.

Accumulation of laboratory wastes for disposal must be performed in accordance with the written waste management plan, which should include all safety factors (chemical incompatibilities being particularly important), regulatory requirements, and factors specific to the disposer and disposal method. A training program should be initiated and maintained to ensure that all laboratory personnel understand the requirements, such as accumulation procedures, safety procedures, and record-keeping. It is especially important that the design conforms to local fire codes. However, sites for accumulation of small quantities of waste over a period of a few days need not be as elaborate as larger waste-storage areas. Thus, modest quantities of single-type wastes can be accumulated temporarily in containers kept in a hood. Such containers should be clearly dated and labeled for content.

Large quantities of flammable wastes should never be accumulated in a closed, unventilated room. A vapor explosion in such a room would be disastrous. An appropriate container for each category of waste in the waste accumulation plan should always be available. The container must be clearly marked to indicate the type of waste it can contain and the hazards associated with this category. It should be dated to indicate when accumulation began.

The accumulation area should be inspected regularly. The inspection frequency can depend on the level of activity and the degree of hazard, but generally should be at least weekly. The inspection should include the following considerations:

1. Adherence to the accumulation plan.
2. Condition of containers.
3. Availability of containers.
4. Dates of containers.
5. Adequate records of container contents.
6. Operation of safety equipment.

Documentation should be kept as to the contents of each container unless the container is labeled to receive, for example, only a single type of waste. The specific data required will depend on the size and complexity of the laboratory operation.

Although the same principles apply to smaller laboratories with simpler or fewer wastes, such laboratories may be able to work with simpler rules and requirements. For example, in a teaching laboratory where students are performing a specific experiment, it should suffice to collect the wastes in containers labeled for each waste and to record the total quantity and date. A similar system can be used in which specific waste solvents are accumulated for recovery. If the wastes are to be placed in a lab pack for landfill disposal or an analogous fiber pack for incineration, the accumulation record should include the information that is on the individual waste containers.

## USING LIQUEFIED PETROLEUM GAS SAFELY

Liquefied petroleum gas (LPG), stored as a liquid under pressure in cylinders, is widely used in homes as cooking gas. The main hazards associated with LPG are fire or explosion, in the case of even minor leakage. In the case of major leakage in confined spaces, asphyxiation due to deficiency of oxygen also may result. Because LPG is colorless and odorless, a distinctive foul odor is added to enable easy detection of a leak. As LPG vapor is heavier than air, these vapors accumulate at lower levels, and a fire or explosion may result.

### Safety When Accepting Door Delivery

- Check that the valve sealing tag is intact and the safety protection cap is in position.
- Check leakage from the valve by applying soap solution.

### Safety When Changing Cylinders

- Extinguish all fires in the room.
- Switch off all electrical appliances.
- Check for leakage from the rubber-tube connections by applying soap solution.
- Never light a matchstick or check the leakage.
- Leave changing of new cylinder to a trained person.
- Do not drag, roll, or drop the cylinders.
- Open windows for free ventilation.
- Preserve the safety protection cap.

### Safe Usage

- Always keep the cylinder in an upright position, away from any heat source, in a well-ventilated place. When moving the cylinder, keep it upright.
- Do not tilt it to draw the last bit of gas; tilting or shaking will obtain no extra gas.
- Position the stove or burner above the level of cylinder.
- To light the burner, open the cylinder valve, hold a lighted matchstick (or gas lighter) over the burner, and only then turn the knob of the burner on.
- To turn off the burner, first close the cylinder valve and then the burner knob.
- When the stove is not in use, keep the cylinder valve closed. Check this every night and whenever you leave the house.
- If the flame goes out during use, do not relight it immediately. First close the cylinder valve and burner knob. Open all doors and windows. Allow time for leaked gas to dissipate. Only then relight the burner.

- Do not use synthetic fabrics (e.g., nylon, terylene) when operating the stove. Wear cotton dresses.
- Keep children away from stove and cylinder.
- Use dry potholders when handling pans on the stove. Do not use trailing towels or aprons.
- Never leave the stove or burner unattended when it is in operation. Cooking materials may overflow on the burners, extinguish the flame, and leakage of gas will occur. Accumulated gas could be ignited.
- Never try to repair or adjust any part of the gas installation or allow untrained persons to do so.
- Do not position a shelf or cabinet above the stove.
- If your stove is near a window, do not use curtains, as they may blow over the burner and catch fire.

### If You Suspect Leakage

- Close burner knobs and cylinder valves, and refix the safety protection cap.
- Extinguish any open flame.
- Do not light a match or bring in other ignited material.
- Open windows for free ventilation.
- Do not touch electrical switches.
- Do not tamper with installation.

Immediately contact your distributor and fire company. Keep phone numbers handy.

# APPENDIXES

## APPENDIX 1 UNITS OF MEASURE USED IN THIS BOOK

| Concept | Description of Unit | Abbreviation |
|---|---|---|
| Dosage of animals or people | Milligrams of compound per kilogram of body weight | mg/kg |
| Concentration of storage in tissue or of residue in food or water | Parts of compound per million parts of tissue, food, or water by weight | ppm |
| Air concentration | Milligrams of compound per cubic meter of air | $mg/m^3$ |
| Formulation concentration | Parts of compound per hundred parts of formulation (weight/volume) | percentage (%) |
| Rate of application to a surface | Milligrams of compound per square meter of area | $mg/m^2$ |

*Industrial Guide to Chemical and Drug Safety*, By T.S.S. Dikshith and Prakash V. Diwan
ISBN 0-471-23698-5 

## APPENDIX 2 CONVERSION TABLE FOR THE UNITS OF MEASURE USED IN THE EVALUATION OF CHEMICALS AND DRUG SAFETY

| **Concentration in Tissue, Food, or Water*** | **ppm** |
|---|---|
| 1 ng/g | 0.001 |
| 1 μg/100 g (μg %) | 0.01 |
| 1 mg/kg | 1 |
| 1 μg/g | 1 |
| 1 μg/100 mg | 10 |
| 1 mg/100 g (mg %) | 10 |
| 1 grain/pound | 142.9 |
| 1 mg/g | 1,000 |
| 1% (concentration) | 10,000 |
| 1 g/pound | 2,204.6 |
| 1 oz/100 lb | 624 |
| 1 oz/60 lb bushel | 1,041.6 |
| **Concentration in Air†** | **$mg/m^3$** |
| 1 μg/L | 1 |
| 1 g/1,000 cu. ft. | 35,315 |
| 1 ppm (of compound in air by volume)‡ | |
| **Concentration in Formulation (Pesticides)** | |
| 1 pound/gallon | 119.8 g/L |
| 1 pound/gallon | 12.0% w/v |
| **Rate of Application (Pesticides)** | |
| 1 pound/acre | 10.4 mg/sq. ft. |
| 1 pound/acre | 112.1 $mg/m^2$ |
| 1 pound/acre | 1.121 kg/ha |
| 1 mg/sq. ft. | 10.8 $mg/m^2$ |

*The concentration [C] of a molar solution expressed as ppm is

$$C = [\text{molecular weight (g/L)}] \times 1{,}000$$

†The concentration [C] of a saturated vapor expressed as $mg/m^3$ at any given temperature is

$$C = 53.8 \times \text{vapor pressure} \times \text{molecular weight}$$

when the vapor pressure at the same temperature is expressed as mm Hg.

‡The is an expression for concentration frequently used in industrial hygiene. It is based on the assumption that the material in question exists as a gas or vapor and expresses the number of

volumes of compound per million volumes of air. The value may be calculated by the formula

$$\text{ppm} = \frac{\text{Observed concentration(mg/liter)} \times 24,450}{\text{Molecular weight of compound}} = \frac{\text{mg/m}^3 \times 24.45}{\text{mol. Wt.}}$$

The figure, 24,450 mL, is the gram molecular volume of a gas at a pressure of 760 mm Hg and a temperature of 25 °C.

## APPENDIX 3 APPROXIMATE DOSES (MG/KG/DAY) OF A COMPOUND FOR CERTAIN ANIMALS ON DIET CONTAINING IT AT DIFFERENT CONCENTRATIONS THAT DO NOT INFLUENCE FOOD CONSUMPTION*

| | | Concentration in the Total Diet Excluding Water (ppm) | | | | | | |
|---|---|---|---|---|---|---|---|---|
| Animal | Gender | 1 | 2 | 5 | 10 | 25 | 50 | 100 |
| Baby chick | M, F | 0.161 | 0.322 | 0.805 | 1.61 | 4.03 | 8.05 | 16.1 |
| Hen | F | 0.058 | 0.116 | 0.290 | 0.58 | 1.45 | 2.90 | 05.8 |
| Adult mouse | M | 0.124 | 0.248 | 0.620 | 1.24 | 3.10 | 6.20 | 12.4 |
| | F | 0.133 | 0.266 | 0.665 | 1.33 | 3.33 | 6.66 | 13.3 |
| Hamster | M, F | 0.083 | 0.166 | 0.416 | 0.833 | 2.08 | 4.16 | 8.33 |
| Weanling rat | M | 0.100 | 0.200 | 0.500 | 1.00 | 2.50 | 5.00 | 10.0 |
| | F | 0.095 | 0.190 | 0.475 | 0.95 | 2.37 | 4.75 | 9.5 |
| Mature rat | M† | 0.045 | 0.090 | 0.220 | 0.45 | 1.12 | 2.25 | 4.5 |
| | F† | 0.053 | 0.106 | 0.260 | 0.53 | 1.31 | 2.62 | 5.3 |
| Adult rhesus monkey | M | 0.016 | 0.032 | 0.080 | 0.16 | 0.40 | 0.80 | 1.6 |
| | F | 0.023 | 0.046 | 0.115 | 0.23 | 0.57 | 1.15 | 2.3 |
| Adult dog | M | 0.019 | 0.038 | 0.095 | 0.19 | 0.47 | 0.94 | 1.9 |
| | F | 0.023 | 0.046 | 0.115 | 0.23 | 0.57 | 1.15 | 2.3 |
| Adult human | M, F | 0.010 | 0.020 | 0.050 | 0.10 | 0.25 | 0.50 | 1.0 |

*Whether a given concentration of a particular compound does food intake must be learned by measurement.

$$\text{Dosage(mg/kg/day)} = \frac{\text{Dosage(mg/day)dietary}}{\text{Body weight(kg)}}$$

$$= \frac{\text{concentration(ppm)} \times \text{food Consumption(kg/day)}}{\text{Body weight(kg)}}$$

†The same values apply to 500-g guinea pigs.

## APPENDIX 4 NORMAL HUMAN VALUES AND STANDARD ASSUMPTIONS

| | |
|---|---|
| Weight, adult male | 70 kg |
| Birth weight | |
| Range | 2.63–4.58 kg |
| Average | 3.36 kg |
| Surface area: 70 kg, 180 cm, adult male | 1.85 $m^2$ |
| Respiratory volume of persons at rest but not asleep | |
| Range | 4.90–12.20 L/person/min |
| Average | 8.732 L/person/min* |
| Average | 8.900 L/person/min† |
| Of worker | 10.0 $m^3$/8-hr workday |
| Of worker | 18.7 $m^3$/24-hr day |
| Food intake adult male | |
| Ordinary food | 1.4 kg/day |
| Dry weight | 0.7 kg/day |
| Baby | 0.6 L/day |
| Water intake, adult | 2.0 L/day |
| Gestation period | |
| From last period | 278 days or 39.7 weeks |
| From fertilization | 268 days or 38.3 weeks |
| Growth of head air | 1.15 cm/month |
| Swallow | 20 ml |

*Observed average.

†Average calculated from the equation cc/min = 2.10 × (weight in grams)$^{1/4}$ [ibid.].

*Source*: Guyton AC. *Am. J. Physiol.*, 1947; **150**:76–77.

## APPENDIX 5 TOXICITY RATING* OF CHEMICALS (GENERAL)

| Toxicity Rate | Dose | For an Average Adult |
|---|---|---|
| Practically nontoxic | >15 g/kg | More than one quart |
| Slightly toxic | 5–15 g/kg | Between a pint and a quart |
| Moderately toxic | 0.5–5 g/kg | Between an ounce and a pint |
| Very toxic | 50–5,000 mg/kg | Between a teaspoonful and an ounce |
| Extremely toxic | 5–50 mg/kg | Between 7 drops and a teaspoonful |
| Super toxic | <5 mg/kg | A taste (less than 7 drops) |

*Probable oral lethal dose for a human adult.

*Source*: Gleason MN, Gosslin RE, Hodge HC, Smith RO. *Clinical toxicology of commercial products: Acute poisoning*. 3rd ed. Baltimore, Md.: Williams & Wilkins; 1969.

## APPENDIX 6 USEPA LABELING TOXICITY CATEGORIES BY HAZARD INDICATOR

| Hazard Indicators | I | II | III | IV |
|---|---|---|---|---|
| Oral $LD_{50}$ | Up to and including 50 mg/kg | From 50 through 500 mg/kg | From 500 through 5,000 mg/kg | Greater than 5,000 mg/kg |
| Inhalation $LD_{50}$ | Up to and including 0.2 mg/L | From 0.2 through 2 mg/L | From 2 through 2 mg/kg | Greater than 20 mg/kg |
| Dermal $LD_{50}$ | Up to and including 200 mg/kg | From 200 through 2000 mg/kg | 2,000 to 20,000 mg/kg | Greater than 20,000 mg/kg |
| Eye effects | Corrosive; corneal opacity not reversible within 7 days | Corrosive; corneal opacity not reversible within 7 days; irritation persisting for 7 days | No corneal opacity; irritation reversible within 7 days | No irritation |
| Skin effects | Corrosive | Severe irritation at 72 hours | Moderate irritation at 72 hours | Mild or slight irritation at 72 hours |

In general, the order of toxicity of various pesticides (in decreasing order) would be insecticides > defoliants > desiccants > herbicides > fungicides. Within the category of insecticides, the general toxicity (in decreasing order are) organophosphates > carbamates > cyclodienes > DDT-relatives > inorganics.

## APPENDIX 7 TOXICITY CRITERIA FOR PESTICIDES

| Hazard Indicators | I | II | III | IV |
|---|---|---|---|---|
| Oral $LD_{50}$ | ≤50 mg/kg* | 50–500 mg/kg | 500–5,000 mg/kg | >5,000 mg/kg |
| Inhalation $LC_{50}$ | ≤0.2 mg/l[†] | 0.2–2.0 mg/L | 2.0–20 mg/L | >20 mg/L |
| Dermal $LD_{50}$ | ≤200 mg/kg* | 200–2,000 mg/kg | 2,000–20,000 mg/kg | >20,000 mg/kg |
| Eye effects | Corrosive; corneal Opacity not reversible Within 7 days | Corrosive opacity reversible within 7 days; irritation Persisting for 7 days | No corneal opacity irritation reversible within 7 days | No irritation |
| Skin effects | Corrosive | Severe irritation at 72 hours | Moderate irritation at 72 hours | Mild or slight irritation at 72 hours |

*Dose expressed as milligrams per kilogram body weight of test animals.
[†]Dose expressed as milligrams per liter of air in the test chamber.
*Source*: Upholt WM. In *Introduction to Crop Production. Fed Regis.* 40, 28,241. Washington DC: Environmental Protection Agency; 1975, 1978.

## APPENDIX 8 HAZARD CLASSIFICATION OF PESTICIDES BY WORLD HEALTH ORGANIZATION

| | $LD_{50}$ for the Rat (mg/kg Body Weight) | | | |
|---|---|---|---|---|
| | Oral Route | | Dermal Route | |
| Class | Solid* | Liquid* | Solid* | Liquid* |
| I Extremely hazardous | 4 or less | 20 or less | 10 or less | 40 or less |
| I Highly hazardous | 5–50 | 20–200 | 10–100 | 40–400 |
| II Moderately hazardous | 50–00 | 200–2,000 | 100–1,000 | 400–4,000 |
| III Slightly hazardous | Over 500 | Over 2,000 | Over 1,000 | Over 4,000 |

*The terms *solid* and *liquid* denote the physical state of the product or formulation under class.

## APPENDIX 9 REGULATORY AGENCIES ASSOCIATED WITH TOXIC SUBSTANCES

| Source | Regulatory Act |
|---|---|
| Air: | A. Clean Air Act<br>— National Ambient Air Quality Standards<br>— Air Toxics(HAPS)<br>— Motor Vehicle Emission Standards |
| Water: | B. Safe Drinking Water Act<br>C. Clean Water Act |
| Food: | Food Quality Protection Act<br>Federal Food, Drug and Cosmetic Act |
| Agrochemicals/ Pesticides | Federal Insecticide Fungicide and Rodenticide Act (FIFRA), United States of America<br>Insecticide Act [Central Insecticide Board (CIB)] Ministry of Agriculture, Government of India, India<br>Control of Pesticides Act, No. 33 of 1980 Parliament of the Democratic Socialist Republic of Sri lanka |
| Hazardous Waste | Resources Conservation and Recovery Act<br>Comprehensive Environmental Response Compensation and Liability Act<br>Superfund Amendments and Reauthorization Act |

Environmental Protection Agency, United States of America; Food and Drug Administration, United States of America, 1999.

## APPENDIX 10 INDUSTRIAL CHEMICALS AND FLASH POINT

| | Flash Point °C |
|---|---|
| Acetaldehyde | −38 |
| Acetone | −18 |
| Acetonitrile | 6 |
| Acetyl chloride | 4 |
| Acrylonitrile | 0 |
| Allyl iodide | <21 |
| Benzene | 11 |
| Butyl alcohols | 24–29 |
| Carbon disulfide | −30 |
| Chloromethane (methyl chloride) | <0 |
| Cyclohexane | −20 |
| 1,2-dichloroethane (ethylene dichloride) | 13 |
| Diethylamine | <−26 |
| Diethyl carbonate | 25 |
| Diethyl ether | −45 |
| Dioxan | 12 |
| Ethanol (ethyl alcohol) | 12 |
| Ethyl acetate | −4.4 |
| Ethyl acrylate | 16 |
| Ethyl chloroformate | 16 |
| Ethyl formate | −20 |
| Hexane | −23 |
| Methanol (methyl alcohol) | 10 |
| 4-Methylpentan-2-one (isobutyl methyl ketone) | 17 |
| Piperidine | 16 |
| Propan-2-ol (isopropyl alcohol) | 12 |
| Pyridine | 20 |
| Tetrahydrofuran | −17 |
| Toluene | 4.4 |
| Triethylamine | −7 |
| Vinyl acetate | −8 |

## APPENDIX 11 INDUSTRIAL GAS CYLINDERS AND SAFETY

| Gases | Distinctive Color | Body Band |
|---|---|---|
| Oxygen ($O_2$) | Black | None |
| Nitrogen ($N_2$) | Gray | Black |
| Carbon dioxide ($CO_2$) | Black | White |
| Ammonia ($NH_3$) | Black | Red & Yellow |
| Freon-12 ($CCl_2F_2$) | Bottom-end Gray<br>Neck-end Violet | None |
| Argon (Ar) | Blue | None |
| Chlorine ($Cl_2$) | Yellow | None |
| Hydrogen ($H_2$) | Red | None |
| Acetylene ($C_2H_2$) | Maroon | None |
| Air | Gray | None |

## APPENDIX 12 HIGHLY POLLUTING INDUSTRIES

- Primary metallurgical manufacturing industries *viz.*, zinc, lead, copper, aluminum, and steel
- Paper, pulp, and newsprint
- Pesticides
- Refineries
- Fertilizers
- Paints
- Dyes and color pigments
- Leather tanning and leather processing
- Rayon and synthetic fiber manufacturing
- Sodium- or potassium cyanide–associated industries
- Basic drug manufacturing
- Foundry industries
- Storage batteries (lead acid type)
- Acids and alkalies
- Plastics fabrication
- Rubber-synthetic manufacturing
- Cement processing and manufacturing
- Asbestos and associated operations
- Fermentation industry
- Electroplating industry

## APPENDIX 13 INDUSTRIES CLASSIFIED BY POLLUTION CONTROL

Industries with high-pollution indices are classified under different categories (e.g., Category **RED**, Category **ORANGE**, and Category **GREEN**). These industries have different capacities based on their production turnover, which classify them as large, medium, and small. These industries need periodical inspections to comply the laws of the pollution control authorities of the concerned country. The inspections are covered under the following laws: The Water (Prevention & Control of Pollution Act 1974; The Air (Prevention & Control of Pollution) Act 1981; The Water (Prevention & Control of Pollution) Cess Act 1977; and The Environment (Protection) Act 1986.

All large-scale industries have been placed under the category RED and the frequency of their inspection ranges from once a month in a 6-month period of time. All medium-scale industries which have been placed under the RED category have a frequency of inspection once in 3 months. All medium-scale industries placed under the ORANGE category will have a frequency of inspection once in 1 year. All medium-scale industries which have been placed under GREEN category have a frequency of inspection once in 2 years. All small-scale industries which have been placed under the RED category have a frequency of inspection once in 6 months. All small-scale industries that have been placed under the ORANGE category have the frequency of inspection once in 1 year; and all small-scale industries which have been placed under the GREEN category have a frequency of inspection once in 2 years.

### CATEGORY: RED

1. Lime manufacture (pending decision on proven pollution control device and Supreme Court's decision on quarrying)
2. Ceramics
3. Sanitary wares
4. Tyres and tubes
5. Refuse incineration
6. Large flour mills
7. Vegetable oils including solvent extracted oils
8. Soap without steam-boiling process
9. Synthetic detergent formulations
10. Steam-generating plants
11. Manufacture of machineries, machine tools, and equipment
12. Manufacture of office and household equipment, and appliances
13. Involving use of fossil fuel combustion
14. Industrial gases (only nitrogen, oxygen, and $CO_2$)
15. Miscellaneous without involving use of fossil fuel combustion
16. Optical glass
17. Petroleum storage and transfer facilities

18. Surgical and medical products including prophylactics and latex products
19. Malted food
20. Manufacture of power-driven pumps, compressors, refrigeration units, and fire fighting equipment
21. Acetylene (synthetic)
22. Glue and gelatine manufacture
23. Metallic sodium
24. Photographic films, papers, and photographic chemicals
25. Plant nutrients (manure)
26. Ferrous and nonferrous metals extraction, refining, casting, forging, alloy making, and processing
27. Dry coal processing, mineral processing, mineral processing industries like ore sintering, beneficiation, and pelletization
28. Phosphate rock-processing plants
29. Cement plants with horizontal rotary kilns
30. Cement plant with vertical shaft kiln technology (pending certification of proven technology on pollution control)
31. Glass and glass products involving use of coal
32. Petroleum refinery
33. Petrochemical industries
34. Manufacture of lubricating oils and greases
35. Synthetic rubber manufacture
36. Coal, oil, nuclear, and wood-based thermal power plants
37. Hydrogenated vegetable oils for industrial purposes
38. Sugar mills industries
39. Kraft paper mills and related industries
40. Coke-oven byproducts and coal tar distillation products
41. Alkalies, caustic soda, and potash manufacturing industries
42. Electrothermal products (artificial abrasives and calcium carbide) industries
43. Phosphorus and associated industries
44. Organic, inorganic acids, and their salts manufacturing
45. Nitrogen compounds, cyanides, cyanamides, and related compounds
46. Manufacturing of explosives, detonators, and fuses-related industries
47. Phthalic anhydride manufacturing industries
48. Processes associated with the manufacturing of chlorinated hydrocarbons, chlorines, fluorine, bromine, iodine, and their compounds
49. Fertilizer industry
50. Paper board and straw board-related industry
51. Manufacturing of pesticides (insecticides, herbicides, and the formulations
52. Manufacturing of basic drugs and formulations
53. Manufacturing of alcohols industrial and potable
54. Manufacturing of leather, tanning, and processing
55. Processes and operations associated with coke making, coal liquification, and fuel gas making industries
56. Industries associated with the manufacturing of fiber glass and processing

57. Refractories
58. Manufacturing of pulp, wood pulp, mechanical, or chemical processing
59. Manufacturing of pigments, dyes, and their intermediates
60. Manufacturing of industrial of industrial carbons, graphite electrodes, anodes, midget electrodes, graphite blocks, graphite crucibles, gas carbons, activated carbon, synthetic diamonds, carbon black, channel black, and lamp black
61. Manufacturing of electrochemicals (other than those covered under alkali group)
62. Manufacturing of paints, enamels, and varnishes
63. Manufacturing of polypropylene
64. Manufacturing of polyvinyl chloride
65. Manufacturing of chlorates, perchlorates, and peroxides
66. Manufacturing of polishes
67. Manufacturing of synthetic resin and plastic products

## CATEGORY: ORANGE

1. Electroplating
2. Galvanizing
3. Manufacture of mirror from sheet glass and photoframing
4. Surgical gauges and bandages
5. Cotton spinning and weaving
6. Wires, pipes, extruded shapes from metals
7. Automobile servicing and repair stations
8. Restaurants
9. Ice cream
10. Mineralized water and soft drink–bottling plants
11. Formulations of pharmaceuticals
12. Dyeing and printing (small units)
13. Laboratory ware
14. Wire drawing (cold process)
15. Bailing straps
16. Steel furniture
17. Fasteners
18. Potassium permanganate
19. Surface coating industries
20. Fragrance, flavors, and food additives
21. Aerated water or soft drink
22. Light engineering industry excluding fabrication, electroplating
23. Small textile industry
24. Plastic industry
25. Chemical industry
26. Readymade garment industry
27. Flour mills

28. Bleaching
29. Degreasing
30. Phosphating
31. Dyeing
32. Pickling
33. Tanning
34. Polishing
35. Cooking of fibers, digesting, desizing of fabric
36. Unhairing, soaking deliming, and bating of hides
37. Washing of fabric
38. Trimming, cutting, juicing, and blanching of fruits and vegetables
39. Washing of equipment and regular floor washing, using considerable cooling water
40. Separating milk and whey
41. Steeping and processing of grain
42. Distillation of alcohol
43. Stilage evaporation
44. Slaughtering of animals, rendering of bones, washing of meat
45. Juicing of sugarcane, extraction of sugar
46. Filtration, centrifugation, distillation pulping, and fermenting of coffee beans
47. Processing of fish
48. DM plant exceeding 20 kilo liters per day capacity
49. Pulp making, pulp processing, and paper making
50. Coking of coal
51. Washing of blast furnace flue gases

## CATEGORY: GREEN

1. Washing of used sand by hydraulic discharge
2. Atta-chakkies
3. Rice millers
4. Ice boxes
5. Dal (pulses) mills
6. Groundnut (peanut) decorticating (dry)
7. Chilling
8. Tailoring and garment making
9. Cotton and woolen hosiery
10. Apparel making
11. Handloom weaving
12. Shoelace manufacturing
13. Gold and silver thread and zari (textile) work
14. Gold and silver smithy
15. Leather footwear and leather products, excluding tanning and hide processing

16. Musical instruments manufacture
17. Sports goods
18. Bamboo and cane products (only dry operations)
19. Cardboard box and paper products (paper and pulp manufacture excluded)
20. Insulation and other coated papers (paper and pulp manufacture excluded)
21. Scientific and mathematical instruments
22. Furniture (wooden and steel)
23. Assembly of domestic electrical appliances
24. Radio assembling
25. Fountain pens
26. Polyethene, plastic, and PVC goods through extrusion or molding
27. Rope (cotton and plastic)
28. Carpet weaving
29. Assembly of air coolers and conditioners
30. Assembly of bicycles, baby carriages, and other small nonmotorized vehicles
31. Electronics equipment (assembly)
32. Toys
33. Candles
34. Carpentry (excluding saw mill)
35. Cold storages (small scale)
36. Oil ginning expelling (no hydrogeneration and no refining)
37. Jobbing and machining
38. Manufacture of steel trunks and suit cases
39. Paper pins and U-clips
40. Block making for printing
41. Optical frames
42. Tyre retreading
43. Power looms and handlooms (without dyeing and bleaching)
44. Printing press
45. Garment stitching and tailoring
46. Thermometer making
47. Footwear (rubber)
48. Plastic processed goods
49. Medical and surgical instruments
50. Electronic and electrical goods
51. Rubber goods industry

## APPENDIX 14 ADVERSE EFFECTS OF AIR POLLUTION ON MATERIALS

| Materials | Air Pollutants | Effects |
|---|---|---|
| Metals | $SO_2$, acid gases | Corrosion, spoilage of surface, loss of metals, tarnishing |
| Building materials | $SO_2$, acid gases particulates | Discoloration, leaching |
| Paint | $SO_2$, $H_2S$ | Discoloration |
| Textiles | $SO_2$, acid gases | Deterioration, reduced tensile strength, and fading |
| Textile dyes | $NO_2$, ozone | Deterioration, reduced tensile strength, and fading |
| Rubber | Ozone, oxidants | Cracking, weakening |
| Leather | $SO_2$, acid gases | Disintegration, powdered surface |
| Paper | $SO_2$, acid gases | Embitterment |
| Ceramics | Acid gases | Change in surface appearance |

## APPENDIX 15 HISTORICAL DEVELOPMENT AND USE OF PESTICIDES

| Period or Era | Chemicals and Location |
|---|---|
| Era of Natural Products | |
| 900 | Arsenites (China) |
| 1690 | Tobacco (Europe) |
| 1800 | Pyrethroids (Caucasus) |
| 1848 | Derris Root (Malaya) |
| Era of Fumigants, Inorganics, Petroleum Products | |
| 1854 | Carbon Disulfide (France) |
| 1867 | Paris Green (United States) |
| 1892 | Lead Arsenate (United States) |
| 1918 | Chloropicrin (France) |
| 1932 | Methyl Bromide (France) |
| Era of Modern Synthetic Insecticides | |
| 1939 | DDT (Germany) |
| 1941 | BHC (France) |
| 1944 | Parathion (Germany) |
| 1945 | Aldrin (United States) |
| 1947 | Demeton (Switzerland) |
| 1958 | Sevin (United States) |
| Era of Hormone Mimics and Pheromones, Rebirth of Botanical Insecticide | |
| 1967 | First Juvenile Hormone Analog (United States) |
| Era of Microbial Insecticides | |
| 1980 | Avermectin Bia |

## APPENDIX 16 CLASSIFICATIONS OF PESTICIDES

1. Insecticides
   - Inorganic (Lead Arsenate)
   - Organic
     — Organochlorine compounds (Lindane)
     — Organophosphorus compounds (Malathion)
     — Carbamate compounds (Sevin)
   - Natural Products
     — Botanicals (Pyrethrum)
     — Microbial (Avermectins)
2. Herbicides
   - Inorganic Chlorates [$NaClO_3$]
   - Organic (Diuron)
3. Fungicides
   - Inorganic (Copper sulfate)
   - Organic (Captan)
4. Rodenticides
   - Inorganic (Barium carbonate)
   - Organic (Methyl fluoroacetate, Warfarin)
5. Fumigants
   - Organic (Methyl Bromide, $CS_2$), Ethylene dibromide (EDB)
6. Others
   - Plant growth regulators (1-naphthylacetic acid)
   - Repellents (Deet, Dimethyl Phthalate)

## APPENDIX 17 PESTICIDES THAT ARE SIGNIFICANTLY POISONOUS TO HUMANS (PARTIAL LISTING)

| Classification | Examples |
|---|---|
| Organophosphorus insecticides | Trichlorfon, Azinphosmethyl, Diazinon, EPN, Parathion, Fenthion, Phosphamidon |
| Carbamate insecticides | Aldicarb, Maneb, Propoxur, Thiram, Zineb, Ziram |
| Chlorinated hydrocarbon | DDT, Dieldrin, BHC, Aldrin, Endrin |
| Insecticides | Lindane, Methoxychlor, Toxaphene |
| Fumigants | $CCl_4$ Acrylonitrite, Chloropicrin, Methyl bromide, Trichloroethane |
| Solvents, oil insecticides | Kerosene, Xylene, Tetralin |
| Inorganic and organometal insecticides | Arsenic trioxide, Lead arsenate Mercuric chloride, Zinc phosphide |
| Insecticides of botanical origin | Nicotine, Pyrethrum, Rotenone |
| Rodenticides | Warfarin, N fluoroacetate |
| Herbicides | Acrolein, Diquat, Paraquat |
| Fungicides | Captafol, Diphenyl hexachlor benzene, 2, 4, 5-T. |

## APPENDIX 18 OBSERVATION OF SIGNS* AND SYMPTOMS† OF INSECTICIDE POISONING

| Organ System | Signs and Symptoms |
|---|---|
| CNS, somatomotor | Twitch, tremor, ataxia, convulsion, rigidity, flaccidity, restlessness, general motor activity, reaction to stimuli, headache, dreams, poor sleep, nervousness, dizziness |
| Autonomic | Miosis, mydriasis, salivation, lacrimation |
| Respiratory | Discharge, rhinorrhea, bradypnea, dyspnea, yawning, constriction of chest, cough, wheezing |
| Ocular | Ptosis, exophthalmos, dimness, lacrimation, conjunctival redness |
| Gastrointestinal | Diarrhea, vomiting |
| General side effects | Temperature, skin texture and color, cyanosis |

*Signs in animals.
†Symptoms in humans.

## APPENDIX 19 ACUTE AND CHRONIC TOXICITY OF REPRESENTATIVE PESTICIDES

| Pesticides | Acute Toxicity | Chronic Toxicity |
|---|---|---|
| Methyl bromide | Eye, skin, respiratory tract irritant, GI tract disturbances, CNS effects (tremors) | CNS depression, visual and speech disturbances, sensory disturbances, kidney damage |
| Carbaryl | Tremors, convulsions, ataxia, nausea, vomiting, excessive salivation, bronchial secretions, dyspnea, lacrimation, teratogenicity | Carcinogen (animal, suspected) |
| Malathion | Tremors, convulsions, ataxia, vomiting, excessive salivations, bronchial secretions, dyspnea, lacrimation, diarrhea | No apparent long-term effects |
| Dieldrin | CNS effects (tremors convulsions, coma), nausea, vomiting | Carcinogen (animal positive), liver damage (glycogen depletion, fatty infiltration, necrosis), kidney damage (fatty changes, necrosis) |
| Rotenone | Skin, eye, lung, irritant, CNS effects (tremors, convulsions) | Fatty changes in liver and kidney |
| 2, 4, 5-T | Teratogenic reproductive effects (decrease in litter size), skin irritation, ataxia, contaminant (TCDD) | No apparent long-term effects |

## APPENDIX 20 BLOOD CHOLINESTERASE INHIBITION (%) AND CLINICAL SIGNIFICANCE

- The development of the characteristic effect(s) of ChE inhibitors is related to the inhibition of enzyme in tissues, rather than in blood.
- Plasma cholinesterase (ChE) falls more rapidly than erythrocyte cholinesterase (AChE)

| Percent Depression | Clinical Significance |
|---|---|
| 40% decrease in plasma cholinesterase | Before first symptoms |
| 80% decrease in plasma cholinesterase | Serious neuromuscular effects |
| 90% decrease in plasma cholinesterase | Emergency treatment |
| 100% decrease in erythrocyte cholinesterase | Dealth |

## APPENDIX 21 CLASSIFICATION OF INSECTICIDES BASED ON THEIR MODE OF ACTION ON THE NERVOUS SYSTEM

| Component | Site Affected | Toxic Signs |
|---|---|---|
| Parasympathetic (muscarinic) | Exocrine glands | Lacrimation, increased salivation |
| | Eyes | Miosis, blurred vision, "bloody tears" |
| | Gastrointestinal tract | Nausea, vomiting, diarrhea |
| | Respiratory tract | Bronchial secretions rhinorrhea, dyspnea |
| | Cardiovascular | Tachycardia, decreased blood pressure |
| | Bladder | Urinary incontinence |
| Parasympathetic and sympathetic (nicotinic) | Cardiovascular | Tachycardia, increased blood pressure |
| Somatic motor (nicotinic) | Skeletal muscles | Fasciculations, ataxia, paralysis |
| Brain (AChE receptors) | CNS | Lethargy, tremors, convulsions, dyspnea, depression of respiratory center, cyanosis |

## APPENDIX 22 TOXIC MANIFESTATIONS OF CHRONIC EXPOSURE TO PESTICIDES AND INSECTICIDES

| | |
|---|---|
| Behavioral | Anxiety and irritability |
| | Depression |
| | Memory deficit |
| | Reduced concentration |
| | Insomnia |
| | Linguistic disturbance |
| Nonbehavioral | Tremor |
| | Ataxia |
| | Paralysis |
| | Paresthesia |
| | Polyneuritis |

## APPENDIX 23 ANTIDOTES FOR ORGANOPHOSPHORUS PESTICIDE POISONING

A Atropinization (blocking acetylcholine action at parasympathetic nerve endings in the CNS)
B Curarization (blocking acetylcholine action at the neuromuscular junction)
C Application of hexamethonium (ganglia protection)
D Application of 2-PAM (reactivation of phosphorylated cholinesterases)
E Application of artificial respiration

## APPENDIX 24 INDUSTRIAL CHEMICALS, INHALATION, AND LUNG DISEASES

| Agents | Occupational Source | Pulmonary Damage |
|---|---|---|
| Asbestos | Mining, construction, ship-building, manufacture of asbestos-containing materials | Asbestos, lung cancer |
| Aluminum dust | Manufacture of aluminum products, fireworks, ceramics, paints, electrical goods, abrasives | Fibrosis |
| Aluminum | Manufacture of abrasives, smelting | Fibrosis initiated from short exposures |
| Ammonia | Ammonia production, manufacture of fertilizers, chemical production, explosives | Irritation |
| Arsenic $Pb_3(AsO_4)_2$ | Manufacture of pesticides, pigments, glass, alloys | Lung cancer, bronchitis, laryngitis |
| Beryllium | Ore extraction, manufacture of alloys, ceramics | Dyspnea, interstitial granuloma, fibrosis, corpulmonale, chronic disease |
| Boron | Chemical process | Acute CNS |
| Cadmium oxide (fume dust) | Welding, manufacture of electrical equipment, alloys, pigments, smelting | Emphysema |
| Carbides of tungsten, titanium, tantalium | Manufacture of cutting edges on tools | Pulmonary fibrosis |
| Chlorine | Manufacture of pulp and paper, plastics, chlorinated chemicals | Irritation |
| Chromium (IV) | Production of chromium compounds, paint pigments, reduction of chromite ore | Lung cancer |
| Coal dust | Coal mining | Pulmonary fibrosis |
| Coke oven emissions | Coke production | Lung cancer (9 times greater than other steel workers) |

| Agents | Occupational Source | Pulmonary Damage |
|---|---|---|
| Hydrogen fluoride | Manufacture of chemicals, photographic film, solvents, plastics | Irritation, edema |
| Iron oxides | Welding, foundry work, steel manufacture, hematite mining, jewelry making | Diffuse fibrosis |
| Kaolin | Pottery making | Fibrosis |
| Manganese | Chemical and metal industries | |
| Nickel | Nickel ore extraction, nickel smelting, electronic electroplating, fossil fuel | Nasal cancer, lung cancer, acute pulmonary edema (NiCO) |
| Osmium tetraoxide | Chemical and metal industry | |
| Oxides of nitrogen | Welding, silo filling, explosive manufacture | Emphysema |
| Ozone | Welding, bleaching flour, deodorizing | Emphysema |
| Phosgene | Production of plastics, pesticides, chemicals | Edema |
| Perchloroethylene | Dry cleaning, metal degreasing, grain fumigating | Edema |
| Silica | Mining, stone cutting construction, farming, quarrying | Silicosis (fibrosis) |
| Sulfur dioxide | Manufacture of chemicals, refrigeration, bleaching, fumigation | |
| Talc | Rubber industry, cosmetics | Fibrosis, pleural sclerosis |
| Tin | Mining, processing of tin | |
| Toluene 2,4-diisocyanate | Manufacture of plastics | Decrement of pulmonary function ($FEV_1$) |
| Vanadium | Steel manufacture | Irritation |
| Xylene | Manufacture of resins, paints, varnishes, other chemicals, general solvent for adhesives | Edema |

## APPENDIX 25 OCCUPATIONAL EXPOSURE TO INDUSTRIAL CHEMICALS AND PULMONARY DISEASES

| Agents | Disease Common Name | Acute Effect |
|---|---|---|
| Asbestos | Asbestosis | |
| Aluminum | Aluminosis | Cough, shortness of breath |
| Aluminum abrasives | Shaver's disease, corundum smelter's lung, bauxite lung | Alveolar edema |
| Ammonia | | Immediate, upper, and lower respiratory tract irritation, edema |
| Arsenic | | Bronchitis |
| Beryllium | Berylliosis | Severe pulmonary edema, pneumonia |
| Boron | | Edema, hemorrhage |
| Cadmium oxide | | Cough, pneumonia |
| Carbides of tungsten, titanium, tantalium | Hard metal disease | Hyperplasia, metaplasia of bronchial epithelium |
| Chlorine | | Cough, hemoptysis, dyspnea, tracheobronchitis, bronchopneumonia |
| Chromium (IV) | | Nasal irritation, bronchitis |
| Coal dust | Pneumoconiosis | |
| Coke oven emissions | | Lung cancer (9 times greater than other steel workers) |
| Cotton dust | Byssinosis | Tightness in chest, wheezing, dyspnea |
| Hydrogen fluoride | | Respiratory irritation, hemorrhagic pulmonary edema |
| Iron oxides | Siderotic lung diseases: Silver finisher's lung, hematite miner's lung, arc welder's lung | Cough |
| Kaolin | Kaolinosis | |
| Manganese | Manganese pneumonia | Acute pneumonia, often fatal |

| Agents | Disease Common Name | Acute Effect |
| --- | --- | --- |
| Nickel | | Pulmonary edema, delayed by 2 days (NiCO) |
| Osmium tetraoxide | | Bronchitis, bronchopneumonia |
| Oxides of nitrogen | | Pulmonary congestion and edema |
| Ozone | | Pulmonary edema |
| Phosgene | | Edema |
| Perchloroethylene | | Pulmonary edema |
| Silica | Silicosis, pneumoconiosis | Silicosis (fibrosis) |
| Sulfur dioxide | | Bronchoconstriction, cough, tightness in chest |
| Talc | Talcosis | |
| Tin | Stanosis | |
| Toluene | | Acute bronchitis, bronchospasm, pulmonary edema |
| Vanadium | | Upper airway irritation and mucous production |
| Xylene | | Pulmonary edema |

## APPENDIX 26 DRUGS AND ADVERSE EFFECTS

| Organ | Drug/Clinical Situation | Adverse Effects |
|---|---|---|
| Heart | Propranolol, doxorubicin, digoxin | Heart failure, dysrhythmia, |
| Brain | L-dopa, Bromocriptine, ethanol (alcohol withdrawal), muscarine receptor antagonists, chlorpromazine | hallucinations, memory impairment, malignant neuroleptic syndrome |
| Eye, ear, taste touch/pain | Ethambutol, chloroquine, aminoglycoside antibiotic, captopril, vincristine | Blindness, deafness, distortion, pain and numbness |
| Locomotor | Beta-blockers, diuretics, prednisolone, phenytoin | Fatigue, gout, osteoporosis, osteomalacia |
| Stomach, pancreas, colon, liver, gallbladder | NSAIDs, asparaginase, ketoprofen, clindamycin, amoxycillin, para-cetamol, phenytoin, octreotide | Peptic ulcer, bleeding, pancreatitis, diarrhea, hepatitis, gallstones |
| Lung | Beta-adrenoceptor antagonists, NSAID | Asthma |
| Kidney | ACEI, NSAID, aminoglycoside antibiotics, analgesic abuse, methysergide, penicillamine, captopril | Acute renal failure, chronic renal failure |
| Genitourinary tract | Cyclophosphamide (hemorrhagic cystitis), thiazide diuretics | Nephrotic syndrome, hematuria, erectile impotence |
| Endocrine or metabolism | Thiazide diuretics, sulphonylureas, amiodarone, chlorpromazine, haloperidol, metoclopramide | Hyperglycemia, thyroid dysfunction, gynecomastia or galactorrhea |

## APPENDIX 27 DRUGS AND RESPIRATORY DISEASES

Abacavir
Abciximab
Acebutolol
Acetaminophen
Acetylcysteine
Acetylsalicylic acid
Acrylate
Acyclovir
Adenosine and derivatives
Adrenaline
Albumin
Albuterol
Allopurinol
Almitrine
Aminoglutethimide
Aminoglycoside antibiotics
Aminorex
Amiodarone
Amitriptyline
Amphotericin B
Ampicillin
Amrinone
Angiotensin-converting enzyme inhibitors (not mentioned elsewhere)
Antidepressants
Anti-inflammatory drugs (nonsteroidal)
Antazoline
Anti-lymphocyte (thymocyte) globulin
Aprotinin
l-Asparaginase
Atenolol
Azapropazone
Azathioprine
Azithromycin
Barbiturates
Beta-blockers
BCG therapy
BCNU
Beclomethasone
Bepridil
Beta-agonists (intravenous in ObGyn)
Betahistine
Betaxolol
Bicalutamide
Bleomycin
Blood transfusions
Bromocriptine
Bucillamine
Bumetanide
Buprenorphine
Busulfan
Cabergoline
Calcium salts
Camptothecin
Captopril
Carbamazepine
Carbimazole
Carmustine (BCNU)
CCNU
Celiprolol
Cephalexin
Cephalosporins
Chlorambucil
Chloroquine
Chlorozotocin (DCNU)
Chlorpromazine
Chlorpropamide
Ciprofloxacin
Cisapride
Clindamycin
Clofazimine
Clofibrate
Clomiphene
Clonidine
Clozapine
Colchicine
Contraceptives (oral)
Contrast media
Cotrimoxazole
Cromoglycate
Curares
Cyclophosphamide

Cyclosporin
Cyproterone acetate
Cytarabine (cytosine arabinoside)
Cytokines
Danazol
Dantrolene
Dapsone
Daunorubicin
Deferoxamine
Desipramine
Dexamethasone
Dexfenfluramine/fenfluramine
Dextropropoxyphene
Diclofenac
Diflunisal
Dihydralazine
Dihydro-5-azacytidine
Dihydroergocristine
Dihydroergocryptine
Dihydroergotamine
Diltiazem
Dimethylsulphoxide
Diphenylhydantoin
Docetaxel
Dothiepin
Doxorubicin
Enalapril
Epinephrine
Epoprostenol
Ergometrine
Ergotamine
Ergots
Erythromycin
Ethambutol
Ethchlorvynol
Etoposide
Etretinate
Febarbamate
Fenbufen
Fenfluramine/dexfenfluramine
Fenoprofen
Fibrinolytics
FK506
Flecainide
Floxuridine
Fludarabine
Fluoresceine
Flutamide
Fluticasone
Fluvastatin
5-Fluorouracil
Fluoxetine
Flurbiprofen
Fosinopril
Fotemustine
Furazolidone
Gemcitabine
Glafenine
Glibenclamide
Gliclazide
G(M)-CSF
Gold salts (aurothiopropanosulfonate)
Gonadotropin
Haloperidol
Heparin
Heroin
Hexamethonium
Hydralazine
Hydrochlorothiazide
Hydrocortisone
Hydroxyquinoleine
Hydroxyurea
Ibuprofen
Ifosfamide
Imipramine
Immunoglobulins (IV)
Indinavir
Indomethacin
Insulin
Interferon-alfa
Interferon-beta
Interferon-gamma
Interleukin-2
Iodine
Irinotecan
Isoflurane
Isoniazide
Isotretinoin
Itraconazole
Ketamine

Ketorolac
Labetalol
Lamoxactam
Latamoxef
Leuprorelin (leuprolide)
Levodopa
Levomepromazine
Lidocaine
Lipids
Lisinopril
Lisuride
Lomustine (CCNU)
Losartan
Loxoprofen
Maprotiline
Mazindol
Mecamylamine
Medroxyprogesterone
Mefloquine
Melphalan
Mephenesin
Mephenytoin
Mercaptopurine
Mesalamine
Metamizole (noramidopyrine)
Metapramine
Metformine
Methadone
Methotrexate
Methyldopa
Methylphenidate
Methylprednisolone
Methysergide
Metoclopramide
Metoprolol
Metronidazole
Miconazole
Midazolam
Minocycline
Minoxidil
Mitomycin C
Mitoxantrone
Montelukast
Morphine (agonists/antagonists)
Moxalactam
Mycophenolate mofetil
Nadolol
Naftidrofuryl
Nalbuphine
Nalidixic acid
Naloxone
Naproxen
Nevirapine
Nicergoline
Niflumic acid
Nilutamide
Niridazol
Nitric oxide (NO)
Nitrofurantoin
Nitroglycerin
Nitrosoureas
Nomifensine
Noramidopyrine (metamizole)
Estrogens
OKT3
Olsalazine
Oxprenolol
Oxyphenbutazone
Paclitaxel
Para-(4)-aminosalicylic acid (PAS)
Paracetamol
Paraffin (mineral oil)
Parenteral nutrition
Paroxetine
Penicillamine
Penicillins
Pentamidine
Pergolide
Perindopril
Phenylbutazone
Phenylephrine
Phenytoin
Pindolol
Piroxicam
Pituitary snuff
Polyethylene glycol
Practolol
Praziquantel
Procainamide
Procarbazine

Propafenone
Propylene glycol
Propoxyphene
Propranolol
Propylthiouracil
Prostacyclin
Prostaglandin F2-alpha
Protamine
Pyrimethamine-dapsone
Pyrimethamine-sulfadoxine
Quinidine
Radiations
Raltitrexed
Retinoic acid
Rifampicin
Risperidone
Ritodrine
Rituximab
Roxythromycin
Salbutamol
Sertraline
Simvastatin
Sirolimus
Sotalol
Steroids
Streptokinase
Streptomycin
Sulfamides-sulfonamides
Sulfasalazine
Sulindac
Tacrolimus
Tamoxifen
Tenidap
Terbutaline
Tetracycline
Thiopental
Tiaprofenic acid
Tiopronin
Ticlopidine
TNF-alpha
Tocainide
Tolazamide
Tolfenamic acid
Tosufloxacin
Tramadol
Trazodone
Triazolam
Trimethoprim-sulfamethoxazole
Trimipramine
Troglitazone
Troleandomycin
l-Tryptophan
d-Tubocurarine
Urokinase
Valproic acid
Valsartan
Vancomycin
Vasopressin
Venlafaxine
Vinblastine
Vindesine
Vinorelbine
Vitamin D
Warfarin
Zafirlukast
Zanamivir
Zomepirac

## APPENDIX 28 SYMPTOMS OF THYROTOXICOSIS AND HYPOTHYROIDISM

| System | Thyrotoxicosis | Hypothyroidism |
|---|---|---|
| Skin and appendages | Warm, moist skin, sweating, heat intolerance, fine, thin hair, Plummer's nails, pretibial dermopathy (Graves' disease) | Pale, cool, puffy skin, dry and brittle hair, brittle nails |
| Eyes, face | Retraction of upper lid with wide stare, periorbital edema, exophthalmos, diplopia (Graves' disease) | Drooping eyelids, periorbital edema, loss of temporal aspects of eyebrows, puffy, nonpitting facies, large tongue |
| Cardiovascular system | Decreased peripheral vascular resistance, increased heart rate, stroke volume, cardiac output, pulse pressure, high-output congestive heart failure, increased inotropic and chronotropic effects, arrhythmias, angina | Increased peripheral vascular resistance, decreased heart rate, stroke volume, cardiac output, pulse pressure, low output congestive heart failure, ECG, bradycardia, prolonged PR interval, flat T wave, low voltage, pericardial effusion |
| Respiratory system | Dyspnea, decreased vital capacity | Pleural effusions, hypoventilation and $CO_2$ retention |
| Gastrointestinal system | Increased appetite, increased frequency of bowel movements, hypoproteinemia | Decreased appetite, decreased frequency of bowel movements, ascites |
| Central nervous system | Nervousness, hyperkinesias, emotional lability | Lethargy, general slowing of mental processes, neuropathies |
| Musculoskeletal system | Weakness and muscle fatigue, increased deep tendon reflexes, hypercalcemia, osteoporosis | Stiffness and muscle fatigue, decreased deep tendon reflexes, increased alkaline phosphatase, LDH, AST |

| System | Thyrotoxicosis | Hypothyroidism |
|---|---|---|
| Renal system | Mild polyuria, increased renal blood flow, increased glomerular filtration rate | Impaired water excretion, decreased renal blood flow, decreased glomerular filtration rate |
| Hematopoietic system | Increased erythropoiesis, anemia1 | Decreased erythropoiesis, anemia1 |
| Reproductive system | Menstrual irregularities, decreased fertility, increased gonadal steroid metabolism | Hypermenorrhea, infertility, decreased libido, impotence, oligospermia, decreased gonadal steroid metabolism |
| Metabolic system | Increased basal metabolic rate, negative nitrogen balance, hyperglycemia, increased free fatty acids, decreased cholesterol and triglycerides, increased hormone degradation, increased requirements for fat- and water-soluble vitamins, increased drug detoxification | Decreased basal metabolic rate, slight positive nitrogen balance, delayed degradation of insulin, with increased sensitivity, increased cholesterol and triglycerides, decreased hormone degradation, decreased requirements for fat- and water-soluble vitamins, decreased drug detoxification |

## APPENDIX 29 IMPORTANT GLOBAL LEGISLATIONS ON DRUGS IN UNITED STATES

| Year | Law | Decisions |
|---|---|---|
| 1906 | Pure Food and Drug Act | Prohibited mislabeling and adulteration of drugs |
| 1909 | Opium Exclusion Act | Prohibited importation of opium |
| 1912 | Amendment to the Pure Food and Drug Act | Prohibited false or fraudulent advertising claims |
| 1914 | Harrison Narcotic Act | Established regulations for use of opium, opiates, and cocaine (marijuana added in 1937) |
| 1938 | Food, Drug, and Cosmetic Act | Required that new drugs be safe as well as pure (but did not require proof of efficacy) Enforcement by FDA |
| 1940 | The Drugs and Cosmetics Act, 1940 and The Drugs and Cosmetics Rules, 1945, Government of India | To regulate manufacture, distribution, and sale of drugs and cosmetics; Also deals with the labeling, packing, and import of drugs |
| 1952 | Durham-Humphrey Act | Vested in the FDA the power to determine which products could be sold without prescription |
| 1962 | Kefauver-Harris Amendments to the Food, Drug, and Cosmetic Act | Required proof of efficacy as well as safety for new drugs and for drugs released since 1938; established guidelines for reporting of information about adverse reactions, clinical testing, and advertising of the new drugs |
| 1970 | Comprehensive Drug Abuse Prevention and Control Act | Outlined strict controls in the manufacture, distribution, and prescribing of habit-forming drugs; established programs to prevent and treat drug addiction |
| 1983 | Orphan Drug Amendments | Amended Food, Drug, and Cosmetic Act of 1938, providing incentive for development of drugs that treat diseases with less than 200,000 patients in the United States |

| Year | Law | Decisions |
|---|---|---|
| 1984 | Drug Price Competition and Patent Restoration Act | Abbreviated new drug applications for generic drugs; required bioequivalence data; patent life extended by amount of time drug delayed by FDA review process; cannot exceed 5 extra years or extend to more than 14 years post-NDA approval |
| 1992 | Expedited Drug Approval Act | Allowed accelerated FDA approval for drugs of high medical need; required detailed postmarketing patient surveillance |
| 1992 | Prescription Drug User Fee Act | Manufacturers pay user fees for certain new drug applications; FDA claims review time for new chemical entities dropped from 30 months in 1992 to 20 months in 1994 |

## APPENDIX 30 QUICK REFERENCE GUIDE TO CHEMICAL AND DRUG TOXICITY

| Chemicals and Drugs | Adverse Effects |
|---|---|
| **Acetanilid; Antipyrine; Phenacetin** | |
| Used to relieve neuralgia and muscular pains, reduce fevers, and in the manufacture of other medicines | Nausea and vomiting, feeble pulse, sub-normal temperature, mental sluggishness, cyanosis, stupor, and collapse |
| **Aconite [and Aconitine]** | |
| A source of alkaloids and one of the most deadly drugs known; used to relieve pain locally, to lower blood pressure, and to reduce fever | Salivation, a tingling sensation on the lips, mouth, and in the throat, nausea and vomiting, followed by collapse |
| **Aminopyrine** | |
| Used to relieve headache, neuralgic pains, lower temperature in fever | Nausea and vomiting, feeble pulse, sub-normal temperature, mental sluggishness, cyanosis, stupor, and collapse |
| **Apomorphine** | |
| Used as a cardiac depressant, emetic, sedative, and hypnotic | Nausea and vomiting, pallor, flow of tears, exhaustion, and collapse |
| **Arnica** | |
| Used as a counterirritant in sore muscles, bruises, sprains, and strains | Pain in the throat and stomach, nausea and vomiting, pallor, weak pulse, subnormal temperature |
| **Aspirin** | |
| Used to reduce fever and relieve pain | Rapid breathing, nausea, vomiting, thirst, headache, irritability, delirium, convulsions, coma |
| **Atropine; Belladonna** | |
| Used to relieve pain and as a respiratory stimulant | Excessive thirst and dryness of the mouth and throat, difficulty in swallowing, dry, flushed skin, dilated pupils, convulsions, coma, and collapse |

| Chemicals and Drugs | Adverse Effects |
|---|---|
| **Barbital, Barbiturates** | |
| Used as sedative and as relief from pain | Overdose causes subnormal temperature, low blood pressure, and cyanosis |
| **Caffeine** | |
| Main use as a stimulant | Headache, restlessness, excitement, mental confusion, pain over the heart, inability to sleep, high blood pressure, strong pulse |
| **Cocaine** | |
| Used to relieve pain | Restlessness, nausea and vomiting, patient may be happy and talkative, pain in the abdomen, convulsions, coma |
| **Codeine** | |
| Used to lessen pain, calm nerves, and induce sleep | Nausea and vomiting, weak pulse, pallor, cold, tired, coma, collapse |
| **Digitalis** | |
| Used as a cardiac stimulant | Nausea and vomiting, pallor, diarrhea, pain in the abdomen, weak pulse, poor vision, headache, dizziness, and collapse |
| **Disulfiram (Antabuse)** | |
| Used to discourage the drinking of alcoholic beverages | Breathlessness, flushing, sweating, rapid heart action, nausea, vomiting, low blood pressure, difficulty in breathing, convulsions |
| **Ergot** | |
| Used to check bleeding from uterus or to contract the uterus in childbirth | Nausea and vomiting, cramp-like pains low in the abdomen, diarrhea, itching and tingling of the skin, weak pulse, heart pains, shortness of breath, muscle spasms and possibly convulsions, and coma before death |

| Chemicals and Drugs | Adverse Effects |
| --- | --- |
| **Morphine; Opium**<br>Used to lessen pain, calm nerves, induce sleep. | Nausea and vomiting, weak pulse, pallor, cold, tired, coma, collapse |
| **Nicotine**<br>Used as a plant spray | Pallor, tremors, palpitations, headaches, dizziness, respiratory paralysis, coma |
| **Strychnine**<br>Used in medicine and in rat poisons. | Dilated pupils, terrified expression, fixed grin, weak and feeble pulse, body arches so that it rests on the head and heels, then relaxes, the body shudders and collapse results |

## APPENDIX 31 INDUSTRIAL CHEMICALS BASED ON THE PRIORITY LIST OF THE EPA, IRC, AND EEC

### I. The Groups of Compounds

A. Organohalogen compounds
  A1. Aliphatic organohalogen compounds
  A2. Chloroethers
  A3. Monocyclic aromatic chlorohydrocarbons
  A4. Bi- and tricyclic chlorohydrocarbons
  A5. Chlorophenols
  A6. Nitroaromatic chlorohydrocarbons
  A7. Aromatic chloroamines and triazine
  A8. Chlorodioxines and furans

B. Organophosphorus compounds

C. Organotin compounds

D. Persistent mineral oils and petroleum hydrocarbons
  D1. Mono- and bicyclic aromatic hydrocarbons
  D2. Polycyclic aromatic hydrocarbons

E. Miscellaneous organic compounds
  E1. Aliphatic hydrocarbons
  E2. Aromatic hydrocarbons
  E3. Pthalate esters
  E4. Nitroaromatic hydrocarbons
  E5. Nitrosamines
  E6. Other organonitrogen compounds

F. Heavy metals and their compounds

G. Cyanide and asbestos

### II. The Individual Groups

A. Organohalogen compounds

| Priority | | List | CAS No. | Name |
|---|---|---|---|---|
| A1. Aliphatic organohalogen compounds | | | | |
| | | EEC | 79-11-8 | Acetic acid, chloro- |
| EPA | IRC | EEC | 309-00-2 | Aldrin |
| EPA | IRC | EEC | 126-99-8 | Butadiene, 2-chloro-1, 3-(Chloroprene) |
| EPA | IRC | EEC | 87-68-3 | Butadiene, hexachloro- |
| | | EEC | 302-17-0 | Chloralhydrate |

| Priority | | List | CAS No. | Name |
|---|---|---|---|---|
| EPA | IRC* | EEC* | 57-74-9 | Chlordane* |
| EPA | IRC | EEC | 608-73-1 | Cyclohexane, hexachloro-(HCH, BHC, mixed isomers) |
| EPA | IRC | EEC | 319-84-6 | Cyclohexane, $\alpha$-Hexachloro- |
| EPA | IRC | EEC | 319-85-7 | Cyclohexane, $\beta$-Hexachloro- |
| EPA | IRC | EEC | 58-89-9 | Cyclohexane, $\gamma$-Hexachloro- (lindane) |
| EPA | IRC | EEC | 319-86-8 | Cyclohexane, $\delta$-Hexachloro- |
| | IRC | EEC | 6108-10-7 | Cyclohexane, $\xi$-Hexachloro- |
| EPA | | | 77-47-4 | Cyclopentadiene, Hexachloro- |
| EPA | IRC | EEC | 60-57-1 | Dieldrin |
| EPA | IRC | EEC | 115-29-7 | Endosulfan |
| EPA | | | 1031-07-8 | Endosulfan sulfate |
| EPA | IRC | EEC | 72-20-8 | Endrin |
| EPA | | | 7421-93-4 | Endrin aldehyde |
| EPA | | | 75-00-3 | Ethane, chloro- (ethyl chloride) |
| | IRC | EEC | 106-93-4 | Ethane, 1,2-Dibromo- |
| EPA | IRC | EEC | 75-34-3 | Ethane, 1,1-Dichloro- (ethylene dichloride) |
| EPA | IRC | EEC | 107-06-2 | Ethane, 1,2-Dichloro- (ethylene dichloride) |
| EPA | IRC | EEC | 71-55-6 | Ethane, 1,1,1-Trichloro- (methyl chloroform) |
| EPA | IRC | EEC | 79-00-5 | Ethane, 1,1,2-Trichloro- |
| EPA | IRC | EEC | 79-34-5 | Ethane, 1,1,2,2-Tetrachloro- |
| EPA | IRC | EEC | 67-72-1 | Ethane, Hexachloro- |
| | | EEC | 76-13-1 | Ethane, 1,12-Trichlorotrifluoro |
| | IRC | EEC | 107-07-3 | Ethanol, 2-chloro- |
| EPA | IRC | EEC | 75-01-4 | Ethene, chloro- (vinyl chloride) |
| EPA | IRC | EEC | 75-35-4 | Ethene, 1,1-Dichloro- (vinylidene chloride) |
| EPA | IRC | EEC | 540-59-0 | Ethene, 1,2-Dichloro[†] |
| EPA | IRC | EEC | 79-01-6 | Ethene, trichloro- |
| EPA | IRC | EEC | 127-18-4 | Ethene, tetrachloro- (perchloro ethylene) |
| EPA | IRC* | EEC* | 67-44-8 | Heptachlor* |
| EPA | IRC | EEC | 102-45-73 | Heptachlor epoxide |
| EPA | | | 74-83-9 | Methane, bromo- (methyl bromide) |
| EPA | | | 74-87-3 | Methane, chloro- (methyl chloride) |

| Priority | | List | CAS No. | Name |
|---|---|---|---|---|
| EPA | IRC | EEC | 75-09-2 | Methane, dichloro- (methylene chloride) |
| EPA | | | 75-25-2 | Methane, tribromo- (bromoform) |
| EPA | IRC | EEC | 67-66-3 | Methane, trichloro- (chloroform) |
| EPA | | | 75-27-4 | Methane, bromodichloro- |
| EPA | | | 124-48-1 | Methane, dibromochloro- |
| EPA | IRC | EEC | 56-23-2 | Methane, tetrachloro- (carbon tetrachloride) |
| EPA | | | 75-71-8 | Methane, dichlorodifluoro |
| EPA | | | 75-69-4 | Methane, trichlorofluoro |
| EPA | IRC | EEC | 78-87-5 | Propane, 1,2-Dichloro- |
| | IRC | EEC | 96-23-1 | Propanol, 1,3-Dichloro-2- |
| | IRC | EEC | 107-05-1 | Propene, 3-Chloro-1- (allyl chloride) |
| EPA | IRC | EEC | 542-75-6 | Propene, 1,3-Dichloro- |
| | IRC | EEC | 78-88-6 | Propene, 2,3-Dichloro- |
| EPA | | | 8001-35-2 | Toxaphene‡ |
| A2. Chloroethers | | | | |
| | IRC | EEC | 106-89-8 | Epichlorohydrin |
| EPA | | | 542-88-1 | Ether, bis(chloromethyl)- |
| EPA | | | 111-44-4 | Ether, bis(2-chloroethyl)- |
| EPA | IRC | EEC | 108-60-1 | Ether, bis(2-chloroisoprophyl)- |
| EPA | | | 110-75-8 | Ether, 2-Chloroethyl vinyl- |
| EPA | | | 101-55-3 | Ether, 2-Bromophenyl phenyl- |
| EPA | | | 7005-72-3 | Ether, 2-Chlorophenyl phenyl |
| EPA | | | 111-91-1 | Methane, Bis(2-chloroethoxy)- |
| A3. Monocyclic aromatic chlorohydrocarbons | | | | |
| EPA | IRC | EEC | 108-90-7 | Benzene, chloro- |
| EPA | IRC | EEC | 95-50-1 | Benzene, 1,2-Dichloro- (o-Dichlorobenzene) |
| EPA | IRC | EEC | 541-73-1 | Benzene, 1,3-Dichloro- (m-Dichlorobenzene) |
| EPA | IRC | EEC | 106-46-7 | Benzene, 1,4-Dichloro- (p-Dichlorobenzene) |
| | IRC | EEC | 87-61-6 | Benzene, 1,2,3-Trichloro- |
| EPA | IRC | EEC | 120-82-1 | Benzene, 1,2,4-Trichloro- |
| | IRC | EEC | 108-70-3 | Benzene, 1,3,5-Trichloro- |
| | IRC | EEC | 95-94-3 | Benzene, 1,2,4,5-Tetrachloro- |

| Priority | | List | CAS No. | Name |
|---|---|---|---|---|
| EPA | IRC | EEC | 118-74-1 | Benzene, hexachloro- |
| | IRC | EEC | 95-49-8 | Benzene, 1-Chloro-2-methyl- (o-Chlorotoluene) |
| | IRC | EEC | 108-41-8 | Benzene, 1-Chloro-3-methyl- (m-Chlorotoluene) |
| | IRC | EEC | 106-43-4 | Benzene, 1-Chloro-4-methyl- (p-Chlorotoluene) |
| | | EEC | 100-44-7 | Benzylchloride ($\alpha$-Chlorotoluene) |
| | | EEC | 98-87-3 | Benzylidenechloride ($\alpha$,$\alpha$-Di-chlorotoluene) |
| | IRC | EEC | 789-02-6 | DDT, o,p′- |
| EPA | IRC | EEC | 50-29-3 | DDT, p,p′- |
| EPA | IRC | EEC | 72-54-8 | DDD, p,p′- |
| EPA | IRC | EEC | 72-55-9 | DDE, p,p′- |

A4. Bi- and tricyclic aromatic chlorohydrocarbons

| Priority | | List | CAS No. | Name |
|---|---|---|---|---|
| | | EEC | 90-13-1 | Naphthalene, 1-Chloro- |
| EPA | | EEC | 91-58-7 | Naphthalene, 2-Chloro- |
| EPA | IRC | EEC | 1336-36-3 | Biphenyls, polychloro- (PCBs) |
| EPA | IRC | EEC | 12674-11-2 | PCB-1016 (Aroclor 1016) |
| EPA | IRC | EEC | 11104-28-2 | PCB-1221 (Aroclor 1221) |
| EPA | IRC | EEC | 11141-16-5 | PCB-1232 (Aroclor 1232) |
| EPA | IRC | EEC | 53469-21-9 | PCB-1242 (Aroclor 1242) |
| EPA | IRC | EEC | 12672-29-6 | PCB-1248 (Aroclor 1248) |
| EPA | IRC | EEC | 11097-69-1 | PCB-1254 (Aroclor 1254) |
| EPA | IRC | EEC | 11096-82-5 | PCB-1260 (Aroclor 1260) |
| | IRC | EEC | | Triphenyls, polychloro- (PCTs) |
| | IRC | EEC | 12642-23-8 | PCT-5442 (aroclor 5442) |
| | IRC | EEC | 11126-42-4 | PCT-5460 (aroclor 5460) |

A5. Chlorophenols

| Priority | | List | CAS No. | Name |
|---|---|---|---|---|
| | IRC | EEC | 94-75-7 | 2,4-D |
| | | EEC | | 2,4-D esters |
| | | EEC | | 2,4-D salts |
| | | EEC | 94-74-6 | MCPA |
| | | EEC | 93-65-2 | Mecoprop |
| | IRC | EEC | 93-76-5 | 2,4,5, -T |
| | | EEC | | 2,4,5, -T esters |
| | | EEC | | 2,4,5, -T salts |
| | | EEC | 120-36-5 | Dichlorprop |

| Priority | | List | CAS No. | Name |
|---|---|---|---|---|
| EPA | IRC | EEC | 95-57-8 | Phenol, 2-Chloro- (o-Chlorophenol) |
| | IRC | EEC | 108-43-0 | Phenol, 3-Chloro- (p-Chlorophenol) |
| | IRC | EEC | 106-48-9 | Phenol, 4-Chloro- (m-Chlorophenol) |
| EPA | IRC | EEC | 120-83-2 | Phenol, 2,4-Dichloro- |
| | IRC | EEC | 95-95-4 | Phenol, 2,4,5-Trichloro- |
| EPA | | EEC | 88-06-2 | Phenol, 2,4,6-Trichloro- |
| EPA | IRC | EEC | 87-86-5 | Phenol, pentachloro- |
| EPA | IRC | EEC | 59-50-7 | Phenol, 4-Chloro-2-methyl (p-Chloro-m-cresol) |
| | | EEC | 95-85-2 | Phenol, 2-Amino-4-chloro- |

A6. Nitroaromatic chlorohydrocarbons

| Priority | | List | CAS No. | Name |
|---|---|---|---|---|
| | | EEC | 89-63-4 | Analine, 4-Chloro-2-nitro- |
| | IRC | EEC | 88-73-3 | Benzene, 1-Chloro-2-nitro (o-Chloronitrobenzene) |
| | IRC | EEC | 121-73-3 | Benzene, 1-Chloro-3-nitro (m-Chloronitrobenzene) |
| | IRC | EEC | 100-00-5 | Benzene, 1-Chloro-4-nitro (p-Chloronitrobenzene) |
| | IRC | EEC | 3209-22-1 | Benzene, 2,3-Dichloronitro- |
| | | EEC | 611-06-3 | Benzene, 2,4-Dichloronitro- |
| | | EEC | 89-61-2 | Benzene, 2,5-Dichloronitro- |
| | | EEC | 601-88-7 | Benzene, 2,6-Dichloronitro- |
| | | EEC | 99-54-7 | Benzene, 3,4-Dichloronitro- |
| | | EEC | 618-62-2 | Benzene, 3,5-Dichloronitro- |
| | | EEC | 97-00-7 | Benzene, 1-Chloro-2,4-dinitro- |
| | | EEC | 3970-40-9 | Toluene, 2-Chloro-3-nitro- |
| | | EEC | 121-68-8 | Toluene, 2-Chloro-4-nitro- |
| | | EEC | 13290-74-9 | Toluene, 2-Chloro-5-nitro- |
| | | EEC | 83-41-1 | Toluene, 2-Chloro-6-nitro- |
| | | EEC | 5367-26-0 | Toluene, 3-Chloro-2-nitro- |
| | | EEC | 38939-88-7 | Toluene, 3-Chloro-4-nitro- |
| | | EEC | 16582-38-0 | Toluene, 3-Chloro-5-nitro- |
| | | EEC | 5367-28-2 | Toluene, 3-Chloro-6-nitro- |
| | IRC | EEC | 89-59-8 | Toluene, 4-Chloro-2-nitro- |
| | | EEC | 89-60-1 | Toluene, 4-Chloro-3-nitro- |
| | | EEC | 1582-09-8 | Trifluralin |

| Priority | List | | CAS No. | Name |
|---|---|---|---|---|
| A7. Aromatic chloroamines and triazine | | | | |
| | IRC | EEC | 95-51-2 | Aniline, 2-Chloro- (o-Chloroaniline) |
| | IRC | EEC | 108-42-9 | Aniline, 3-Chloro- (m-Chloroaniline) |
| | IRC | EEC | 106-47-8 | Aniline, 4-Chloro- (p-Chloroaniline) |
| | IRC | EEC | 1746-81-2 | Monolinuron |
| | IRC | EEC | 608-27-5 | Aniline, 2,3-Dichloro- |
| | IRC | EEC | 554-00-7 | Aniline, 2,4-Dichloro- |
| | IRC | EEC | 95-82-9 | Aniline, 2,5-Dichloro- |
| | IRC | EEC | 608-31-1 | Aniline, 2,6-Dichloro- |
| | IRC | EEC | 95-76-1 | Aniline, 3,4-Dichloro- |
| | IRC | EEC | 626-43-7 | Aniline, 3,5-Dichloro- |
| | | EEC | 709-98-8 | Propanil |
| | IRC | EEC | 330-55-2 | Linuron |
| | | | See class A5 | Phenol, 2-Amino-4-chloro- |
| | | | See class A6 | Aniline, 4-Chloro-2-nitro- |
| | | EEC | 87-60-5 | Toluidine, 3-Chloro-o- |
| | | EEC | 95-79-4 | Toluidine, 5-Chloro-o- |
| | | EEC | 87-63-8 | Toluidine, 6-Chloro-o- |
| | | EEC | 29027-17-6 | Toluidine, 2-Chloro-m- |
| | | EEC | 7149-75-9 | Toluidine, 4-Chloro-m- |
| | | EEC | | Toluidine, 5-Chloro-m- |
| | | EEC | 95-81-8 | Toluidine, 6-Chloro-m- |
| | | EEC | 615-65-6 | Toluidine, 2-Chloro-p- |
| | | EEC | 95-74-9 | Toluidine, 3-Chloro-p- |
| | | | | Trifluralin: see class A5. |
| | | EEC | 1331-47-1 | Benzidine, dichloro- |
| | | EEC | 84-68-4 | Benzidine, 2,2′-dichloro- |
| EPA | | EEC | 91-94-1 | Benzidine, 3,3′-Dichloro- |
| | | EEC | 63390-11-4 | Benzidine, 3,5′-Dichloro- |
| | | EEC | 1698-60-8 | Pyrazo |
| | | EEC | 122-34-9 | Simazine |
| | IRC | EEC | 108-77-0 | Triazine, 2,4,6-Trichloro-1,3,5- (cyanuric acid chloride) |

*Has been removed from priority list.

†EPA: 1,2-*trans*-Dichloroethene; IRC and EEC; *cis* or *trans* isomer not specified.

‡Technical mixture of mostly polychlorobornanes, polychlorobornenes, or polychlorotricyclenes.

B. Organophosphorus compounds

| Priority | List | CAS No. | Name |
|---|---|---|---|
| IRC | EEC | 86-50-0 | Azinphos-methyl |
| IRC | EEC | 2642-71-9 | Azinphos-ethyl |
| IRC | EEC | 56-72-4 | Coumaphos |
| | EEC | 8065-48-3 | Deweton (mixture of Demeton-O- with Demeton-S) |
| | EEC | 298-03-3 | Demeton-O- |
| | EEC | 126-75-0 | Demeton-S- |
| | EEC | 919-86-8 | Demeton-S-methyl |
| | EEC | 17040-19-6 | Demeton-S-methylsulfon |
| | EEC | 62-73-7 | Dichlorvos |
| | EEC | 60-51-5 | Dimethoate |
| IRC | EEC | 298-04-4 | Disulfoton |
| | EEC | 122-14-5 | Fenitrothion |
| | EEC | 55-38-9 | Fenthion |
| | EEC | 121-75-5 | Malathion |
| IRC | EEC | 10265-92-6 | Methamidophos |
| | EEC | 7786-34-7 | Mevinphos |
| | EEC | 1113-02-6 | Omethoate |
| | EEC | 301-12-2 | Oxydemeton-methyl |
| IRC | EEC | 56-38-2 | Parathion |
| IRC | EEC | 298-00-0 | Parathion, methyl- |
| | EEC | 126-73-8 | Phosphate, tributyl- |
| | EEC | 14816-18-3 | Phoxim |
| | EEC | 24017-47-8 | Triazophos |
| | EEC | 52-68-6 | Trichlorfon |

C. Organotin compounds

| Priority | List | CAS No. | Name |
|---|---|---|---|
| | EEC | 818-08-6 | Tin oxide, dibutyl- |
| | EEC | 683-18-1 | Tin dichloride, dibutyl- |
| | EEC | | Tin salts, other dibutyl- |
| IRC | EEC | 56-35-9 | Tin oxide, tributyl- |
| | EEC | 1461-25-2 | Tin, tetrabutyl- |
| IRC | EEC | 639-58-7 | Tin chloride, triphenyl- (fentin chloride) |
| IRC | EEC | 76-87-9 | Tin hydroxide, triphenyl- (fentin hydroxide) |
| IRC | EEC | 900-95-8 | Tin acetate, triphenyl- (fentin acetate) |

D. Persistent mineral oils and petroleum hydrocarbons

| Priority | | List | CAS No. | Name |
|---|---|---|---|---|
| D1. Mono- and bicyclic aromatic hydrocarbons | | | | |
| EPA | IRC | EEC | 71-43-2 | Benzene |
| EPA | | EEC | 108-88-3 | Toluene |
| | | EEC | 1330-20-7 | Xylene (isomers) |
| | | EEC | 95-47-6 | Xylene, o- |
| | | EEC | 108-38-3 | Xylene, m- |
| | | EEC | 106-42-3 | Xylene, p- |
| EPA | | EEC | 100-41-4 | Benzene, ethyl- |
| | | EEC | 98-82-8 | Benzene, isopropyl- |
| | | EEC | 92-52-4 | Biphenyl |
| D2. Polycyclic aromatic hydrocarbons (PAHs) | | | | |
| EPA | | | 83-32-9 | Acenaphthene |
| EPA | | | 208-96-8 | Acenaphthylene |
| EPA | | EEC | 120-12-7 | Anthracene |
| EPA | | | 56-55-3 | Anthracene, benzo[a]-(1,2-Benzanthracene) |
| EPA | | | 53-80-3 | Anthracene, dibenzo[a,h]-(1,2,5,6-Dibenzanthracene) |
| EPA | | | 218-01-9 | Chrysene |
| EPA | IRC* | EEC* | 206-44-0 | Fluoranthene* |
| EPA | IRC* | EEC* | 205-99-2 | Fluoranthene*, 3,4-Benzo(Benzo[b]-fluoranthene) |
| EPA | IRC* | EEC* | 207-08-9 | Fluoranthene*, 11,12-Benzo(Benzo[k]-fluoranthene) |
| EPA | | | 86-73-7 | Fluorene |
| EPA | IRC | EEC | 91-20-3 | Naphthalene |
| EPA | IRC* | EEC* | 191-24-2 | Perylene*, 1,12-Benzo-(Benzo[g,h,I]-perylene) |
| EPA | | | 85-01-8 | Phenantrene |
| EPA | | | 129-00-0 | Pyrene |
| EPA | IRC* | EEC* | 50-32-8 | Pyrene*, 3,4-Benzo-(Benzo[a]-pyrene) |
| EPA | IRC* | EEC* | 193-39-5 | Pyrene*, Indeno [1,2,3-cd]- |
| | 6 PAHs according to Borneff | | | |
| EPA | | | 107-02-8 | Acrolein (2-Propenal) |

E. Miscellaneous organic compounds

| Priority | List | | CAS No. | Name |
|---|---|---|---|---|
| E1. Aliphatic hydrocarbons | | | | |
| E2. Aromatic hydrocarbons | | | | |
| EPA | | | 108-95-2 | Phenol |
| EPA | | | 105-67-9 | Phenol, 2,4-Dimethyl- |
| E3. Phtalate esters | | | | |
| EPA | | | 131-11-3 | Phtalate, dimethyl- |
| EPA | | | 84-66-2 | Phtalate, diethyl- |
| EPA | | | 84-74-2 | Phtalate, di-n-butyl- |
| EPA | | | 117-84-0 | Phtalate, di-n-octyl- |
| EPA | | | 117-81-7 | Phtalate, bis(2-ethylhexyl)- |
| EPA | | | 85-68-7 | Phtalate, butylbenzyl- |
| E4. Nitroaromatic hydrocarbons | | | | |
| EPA | | | 98-95-3 | Benzene, nitro- |
| EPA | | | 88-75-5 | Phenol, 2-Nitro- (o-Nitrophenol) |
| EPA | | | 100-02-7 | Phenol, 4-Nitro- (p-Nitrophenol) |
| EPA | | | 51-28-5 | Phenol, 2,4-Dinitro- |
| EPA | | | 121-14-2 | Toluene, 2,4-Dinitro- |
| EPA | | | 606-20-2 | Toluene, 2,6-Dinitro- |
| EPA | | | 534-52-1 | Cresol, 2,6-Dinitro-o- |
| E5. Nitrosamines | | | | |
| EPA | | | 65-75-9 | Nitrosamine, dimethyl- |
| EPA | | | 621-64-7 | Nitrosamine, di-n-propyl- |
| EPA | | | 86-30-6 | Nitrosamine, diphepyl- |
| E6. Other organonitrogen compounds | | | | |
| EPA | | | 107-13-1 | Acrylonitrile (2-Propenenitrile) |
| | | EEC | 124-40-3 | Dimethylamine |
| | | EEC | 109-89-7 | Diethylamine |
| EPA | IRC | EEC | 92-87-5 | Benzidine |

| Priority | | List | CAS No. | Name |
|---|---|---|---|---|
| EPA | | | 122-66-7 | Hydrazine, 1,2-Diphenyl- (Hydrazobenzene) |
| EPA | | | 78-59-1 | Isophorone |
| F. Heavy metals and their compounds[†] | | | | |
| EPA | | | 7440-36-0 | Antimony and compounds (total) |
| EPA | IRC | EEC | 7440-38-2 | Arsenic (total) |
| EPA | | | 7440-41-7 | Beryllium (total) |
| EPA | IRC | EEC | 7440-43-9 | Cadmium (total) |
| EPA | | | 7440-47-3 | Chromium (total) |
| EPA | | | 7440-50-8 | Copper (total) |
| EPA | | | 7439-92-1 | Lead (total) |
| EPA | IRC | EEC | 7439-97-6 | Mercury (total) |
| EPA | | | 7440-02-0 | Nickel (total) |
| EPA | | | 7782-49-2 | Selenium (total) |
| EPA | | | 7440-22-4 | Silver (total) |
| EPA | | | 7440-28-4 | Thallium (total) |
| EPA | | | 7440-66-6 | Zinc (total) |
| G. Cyanide and asbestos | | | | |
| EPA | | | 420-05-3 | Cyandie |
| EPA | | | | Asbestos (Fibrous) |

*Has been removed from priority list.

[†]According to EPA: The term "compounds" shall include organic and inorganic compounds; IRC: Arsenic and its mineral compounds; EEC: Arsenic and its inorganic compounds.

## APPENDIX 32 QUICK REFERENCE GUIDE TO CHEMICAL AND DRUG TOXICITY

| Chemicals | Adverse Effects |
|---|---|
| **Acid, Acetic** | |
| Used to make acetates, acetate plastics, acetate rayon, and as a solvent | Skin is yellow where it comes in contact with the acid; burns on the lips and mouth; pain in the throat and stomach; difficulty in swallowing; nausea and vomiting; feeble pulse; diarrhea and collapse |
| **Acid, Boric** | |
| Used as an eyewash and in external ointments | Nausea, diarrhea, headache, cold sweat, subnormal temperature, rash, collapse |
| **Acid, Hydrochloric** | |
| Used extensively in industry and the laboratory | The lips and mouth on contact with the acid are usually white at first, but later turn brown; pain in the throat and stomach; difficulty swallowing; nausea and vomiting; feeble pulse; diarrhea and collapse |
| **Acid, Hydrocyanic** | |
| Gas used as a fumigant for citrus trees, and in ships and buildings against rodents and vermin; salts used for case-hardening steel | Possesses a peculiar peach blossom odor; nausea and vomiting; feeble pulse; shallow breathing; dyspnea; cyanosis; convulsions and collapse |
| **Acid, Hydrofluoric** | |
| Gas or liquid used for etching glass; also used in the manufacture of fluorides | Skin is yellow where it comes in contact with the acid; burns on the lips and mouth; pain in the throat and stomach; difficulty swallowing; nausea and vomiting; feeble pulse; diarrhea and collapse |

| Chemicals | Adverse Effects |
|---|---|
| **Acid, Nitric** | |
| Used extensively as a nitrating agent in making explosive and fertilizers; also used as an oxidizing agent | Stains on the lips and mouth are first white, later turning to a deep yellow. |
| **Acid, Oxalic** | |
| Used as an industrial bleach and as an oxidation-reduction standard in the laboratory | |
| **Acid, Phosphoric** | |
| Used in manufacture of many phosphates; also in the engraving and lithography trades | The adverse effects are similar to acetic acid |
| **Acid, Sulfuric** | |
| The most widely used acid in chemical industry | |
| **Alcohol, Ethyl** (grain alcohol) | |
| Used in beverages, medicines, and extracts | Effects vary; some patients become quarrelsome, some sentimental, others fall asleep; nausea and vomiting; patient enters a stage of depression. |
| **Alcohol, Isopropyl** | |
| A substitute for rubbing alcohol | |
| **Alcohol, Methyl** (wood alcohol) | |
| Used as solvent for shellacs and resins, in the manufacture of dyes and varnishes, as an antifreeze and a fuel | Initial symptoms like those of ethyl alcohol; later, nausea and vomiting; dizziness, headache; dilated pupils; delirium; blindness |
| **Ammonium Hydroxide** (ammonia water) | |
| Used in cleaning and bleaching, removing stains; has a wide variety of uses about the home | Burns on the lips and mouth; severe pains in the throat and stomach; diarrhea, weak pulse; pallor and collapse |

| Chemicals | Adverse Effects |
|---|---|
| **Antimony Trichloride**<br>Used in medicine and in the manufacture of alloys and fireworks<br>**Antimony Potassium Tartrate**<br>Used as an emetic and expectorant | Metallic taste in the mouth; pains in the abdomen; nausea and vomiting; vomitus is blood-stained; spasms of the fingers, arms and legs, followed by collapse |
| **ANTU**<br>Used as a rodenticide | Sharp drop in temperature and pulmonary edema |
| **Arsenic**<br>Used in hardening metals, in alloys, rat poisons, flypaper, for trees and garden sprays and in dyes | Pain in the throat and stomach; nausea and vomiting; pallor; weak pulse; abdominal cramps; thirst; coma; convulsions and collapse |
| **Barium Acetate**<br>Used as a mordant for printing fabrics<br>**Barium Carbonate**<br>Used in rat poisons, paints, and enamels<br>**Barium Chloride**<br>Uses are similar to those of barium carbonate<br>**Barium Sulfide**<br>Used as a depilatory, in luminous paints and vulcanizing rubber<br>**Barium Sulfite**<br>Has a variety of uses in industry | Nausea and vomiting; abdominal cramps; diarrhea; salivation; paralysis of the arms and legs; pallor; weak pulse |

| Chemicals | Adverse Effects |
|---|---|
| **Benzene Hexachloride** | |
| Used as an insecticide | Vomiting; diarrhea; convulsions; difficulty breathing |
| **Bismuth Compounds** | |
| Used medicinally many times as bismuth dressings | Nausea and vomiting; salivation; a blue line at the junction of the teeth and gums; swelling of the gums, tongue, and throat |
| **Borates** | |
| Used in cleansers, soaps, and detergents | Nausea; diarrhea; headache; cold sweat; subnormal temperature; rash; collapse |
| **Bromides** | |
| Like sodium bromide, potassium bromide, ammonium bromide, used as drugs | Mental confusion; nausea; vomiting; stomach pains; delirium; coma |
| **Cadmium** | |
| Used for plating other metals and alloys | Nausea; Vomiting; diarrhea; headache; stomach pains; salivation |
| **Calcium Hydroxide and Calcium Oxide** | |
| Used in plasters, cements, mortars, water paints, dehairing hides, and as an insecticide | Pain in the throat and stomach; nausea and vomiting; thirst; pallor; weak pulse; collapse |
| **Carbon Dioxide** | |
| Used in beverages and fire extinguishers, as solid dry ice, and as a refrigerant | Headache; unconsciousness; failure of respiration and circulation |
| **Carbon Monoxide** | |
| Present in automobile exhaust and in some industrial gases | Headache; bluish red patches on body; unconsciousness; failure of respiration and circulation |
| **Carbon Tetrachloride** | |
| Used in the extinguishers, dry cleaning, and as a solvent | Nausea and vomiting; headache; dizziness; pallor; weak pulse; subnormal temperature |

| Chemicals | Adverse Effects |
|---|---|
| **Chloral Hydrate** | |
| Used to induce sleep; also known as "knockout drops" | Drowsiness; lassitude; cold hands and feet; nausea and vomiting; headache; stupor; heart failure |
| **Chlordane, Dieldrin, DDT, Heptachlor** | |
| Used as an insecticide | Overexcitability; tremors; convulsions; nausea; vomiting; weakness; depression; coma |
| **Chlorine and Chlorine Water** | |
| Used to disinfect and deodorize, a bleach for wood, paper, pulp, cotton, and many other products | Pain in the throat and stomach; nausea and vomiting; weak pulse; pallor; difficulty breathing |
| **Chloroform** | |
| Used as an anesthetic, analgesic, antiseptic, and solvent | Slow, weak pulse becoming ever slower; pallor; diluted pupils; paralysis of the heart |
| **Copper Acetate or Cupric Acetate** | |
| Used in the manufacture of pigments, fungicides, algacides, and insecticides, in dyeing and printing fabrics | |
| **Copper Acetoarsenite** | |
| Used as a pigment, an insecticide, and a wood preservative | |
| **Copper Arsenite or Cupric Arsenite** | |
| Its uses are similar to those of copper acetoarsenite and cupric acetate | Nausea and vomiting; pallor; diarrhea; symptoms of collapse; heart failure |
| **Copper Sulfate or Cupric Sulfate** | |
| Its uses are similar to those of copper acetoarsenite, cupric acetate, and cupric arsenite | |

| Chemicals | Adverse Effects |
|---|---|
| **Creolin, Creosote, and Cresols** | |
| Water emulsion of phenolics (cresols); used as disinfectants, germicides, and deodorants | Like phenol, these chemicals cause severe burns |
| **2,4-D** | |
| Used as a weed killer | Irritation of eyes; intestinal disturbance; muscle stiffness; paralysis; coma |
| **Cyanides** | |
| Potassium, sodium, and other salts | Possesses a peculiar peach blossom odor; Nausea and vomiting; feeble pulse; shallow breathing; dyspnea; cyanosis; convulsions and collapse |
| **Dinitro-o-Cresol** | |
| Used as a selective weed killer and insecticide | Thirst; fatigue; excessive sweating; nausea; vomiting; abdominal pains; high temperature; difficulty breathing; restlessness; convulsions; prostration |
| **Dinitrophenol** | |
| Used as an insecticide and weed killer | |
| **Ether** | |
| Used as a general anesthetic, a stimulant, a solvent, and cleaning agent | Slow weak pulse becoming ever slower; pallor; diluted pupils; paralysis of the heart |
| **Fluorides** | |
| Used as an insecticide | Burning cramp-like pains in the abdomen; grayish-blue skin; weak pulse; pallor; collapse |
| **Fluoroacetates** | |
| Used as rat poison | Nausea; vomiting; mental uneasiness; epileptiform convulsions; uneven heart beat and respiration; exhaustion; coma |

| Chemicals | Adverse Effects |
|---|---|
| **Formaldehyde** | |
| Used in embalming fluid, for hardening, films, as a germicide, antiseptic, and deodorant | Nausea and vomiting; clammy skin; weak pulse; pallor; burning in the mouth and throat; collapse |
| **Gasoline** | |
| Has many uses, the greatest being as fuel | Nausea and vomiting; headache; giddiness; affected vision; in general, symptoms greatly resemble drunkenness |
| **Hydrogen Peroxide** | |
| Used in medicine and also as a bleaching agent, an oxidizing agent, an antiseptic, and a catalyst | Nausea and vomiting; pallor; weak pulse |
| **Hydrogen Sulfide** | |
| Used as a reducing agent | Nausea vomiting; greenish face; weak pulse; coma and respiratory failure |
| **Iodoform** | |
| Used as an antiseptic | Headache; rapid; weak pulse; pallor; dizziness; may attempt suicide; collapse |
| **Lead Salts** | |
| Have many uses in industry | Headache; metallic taste in mouth and throat; nausea and vomiting; blue line on the gums; constricted throat; diarrhea; anemia and paralysis may appear |
| **Lye or Sodium Hydroxide** | |
| Used extensively in industry and the home | Burns on the lips and mouth; severe pains in the throat and stomach; diarrhea; weak pulse; pallor; collapse |

| Chemicals | Adverse Effects |
|---|---|
| **Mercuric Oxide, Red**<br>Used in medicine and industry | |
| **Mercuric Oxide, Yellow**<br>Used in medicine and industry | Metallic taste in the mouth; nausea and vomiting; thirst; diarrhea; weak pulse; slow, shallow breathing; collapse |
| **Mercurous Chloride**<br>A powerful antiseptic | |
| **Mercury Bichloride**<br>Has many uses in medicine; also used in various industries | |
| **Naphthalene**<br>In mothballs, in dye, resin, and plastic industries | Restlessness; depression; twitching; the urine is brown to black; weak pulse; pallor; coma; snoring |
| **Phenol**<br>Used as an antiseptic, disinfectant, and deodorant | Whitish burns on the mouth; pains in the throat and stomach; nausea and vomiting; dizziness; pallor; weak pulse; shallow breathing; depression; unconsciousness |
| **Phosphorus, Red**<br>Its uses are similar to those of white phosphorus | |
| **Phosphorus, White**<br>Used in fireworks, poisons for mice and rats | Nausea and vomiting; a garlic taste; headache; pallor; weak pulse; diarrhea; vomitus luminous in dark; collapse |
| **Picric Acid**<br>Used in matches, explosives, in the leather industry, and in the manufacture of textile mordant | Skin yellow where it contacts the acid; weak pulse; pallor; nausea and vomiting; convulsions; collapse |

| Chemicals | Adverse Effects |
|---|---|
| **Potassium Carbonate** | |
| Used in the manufacture of soap, glass, and pottery | Nausea and vomiting; pain in the throat and stomach; weak pulse; pallor; collapse |
| **Potassium Chlorate** | |
| Used in the manufacture of matches and fireworks | Nausea and vomiting; pain in the throat and stomach; diarrhea; jaundice; weak pulse; pallor; cyanosis; coma; collapse |
| **Potassium Permanganate** | |
| Used for bleaching resins, waxes, oils, and fats | Nausea and vomiting; rapid, weak pulse; pallor; cold, clammy skin; collapse |
| **Silver Nitrate** | |
| Used in the manufacture of indelible inks, silver salts, and for re-silvering mirrors | Nausea and vomiting, the vomitus being black; pain in the throat and stomach; weak pulse; pallor; coma; collapse |
| **Sodium Carbonate** | |
| Used in manufacture of soap, glass, sodium salts, as a detergent and water softener | Nausea and vomiting; pain in the throat and stomach; weak pulse; pallor; collapse |
| **Sodium Fluoride** | |
| Used as an insecticide | Burning cramp-like pains in the abdomen; grayish-blue skin; weak pulse; pallor; collapse |
| **Sodium Fluoroacetate [1080]** | |
| Used as rat poison | Nausea; vomiting; mental uneasiness; epileptiform convulsions; uneven heart beat and respiration; exhaustion; coma |
| **Sodium Hydroxide** | |
| Used in the manufacture of paper and soap, in oil refining, and other industries | Burns on the lips and mouth; severe pains in the throat and stomach; diarrhea; weak pulse; pallor; collapse |

| Chemicals | Adverse Effects |
|---|---|
| **Sodium Nitrate** | |
| Used in the manufacture of diazo dyes | Nausea and vomiting; flushed face; violent then lessened heart action; dilated pupils; pallor; collapse |
| **Thallium Salts** | |
| Used in rat poisons and ant powders | Severe abdominal pains; purplish gums; foul breath; salivation; respiratory failure |
| **Thiocyanates** | |
| Used as an insecticide | Respiratory difficulty and convulsions |
| **Toxaphene** | |
| Used as an insecticide | Convulsions sometimes preceded by nausea and vomiting; weakness; lassitude; amnesia |
| **White Lead** | |
| Used in putty and pigments | Metallic taste; dry throat; nausea and vomiting; diarrhea; leg cramps; blue line on gums; pallor; weak pulse; anemia; paralysis |
| **Zinc Acetate; Zinc Chloride; Zinc Sulfate** | |
| Used in medicine and many industries | Metallic taste; pain in the stomach; salivation; nausea and vomiting, the vomitus of bloody material; purging; pallor; collapse |
| **Zinc Phosphide** | |
| Used as rat poison | Difficulty breathing; nausea; vomiting; stomach pains; diarrhea; slow action of heart; circulatory collapse |

## APPENDIX 33 INDUSTRIAL CHEMICALS AND GLOBAL DISASTERS

Global productivity of different chemicals now ranges to several hundred million tons every year. With the ambitious growth of chemical industries to meet national and domestic needs, innumerable compounds are being manufactured. In fact, large volumes of chemicals are produced, stored, and transported often with less care and precautions. A number of chemicals are known to be toxic and potentially hazardous to humans and the environment. In fact, several instances indicate that chemical accidents have occurred all over the world. It is now clear in all most all cases that negligence, improper management, and lack of proper knowledge for the disposal of hazardous chemicals have lead to such disasters where human health and environmental safety were in jeopardy. The ecological and environmental catastrophe from Schweizerhalle, Switzerland in 1986 is an example. The Rhine River was polluted with large quantities of chemicals and resulted in contamination of a huge fish population in the river in addition to causing ecological disturbances. Although some chemical disasters provide details of human injuries and tragedies, a number of cases fail to provide substantial documentary evidence. This lack has failed to educate the public when a similar chemical disaster occurs in another part of the globe. The following list provides a few of the major global disasters linked with different industrial chemicals.

| Year | Location | Chemical | Death | Injuries |
|---|---|---|---|---|
| 1958 | Kerala, India | Parathion poisoning and food contamination | Large number | — |
| 1966 | Corpus Cristi, Texas | Malathion poisoning | — | — |
| 1969 | Columbia, Egypt, Iran, Malaysia, Mexico | Parathion poisoning | — | — |
| 1972 | Iraq | Organic mercury poisoning in seed grains | — | — |
| | Love Canal, Niagara Falls, Italy | Dioxin contamination | — | — |
| 1974 | Decatur, Illinois | Propane explosion | 7 | 152 |
| 1974 | Flixborough, United Kingdom | Explosion in caprolactum plant | 28 | 89 |

| Year | Location | Chemical | Death | Injuries |
|---|---|---|---|---|
| 1975 | Beek, The Netherlands | Propylene explosion | 14 | 107 |
| 1976 | Pakistan | Malathion poisoning among workers | — | — |
| 1976 | Seveso, Italy | Dioxin/TCDD | — | 193 |
| 1977 | Chicago, Illinois | Hydrogen sulfide release | 8 | 29 |
| 1978 | Santacruz, Mexico | Methane fire | 52 | — |
| 1978 | Xilatopec, Mexico | Gas explosion in transit | 100 | 150 |
| 1978 | Los Alfaques | Propylene transfer | 216 | 200 |
| 1979 | Three Mile Island, Pennsylvania | Nuclear reactor accident | | |
| 1979 | Novosibirsk, Soviet Union | Chemical plant accident | 300 | — |
| 1980 | Sommerville, | $PCl_3$ accident | — | 300 |
| 1981 | Tacoa, Venezuela | Oil explosion | 145 | — |
| 1982 | Taff, U.S.A | Acrolein explosion | | |
| 1983 | Denver, Colorado | Leakage of 20,000 gallons of Nitric acid | — | — |
| 1984 | Arkansas | Ethylene oxide leakage | — | — |
| 1984 | Callao, Peru | Pipe breakage of tetraethyl | — | — |
| 1984 | Sao Paulo, Brazil | Petro pipeline explosion | 508 | — |
| 1984 | Middle Port, New York | Leakage of MIC | — | — |
| 1984 | Ixhuatepec, Mexico | LPG tank explosion | 452 | 4248 |
| 1984 | Matamoras, Mexico, | Leakage of ammonia from a fertilizer plant | — | — |
| 1984 | Bhopal, India | Leakage in pesticide plant | 2,600* 16,000† | A large population |
| 1985 | Trichur, Kerala, India | Leakage of chlorine gas in a textile plant | — | — |

| Year | Location | Chemical | Death | Injuries |
|---|---|---|---|---|
| 1985 | Cubatao, Brazil | Leakage of ammonia from a fertilizer plant | — | — |
| 1985 | Westmalle, Belgium | Release of cloud of chlorine gas | — | — |
| 1985 | Karlskoga, Sweden | Leakage of sulfuric acid vapor | — | — |
| 1986 | Chernobyl, Soviet Union | Nuclear reactor accident | 325 | 300 |
| 1986 | Devnya, Bulgaria | Fire in a chemical complex | 17 | 19 |
| 1988 | North Sea, United Kingdom | Pipe alpha oiling explosion | 166 | |
| 1989 | Prince William Sound, Alaska | Exxon Valdez spill; Millions of tons of oil spill | | |
| 1991 | Kuwait, Gulf War | Petroleum oils | | A large population of adults and children |
| | West Virginia, U.S.A | Toxic gas in a pesticide plant | — | — |

*Fatal disasters on the night December 2, 1984.

†Subsequent human tragedies related with MIC.

## APPENDIX 34 HISTOLOGIC CLASSIFICATION OF TUMORS

| Benign Tumor | Malignant | Tissue of Original |
|---|---|---|
| Papilloma | Squamous | Epithelial tissue |
| | Carcinoma | Surface epithelium |
| | | Glandular epithelium |
| Adenoma | Adenocarcinoma | Connective tissue |
| | | Fibroblast |
| Osteoma | Osteogenic sarcoma | Bone |
| Lipoma | Liposarcoma | Fat |
| Leiomyoma | Leiomyosarcoma | Muscle tissue |
| Rhabdomyoma | Lymphangiosarcoma | Smooth muscle |
| | | Striated muscle |
| | Lymphosarcoma | Lymphoid and hemopoietic tissue |
| | Lymphatic leukemia | Lymphocytes |
| | Myeloid leukemia | Myeloid cells |

## APPENDIX 35 HUMAN FETUS, DRUGS, AND CHEMICAL TOXICITY

| | |
|---|---|
| 1. Alcohols | Muscular hippotonia |
| 2. Antibacterials | Deposition in bones |
| Streptomycin | Discoloration of teeth |
| Tetracyclines | Inhalation of bone growth |
| 3. Sulfonamides | 8$^{th}$ nerve damage (anemia) |
| 4. Novobiocin | Hyperbilirubinemia |
| 5. Erythromycin | Liver damage |
| 6. Nitrofurantoin | Hemolysis |
| 7. Anticoagulants<br>8. Coumarin<br>9. Sodium warfarin | Hemorrhage, death |
| 10. Antidiabetics | |
| Tolbutamide | Thrombocytopenia |
| Chlorpropamide | Prolonged hypoglycemia |
| Phenoformin | Lactic acidosis |
| Insulin (shock) | Fatal |
| 11. Ammonium chloride | Acidosis |
| 12. Antihistamines | Infertility |
| 13. Antithyroid drugs | Hypothyroidism |
| 14. Barbiturates | Coagulation defects |
| Diphenyl hydantoin | Withdrawal syndrome |
| 15. Phenobarbital excess | Neonatal bleeding, death |
| 16. Diazepam | Hypothermia |
| 17. Sedatives | Behavioral changes |
| 18. Meprobamate | Retarded development |
| 19. Meperidine | Neonatal depression |
| 20. Primidone | Withdrawal symptoms |
| 21. Heroin, morphine | Withdrawal syndrome |
| 22. Reserpine | Nasal congestion, lethargy, respiratory depression, brachycardia |
| 23. Phenothiazines | Hyperbilirubinemia, depression, hypothermia |
| 24. Magnesium sulfate | Central depression, Neuromuscular block |
| 25. Chloroquine | Death |
| 26. Solvents | Newborn depression |
| 27. Salicylates in large amounts | Bleeding |

## APPENDIX 36 LIPOPROTEIN-RELATED DISORDERS

| Disorder | Manifestations | Single Drug* | Drug Combination |
|---|---|---|---|
| Primary chylomicronemia (familial lipoprotein lipase or cofactor deficiency) | Chylomicrons, VLDL* | Dietary management | Niacin plus fibrate |
| Familial hypertriglyceridemia, severe or moderate | VLDL, chylomicrons increased VLDL increased; chylomicrons may be increased | Niacin, fibrate | Niacin plus fibrate |
| Familial combined hyperlipoproteinemia | VLDL increased | Niacin, fibrate | |
| LDL increased | Niacin, reductase inhibitor, resin | Niacin plus resin or reductase inhibitor | |
| VLDL, LDL increased | Niacin, reductase inhibitor | Niacin plus resin or reductase inhibitor | |
| Familial dysbetalipoproteinemia | VLDL remnants; chylomicron remnants increased | Fibrate, niacin | Fibrate plus niacin or niacin plus reductase inhibitor |
| Familial hypercholesterolemia Heterozygous | LDL increased | Resin, reductase inhibitor, niacin | Two or three individual drugs |
| Homozygous | LDL increased | Niacin, atorvastatin | Resin plus niacin plus reductase inhibitor |
| Lp(a) hyperlipoproteinemia | Lp(a) increased | Niacin | |

*Single-drug therapy should be evaluated before drug combinations are used.

*VLDL — very low density lipoprotein.

## APPENDIX 37 DRUG EFFECTS AND THYROID FUNCTION

| Drug Effect | Drugs |
|---|---|
| Change in thyroid hormone synthesis | |
| Inhibition of Thyrotropin Releasing Hormone (TRH) or Thyroid Stimulating Hormone (TSH) secretion without induction of hypothyroidism | Dopamine, levodopa, corticosteroids, somatostatin |
| Inhibition of thyroid hormone synthesis or release with the induction of hypothyroidism (or occasionally hyperthyroidism) | Iodides (including amiodarone), lithium, aminoglutethimide |
| *Alteration of thyroid hormone transport and serum total $T_3$ and $T_4$ levels, but usually no modification of $FT_4$ or TSH* | |
| Increased Thyroid Binding Globulin (TBG) | Estrogens, tamoxifen, heroin, methadone, mitotane |
| Decreased Thyroid Binding Globulin (TBG) | Androgens, glucocorticoids |
| Displacement of $T_3$ and $T_4$ from TBG with transient hyperthyroxinemia | Salicylates, fenclofenac, mefenamic acid, furosemide |
| *Alteration of $T_4$ and $T_3$ metabolism with modified serum $T_3$ and $T_4$ levels but not $FT_4$ or TSH levels* | |
| Induction of increased hepatic enzyme activity | Phenytoin, carbamazepine, phenobarbital, rifampin, rifabutin |
| Inhibition of 5′-deiodinase with decreased $T_3$, increased $rT_3$ | Iopanoic acid, ipodate, amiodarone, beta-blockers, corticosteroids, propylthiouracil, cholestyramine, |
| Other interactions | |
| Interference with $T_4$ absorption | colestipol, aluminium hydroxide, sucralfate, ferrous sulfate, some calcium preparations |
| Induction of autoimmune thyroid disease with hypothyroidism or hyperthyroidism | Interferon-$\alpha$, interleukin-2 |

## APPENDIX 38 NEW DRUGS, CATEGORY, AND THEIR USES IN INDIA

| | |
|---|---|
| 5-Aminosalicyclic acid | Ulcerative colitis |
| Abciximab | Adjunct to ptea |
| Aclarubicin injection | Anticancer |
| Adenosine | Anti-epileptic |
| Alendronate sodium | Postmenopausal osteoporosis |
| Alprazolam | Anxiolytic agent |
| Altretamine | Anticancer |
| Ambroxol hydrochloride | Mucolytic |
| Amifostine | Chemoprotective agent against the serious oxicites associated with intensive regimens of platinum and alkylatin agent chemotherapy |
| Amineptine | Antidepressant |
| Amiodrone hydrochloride | Cardiac |
| Amitraz | Veterinary |
| Amlodipine | Antihypertensive |
| Amrinone lactate | Intropic agent |
| Amsacrine injection | Anticancer |
| Arteether | Antimalarial |
| Artesunate | Malaria |
| Arthmether injection | Antimalarial |
| Astemizole | Antiallergic |
| Atorvastatin | Antihyperlipidemia |
| Azelastine hydrochloride | Antiallergic rhinitis |
| Azithromycin | Antibacterial |
| Bacillocid special | Disinfectant surface |
| Baclofen | Muscle relaxant |
| Bambuterol | Bronchodilator |
| Basiliximab injection | Immunosuppressant |
| Benazepril hydrochloride | Antihypertensive |
| Benidipine hydrochloride | Hypertension; angina pectoris |
| Benoxinate hydrochloride | Ophthalmic |
| Benzyadmine | Topical analgesic |
| Betaxolol hydrochloride | Ophthalmic |
| Bezafibrate | Lipid-lowering agent |
| Bifonazole | Antifungal |
| Biphenyl acetic acid | NSAID |
| Bisoprolol fumarate | Antihypertensive |
| Botulinum toxin type AIM injection | Neurotoxin acting |
| Brimonidine tartrate | Ophthalmic |
| Budesonide | Antiasthmatic |

| | |
|---|---|
| Bulaquine compound | Antimalarial |
| Bupropion SR tablets (150 mg) | Smoking cessation |
| Buspirone | Anxiolytic |
| Butafosfane + cyanocobalmin injection | Veterinary drug |
| Butalex | Veterinary drug |
| Butenafine hydrochloride cream (1%) | Antifungal |
| Cabapentin | Antiepileptic |
| Calcitriol sofl gelatin capsule | Metabolic bone disease |
| Calcitrol injection | Vitamin $D_3$ biologically active for chronic venours, for vascular disorders |
| Calcium dobesilate | |
| Carboplatin injection | Anticancer |
| Carteolol hydrochloride | Antihypertension and glaucoma |
| Carvedilol | Antihypertensive agent |
| Cefaclor | Antibiotic |
| Cefadroxyl | Antibiotic |
| Cefixime | Antibiotic |
| Cefoperazone sodium injection (IM/IV) | — |
| Cefpirome | Antiinfective, cephalosporin antibiotic |
| Cefpodoxime proxetil | Antibacterial for bronchitis, cephalosporin antibiotic |
| Ceftazidime pentahydrate | Antibiotic |
| Ceftibuten (finished formulation) | URTI, LRTI, UTI |
| Ceftizoxime Sodium Sterile | Cephalosporin |
| Ceftriaxone sodium | Antibiotic |
| Cefuroximeaxetil | Antibiotic |
| Cefuroxime | Antibiotic |
| Celecoxib | NSAID |
| Nevirapine | Anti–HIV |
| Centchroman | Oral contraceptive |
| Centropazine | Antidepressant |
| Cephacetril sodium | Veterinary drug |
| Cerivastatin | Lipid-lowering Agent |
| Cetirizine | Seasonal rhinitis; perennial allergic rhinitis; pruritus |
| Chandonium iodide | Nonsteroidal neuromuscular blocking agent |
| Chelated amino acid–based minerals | Trace mineral deficiencies |
| Ciprofloxacin hydrochloride | Antibacterial |
| Cisapride | Gastrointestinal disorders |
| Clarithromycin | Antibiotic |

| | |
|---|---|
| Clobazam | Antiepileptic |
| Clomipramine hydrochloride | Antidepressant |
| Clonazepam | Antiepileptic |
| Clopidogrel | Anti-Platelet aggregation inhibitor |
| Closantel | Veterinary drug |
| Clozapine | Schizophrenia |
| Colfosceric palmitate | Respiratory distress syndrome |
| Coligen | Veterinary drug |
| Collagen with gentamicin implant | Antiinfective |
| Collagenase | — |
| Complex 15 cream | Phospholipid |
| Creatine phosphate | Coadjuvant for cardiac muscles |
| Culture rabies vaccine, lofepramine | Antidepressant |
| Cumophos | Veterinary drug |
| Cyclosporin oral soln | Immunosuppressant |
| Cypermethrin high cis | For control of ticks flies lice, and mites in cattle, sheeps, dogs, poultry |
| Danofloxacin mesylate injection | Veterinary drug |
| Desogestrel (to be combined with ethinyl estradiol) | Oral contraceptive |
| Desonide | Corticosteroid |
| Dexfenfluramine hydrochloride | Antiobesity |
| Diazinon | Veterinary drug |
| Dinoprostone (PGE2) | Induction of labor |
| Dipivefrin hydrochloride | Antiglaucoma |
| Disodium Pamidronate injection | Anticancer |
| Dobutamine | Inopathic agent |
| Docetaxel injection | Anticancer |
| Doneprazal hydrochloride | Alzheimer's disease |
| Dovamectin | Veterinary drug |
| Doxazosim mesylate | Antihypertensive |
| DPT-ACT HIB vaccine | Vaccine |
| Eeroplast (fitrin adhesive glue) | Surgical |
| Efavirenz 50/100/200 mg* | Anti–HIV |
| Enalapril maleate | All grades of hypertension |
| Enrofloxacin | Veterinary drug |
| Eptifibatide IV injection | Acute coronary syndrome |
| Equino influenza and equine | Vaccine |
| Erythropoietin (rDNA) | Anemia due to chronic renal failure |
| Esmolol hydrochloride | Cardioselective beta-blocker |
| Estradiol TTS 25, 50, 100 mcg/day | Estrogen |
| Ethamsylate | Antihemostatic |
| Etidronate disodium | Anticancer |

| | |
|---|---|
| Etodolac | NSAID |
| Midazolam | Anesthetic general |
| Etoposide capsules and injection | Anticancer |
| Famotidine | Antiulcer |
| Farvovirus vaccine | Vaccine |
| Felodipine | Antihypertensive |
| Fenifibrate | Antihyperlipidemia |
| Fexofenadine hydrochloride | Antihistamine |
| Fiararubicin hydrochloride injection | Anticancer |
| Filgrastim injection | Anticancer |
| Finasteride | Benign prostatic hypertrophy |
| Flavophospholipol flavomycin | Antibiotic Veterinary |
| Flecainide acetate | Antiarrhythmic |
| Fluconazole | Systemic antifungal |
| Fludarabine phosphate IV injection | Anticancer |
| Flumethrin | Veterinary drug |
| Flunarizine hydrochloride | Antimigraine |
| Flunisolide | Allergic rhinitis |
| Fluoxetine hydrochloride | Antidepressant |
| Flurometholone | In noninflammatory ocular condition |
| Fluticasone propionate | Topical steroid |
| Fluvoxamine maleate | Antidepressant |
| Formestane depot injection | Anticancer |
| Fortison liquid/power pepti-2000-liquid powder | Nutritional preparation |
| Fotemustine IV injection (finished formulation) | In the treatment of disseminated malignant melanoma including cerebral localization and primary malignant — Cerebral tumors |
| Fucidic acid hemihydrate | Ophthalmic |
| Gadodiamide IV injection | Contrast media |
| Gadopentetic acid dimelumine salt | Contrast media |
| Ganciclovir | Antiviral |
| Gemcitabine injection | Anticancer |
| Gemfibrozil | Lipid-lowering agent |
| Genotropin injection | Growth hormone |
| Ginkgo biloba extract | Herbal extract |
| Gliclazide | Antidiabetic |
| Glimepride | Antidiabetic |
| Glucomannan | Antidiabetic (herbal) |
| Goserelin acetate | Anticancer |
| Hemophillus type b conjugate vaccine | Influenzae vaccine |

| | |
|---|---|
| Haloperidol, decanoate | Chronic schizophrenia |
| Haltrexone | Narcotic and alcohol dependence |
| Hepatitis-A vaccine | Vaccine |
| Horse antihuman lymphocyte | Aplastic anemia |
| Idarubicin hydrochloride | Anticancer |
| Idebenone | Cerebral arteriosclerosis |
| Immucyst (bacillus calmette guerin vaccine) | Anticancer |
| Indinavir sulfate* | Anti-AIDS |
| Influenza vaccine inactivated | Prophylaxis against influenza Inhibitor agent |
| Interferon alpha injection (recombinant) | Anticancer |
| Iopramide (ultravist) | Contrast media |
| Ioversol injection | Contrast media |
| Ipratropium bromide | Anticholinergic bronchodilator |
| Irbesartan | Antihypertensive |
| Irinotecan hydrochloride trihydrate | Anticancer |
| Isepamicin | Aminoglycoside antibiotic |
| Isoflurane | Anesthesia |
| Itraconazole | Antifungal |
| Itrolan (isovist) | Contrast media |
| Ketorolac tromethamine antiinflammatory | Nonsteroidal, analgesic |
| Ketotifen | Antiasthmatic drug |
| Korsolex concentrate | Instrument disinfectant |
| Lacidipine | Antihypertensive |
| Lamivudine | Anti-AIDS |
| Lansoprazole | Antiulcer |
| Latonoprost | Ophthalmic |
| L-Carnitine | Vitamin deficiency |
| Letrozole | Anticancer |
| Levobunolol hydrochloride | Beta-blocker |
| Levofloxacin | Antibacterial |
| Levovist | Contrast media |
| Lispra (rDNA) injection | |
| Levprolide acetate injection | Anticancer |
| Lisinopril | Antihypertensive |
| Lomefloxacin | Antibacterial |
| Loratadine | Long-acting antihistamine |
| Losartan potassium | Antihypertensive |
| Lovastatin | Cardiovascular |
| Loxzpine | Antidepressant |
| Madramicin concentrate | Veterinary drug |

| | |
|---|---|
| Mefloquine hydrochloride | Antimalarial |
| Melagenine lotion | Leukoderma |
| Melatonin | Hormone |
| Melitracin hydrochloride | Antidepressant |
| Meloxicam | NSAID |
| Meparticin | Benign prostatic hyperplasia |
| Milrinone Lactate injection | Congestive heart failure |
| Minocycline | Antibiotic |
| Minoxidil | Male-pattern baldness |
| Misoprostol | Antiulcer |
| Mitoxantrone hydrochloride | Anticancer |
| Mitrazapine | Antidepressant |
| Mometasone furoate | Topical corticosteroid |
| Monocomponent insulin | Antidiabetic |
| Monohydrate sodium, hyaluronate soln | Aid to intraocular surgery |
| Mosapride | Prokinetic agent |
| Moxidectin (10%) intramuscular | Veterinary drug |
| Moxifloxacin | Antibacterial |
| Mupirocin | Antibacterial skin infection |
| Mycophenolate mofetil | Immunosuppressive agent |
| Nabumetone | NSAID |
| Nafarelin acetate | GnRH antagonist |
| Naloxone hydrochloride | Narcotic antagonist |
| N-butyl-2-cyano crylate | Antibacterial |
| Nefopam | Analgesic |
| Nelfinavir* | Anti–HIV |
| Nicardipine hydrochloride | Antihypertensive |
| Nicorandil | Antiangina |
| Nicotine transdermal therapeutic | Antismoking |
| Nilutamide | Anticancer |
| Nimesulide | NSAID |
| Nimodipine | Cerebral protective agent |
| Nitrendipine | Antihypertensive |
| Ofloxacin | Antibacterial |
| Olanzapine | Antipsychotic |
| Omeprazole | Antiulcer |
| Ondanseteron hydrochloride injection | Antiemetic |
| Oral polio vaccine (verocell) | Polio vaccine oral |
| Orthoclone sterile ($OKT_3$) | Immunosuppressant |
| Oxaliplatin injection | Anticancer |
| Oxiconazole | Antifungal |
| Oxybutynin chloride | Antimuscarinic agent |
| Paclitaxel injection | Anticancer |

| | |
|---|---|
| Pantoprazole tablets | Antiulcer agent |
| Paspat injection | Bronchial asthma allergic |
| PCEC vaccine (rabies HEP flurystrain) | Vaccine |
| Pefloxacin mesylate | Antibacterial |
| Penfluridol | Neuroleptic agent |
| Perindopril | Antihypertensive |
| Permethrin | Antilice/Antiscabies |
| Pipecuronium bromide | Nondepolarizing muscle-relaxing agent |
| Piperacillin + tazobactum intravenous injection | Antibiotic |
| Piperacillin sodium | Antibiotic |
| Piribedil | Idiopathic Parkinson's disease |
| Piroxicam betacyclodextrin | NSAID |
| Plasminogen activator injection (actilyse), desmopressine intranasal | Central or neurogenic diabetes |
| Podophyllotoxin | Anogenital/nonfacial warts |
| Polysaccharide pneumococcal vaccine | Pneumonia |
| Polyvinyl alcohol with povidone and chlorbutanol eye drops | Antibacterial |
| Prazosin | Antihypertensive |
| Prednicarbate | Topical corticosteroid |
| Prenoxdiazine hydrochloride (Libexin) | Antitissue |
| Promegestone | Menstrual irregularities and dysmenorrhea |
| Propfol injection | Anesthetic |
| Propoxur + flumethrin | Veterinary drug |
| Purified chick embryo cell tissue | Vaccine (human use) |
| Purified vicapsular polysaccharide of salmonella typhi | Typhoid vaccine |
| Rabbit Antihuman thymocyte immunoglobulin intravenous | Aplastic anemia |
| Rabies vaccine inactivated (Vero-Cell) | Vaccine |
| Globulin, Ramipril hydrochloride | ACE inhibitor |
| Recombinant human tissue type | Fibrinolytic therapy |
| Reloxifen | Menopausal osteoporosis |
| Repaglinide | Antidiabetic |
| Retrodrine hydrochloride | Toxemia of pregnancy |
| Rhinopneumonitis vaccine | Veterinary drug |
| Antirabies vaccine (killed) | Vaccine, veterinary drug |
| Ribavirin (virazole) | Antiviral |

| | |
|---|---|
| Rilmenidine dihydrogen phosphate | Antihypertensive |
| Riluzole tablets | Motor neuron disease |
| Risperidone | Antipsychotic |
| Ritonavir capsules | Anti–HIV |
| Rivastigmine caps | Acetylcholine esterase |
| Rocuromiem bromide injection | Muscle relaxant |
| Rofecoxib | NSAID |
| Rosiglitazone maleate | Antidiabetic |
| Rowachol | For hepatobiliary disorders |
| Roxatidine acetate hydrochloride | Antiulcer |
| Roxithromycin | Antibacterial |
| Saccharomyces boulardi | Irritable bowel syndrome |
| Saccharomyces boulardil | Yeast |
| Salinomycin | Veterinary drug |
| Salmetrol xinofate | Bronchodilator |
| Salmon calcitonin injection | Anticancer |
| Saquinavir capsules | Anti–HIV |
| Secnidazole | Amoebiasis, bacterial and richemonal, vaginitis |
| Selegiline hydrochloride | Antiparkinsonism |
| Semduramycin sodium | Veterinary drug |
| Serenda repcus, serenovarepens | Benign prostatic hypertrophy |
| Sertraline hydrochloride | Antidepressant |
| Sevoflurene inhalation | Anesthetic |
| Sibutramine hydrochloride | Antiobesity |
| Sildenafil citrate | Erectile dysfunction |
| Silymarin | Heptoprotective |
| Simvastatin | Antihypercholesterolemia |
| Sodium enoxaparin | Thromboembolic disorders |
| Sodium picosulfate | Laxative |
| Solution/nasal spray/injection | Insipidu, bleeding disorders, nocturnal enuresis |
| Somatostatin intravenous infusion | Antihemorrhage |
| Somatropine (rDNA) injection | Growth hormone |
| Sorbister capsules | Nephrologic |
| Sparfloxacin | Community-acquired pneumonia, acute exacerbation of chronic bronchitis, other lower gonococcal and nongonococcal burethritis in complicated UTI |
| Spenglersan | Immunomodulator |
| Stavudine | Anti–HIV |
| Sterillium rubin disinfectant | Disinfectant for hands |
| Stronger neominophagen-C | Hepatoprotective |
| Sulbutiamine | Astheniatic |

| | |
|---|---|
| Sulpha chloropyrazine sodium | Veterinary drug |
| Sulpride | Antidepressant |
| Sultamicilline | Antimicrobial agent |
| Sumartriptan tablets/injection | Antimigraine |
| Tacrine hydrochloride | Dementia in patients with alzheimers |
| Nitroscanate | Anthelmentic, veterinary drug |
| Taurolidine 2% or 0.5% | Antimicrobial |
| Teicoplanin for intravenous/instramuscular | Antiriotic |
| Temozolamide | Anticancer |
| Tenoxicam | NSAID, giaridiasis |
| Teratolol hydrochloride | Antihypertensive |
| Terazosin hydrochloride | Antihypertensive |
| Terconazole | Antifungal |
| Terfenadine | Antihistamine |
| Terlipressin Acetate intravenous injection | Anticoagulant |
| Tetra chlorodecaoxide | Immunomodulator |
| Tetrabenazine | Presynaptic monoamine depleting agent |
| Thiocochicoside | Antispasmodic |
| Thymosin alfa 1 injection[†] | Anti–hepa-B |
| Tiaprost trometamon injection | Veterinary drug |
| Ticlopidine | Antiplatlet aggregation |
| Tiobolone | Natural or surgical menopause |
| Tizenidine hydrochloride | Muscle relaxant |
| Tolfenamic acid | NSAID |
| Toltrazuril | Veterinary drug |
| Topiramate | Antiepileptic |
| Topotecan hydrochloride | Anticancer (ovarian) |
| Tramadol | Analgesic |
| Trapidil | Adjunct in antioplasty |
| Triclabendazole | Broad-spectrum flukicide, veterinary drug |
| Triflusalcaps 30 mg[†] | Prevention of thromboembolism |
| Trinidazole | Antiprotozoal |
| Tripotassium dicitrato bismuthate | Antiulcer |
| Triptorelin injection | Anticancer |
| Tromantadine hydrochloride | Antiviral |
| Vancomycin hydrochloride injection | Antibacterial |
| Vindesine sulfate injection | Anticancer |
| Vinorelbin tartrate injection* | Anticancer |
| Xylazine hydrochloride | Veterinary drug |
| Zalcitabine | Anti-AIDS |
| Zidovudine capsule and injection | Treatment of HIV infections |
| Zolpidem | Insomnia |

| | |
|---|---|
| Zopiclone | Short-term treatment of insomnia |
| Zuclopenthixole acetate | Antimaniac and antischizophrenia |

*Anticancer/Anti–AIDS or immunosuppressive drugs.
†Drugs referred to ICMR and experts.

*Source*: Drugs cleared from the Drug Controller General India, and periodically Notified in the Official Gazette Bulletins, Ministry of Health, New Delhi, Government of India, India.

# GLOSSARY

**Abatement**: Reducing the degree or intensity of, or eliminating, pollution.

**Abiotic**: Nonbiologic.

**Absorbance**: The logarithm to the base of 10 of the reciprocal of transmittance.

**Absorbate**: Substance/or chemical that has been retained by the process of absorption.

**Absorbent**: Material in which absorption occurs.

**Absorption**: Movement of substances into the blood vascular system or into the tissues of the organism.

**Acclimatization**: The physiologic and behavioral adjustments of an organism to changes in its environment.

**Accretion**: A phenomenon consisting of the increase in size of particles by the process of external additions.

**Action level**: A level of the chemical similar to a tolerance level, except it is not established through formal regulatory proceedings. It is an informal judgment by a regulatory agency on what amount of a chemical should be allowed in food products.

**Active ingredient**: In any pesticide product, the component that kills, or otherwise controls, target pests. Pesticides are regulated primarily on the basis of active ingredients.

**Acute dermal toxicity**: Adverse effects occurring within a short time of dermal application of a singular dose of a test chemical.

**Acute exposure**: A single exposure to a toxic substance which may result in severe biologic harm or death. Acute exposures are usually characterized as lasting no longer than a day, as compared with longer, continuing exposure over a period of time.

**Acute inhalation toxicity**: Adverse effects produced by a test chemical following a single uninterrupted exposure inhalation or respiratory route over a short period of time (24 hours or less) and the chemical capable of being inhaled.

**Acute oral toxicity**: Adverse effects produced within a short time of oral administration of a single dose of a test chemical or multiple doses given within 24 hours.

*Industrial Guide to Chemical and Drug Safety*, By T.S.S. Dikshith and Prakash V. Diwan
ISBN 0-471-23698-5 

**Acute test**: A test lasting a short time (e.g., 14 days).

**Acute toxicity**: Thc capacity of a substance to cause adverse health effects or poisonous effect or death as a result of a single or short-term exposure.

**Adenoma**: A benign tumor of glandular tissue; can be precancerous in cases such as polyps in the colon.

**ADI**: **A**cceptable **D**aily **I**ntake; the amount of a specific food additive or contaminant (e.g., pesticide) thought to be the maximum level that should be consumed on a daily basis. ADI values are normally determined by experts of WHO and FAO Codex Alimentarius Committee.

**Adsorbate**: Chemical that has been retained by the process of adsorption.

**Adsorbent**: A solid material on the surface of which adsorption takes place.

**Adsorption**: A physical process in which molecules of gas of dissolved chemicals or liquids adhere in an extremely thin layer to the surfaces of solid bodies with which they are in contact.

**Aerosol**: Suspension of tiny particles of a solid, liquid, or gaseous matter.

**Aetiology**: The science of cause or origin of a disease.

**Agglomeration**: A process of contact and adhesion whereby the particles of a dispersion form clusters of increasing size.

**Air pollution**: Contamination of atmospheric air with substances or chemicals not considered unsuitable for health.

**Ambient air**: Air surrounding on all sides.

**Aflotoxin**: Toxins produced by common molds (e.g., *Aspergillus flavus* and species in different foods).

**Agent orange**: Refer to 2,4, 5-T, 2,4,-D, Dioxin (in Chapter).

**Ames test**: A method of an experiment performed using bacteria as a test system to determine the mutagenic potential of a substance or chemical.

**Allergy**: An altered immune response to a specific substance on reexposure.

**Amino acids**: Building blocks of protein by cells; there are about 20 amino acids.

**Anemia**: A condition suggesting lack of red blood cells (RBC).

**Angioedema**: A reaction in the skin and underlying tissue showing swelling and red blotches.

**Antigen**: A foreign substance that provokes immune response when introduced into the body and the body reacts by making antibodies.

**Antagonistic**: Reduction of the effect of one chemical by the other when interacted.

**Anthropogenic**: Effects produced as a result of human activities.

**Aphotic zone**: The deeper part of lakes, sea, and ocean where light does not penetrate.

**Aphytic zone**: Part(s) of the lake floor where vegetation is not available.

**Application factor**: Refers to number used to estimate concentration of a substance/chemical that will not produce significant adverse effects/harm to a population during chronic exposure. The factor is based on the formula:

$$\text{Application factor} = \frac{\text{MATP}}{96 - \text{h LC50}}$$

(MATC - maximum allowable toxicant concentration)

**Aquaculture**: Breeding and rearing of fish in captivity; also termed pisciculture.

**Aquatic organism**: Organism(s) related to living water bodies.

**Aqueous**: Related to watery solution.

**Arboreal**: Related to plants.

**Aromatic**: Technical term describing a compound or chemical that contains one or more benzene rings.

**Aromatic amines**: Petrochemical compounds with a pungent odor (are known to produce cancer).

**Ash**: Mineral content of a product that remains after complete combustion.

**Atmosphere, an**: A unit of pressure equal to the pressure exerted by a vertical column of mercury 760 mm high at a temperature of 0°C and under standard gravity.

**Atmosphere, the**: The gaseous envelop surrounding a planet; the Earth's atmosphere is surrounded by a mass of air largely composed of oxygen (20.9%) and nitrogen (79.1%) by volume and carbon dioxide (0.03%) and traces of noble gases, water vapor, organic matter, and suspended solid particles.

**Atmospheric dispersion**: The mechanism of dilution of gaseous or smoke pollution leading to progressive decrease of pollutants.

**Autoimmunity**: A condition in which the immune responses of an animal are directed against its own tissues.

**Autotrophic**: Refers to organisms that produce their own organic constituents from inorganic compounds using sunlight for energy or by oxidation process.

**Basal diet**: Ration for adults, starter ration for young; appropriate to the species and should meet the standard nutritional requirement.

**Base pair mutagens**: Chemicals/agents that produce a base change in the DNA.

**Benign**: A condition of growth that is harmless.

**Benign tumor**: Slow-growing set of cells with abnormal look of a tumor.

**Bioaccumulation**: A process whereby a living organism acquires and stores chemicals through bioconcentration after ingestion or exposure.

**Bioaccumulants**: Substances that increase in concentration in living organisms as they take in contaminated air, water, or food because the substances are very slowly metabolized or excreted. (*See* Biological magnification.)

**Bioaccumulation factor**: The ratio of concentration of a chemical in an organism to its concentration in the food.

**Bioassay**: The quantitative measurement of a chemical's effects on the organism under standard conditions.

**Bioavailability**: Availability or presence of a chemical or metabolite in the body of the animal.

**Biochemical**: A substance or chemical produced by a living organism or system.

**Biochemical oxygen demand (BOD)**: Amount of oxygen used for biochemical oxidation by a unit volume of water at a given temperature and for a given period of time; BOD is used for the measurement of degree of water pollution.

**Biocide**: General term for any substance that kills or inhibits the growth of microorganisms (mold, slime, bacterium, fungus).

**Bioconcentration**: A process whereby living organisms acquire chemicals from water through gills or integument and store it in their bodies at concentration(s) higher than in the environment.

**Biodegradable**: Capability of an organism or biologic system to break the chemical into simpler chemicals.

**Biological oxygen demand (BOD)**: An indirect measure of the concentration of biologically degradable material present in organic wastes. It usually reflects the amount of oxygen consumed in 5 days by biologic processes breaking down organic waste.

**Biomagnification**: A phenomenon where the bioaccumulated chemical (usually a toxicant) increases in concentration as they pass upward through two or more trophic levels.

**Biotechnology**: Application of living organisms to produce new products or substances.

**Breathing zone**: The location in the atmosphere at which individual animals or humans breathe.

**Cancer**: Injurious malignant growth of cells and tissue of potentially unlimited size invading local tissues and spreading to distant areas of the body.

**Carcinogen**: Any substance or chemical that can cause cancer in animals or humans.

**Carcinogenesis**: A biologic process involving the transformation of a normal cell into a cancer cell.

**Carcinoma**: A malignant tumor of the cells that involves lung, gut, skin, and epithelial tissues; ranges to approximately 90% of all types of cancer.

**CCPR**: Codex Committee on Pesticide Residues.

**Cell**: The smallest structural unit of all living organisms.

**Chemical oxygen demand (COD)**: A measure of the oxygen required to oxidize all compounds or organic and inorganic matter in a sample of water. COD is expressed as ppm of oxygen taken from a solution of boiling potassium dichromate for 2 hours. COD is used to assess the strength of sewage and waste.

**Chlorinated hydrocarbons**: 1. Chemicals containing only chlorine, carbon, and hydrogen. These include a class of persistent, broad-spectrum insecticides that linger in the environment and accumulate in the food chain. Among them are DDT, aldrin, dieldrin, heptachlor, chlordane, lindane, endrin, Mirex, hexachloride, and toxaphene. Other examples include TCE, used as an industrial solvent. 2. Any chlorinated organic compounds, including chlorinated solvents such as dichloromethane, trichloromethylene, and chloroform.

**Chlorofluorocarbons (CFCs)**: A family of inert, nontoxic, and easily liquefied chemicals used in refrigeration, air conditioning, packaging, insulation, or as solvents and aerosol propellants. Because CFCs are not destroyed in the lower atmosphere, they drift into the upper atmosphere where their chlorine components destroy the ozone.

**Cholinesterase**: The enzyme found in animals that regulates nerve impulses by the inhibition of acetylcholine; cholinesterase inhibition in animals indicates a variety of acute symptoms such as nausea, vomiting, blurred vision, stomach cramps, and rapid heart rate.

**Chromatid—type aberration**: The damage expressed as breakage of single chromatids at the same locus.

**Chromosome**: Structure found in the nucleus of a cell; these structures bear the DNA (deoxyribonucleic acid) from which genes are made. Genes carry the genetic code of the organism.

**Chromosome-type aberration**: The damage expressed in both sister chromatids at the same locus.

**Chronic toxicity**: The capacity of a substance to cause adverse effects or harmful effects in the organism after long-term exposure.

**Co-carcinogen**: A substance or agent that assists carcinogens to cause cancer.

**Code of federal regulations (CFR)**: Document that codifies all rules of the executive departments and agencies of the federal government. It is divided into 50 volumes, known as titles. Title 40 of the CFR (referenced as 40 CFR) lists all environmental regulations.

**Complete carcinogens**: Substances or chemicals that will both initiate and promote cancer.

**Condensation**: The process of converting a chemical in the gaseous phase to a liquid or solid state by decreasing temperature, increasing pressure, or both.

**Congenital**: Related to a condition that exists before birth.

**Contact dermatitis**: Skin outbreaks caused by direct contact with a substance.

**Contaminant**: A substance or chemical added by humans or by natural activities in which sufficient concentrations may render the living or working atmosphere unacceptable.

**Contamination**: Introduction into water, air, and soil of microorganisms, chemicals, toxic substances, wastes, or wastewater in a concentration that makes the medium unfit for its next intended use. Also applies to surfaces of objects, buildings, and various household and agricultural use products.

**Convulsion**: Abnormal and involuntary jerks and shaking of the body.

**Corrosive**: Any liquid or solid that causes visible destruction or irreversible alteration of skin tissue at the place of contact.

**Cost-benefit analysis**: A quantitative evaluation and decision-making technique where comparisons are made between the costs of a proposed regulatory action on the use of a substance or chemical with the overall benefits to society of the proposed action; often converting both the estimated costs and benefits into health and monetary units.

**CPSC**: Consumer Product Safety Commission.

**Cumulative toxicity**: Adverse effects caused by substances or chemicals after repeated doses, prolonged exposure, or increased concentration of the chemical or metabolites in susceptible tissues of animals.

**Cytogenetics**: A discipline of science linking the study of heredity with that of the physical appearance of the chromosomes.

**Cytolysis**: The phenomenon in which the cell undergoes destruction particularly due to the disintegration of cell membrane.

**Decible**: The unit used for the measurement of the intensity of sound on a logarithmic scale, based on measurements of sound intensity in watts per square meter and related to a reference. For instance, 10 watts/m$^2$ is the intensity of the quietest sound perceptible to human ear.

**Delney Clause**: An amendment enacted in 1958 to the Pure Food Act that prohibits the addition to food of all detectable amounts of a carcinogen.

**Density**: The mass per unit volume of a substance or chemical.

**Dermal corrosion**: Production of irreversible tissue damage on the skin following the application of a test substance or chemical.

**Dermal irritation**: An irreversible inflammatory change on the skin following exposure to a substance or chemical.

**Dermatitis**: Inflammation of the skin due to chemical reaction on the body.

**Detergents**: Group of synthetic, organic, water-soluble cleansing agents. Detergents are not prepared from fats and oils and are not inactivated by hard water.

**Desorption**: The process of freeing from an absorbed state.

**Detergent**: A surface active agent which is used to remove dirt or grease from a surface.

**Dioxin**: Any of a family of compounds known chemically as dibenzo-p-dioxins. Concern about them arises from their potential toxicity as contaminants in commercial products. Tests on laboratory animals indicate that it is one of the more toxic anthropogenic (man-made) compounds.

**Dissolved oxygen (DO)**: In-oxygen molecules that are dissolved water and the unit of expression is ppm.

**Dispersoids**: The particles of a dispersion.

**Disposals**: Final placement or destruction of toxic, radioactive, or other wastes; surplus or banned pesticides or other chemicals; polluted soils; and drums containing hazardous materials from removal actions or accidental releases. Disposal may be accomplished through use of approved secure landfills, surface impoundments, land farming, deep-well injection, ocean dumping, or incineration.

**Diural**: Recurring daily.

**DNA**: Deoxyribonucleic acid (*See* Chromosomes).

**Dominant lethal mutation**: The mutation occurring in a germ cell which does not cause disfunction of the gamete but is lethal to the fertilized egg or developing embryo.

**Dosage/Dose**: 1. The actual quantity of a chemical administered to an organism or to which it is exposed. 2. The amount of a substance that reaches a specific tissue (e.g., the liver). 3. The amount of a substance available for interaction with metabolic processes after crossing the outer boundary of an organism.

**Dosage**: A composite term indicating the dose (size), its frequency, number of doses, and the duration of dosing.

**Dose**: Quantity or volume of a substance or chemical administered once to an organism; dose is expressed as mg/g of test substance or chemical per unit weight of animal (mg/kg body weight).

**Dose-response relationship**: The quantitative relationship between the amount of exposure to a substance and the extent of toxic injury or disease produced.

**Dust**: Consists of small, very fine, solid particles in the air; the sources of dust could be from the following: Automotive paint; Batteries; Brake fluid; Engine degreaser; Epoxies and adhesives; Flea powder; Gasoline; Herbicides; Insecticides; Mothballs; Oil-based paints; Paint stripper; Photographic chemicals; Polishes containing nitrobenzenes; Wood preservatives.

**EC50**: The calculated concentration of a substance or chemical that would kill 50% of an exposed population of animals.

**Ecology**: Branch of science dealing with interrelationship(s) of organisms and their environment.

**Ecosystem**: A complex system in which different biologic communities and their nonliving environmental surroundings function independently and interact.

**Edaphic factor**: A composite of physical and biologic characteristics of the soil that disturbs the ecosystem.

**Edema**: Swelling of tissues normally associated with the collection of fluids.

**Effluent standard**: The maximum amount of a specified pollutant permitted in effluents.

**Elute**: To remove sorbed substance or chemical from a sorbent by means of a fluid.

**Emission**: Pollution discharged into the atmosphere from smokestacks, other vents, and surface areas of commercial or industrial facilities; from residential chimneys; and from motor vehicles, locomotives, or air craft exhausts.

**Emission mixture**: The total amount of a substance or chemical discharged into the air from a stack. Vent or discrete source.

**Embryotoxic**: Adverse effects or bad effects on the growing embryo.

**Environment**: The physical, chemical, and biotic conditions surrounding an organism or animal.

**Environmental impact statement**: A document required of federal agencies by the National Environmental Policy Act for major projects or legislative proposals significantly affecting the environment. A tool for decision-making, it describes the positive and negative effects of the undertaking and cites alternative actions.

**Enzyme**: A protein produced by living cells; enzymes regulate the rate of chemical reaction without being altered in the process.

**Epidemiology**: Discipline of science studying the distribution of disease, or other health-related states and events in human populations, as related to age, gender, occupation, ethnicity, and economic status to identify and alleviate health problems and promote better health.

**Estuary**: Region of interaction between rivers and near-shore ocean waters, where tidal action and river flow mix fresh and salt water. Such areas include bays, mouths of rivers, salt marshes, and lagoons. These brackish water ecosystems shelter and feed marine life, birds, and wildlife.

**Et 50**: The calculated time to kill 50% of an exposed population to a specific concentration of a substance or chemical.

**Excretion**: The physiologic process in which the chemical or metabolite is eliminated from the body through urine, feces, sweat, and or exhaled gas.

**Eye corrosion**: The production of an irreversible tissue damage in the eye(s) following the anterior surface contact of a substance or chemical.

**Eye irritation**: The production of an irreversible change in the eye(s).

**Fetotoxic**: Adverse effect or bad effect to the fetus.

**FFDCA**: Federal Food, Drug, and Cosmetic Act.

**Fiber**: A fiber is a solid particle, and normally its length is at least three times its width; hazard of a particle depends on the size of the fiber.

**FIFRA**: Federal Insecticide Fungicide Rodenticide Act.

**Flue gas**: The air coming out of a chimney after combustion in the burner it is venting. Flue gas includes, nitrogen oxides, carbon oxides, water vapor, sulfur oxides, particles, and many chemical pollutants.

**Fluorocarbons (FCs)**: Organic compounds analogous to hydrocarbons in which one or more hydrogen atoms are replaced by fluorine. FCs were once used in the United States as a propellant for domestic aerosols and are now found mainly in coolants and some industrial processes. FCs containing chlorine are called chlorofluorocarbons (CFCs). These are believed to be modifying the ozone layer in the stratosphere and are responsible for allowing more harmful solar radiation to reach the Earth's surface.

**Foam**: A gas in liquid dispersion.

**Food chain**: Order of organisms in which each organism feeds on the member below it; the lowest order could be an alga or a plant.

**Fog**: Visible aerosol in which the dispersed phase is liquid.

**FQPA**: Food Quality Protection Act.

**Fume**: Very fine and small solid particles generated by condensation of vapors from the gaseous state, generally after volatilization from melted chemicals at high temperature; particle diameter is generally less than 1 μ.

**Fumigants**: Toxic compounds in vapor form used to destroy rodents, insects, and infectious organisms.

**Gas**: One of the three states of aggregation of matter having no independent shape or volume matter but with a property of indefinite expansion.

**Gene**: The part of DNA molecule that carries the information defining the sequence of amino acids in a specific polypeptide chain; in simple words, the chromosome that carries a particular inherited characteristic.

**Genome**: A term for all the genes in a cell.

**Genotoxic**: Substance or chemical causing damage to the genetic material of living organisms.

**Global warming**: An increase in the near surface temperature of the Earth. Global warming has occurred in the distant past as the result of natural influences, but the term is most often used to refer to the warming predicted to occur as a result of increased emissions of greenhouse gases. Scientists generally agree that the Earth's surface has warmed by about 1°F in the past 140 years. The Intergovernmental Panel on Climate Change (IPCC) recently concluded that increased concentrations of greenhouse gases are causing an increase in the Earth's surface temperature and that increased concentrations of sulfate aerosols have led to relative cooling in some regions, generally over and downwind of heavily industrialized areas.

**GRAS**: A level of substance or chemical generally recognized as safe (the term refers to food additives and related substances).

**Generally recognized as safe (GRAS)**: Designation by the USFDA that a chemical or substance (including certain pesticides) added to food is considered safe by experts, and so is exempted from the usual FFDCA food-additive tolerance requirements.

**Greenhouse effect**: The warming of the Earth's atmosphere attributed to a buildup of carbon dioxide or other gases; some scientists think that this buildup allows the sun's rays to heat the Earth, while making the infrared radiation atmosphere opaque to ultraviolet? radiation, thereby preventing a counterbalancing loss of heat.

**Growth hormone (GH)**: A protein produced by the pituitary gland which promotes the growth of the whole body.

**GUP**: General use pesticide.

**Habitat**: The place where a population (e.g., human, animal, plant, microorganism) lives, and its surroundings, both living and nonliving.

**Half-life**: In simple terms, it denotes the time required for the elimination of one half of the total dose of a chemical from the body. For instance, the biochemical half-life of DDT in the environment is 15 years. Similarly, the time required for half of the atoms of a radioactive element to undergo self-transmutation or decay (half-life of radium is 1620 years).

**Hazard**: Ability of a substance to cause injury; probability that a substance (chemical or physical agent) can cause injury or adverse effect to animals and humans under a set condition; hazard is the inverse function of safety.

**Hazard $\propto$ (Toxicity x Bioavailability)**:

**Hazard evaluation**: A component of risk evaluation that involves gathering and evaluating data on the types of health injuries or diseases that may be produced by a chemical and on the conditions of exposure under which such health effects are produced.

**Hazardous ranking system**: Evaluation of risks to public health and the environment related with rejected or uncontrolled hazardous waste sites and the principles of screening thereof by USEPA. The scorings are based on the potential of hazardous substances spreading from the site through the air, surface water, or groundwater, and on other factors such as density and proximity of human population. Scorings of this kind help to decide the site(s) on the National Priorities List and related ranking.

**Hazardous substance**: Kind of substance or material, (e.g., corrosive, ignitable, explosive, or chemically reactive) that may pose a threat to human health and environment safety. Normally, the USEPA designates typical hazardous substance(s).

**Heterotrophic**: Organisms which require ready-made organic food materials from which they can produce most of their own constituents.

**Homeostasis**: An inherent tendency in an organ toward maintaining physical and psychologic stability; the maintenance of constancy within a biologic system either in terms of interaction between the organisms of a community or between the internal environment of an organism or individual.

**Immunity**: The ability of an organism to combat infection(s) by parasites.

**Immunotoxic**: Substances harmful to the immune system.

**Induction period**: The length of time (at least 1 week) following a sensitization during which period a hypersensitive state is developed by the animal.

**Initiator**: Substance or chemical that starts the process of tumor formation by causing permanent damage to the DNA.

**In Vitro**: Studies performed in isolation from living organisms; a process occurring in a test tube.

**In Vivo**: Studies performed with in the living organism; a process occurring in the intact body of an organism.

**Irritant**: A substance that can cause an inflammatory reaction to eye, skin, or respiratory system.

**JMPR**: Joint Meeting on Pesticides Residues (FAO/WHO); JMPR advises the CCPR, which is responsible for recommending the Codex MRL.

**Landfills**: 1. Sanitary landfills are disposal sites for nonhazardous solid wastes spread in layers, compacted to the smallest practical volume, and covered by material applied at the end of each operating day. 2. Secure chemical landfills are disposal sites for hazardous waste, selected and designed to minimize the chance of release of hazardous substances into the environment.

**Latency**: The time period between exposure and the first appearance of the effect.

**$LC_{50}$**: Lethal concentration, 50%; a statistically derived concentration of a substance or chemical in air that will kill 50% of the test animals that inhale it. $LC_{50}$ is expressed as mg/L (of air), or as parts per million (ppm).

**$LD_{50}$**: Lethal dose, 50%; a statistically derived dose after a single administration of the test chemical or substance that will kill 50% of test animals. The chemical may be administered by mouth (ingestion), by skin application (dermal), by inhalation, or by injection (systemic). The term often suggests a coarse measure of acute toxicity.

**Lethal mutation**: A mutation in the genome that causes death to the carrier.

**Leukemia**: Abnormal number of leukocytes produced in the bone marrow; covers about 4% of all cancers in humans.

**Leukoplakia**: White patches in the oral cavity (mouth) of humans that becomes cancerous; covers about 3% of patients.

**Lymphoma**: Development of cancer from leukocytes of the lymph nodes, spleen, CNS, or bowel; covers about 5% of all cancers.

**Malformation**: An abnormal or defective form normally caused by a toxic substance or chemical.

**Malignant**: A very injurious or deadly spreading (term generally associated with cancer).

**Malignant tumor**: Cancer that spreads into different neighboring organs of the body.

**Maximum allowable concentration (MAC)**: A concentration of a substance or chemical calculated from long-term studies indicating that no significant adverse effect or harm would be produced to an exposed population.

**Maximum residue level (MRL)**: Comparable to a U.S. tolerance level, the maximum residue level is the enforceable limit on food pesticide levels in some countries. Levels are set by the Codex Alimentarius Commission, a United Nations agency managed and funded jointly by the World Health Organization and the Food and Agriculture Organization.

**Median tolerance limit (TLM)**: A statistically calculated concentration of a substance or chemical where 50% of the exposed population survive a specific period of time, (e.g., TLM 24 hours, TLM 72 hours).

**Melanoma**: Cancer of the melanocytes or pigment cells of the skin tissue.

**Metabolism**: The physiologic changes comprising of the whole gamut of biochemical reactions when a substance or chemical undergoes in an organism; ***anabolism***, which is changes leading to the synthesis of products and ***catabolism***, which is changes leading to the breakdown of products.

**Metabolite**: Chemical constituents formed in the body of an animal as a result of metabolism.

**Micronuclei**: Small particles of genetic material and consisting of accentric fragments of chromosomes or the entire chromosome that lag behind at anaphase phase of cell division; after telophase, the fragments may not be included in the nuclei of daughter cells, but form single or multinuclei in the cytoplasm.

**Mist**: Liquid normally water in the form of particles suspended in the atmosphere at or near the surface of the Earth.

**Mitotic crossing-over**: A phenomenon in which exchange of DNA segments between genes (more generally between gene and its centromere) occurs to produce reciprocal products.

**Mitotic gene conversion**: A phenomenon in which the unilateral transfer of DNA sequence information occurs within a gene and results in nonreciprocal products.

**Monitoring**: Periodic, regular, or continuous surveillance or testing to determine the level of compliance with statutory requirements or pollutant levels of substances or chemicals in various media, humans, plants, and animals.

**Montreal protocol**: Treaty signed in 1987 that governs stratospheric ozone protection and research, and the production and use of ozone-depleting substances. It provides for the end of production of ozone-depleting substances such as CFCS. Under the protocol, various research groups continue to assess the ozone layer. The multilateral fund provides resources to developing nations to promote the transition to ozone-safe technologies.

**Morbidity**: Rate of disease incidence in a population.

**Mordants**: Chemicals that are insoluble compounds that serve to fix a dye, usually a weak dye.

**Mortality**: Rate of death.

**MRL**: Maximum residue level (a level much below ADI).

**MSDS**: Material safety data sheet.

**Mutagens**: Substances or chemicals causing mutation.

**Mutagenic**: Capability of a substance or chemical to cause genetic damage such as mutations.

**Mutagenesis**: The production of mutation occurring in the genetic store of organisms because of toxic chemicals.

**Mutation**: Any heritable change occurring in the genetic material.

**Mycotoxins**: Toxic substance produced by fungus.

**Myeloma**: Cancer beginning in plasma cells of bone marrow.

**Nasopharyngeal carcinoma**: Cancer of the throat.

**Necrosis**: Mass death of area(s) of tissue(s) surrounded by otherwise healthy tissue.

**Neoplasm**: Formation of new tissues associated with disease (term denoting tumors).

**Nephrotoxic**: Substances or chemicals causing injury to kidneys.

**NIOSH**: **N**ational **I**nstitute of **O**ccupational **S**afety and **H**ealth. U.S. Congress set up this institute in 1970 to play a key role in the protection of the health of occupational workers. The agency conducts occupational health research, inspects industries and manufacturing plants at the request of the employers and workers, and collects data for their own studies, to recommend standards for safe exposure to hazardous substances.

**Nitric oxide (NO)**: A gas formed by combustion under high temperature and high pressure in an internal combustion engine; it is converted by sunlight and photochemical processes in ambient air to nitrogen oxide. NO is a precursor of ground-level ozone pollution, or smog.

**Nitrogen oxide (NOx)**: The result of photochemical reactions of nitric oxide in ambient air; a major component of photochemical smog. It is a product of combustion from transportation and stationary sources and a major contributor to the formation of ozone in the troposphere and to acid deposition.

**NOEC**: **N**o **O**bserved **E**ffect **C**oncentration. The highest tested concentration of a test substance or chemical at which no statistically significant lethal or other adverse effects are observed in animals.

**NOEL**: **N**o **O**bserved **E**ffect **L**evel. Normally a maximum or highest test dose level of a substance or chemical that has not produced any statistically significant adverse effect in animals.

**Non-Hodgkin's lymphomas**: Cancer of the lymphatic tissue.

**Odor**: The property of a chemical that affects the sense of smell, scent, or perfume.

**OECD**: **O**rganization for **E**conomic **C**ooperation and **D**evelopment. The OECD is an international agency that supports programs designed to facilitate trade and development.

**OECD Guidelines**: Testing guidelines prepared by the Organization of Economic and Cooperative Development of the United Nations. They assist in preparation of protocols for studies of toxicology and environmental fate.

**OPIDN**: **O**rgano**p**hosphorus-**i**nduced **d**elayed **n**eurotoxicity.

**Organic chemicals or compounds**: Naturally occurring (animal-, plant-produced, or synthetic) substances containing mainly carbon, hydrogen, nitrogen, and oxygen.

**Organism**: An individual microorganism, plant, or animal carrying out all its functions independently.

**OSHA**: **O**ccupational **S**afety and **H**ealth **A**dministration. An agency in the U.S. Department of Labor that establishes workplace safety and health regulations. In workplace crisis, NIOSH and OSHA work in tandem to identify the details of occupational health and recommend solutions. According to the Law, NIOSH documents should be used by OSHA to help health and safety regulations for the set industry.

**Osteogenic sarcoma**: Cancer associated with bone structures.

**Oxidation**: The chemical addition of oxygen to break down pollutants or organic waste (e.g., destruction of chemicals such as cyanides, phenols, and organic sulfur compounds in sewage by bacterial and chemical means).

**Oxidizer**: A substance that causes oxygen to combine with another substance; oxygen and hydrogen peroxide are examples of oxidizers.

**Ozone depletion**: Destruction of the stratospheric ozone layer that protects the Earth from harmful effects of ultraviolet radiation. Depletion of the ozone layer is due to the breakdown of certain chlorine- or bromine-containing compounds (chlorofluorocarbons or halons), which break down when they reach the stratosphere and then catalytically destroy ozone molecules.

**Ozone layer**: The protective layer in the atmosphere, about 15 miles above the ground, that absorbs some of the sun's ultraviolet rays, thereby reducing the amount of potentially harmful radiation that reaches the Earth's surface.

**PAHs**: **P**olycyclic **a**romatic **h**ydrocarbons.

**Particle**: Very small but discrete mass of a solid or liquid matter.

**Particle concentration**: Concentration expressed in terms of number of particles per unit volume of air or other gas.

**Particle size distribution**: The relative percentage by weight or in number of each different-sized fraction of particulate matter.

**Pest**: An insect, rodent, nematode, bacteria, virus, fungus, weed, and other organism that compete with humans for food or other resources. These are not beneficial but injurious to man and environment.

**pH**: The scale to measure acidity and alkalinity of a medium (e.g., water, soil, body fluid). pH is the hydrogen ion concentration. A truly neutral solution is neither acidic nor alkaline; pH of water is 7.0. pH 0–2 Strongly acidic; pH 3–5 Weakly acidic; pH 6–8 Neutral; pH 9–11 Weakly basic; pH 12–14 Strongly basic.

**Pharmacokinetics**: Term suggesting the quantitative uptake of drugs by the body, biotransformation, distribution, metabolism, and excretion from the body of animals or humans.

**Photocarcinogenesis**: Carcinogenic effects associated with exposure to ultraviolet light.

**Photochemical reaction**: Any chemical reaction that is initiated as a result of absorption of light.

**Photochemical smog**: Type of air pollution due photochemical reaction in the atmosphere.

**Pneumoconiosis**: Health conditions characterized by permanent deposition of substantial amounts of particulate matter in the lungs and by the tissue reaction to its presence; can range from relatively harmless forms of sclerosis to the destructive or fatal fibrotic effect of silicosis.

**Pollution**: Generally, the presence of a substance in the environment that, because of its chemical composition or quantity, prevents the functioning of natural processes and produces undesirable environmental and health effects. Under the Clean Water Act, for example, the term has been defined as the man-made or man-induced alteration of the physical, biological, chemical, and radiological integrity of water and other media.

**Pollution prevention**: 1. Identifying areas, processes, and activities that create excessive waste products or pollutants reduce or prevent them through alteration or elimination of a process. Such activities, consistent with the Pollution Prevention Act of 1990, are conducted across all USEPA programs and can involve cooperative efforts with such agencies as the Departments of Agriculture and Energy. 2. The USEPA has initiated a number of voluntary programs in which industrial, commercial, or "partners" join with USEPA in promoting activities that conserve energy, conserve and protect water supply, reduce emissions, or find ways of using them as energy resources, and reduce the waste stream. Among these are Agstar, to reduce methane emissions through manure management; Climate Wise, to lower industrial greenhouse-gas emissions and energy costs; Coalbed Methane Outreach, to boost methane recovery at coal mines; Design for the Environment, to foster including environmental considerations in product design and processes; Energy Star programs, to promote energy efficiency in commercial and residential buildings, office equipment, transformers, computers, office equipment, and home appliances; Environmental Accounting, to help businesses identify environmental costs and factor them into management decision-making; Green Chemistry, to promote and recognize cost-effective breakthroughs in chemistry that prevent pollution; Green Lights, to spread the use of energy-efficient lighting technologies; Indoor Environments, to reduce risks from indoor air pollution; Landfill Methane Outreach, to develop landfill gas-to-energy projects; Natural Gas Star, to reduce methane emissions from the natural gas industry; Ruminant Livestock Methane, to reduce methane emissions from ruminant livestock; Transportation Partners, to reduce carbon dioxide emissions from the transportation sector; Voluntary Aluminum Industrial Partnership, to reduce perfluorocarbon emissions from the primary aluminum industry; WAVE, to promote efficient water use in the lodging industry; and Wastewise, to reduce business-generated solid waste through prevention, reuse, and recycling.

**Polymer**: A natural or synthetic chemical structure in which two or more like molecules are joined to form a more complex molecular structure (e.g., polyethylene in plastic).

**Polyvinyl chloride (PVC)**: A tough, environmentally indestructible plastic that releases hydrochloric acid when burned.

**Polyploidy**: A condition in which number of chromosomes in a cell(s) is more than normal numbers.

**Potentiation**: The ability of one substance or chemical to increase the effect of another.

**ppb**: The abbreviation for parts per billion; parts of a chemical or substance per billion parts of air or water. Ppb is 1,000 times smaller than ppm.

**ppm**: The abbreviation for parts per million; parts of a substance or chemical per million parts of air or water.

**Precision**: A degree of agreement of repeated measurements of the same property expressed in terms of dispersion of test results about the mean results obtained by repetitive testing of homogeneous sample under specified conditions. The precision of a test method is expressed quantitatively as the standard deviation computed from the results of a series of controlled determinations.

**Procarcinogens**: Substances or chemicals that are not carcinogenic by nature by themselves, but on body metabolism can become carcinogens.

**Promoters**: Substances, chemicals, or agents that do not cause cancer, but enhance the incidence of cancer on initiated cells.

**Protocol**: A series of formal steps for conducting a test.

**PVC**: Polyvinyl chloride.

**Quantitative risk assessment (QRA)**: A process that relies on mathematical modeling and estimations usually derived from animal test results and the probability of risk for a substance or chemical at low doses to which human population is normally exposed to.

**Radionuclide**: A nuclide with radioactive property.

**RANDON**: A naturally occurring radioactive inert gas formed by radioactive decay of radium atoms in soil and rocks and that cannot be seen, smelled, or tasted.

**RDA**: Term suggesting recommended daily allowance; the National Academy of Sciences sets the required nutrient values for the healthy people in the United States. The values consider the needs of all individuals.

**RDI**: Recommended daily intake.

**Reactivity**: The ability of a substance or chemical to undergo a chemical reaction, such as combining with another substance; highly reactive substances are often hazardous.

**Recombinant DNA**: New DNA that is formed by combining pieces of DNA from different organisms or cells.

**Reduction**: The process of removing oxygen from a compound or the addition of hydrogen.

**Registration**: A process involving the scientific, legal, and administrative rules and regulations through which a governmental body (United States Environmental Protection Agency), Central Insecticide Board (CIB, India,) and similar bodies in other countries evaluate the ingredients of pesticide formulations. It may be stated further as a formal listing with regulatory bodies of a new pesticide before it can be sold or distributed. The Federal Insecticide, Fungicide, and Rodenticide Act (USEPA) or Insecticide Act (India) or similar bodies are responsible for registration (premarket licensing) of pesticides based on data demonstrating no unreasonable adverse effects on human health or the environment when applied according to approved label directions.

**Reproductive toxicity**: Effect(s) that may alter the normal reproductive process of an animal, (e.g., loss of fertility).

**Residue**: The pesticide remaining in the product after natural or other processes.

**Risk**: The predicted or actual frequency of occurrence of adverse effects of a substance; probability that harm will occur in a given set of conditions by the use of substances or chemicals.

**Risk assessment**: A process that evaluates the probability that harm will occur in a given set of conditions.

**Risk/Benefit**: The relation between the risks and benefits of a procedure or treatment.

**RUP**: Restricted use pesticide

**Safe water**: Water that does not contain harmful bacteria, toxic materials, or chemicals, and is considered safe for drinking even if it may have taste, odor, color, and certain mineral problems.

**Safety**: Practical certainty or very high probability that injury will not occur from exposure to a hazard; the reverse function of the sum total of toxicity and bioavailability; safety may be expressed as follows:

$$\text{Safety} = \frac{1}{\text{Toxicity} \times \text{Bioavailability}}$$

**Safety factor**: A number, normally 100 or 1,000, which is divided into the highest no-observed-effect dose arrived at animal studies to set an acceptable daily intake level for humans.

**Sampling**: A process consisting of the withdrawal or isolation of a fraction part of a whole. In the analysis of gas or polluted air, the separation of a portion of ambient atmosphere with or without the simultaneous isolation of the selected component.

**Sarcoma**: Tumors of connective tissues, muscles, fat cells, bones, and related organs.

**Sediments**: Soil, sand, and minerals washed from land into water, usually after rain. They pile up in reservoirs, rivers, and harbors, destroying fish and wildlife habitat, and clouding the water so that sunlight cannot reach aquatic plants. Careless farming, mining, and building activities will expose sediment materials, allowing them to wash off the land after rainfall.

**Semistatic test**: A test using aquatic organisms in which the outflow of the solution is made possible but with occasional batch-wise renewal of the test solution for a prolonged period of time (e.g., 24 hours).

**Sewage**: The waste and wastewater produced by residential and commercial sources and discharged into sewers.

**Sex-linked genes**: Set of genes located on the sex chromosomes (X and Y) of animals or humans.

**Sister chromatid exchange**: Genetic process leading to the formation of reciprocal exchange of DNA between the two DNA molecules of a replicating chromosome.

**Skin sensitization**: Allergic contact dermatitis. An immunologically modulated cutaneous reaction to a substance or chemical by the animal. Response by humans to skin sensitization are observed in the form of erythema, edema, papules, pruritis, bullae, and vesicles.

**Solubility**: The amount of mass of a substance or chemical that will dissolve in a unit volume of solution; aqueous solubility is the maximum concentration of a chemical that will dissolve in pure water at a reference temperature.

**Solution**: A mixture in which the components are uniformly dispersed.

**Solvent**: Usually a liquid capable of dissolving or dispersing one or more substances; certain chemicals commonly called solvents can dissolve many different chemicals (e.g., water, ethanol, acetone, hexane, toluene).

**Stack**: A chimney, smokestack, or vertical pipe that discharges used air.

**Stack effect**: Air, as in a chimney, that moves upward because it is warmer than the ambient atmosphere.

**Stack effect**: Flow of air resulting from warm air rising, creating a positive pressure area at the top of a building and a negative pressure area at the bottom. This effect can overpower the mechanical system and disrupt building ventilation and air circulation.

**Static test**: A test using aquatic organisms in which there is no flow of the test solution. The test solution remains unchanged through the period of test.

**Subchronic delayed neurotoxicity**: The repeated daily administration of a test substance or chemical causing adverse effects, such as delayed onset of locomotor ataxia.

**Subchronic toxicity**: Appearance of adverse effects as a result of repeated daily dosing of a test substance or chemical to experimental animals through oral, dermal, or respiratory route for a period not exceeding 10% of its life span.

**Smog**: Refers to extensive atmospheric contamination by aerosols (the term is derived from smoke and fog).

**Smoke**: Small gas-borne particles created from incomplete combustion of a variety of materials; essentially consists of carbon and other combustible materials.

**Soot**: Agglomeration of particles of carbon impregnated with tar formed in the incomplete combustion of carbonaceous material.

**Sorbent**: A solid/liquid medium in or on which materials are retained by absorption or adsorption.

**Synergism**: A phenomenon in which exposure to more than one chemical can result in health effects greater than expected when the effects of exposure to each chemical are added together. In simple terms: $1 + 1 = 3$. If chemicals have synergistic properties, the potential hazards of the chemicals should be reevaluated, considering their synergistic properties.

**Teratogenic**: Organisms showing the formation of noninheritable birth defects.

**Teratogen**: A substance or chemical capable of causing birth defects or abnormalities.

**Terotogenicity**: The biologic process where substances or chemicals produce permanent structural and functional abnormalities to the fetus during the period of embryonic development.

**Threshold**: The lowest dose of a substance or chemical at which a specified measurable effect(s) is observed and below which it is not observed.

**Threshold level**: Time-weighted average pollutant concentration values, exposure beyond which is likely to adversely affect human health. (*See* Environmental exposure)

**Threshold limit value (TLV)**: The concentration of an airborne substance to which an average person can be repeatedly exposed without adverse effects. TLVs may be expressed in three ways: 1) TLV-TWA (Time weighted average), based on an allowable exposure averaged over a normal 8-hour workday or 40-hour workweek; 2) TLV-STEL (Short-term exposure limit) or maximum concentration for a brief specified period of time, depending on a specific chemical (TWA must still be met); and 3) TLV-C (Ceiling Exposure Limit) or maximum exposure concentration not to be exceeded under any circumstances. TWA must still be met.

**Tolerance**: The amount of substance or chemical legally permitted in food products; tolerance limits are established through formal regulatory procedures; for example, with permissible residue levels for pesticides in raw agricultural produce and processed foods, when a pesticide is registered for use on a food or a feed crop, a tolerance must be established. The USEPA establishes tolerance levels, which are enforced by the Food and Drug Administration and the Department of Agriculture. Use of products indicating no set tolerance values is considered illegal.

**Toxic**: Harmful to living organisms.

**Toxic cloud**: Airborne plume of gases, vapors, fumes, or aerosols containing toxic materials.

**Toxicity**: The intrinsic capability of a chemical or substance to cause injury or adverse effects to animals, humans, or plants.

**Toxicity**: The degree to which a substance or mixture of substances can harm humans or animals. Acute toxicity involves harmful effects in an organism through single or short-term exposure. Chronic toxicity is the ability of a substance or mixture of substances to cause harmful effects over an extended period, usually on repeated or continuous exposure, sometimes lasting for the entire life of the organism. Subchronic toxicity is the ability of the substance to cause effects for more than 1 year but less than the lifetime of the organism.

**Toxicokinetics**: Quantitative uptake of xenobiotics by the body, its biotransformation, distribution, metabolism, and elimination from the body.

**Toxicological profile**: The examination or summary and interpretation of a hazardous substance or chemical to determine levels of exposure and the associated health effects.

**TWA (Time Weight Average)**: The average concentration of a chemical or substance in air over the total exposure time. Usually an 8-hour workday.

**Ultraviolet (UV) rays**: Radiation from the sun that can be useful or potentially harmful. UV rays from one part of the spectrum (UV-A) enhance plant life. UV rays from other parts of the spectrum (UV-B) can cause skin cancer or other tissue damage. The ozone layer in the atmosphere partly shields us from ultraviolet rays reaching the Earth's surface.

**Vapor**: The gaseous phase of matter normally from liquid or solid material and observed at normal temperature; most organic solvents evaporate and produce vapors.

**Vinyl chloride**: A chemical compound, used in producing some plastics, which is believed to be oncogenic.

**Virtually safe dose (VSD)**: The dose of chemical corresponding to the level of risk determined and accepted by regulatory agencies; the dose-to-risk relationship is based on a chemical dose-response curve.

**VOCs**: **V**olatile **o**rganic **c**ompounds.

**Volatility**: The tendency of a liquid to evaporate into a gas form or vapor. Organic solvents on inhalation are in the form of vapors.

**Vulnerable zone**: An area over which the airborn concentration of a chemical accidentally released could reach the level of concern.

**Wastewater**: The spent or used water from a home, community, farm, or industry that contains dissolved or suspended matter.

**Water pollution**: The presence in water of enough harmful or objectionable material to damage the water's quality.

**Xenobiotics**: Substances or chemicals that are pharmacologically and toxicologically active but foreign to organisms (animals, humans and plants).

**Zero air**: Atmospheric air purified to contain less than 0.1 ppm total hydrocarbons.

# INDEX



*Industrial Guide to Chemical and Drug Safety*, By T.S.S. Dikshith and Prakash V. Diwan
ISBN 0-471-23698-5